PRINCIPLES
OF
SOLAR
ENGINEERING

CONTRIBUTING AUTHORS

PROFESSOR JEFF MOREHOUSE
Mechanical Engineering Department
University of South Carolina
Columbia, South Carolina

ROBERT C. BROWN
Center for Sustainable Environmental Technologies
Departments of Mechanical and Chemical Engineering
Iowa State University
Ames, Iowa

PRINCIPLES OF SOLAR ENGINEERING

Second Edition

D. YOGI GOSWAMI
Professor and Director
Solar Energy & Energy Conservation Laboratory
University of Florida

FRANK KREITH
Consulting Engineer and ASME Fellow,
National Conference of State Legislatures

JAN F. KREIDER
University of Colorado

ALERE FLAMMAM

TAYLOR & FRANCIS

· Founded 1798 ·

Published in 2000 by
Taylor & Francis Group
270 Madison Avenue
New York, NY 10016

Published in Great Britain by
Taylor & Francis Group
2 Park Square
Milton Park, Abingdon
Oxon OX14 4RN

International Standard Book Number-10: 1-56032-714-6 (Hardcover)
International Standard Book Number-13: 978-1-56032-714-1 (Hardcover)
Library of Congress Card Number 99-31349

This book contains information obtained from authentic and highly regarded sources. Reprinted material is quoted with permission, and sources are indicated. A wide variety of references are listed. Reasonable efforts have been made to publish reliable data and information, but the author and the publisher cannot assume responsibility for the validity of all materials or for the consequences of their use.

Library of Congress Cataloging-in-Publication Data

Goswami, D. Yogi.
    Principles of solar engineering / D. Yogi Goswami, Frank Kreith, Jan F. Kreider.— 2nd ed.
        p.   cm.
        Kreith's name appears first on the previous ed.
        Includes bibliographical references and index.
        ISBN 1-56032-714-6 (alk. paper)
    1. Solar energy. I. Kreith, Frank. II. Kreider, Jan F., 1942–   .
    III. Kreith, Frank. Principles of solar engineering. IV. Title.
TJ810.K73   1999
631.47—dc21   99-31349

Taylor & Francis Group
is the Academic Division of Informa plc.

**Visit the Taylor & Francis Web site at
http://www.taylorandfrancis.com**

CONTENTS

PREFACE

During the 20 years that have elapsed since the first edition of *Principles of Solar Engineering,* there have been a number of significant advances in solar energy applications. The new applications include solar detoxification, which utilizes solar photochemical processes to treat contaminated water, air and soil; and solar photovoltaic (PV) conversion, which has matured markedly. PV systems are now more efficient with costs at about $5/watt compared to $30/watt in 1980. For this reason, photovoltaics are now cost effective in many stand-alone applications all over the world. New advances in manufacturing have reduced the energy payback time to two-and-a-half to five years, while lifetimes have increased to more than 25 years. Passive solar heating is now a standard feature in modern, energy efficient homes built in sunny, cold parts of the country. Solar water-heating systems are reliable, efficient and long lasting.

Solar thermal power conversion has been demonstrated successfully at a level of hundreds of megawatts. A large solar thermal power plant using concentrating parabolic troughs has been operating reliably for more than ten years, producing 354 MW under design conditions. Central receiver power has been demonstrated at a level of 10 MW. Solar powered Stirling engine systems now use advanced concepts, including free piston designs and stretched membrane dishes. Finally, solar resource assessment is advancing with new measurement and estimation procedures using satellite data.

Recent solar energy books have not covered all of the sustainable, renewable technologies in depth. Therefore, the authors have included engineering level detail for all included technologies. Over the years, our students at Colorado and Florida and many other schools have requested that we produce this revision. The new edition of *Principles of Solar Engineering* is designed to fill the gap. The second edition is nearly completely new, using less than 20% of the material from the first edition. Key additions to the book include:

modern methods of solar resource assessment including satellite measurements;

new developments in concentrating solar thermal collectors;

a new chapter on methods for passive heating, cooling and daylighting;

the latest developments in solar cooling and dehumidification;

a thorough treatment of solar thermal power and industrial process heat;

a new chapter on photovoltaics with a thorough treatment of fundamentals, design applications and manufacturing;

a new chapter on solar photochemical applications (the first in a solar energy textbook);

and a new chapter on capturing solar energy through biomass. Biomass is expected to make a major global contribution in the future, for both stand-alone biomass power systems and hybrid solar-biomass power systems.

The new edition has been revised to make it useful all over the world. Examples and homework problems are provided that cover latitudes in northern and southern hemispheres. We include the most current worldwide solar radiation data (including addresses on the World Wide Web). Applications and examples have been chosen with a wide variety of countries in mind.

The book contains many homework problems and is designed to be both a textbook in engineering schools and a reference for renewable energy scientists, engineers, architects, and building construction professionals. We assume that users of this book have a basic background in physics, chemistry, mathematics, thermodynamics, and heat transfer. Specific information needed from thermodynamics and heat transfer is integrated into chapters but in such a way that instructors can pass over it if students have sufficient background.

Because of this book's comprehensiveness, a single course cannot cover all of the material in one-semester. Depending on the scientific and professional backgrounds of students, a one-semester undergraduate or graduate course can be tailored to cover a variety of emphases.

The authors would like to thank Dr. Jeffrey Morehouse for writing Chapter 7 "Passive Methods for Heating, Cooling, and Daylighting" and Dr. Robert Brown for writing Chapter 11 "Capturing Solar Energy through Biomass." Their up to date contributions have provided valuable breadth and depth to the new edition.

Finally, D. Yogi Goswami, who was responsible for most of the revisions, would like to dedicate this edition to the memory of his late father, Mr. G. L. Goswami, a Sanskrit scholar, an author and a teacher, who was a constant source of inspiration, but who passed away during the course of this revision.

D. Yogi Goswami
Frank Kreith
Jan F. Kreider

DEDICATION

Dedicated to the students of sustainability and renewable energies in the 21st Century.

INTRODUCTION TO SOLAR ENERGY CONVERSION

The National Science Foundation in testimony before the Senate Interior Committee in 1972 stated that "Solar energy is an essentially inexhaustible source potentially capable of meeting a significant portion of the nation's future energy needs with a minimum of adverse environmental consequences. . . . The indications are that solar energy is the most promising of the unconventional energy sources. . . ." Despite this encouraging assessment of the potential of solar energy, considerable technical and economic problems must be solved before large-scale utilization of solar energy can occur. The future of solar power development will depend on how we deal with a number of serious constraints, including scientific and technological problems, marketing and financial limitations, and political and legislative actions favoring conventional and nuclear power. In addition, the education of engineers will have to change its focus from non-renewable fossil-fuel technology to renewable power sources. There appears to be general agreement that the most significant of the renewable energy sources is solar radiation. Hence, it is the objective of this book to present the basic technical background necessary for the design and economic analysis of solar energy utilization systems.

This book assumes that the reader is familiar with traditional thermodynamics, basic heat transfer, and fluid mechanics, and has a knowledge of calculus and ordinary differential equations. Some elements of radiation, fluid mechanics, and heat transfer specific to solar engineering are presented in the text. The design and analysis of solar utilization schemes are approached from a systems-analysis viewpoint, which combines technical design with economic analysis. There is no single solution to a given task in solar energy utilization, and each problem must be analyzed separately from fundamental principles. Whereas with systems utilizing conventional fuel sources such as coal or gas one can specify the heat and power requirements and the manufacturer can meet these requirements with a device guaranteed to meet specifications, in the design of a solar system it is necessary to match the available power source to the task at hand. It is, therefore, not possible to make a general case for or against the utilization of solar energy. Instead, the engineer will have to form a judgment on the basis of the task at hand, the systems and resources available to achieve a technical solution, and the economics involved.

Thermal Conversion

Thermal conversion is a technological scheme that utilizes a familiar phenomenon. When a dark surface is placed in sunshine, it absorbs solar energy and heats up. Solar energy collectors working on this principle consist of a sun-facing surface which transfers part of the energy it absorbs to a working fluid in contact with it. To reduce heat losses to the atmosphere and to improve its efficiency, one or two sheets of glass are usually placed over the absorber surface. This type of thermal collector suffers from heat losses due to radiation and convection. Such losses increase rapidly as the temperature of the working fluid increases. Improvements, such as the use of selective surfaces, evacuation of the collector to reduce heat losses, and special kinds of glass, are used to increase the efficiency of these devices.

The simple thermal-conversion devices described above are called flat-plate collectors. They are available today for operation over a range of temperatures up to approximately 365 K (200°F). These collectors are suitable mainly for providing hot service water and space heating and possibly are also able to operate absorption-type air-conditioning systems.

The thermal utilization of solar energy for the purpose of generating low-temperature heat is at the present time technically feasible and economically viable for producing hot water and heating swimming pools. In some parts of the world thermal low-temperature utilization is also attractive for heating and cooling buildings.

The generation of higher working temperatures, as needed, for example, to operate a conventional steam engine, requires the use of focusing devices in connection with a basic absorber-receiver. Operating temperatures as high as 4000 K (6740°F) have been achieved, and the generation of steam to operate pumps for irrigation purposes has also proved technologically feasible. At the present time, focusing devices for the generation of steam to produce electric power are under construction in different regions of the world, and cost estimates suggest that the cost of solar power in favorable locations will be no more than that of nuclear power when the development of these plants has been completed.

Photovoltaic Conversion

The conversion of solar radiation into electrical energy by means of solar cells has been developed as a part of satellite and space-travel technology. The theoretical efficiency of solar cells is about 30 percent, and, in practice, efficiencies as high as 25 percent have been achieved with silicon photovoltaic devices. Overall system efficiencies are in the range of 10 to 14 percent. The technology of photovoltaic conversion is well developed, but large-scale application is hampered by the high price of the cells.

Photovoltaic solar cells, as opposed to conventional collectors, which convert solar radiation into heat, utilize energetic photons of the incident solar radiation to produce electricity directly. Thus, this technique is often referred to as *direct solar conversion*. Conversion efficiencies of thermal systems are limited by collector temperatures, whereas the conversion efficiency of photo cells is limited by other factors.

Biological Conversion

Biological conversion of solar energy by means of photosynthesis is a natural process that has been studied by scientists for many decades. This form of solar energy utilization has been of the greatest importance by far to the human race. It provides a small but vital part of our energy consumption in the form of food and for thousands of years has served our ancestors in the form of wood as the only source of heat. Last, but not least, it is this process that in the course of millions of years produced our fossil fuels, which currently provide most of our energy.

In principle, it might be possible to cultivate appropriate plants exclusively for the purpose of generating power either by direct bioconversion or by pyrolytic conversion into liquid or gaseous fuel. Unfortunately, the photosynthetic yield in agricultural application is only of the order of 1 percent. In temperate climates, even under favorable conditions, the average annual harvest is of the order of 20 metric tons of organic dry material per hectare. However, cultivation of special plants such as sugarcane can achieve annual mean conversion efficiencies of the order of 2.5 percent, and recycling processes by which the waste products of foods or animals are used to produce methane, which can also be converted into methanol, are within the realm of possibility. Bioconversion is expected to provide an increasing amount of total energy needs in the future.

Wind Power

The utilization of wind power has been widespread since medieval times. Windmills were used in rural United States to power irrigation pumps and drive the small electric generators used to charge the batteries that provided electricity during the last century. A windmill or wind turbine converts the kinetic energy of moving air into mechanical motion, usually in the form of a rotating shaft. This mechanical motion can be used to drive a pump or to generate electric power.

The technology of wind-energy conversion is well developed and will not be treated in this text. The energy content of the wind increases with the third power of the wind velocity, and wind-power installations are economical in regions where winds of sufficient strength and regularity occur. The construction of wind-power installations does not require any new technologies, and cost estimates in favorable regions of the world are fairly close to those of fossil fuel energy sources. However, in order to produce appreciable amounts of power, installations have to be fairly large. Rotor diameters of 50 m and heights of 100 m are dimensions typically suggested for wind generators, and such giant windmills placed at regular distances might not be aesthetically pleasing. Moreover, wind velocities decrease considerably at night and vary with the weather. Consequently, for reliable power from the wind energy must be stored. These are some of the reasons why wind power has not yet found large-scale applications even in favorable regions. However, in some parts of the world wind power will make an appreciable contribution to the overall energy demand.

Ocean Energy Conversion

Almost 71 percent of the world's surface is covered by oceans. Oceans serve as tremendous storehouses of solar energy because of the temperature differences produced by the sun as well as the kinetic energy stored in the waves. There are a number of places in the ocean where temperature differences of the order of 20–25 K exist at depths of less than 1000 m, and these temperature differences could be used to operate low-pressure heat engines. Although the thermodynamic efficiency of a heat engine operating on such a small temperature difference is low, the available amount of thermal energy is very large. However, putting this energy-conversion method into practice requires the development of efficient and cheap heat exchangers that can withstand the rough marine conditions. Since heat-exchange equipment is the most expensive part of any ocean thermal conversion scheme, the cost of using the temperature gradients in the ocean for practical solar energy utilization depends largely on this development.

The second method of utilizing the oceans for energy generation is through ocean waves. This approach is being studied in Japan and the United Kingdom. Prototype installations have been built, but cost estimates for ocean energy utilization methods are not encouraging and will not be treated in this text.

AVAILABILITY OF SOLAR ENERGY

The amount of solar radiant energy falling on a surface per unit area and per unit time is called *irradiance*. The mean extraterrestrial irradiance normal to the solar beam on the outer fringes of the earth's atmosphere is approximately 1.35 kW/m². Since the earth's orbit is elliptical, the sun-to-earth distance varies slightly with time of year, and the actual extraterrestrial irradiance varies by ±3.4 percent during the year. The angle subtended by the sun when viewed from the earth is only 0.0093 rad (approximately 32 min of arc), and the direct beam radiation reaching the earth is therefore almost parallel. Although the brightness of the solar disc decreases from center to edge, for most engineering calculations the disc can be assumed to be of uniform brightness. The radiant energy from the sun is distributed over a range of wavelengths, and the energy falling on a unit surface area per unit time within a particular spectral band is known as the *spectral irradiance*; its value is usually expressed in watts per square meter per nanometer of bandwidth. The extraterrestrial spectral irradiance is shown in Fig 1.2. It can be approximated by the spectrum of a black body at 5800 K. In the upper part of Fig. 1.1, the wave bands typically useable for different solar applications are shown,

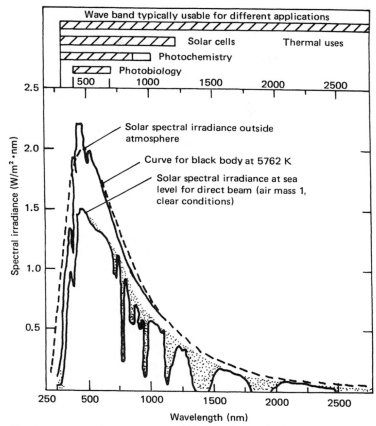

Figure 1.1 Spectral irradiance curves for direct sunlight extraterrestrially and at sea level with the sun directly overhead. Shaded areas indicate absorption due to atmospheric constituents, mainly H_2O, CO_2, and O_3. Wavelengths potentially utilized in different solar energy applications are indicated at the top.

and the lowest curve shows the spectral direct-beam irradiance at sea level on earth under clear sky conditions with the sun overhead.

The earth and its atmosphere receive continuously 1.7×10^{17} W of radiation from the sun. A world population of 10 billion with a total power need per person of 10 kW would require about 10^{11} kW of energy. It is thus apparent that if the irradiance on only 1 percent of the earth's surface could be converted into useful energy with 10 percent efficiency, solar energy could provide the energy needs of all the people on earth. This figure is often quoted by solar energy enthusiasts, but unfortunately the nature of this energy source has technical problems and economic limitations that are not apparent from this macroscopic view of the energy budget. The principal limitations are that the solar energy received on earth is of small flux density, is intermittent, and falls mostly in remote places. The implications of these factors will be discussed.

LIMITATIONS OF SOLAR ENERGY

The first problem encountered in the engineering design of equipment for solar energy utilization is the low flux density which makes necessary large surfaces to collect solar energy for large-scale utilization. Also, the larger the surfaces the more expensive the delivered energy becomes. When the sun is directly overhead on a cloudless day, 10 m^2 of surface could theoretically provide energy at 10 percent efficiency of collection at the rate of 1 kW. Several factors reduce this amount in practice. One of these is losses in the atmosphere. The solar spectrum is substantially modified in passing through the earth's atmosphere, and approximately 25–50 percent of its energy is lost by scattering and absorption. Even on a cloud-free day with unpolluted skies, about 30 percent of the incident energy is lost. Some is scattered back into space by air molecules, and some is absorbed by ozone, water vapor, and carbon dioxide in the atmosphere. Even on a clear day, the diffuse energy from the sky that cannot be collected efficiently accounts for about 20 percent of the total irradiance on a horizontal surface, and this percentage increases substantially in locations where there are natural clouds and anthropogenic pollution. Moreover, as shown in Fig. 1.2, as a result of almost total absorption of solar energy by ozone at wavelengths below 300 nm and by carbon dioxide at wavelengths beyond 2500 nm the irradiance on the earth's surface is effectively limited to wavelengths between 300 and 2500 nm. In that range, only the solar energy falling between 400 and 700 nm, or about half of the total earth irradiance, can be used by living plants photosynthetically.

The total solar energy reaching the earth is made up of two parts: the energy in the direct beam and the diffuse energy from the sky. Although plants can use direct and diffuse solar energy, most manmade solar collectors can convert only direct energy efficiently. The amount of direct energy depends on the cloudiness and the position of the sun and is obviously greatest on clear days. Some solar radiation falling on clouds is diffused by scattering, but clouds do not absorb all of the energy. The effect of clouds is mainly to increase the percentage of diffuse energy in the total energy reaching the surface, and diffuse irradiance in summer months with high sun and broken clouds can be as high as 400 W/m^2. Thick clouds let less energy pass than do thin clouds, and they scatter proportionally more of the total energy back into space.

The second practical limitation that is not apparent from the macroscopic energy view is that most of the solar energy falls on remote areas and would therefore require some means of transportation to be useful to the industrialized nations. The annual mean surface global irradiance on a horizontal plane as mapped by Budyko [7] is shown in Fig. 1.3. The mean amount of energy available on a horizontal plane is greatest in the continental desert areas around latitudes 25°N and 25°S of the equator and falls off toward both the equator and the poles. The highest annual mean irradiance is 300 W/m^2 in the Red Sea area. Clouds reduce the mean global irradiance considerably

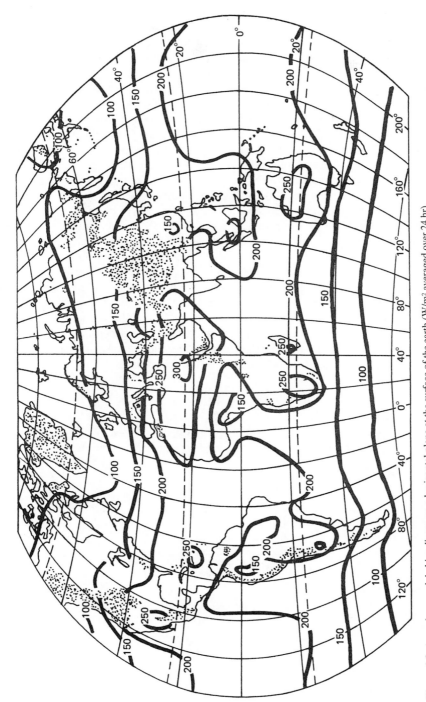

Figure 1.2 Annual mean global irradiance on a horizontal plane at the surface of the earth (W/m² averaged over 24 hr).

7

in equatorial regions, and the annual mean irradiance for Singapore is almost the same as that for Paris. However, whereas in northern climates the mean horizontal surface global irradiance varies from season to season, it remains relatively constant in equatorial regions. Typical values of mean annual horizontal surface irradiance are: Australia, about 200 W/m²; United States, 185 W/m²; United Kingdom, 105 W/m².

Although the global distribution of solar energy does not favor the industrialized parts of the world, it may be of help to industrially developing countries located in the favorable radiation belt. Table 1.1 shows the total solar electric energy potential for the solar energy-rich areas of the world. It is important to note that many of these parts of the world, for example, Saudi Arabia, central Australia, and parts of South Africa and India, are flat deserts that are practically unuseable for agriculture. At the same time, it should be noted that these are also regions with little or no water, which can create special problems in the generation of thermal electric power.

The third limitation of solar energy as a large-scale source of power and heat is its intermittency. Solar energy has a regular daily cycle due to the turning of the earth around its axis, a regular annual cycle due to the inclination of the earth axis with the plane of the ecliptic and due to the motion of the earth around the sun, and is also unavailable during periods of bad weather. These daily and seasonal variations in solar radiation, exacerbated by variation due to weather, introduce special problems in storage and distribution of this energy, which are entirely different from those problems involved in the utilization of conventional sources such as coal and oil. In assessing the prospects for using solar energy, it therefore becomes particularly important to ascertain for which application the diffuse cyclic and intermittent nature of the source will not introduce insurmountable technical and economic problems.

Table 1.1. Solar-electric energy from earth's high-insolation areas[a]

Desert	Nominal areas km²	Nominal annual thermal energy flux GW · hr(th)/km²	Percentage of area assumed usable	Electric energy extracted at 25% efficiency (GW · hr/yr)
North Africa	7,770,000	2,300	15[b]	670,000,000
Arabian Peninsula	1,300,000	2,500	30[c]	244,000,000
Western and central Australia	1,550,000	2,000	25	194,000,000
Kalahari	518,000	2,000	50	129,000,000
Thar (northwest India)	259,000	2,000	50	65,000,0000
Mojave, southern California	35,000	2,200	20	3,900,000
Vizcaino, Baja, California (Mexico)	15,500	2,200	25	2,100,000
Total/Average	11,447,500	2,190 ave.	31 ave.	1,308,000,000

[a]From [10].
[b]Parts of Arabian and Libyan deserts.
[c]About 60 percent of Rub' al Khali desert.

Energy Storage

In addition to the technical problems already noted, widespread use of solar energy systems requires satisfactory means of storing the energy once it has been collected. Figure 1.3 shows a comparison of the energy storage densities of various storage media. Whereas considerable progress has been made in developing systems for extracting energy from the sun's radiation, the state of development of storage systems is less satisfactory. For low-temperature storage, such as is necessary in heating and cooling buildings, sensible heat storage in water or rocks is a reasonable solution. For electrical energy production systems, however, either thermal storage at elevated temperature or electrochemical battery storage is currently being used. Table 1.2 lists some commercially available batteries in the U.S.A. in 1993. Hydro storage is currently available and widely used. It could be useful in connection with solar power plants, but

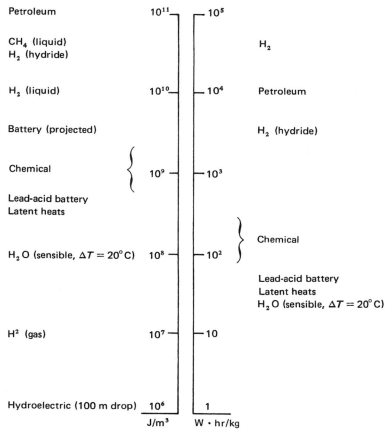

Figure 1.3 Mass and volumetric energy densities for typical energy-storage media. Redrawn from [24].

Table 1.2. Electric storage batteries

Manufacturer and Model	Model Number	Shallow/ Deep Cycle (S/D)	Nominal Capacity (Ah)	Nominal Voltage (V)	Depth of Discharge (%)	Life (Cycles)	Energy Delivered (kWh)	Cost ($/kWh)
GNB Absolute	638	S	42	6	50	1,000	126	0.50
	1260	S	59	12	50	1,000	359	0.32
	6-35A09	S	202	12	50	3,000	3,636	0.18
	3-75A25	S	1,300	6	50	3,000	11,700	0.10
Exide Tubular Modular	6E95-5	S	192	12	15	4,100	1,417	0.42
					20	3,900	1,797	0.33
	6E120-9	S	538	12	15	4,100	3,970	0.23
					20	3,900	5,036	0.19
	3E120-21	S	1,346	6	15	4,100	4,967	0.19
					20	3,900	6,299	0.15
Delco-Remy Photovoltaic	2000	S	105	12	10	1,800	227	0.34
					15	1,250	236	0.33
					20	850	214	0.36
Globe Solar Reserve Gel Cell	3SRC-125G	S	125	6	10	2,000	150	1.21
	SRC-250G	S	250	2	10	2,000	100	1.00
	SRC-375G	S	375	2	10	2,000	150	1.00
Globe	GC12-800-38	S	80	12	20	1,500	288	0.35
		D	80	12	80	250	240	0.42
GNB Absolute	638	D	40	6	80	500	96	0.65
	1260	D	56	12	80	500	269	0.42
	6-35A09	D	185	12	80	1,500	2,664	0.25
	3-75A25	D	1,190	6	80	1,500	8,568	0.14
Surrette	CH-375	D	375	6	80	1,400	2,520	0.09
	NS-29	D	490	6	80	1,400	3,293	0.16
	NS-33	D	564	8	80	1,400	5,053	0.16
Exide Tubular Modular	6E95-5	D	180	12	80	1,800	3,110	0.19
	6E120-9	D	360	12	80	1,800	6,221	0.15
	3E120-21	D	1,250	6	80	1,800	10,800	0.09

Source: From Strong, S.J. and W.G. Scheller, *The Solar Electric House: Energy for the Environmentally Responsive, Energy-Independent Home*, Sustainability Press, Still River, MA, 1993, with permission.

would require siting the power plants in areas where hydro storage capability exists. Compressed-air systems in combination with gas turbines are technologically possible, as is thermal energy storage in liquid metals or molten salts.

Another possibility is using solar energy to generate hydrogen and storing the energy in the gaseous or liquid phase. There appear to be no technical difficulties to the large-scale production, storage, and subsequent use of hydrogen in either liquid or gaseous form, but the hydrogen production efficiency is relatively low and, consequently, the cost of hydrogen storage and delivery systems is quite high at the present time.

THE ECONOMICS OF SOLAR SYSTEMS

Although solar energy is essentially free, there is a definite cost associated with its utilization. In its simplest form, neglecting interest charges on capital, one could calculate the cost of solar energy in the following manner. Assume that a solar system will have a life of T years and its initial cost is C_0 dollars. If, during its life, the system will on the average receive every year an amount of energy Q, the unit cost of energy, neglecting interest charges, is equal to the cost of the installation divided by the total energy delivery during its lifetime. For example, if a solar energy collector costs $100/m² surface, has an expected life of 20 yr, and is installed in a part of the country where the mean annual horizontal surface irradiance is 200 W/m² averaged over 24 hr, the cost of solar energy C_s will be equal to

$$C_s = \frac{C_0}{Q \times T}$$

$$= \frac{100\ (\$/m^2) \times 1000(\text{W/kW})}{200\ (\text{W/m}^2) \times 24\ (\text{hr/day}) \times 365\ (\text{days/yr}) \times 20\ \text{yr}}$$

$$= \$0.00285/\text{kW}\cdot\text{hr} \tag{1.1}$$

It is obvious, however, that no collector will perform at 100 percent efficiency. Consequently, the collector efficiency enters the economics, because according to thermodynamic laws only a fraction of the incident energy can be converted into useful heat. Assuming that the efficiency of the collector η_c is 50 percent, the cost of solar energy will be twice that calculated in the previous example and will be given by

$$C_s = \frac{C_0}{QT\eta_c} = \$0.0057/\text{kW}\cdot\text{hr} \tag{1.2}$$

Note that installing the very same collector in another part of the country where the mean irradiance is only 100 W/m² would again double the cost of the solar energy, although the system cost would be the same.

REFERENCES AND SUGGESTED READINGS

1. Budyko, M. I. 1958. *The Heat Balance of the Earth's Surface,* 259. N. Stepanov trans. Washington, D.C.: U.S. Department of Commerce.
2. Daniels, F. 1967. Direct use of the sun's energy. *Am. Sci.,* 55, 1, 15–47.
3. Ehricke, K. A. 1973. The Power Relay Satellite Concept of the Framework of the Overall Energy Picture. *N. Am. Aerospace Rockwell Int. Rept. E-73-12-1,* December.
4. EIA. 1993. Annual Energy Outlook 1993, International Energy Outlook, Energy Information Administration, Office of Integrated Analysis and Forecasting, U.S. DOE, DOE/EIA-0383(93), Washington, D.C.
5. IEA. 1994. World Energy Outlook, Economic Analysis Division, International Energy Agency, Paris.
6. U.S. DOE. 1991. National Energy Strategy–Powerful Ideas for America, 1991. National Technical Information Service, U.S. Department of Commerce, Springfield, VA.
7. Offenhartz, P. O. D. 1976. Classical Methods of Storing Thermal Energy. In *Sharing the Sun–Solar Technology in the Seventies,* vol. 8, ed. K. W. Böer. Winnipeg: Solar Energy Society of Canada.
8. Press, W. H. 1976. Theoretical maximum for energy from direct and diffuse sunlight: *Nature,* 264: 734–735.
9. NSF/NASA Solar Energy Panel. 1972. *Solar Energy as a National Resource.* Washington, D.C.: U.S. Government Printing Office.

FUNDAMENTALS OF
SOLAR RADIATION

2

In houses with a south aspect, the sun's rays penetrate into the porticos in winter, but in summer the path of the sun is right over our heads and above the roofs so that there is shade.

Socrates, ca. 400 B.C.

2.1 THE PHYSICS OF THE SUN AND ITS ENERGY TRANSPORT

The nature of energy generation in the sun is still an unanswered question. Spectral measurements have confirmed the presence of nearly all the known elements in the sun. However, 80 percent of the sun is hydrogen and 19 percent helium. Therefore, the remaining 100-plus observed elements make up only a tiny fraction of the composition of the sun. It is generally accepted that a hydrogen-to-helium thermonuclear reaction is the source of the sun's energy. Yet, because such a reaction has not been duplicated in the laboratory, it is unclear precisely what the reaction mechanism is, what role the turbulent flows in the sun play, and how solar prominences and sunspots are created.

The nature of the energy-creation process is of no importance to terrestrial users of the sun's radiation. Of interest is the amount of energy, its spectral and temporal distribution, and its variation with time of day and year. These matters are the main subject of this chapter.

The sun is a 13.9×10^5 km diameter sphere comprised of many layers of gases, which are progressively hotter toward its center. The outermost layer, that from which energy is radiated into the solar system, is approximately at an equivalent black-body temperature of 5760 K (10,400° R). The center of the sun, however, may be at 20×10^6 K. The rate of energy emission from the sun is 3.8×10^{23} kW, which results from the conversion of 4.3×10^9 g/sec (4.7×10^6 ton/sec) of mass to energy. Of this total, only a tiny fraction, approximately 1.7×10^{14} kW, is intercepted by the earth, which is located about 150 million km from the sun (Fig. 2.1).

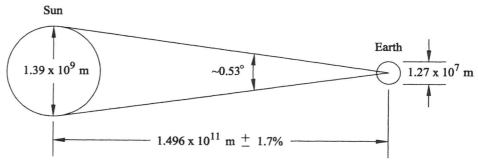

Figure 2.1. Relationship between the sun and the earth.

Solar energy is the world's most abundant permanent source of energy. The amount of solar energy intercepted by the planet earth is 5000 times greater than the sum of all other inputs (terrestrial nuclear, geothermal and gravitational energies, and lunar gravitational energy). Of this amount, 30 percent is reflected to space, 47 percent is converted to low-temperature heat and reradiated to space, and 23 percent powers the evaporation/precipitation cycle of the biosphere. Less than 0.5 percent is represented in the kinetic energy of the wind and waves and in photosynthetic storage in plants.

Total terrestrial radiation is only about one-third of the extraterrestrial total during a year, and 70 percent of that falls on the oceans. However, the remaining 1.5×10^{17} kW · hr that falls on land is a prodigious amount of energy—about 6000 times the total energy usage of the United States in 2000. However, only a small fraction of this total can be used because of physical and socioeconomic constraints as described in Chapter 1.

2.2 THERMAL RADIATION FUNDAMENTALS

The material presented in this section has been selected from textbooks on heat transfer and radiation [for example, Refs. 24, 38, 39, 69]. It provides the background needed to understand the nature of solar radiation for the engineering analysis of solar energy systems.

To begin then, all radiation travels at the speed of light which is equal to the product of the wavelength and the frequency of radiation. The speed of light in a medium equals the speed of light in a vacuum divided by the refractive index of the medium through which it travels:

$$c = \lambda\mu = \frac{c_0}{n}, \tag{2.1}$$

where λ = wavelength in m (or μm, 1 μm = 10^{-6}m)
μ = frequency in $\sec^{-1}$,

c = speed of light in a medium (m/sec),
c_0 = speed of light in a vacuum (m/sec), and
n = index of refraction of the medium.

Thermal radiation is one kind of electromagnetic energy and all bodies emit thermal radiation by virtue of their temperature. When a body is heated, its atoms, molecules, or electrons are raised to higher levels of activity called excited states. However, they tend to return to lower energy states, and in this process energy is emitted in the form of electromagnetic waves. Changes in energy states result from rearrangements in the electronic, rotational, and vibrational states of atoms and molecules. Since these rearrangements involve different amounts of energy changes and these energy changes are related to the frequency, the radiation emitted by a body is distributed over a range of wavelengths. A portion of the electromagnetic spectrum is shown in Fig. 2.2. The wavelengths associated with the various mechanisms are not sharply defined; thermal radiation is usually considered to fall within the band from about 0.1 to 100 μm, whereas solar radiation has most of its energy between 0.1 and 3 μm.

For some problems in solar energy engineering, the classical electromagnetic wave theory is not suitable. In such cases, for example in photovoltaic or photochemical processes, it is necessary to treat the energy transport from the point of view of quantum mechanics. In this view, energy is transported by particles or *photons* which are treated as energy units or quanta rather than waves. The energy of a photon, E_p, of frequency v_p is

$$E_p = h\mu_p, \tag{2.2}$$

where h = Planck's constant (6.625×10^{-34} J · sec).

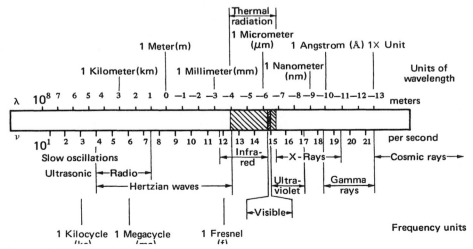

Figure 2.2. Electromagnetic radiation spectrum.

2.2.1 Black Body Radiation

The energy density of the radiation emitted at a given wavelength (monochromatic) by a perfect radiator, usually called a black body, is given according to the relation

$$E_{b\lambda} = \frac{C_1}{(e^{C_2/\lambda T} - 1)\lambda^5 n^2},\tag{2.3}$$

where $C_1 = 3.74 \times 10^8 \text{W} \cdot \mu\text{m}^4/\text{m}^2$ (1.19×10^8 Btu $\cdot \mu\text{m}^4/\text{hr} \cdot \text{ft}^2$),
       $C_2 = 1.44 \times 10^4 \, \mu\text{m} \cdot \text{K}$ ($2.59 \times 10^4 \, \mu\text{m} \cdot °\text{R}$), and
       n = Refractive index of the medium = 1.0 for vacuum; n is taken to be approximately equal to 1 for air.

The quantity $E_{b\lambda}$ has the units of W/m$^2 \cdot \mu$m (Btu/hr $\cdot$ ft$^2 \cdot \mu$m) and is called the monochromatic emissive power of a black body, defined as the energy emitted by a perfect radiator per unit wavelength at the specified wavelength per unit area and per unit time at the temperature T.

The total energy emitted by a black body, E_b, can be obtained by integration over all wavelengths. This yields the Stefan-Boltzmann law

$$E_b = \int_0^\infty E_{b\lambda} \, d\lambda = \sigma T^4,\tag{2.4}$$

where σ = Stefan-Boltzmann constant
     = 5.67×10^{-8} W/m$^2 \cdot$ K^4 (0.1714×10^{-8} Btu/hr $\cdot$ ft$^2 \cdot$ R^4),
   T = absolute temperature in K (or R = 460 + °F).

The concept of a black body, although no such body actually exists in nature, is very convenient in engineering because its radiation properties can readily be related to those of real bodies.

2.2.2 Radiation Function Tables

Engineering calculations of radiative transfer are facilitated by the use of radiation function tables, which present the results of Planck's law in a more convenient form than Eq. (2.3). A plot of the monochromatic emissive power of a black body as a function of wavelength as the temperature is increased is given in Fig. 2.3. The emissive power shows a maximum at a particular wavelength. These peaks, or inflection points, are uniquely related to the body temperature. By differentiating Planck's distribution law [Eq. (2.3)] and equating to zero, the wavelength corresponding to the maximum value of $E_{b\lambda}$ can be shown to occur when

$$\lambda_{\max} T = 2897.8 \mu m \; K(5215.6 \mu m \; R).\tag{2.5}$$

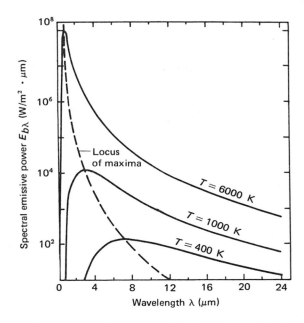

Figure 2.3. Spectral distribution of black-body radiation.

Frequently one needs to know the amount of energy emitted by a black body within a specified range of wavelengths. This type of calculation can be performed easily with the aid of the radiation functions mentioned previously. To construct the appropriate radiation functions in dimensionless form, note that the ratio of the black body radiation emitted between 0 and λ and between 0 and ∞ can be made a function of the single variable (λT) by using Eq. (2.3) as shown below (for $n = 1$):

$$\frac{E_{b,0-\lambda}}{E_{b,0-\infty}} = \frac{\int_0^\lambda E_{b\lambda}\, d\lambda}{\sigma T^4} = \int_0^{\lambda T} \frac{C_1 d(\lambda T)}{\sigma (\lambda T)^5 \, (e^{C_2/\lambda T} - 1)}. \tag{2.6}$$

The above relation is plotted in Fig. 2.4 and the results are also shown in tabular form in Table 2.1. In this table, the first column is the ratio of λ to λ_{max} from Eq. (2.5), and the third column the ratio of $E_{b,0-\lambda}$ to σT^4 from Eq. (2.6). For use on a computer Eq. (2.6) can be approximated by the following polynomials:

$$v \geq 2 \quad \frac{E_{b,0-\lambda}}{\sigma T^4} = \frac{15}{\pi^4} \sum_{m=1,2,\dots} \frac{e^{-mv}}{m^4} \{[(mv + 3)mv + 6]mv + 6\}, \text{ and} \tag{2.7a}$$

$$v < 2 \quad \frac{E_{b,0-\lambda}}{\sigma T^4} = \frac{15}{1 - \pi^4} v^3 \left(\frac{1}{3} - \frac{v}{8} - \frac{v^2}{60} - \frac{v^4}{5040} + \frac{v^6}{272,160} - \frac{v^8}{13,305,600} \right), \tag{2.7b}$$

where $v = C_2/\lambda T$.

Table 2.1. Thermal radiation functions[a]

$\lambda/\lambda_{\max}$	$\dfrac{E_{b\lambda}}{E_{b\lambda,\max}}$	$\dfrac{E_{b\lambda,0-\lambda}}{\sigma T^4}$	$\lambda/\lambda_{\max}$	$\dfrac{E_{b\lambda}}{E_{b\lambda,\max}}$	$\dfrac{E_{b\lambda,0-\lambda}}{\sigma T^4}$	$\lambda/\lambda_{\max}$	$\dfrac{E_{b\lambda}}{E_{b\lambda,\max}}$	$\dfrac{E_{b\lambda,0-\lambda}}{\sigma T^4}$
0.00	0.0000	0.0000	1.50	0.7103	0.5403	2.85	0.1607	0.8661
0.20	0.0000	0.0000	1.55	0.6737	0.5630	2.90	0.1528	0.8713
0.25	0.0003	0.0000	1.60	0.6382	0.5846	2.95	0.1454	0.8762
0.30	0.0038	0.0001	1.65	0.6039	0.6050	3.00	0.1384	0.8809
0.35	0.0187	0.0004	1.70	0.5710	0.6243	3.10	0.1255	0.8895
0.40	0.0565	0.0015	1.75	0.5397	0.6426	3.20	0.1141	0.8974
0.45	0.1246	0.0044	1.80	0.5098	0.6598	3.30	0.1038	0.9045
0.50	0.2217	0.0101	1.85	0.4815	0.6761	3.40	0.0947	0.9111
0.55	0.3396	0.0192	1.90	0.4546	0.6915	3.50	0.0865	0.9170
0.60	0.4664	0.0325	1.95	0.4293	0.7060	3.60	0.0792	0.9225
0.65	0.5909	0.0499	2.00	0.4054	0.7197	3.70	0.0726	0.9275
0.70	0.7042	0.0712	2.05	0.3828	0.7327	3.80	0.0667	0.9320
0.75	0.8007	0.0960	2.10	0.3616	0.7449	3.90	0.0613	0.9362
0.80	0.8776	0.1236	2.15	0.3416	0.7565	4.00	0.0565	0.9401
0.85	0.9345	0.1535	2.20	0.3229	0.7674	4.20	0.0482	0.9470
0.90	0.9725	0.1849	2.25	0.3053	0.7777	4.40	0.0413	0.9528
0.96	0.9936	0.2172	2.30	0.2887	0.7875	4.60	0.0356	0.9579
1.00	1.0000	0.2501	2.35	0.2731	0.7967	4.80	0.0308	0.9622
1.05	0.9944	0.2829	2.40	0.2585	0.8054	5.00	0.0268	0.9660
1.10	0.9791	0.3153	2.45	0.2447	0.8137	6.00	0.0142	0.9790
1.15	0.9562	0.3472	2.50	0.2318	0.8215	7.00	0.0082	0.9861
1.20	0.9277	0.3782	2.55	0.2197	0.8290	8.00	0.0050	0.9904
1.25	0.8952	0.4081	2.60	0.2083	0.8360	9.00	0.0033	0.9930
1.30	0.8600	0.4370	2.65	0.1976	0.8427	10.00	0.0022	0.9948
1.35	0.8231	0.4647	2.70	0.1875	0.8490	20.00	0.0002	0.9993
1.40	0.7854	0.4911	2.75	0.1780	0.8550	40.00	0.0000	0.9999
1.45	0.7477	0.5163	2.80	0.1691	0.8607	50.00	0.0000	1.0000

[a]
λ = wavelength in μm.

$\lambda_{\max}$ = wavelength at $E_{b\lambda,\max}$ in μm = $2898/T$.

$E_{b\lambda}$ = monochromatic emissive power in $W/m^2 \cdot \mu$m

　　　 = $374.15 \times 10^6/\lambda^5[\exp(14{,}387.9/\lambda T)-1]$,

$E_{b\lambda,\max}$ = maximum monochromatic emissive power in $W/m^2 \cdot \mu$m

　　　 = $12.865 \times 10^{-12} T^5$,

$E_{b\lambda,0-\lambda} = \displaystyle\int_0^\lambda E_{b\lambda}\, d\lambda$,

$\sigma T^4 = E_{b\lambda,0-\infty} = 5.670 \times 10^{-8} T^4\ W/m^2$, and

T = absolute temperature in K.

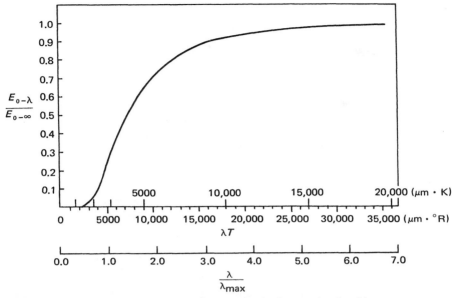

Figure 2.4. Fraction of total emissive power in spectral region between $\lambda = 0$ and λ as a function of λT and λ/λ_{max}.

2.2.3 Intensity of Radiation and Shape Factor

The emissive power of a surface gives the total radiation emitted in all directions. To determine the radiation emitted in a given direction we must define another quantity, the radiation intensity I. This quantity is defined as the radiant energy passing through an imaginary plane in space per unit area per unit time and per unit solid-angle perpendicular to the plane as shown in Fig. 2.5. I is defined by the relation

$$I = \lim_{\substack{dA' \to 0 \\ d\omega \to \infty}} \frac{dE}{dA' d\omega} \tag{2.8}$$

Radiation intensity has both magnitude and direction. It can be related to the radiation flux, defined as the radiant energy passing through an imaginary plane per unit area per unit time in all directions. Note that, whereas for the intensity, the area dA' is perpendicular to the direction of the radiation, for the flux the area dA is at the base in the center of a hemisphere through which all of the radiation passes. Recalling that the definition for the solid angle between dA' and dA is $d\omega = dA'/r^2$, the radiation flux q_r emanating from dA can be obtained by integrating the intensity over the hemisphere. As shown in Fig. 2.5, the unit projected area for I is $dA \cos \theta$ and the differential area dA' on the hemisphere is $r^2 \sin \theta\, d\theta\, d\phi$; thus

$$q_r = \int_0^{2\pi} \int_0^{\pi/2} I \cos\theta \, \sin\theta \, d\theta \, d\phi. \tag{2.9}$$

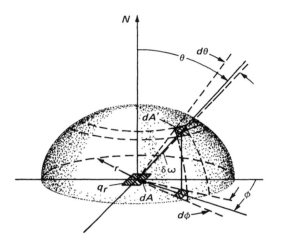

Figure 2.5. Schematic diagram illustrating radiation intensity and flux.

If the area dA is located on a surface, the emissive power E can also be obtained from Eq. (2.9). For the special case of a diffuse surface, for which I is the same in all directions, Eq. (2.9) gives

$$q_r = \pi I. \tag{2.10}$$

Since all black surfaces are diffuse,

$$E_b = \pi I_b \tag{2.11}$$

Equation (2.11) can, of course, also be written for monochromatic radiation as:

$$E_{b\lambda} = \pi I_{b\lambda} \tag{2.12}$$

In the evaluation of the rate of radiation heat transfer between two surfaces, not only their temperatures and their radiation properties but also their geometric configurations and relationships play a part. The influence of geometry in radiation heat transfer can be expressed in terms of the ***radiation shape factor*** between any two surfaces 1 and 2 defined as follows:

F_{1-2} = fraction of radiation leaving surface 1 that reaches surface 2, and
F_{2-1} = fraction of radiation leaving surface 2 that reaches surface 1.

In general, F_{m-n} = fraction of radiation leaving surface m that reaches surface n. If both surfaces are black, the energy leaving surface m and arriving at surface n is $E_{bm}A_mF_{m-n}$ and the energy leaving surface n and arriving at m is $E_{bn}A_nF_{n-m}$. If both surfaces absorb all the incident energy, the net rate of exchange $q_{m \rightleftarrows n}$ will be

$$q_{m \Leftrightarrow n} = E_{bm} A_m F_{m-n} - E_{bn} A_n F_{n-m} \tag{2.13}$$

If both surfaces are at the same temperature, $E_{bm} = E_{bn}$ and the net exchange is zero, $q_{m \rightleftarrows n} = 0$. This shows that the geometric radiation shape factor must obey the reciprocity relation

$$A_m F_{m-n} = A_n F_{n-m} \tag{2.14}$$

The net rate of heat transfer can therefore be written in two equivalent forms:

$$q_{m \Leftrightarrow n} = A_m F_{m-n}(E_{bm} - E_{bn}) = A_n F_{n-m}(E_{bm} - E_{bn}). \tag{2.15}$$

The evaluation of geometric shape factors is in general quite involved. For a majority of solar energy applications, however, only a few special cases are of interest. One of these is a small convex object of area A_1 surrounded by a large enclosure A_2. Since all radiation leaving A_1 is intercepted by A_2, $F_{1-2} = 1$ and $F_{2-1} = A_1/A_2$.

Another case is the exchange of radiation between two large parallel surfaces. If the two surfaces are near each other, almost all of the radiation leaving A_1 reaches A_2 and vice versa. Thus, $F_{1-2} = F_{2-1} = 1.0$, according to the definition of the shape factor. A third case of importance is the exchange between a small surface ΔA_1 and a portion of space A_2, for example, the exchange between a flat-plate solar collector tilted at an angle β from the horizontal and the sky it can see. For this situation we refer to the definition of radiation flux (see Fig. 2.5). The portion of the radiation emitted by ΔA_1 that is intercepted by the surrounding hemisphere depends on the angle of tilt. When the surface is horizontal, $F_{1-2} = 1$; when it is vertical, $F_{1-2} = 1/2(\beta = 90°)$. For intermediate values it can be shown that [69]:

$$F_{1-2} = \frac{1}{2}(1 + \cos\beta) = \cos^2\left(\frac{\beta}{2}\right). \tag{2.16}$$

If the diffuse sky radiation is uniformly distributed and assumed to be black, then a small black area A_1 receives radiation at the rate

$$A_1 F_{1-sky} E_{sky} = \frac{A_1}{2}(1 + \cos\beta)\sigma T_{sky}^4 \tag{2.17}$$

whereas the net radiation heat transfer is given by

$$q_{sky \Leftrightarrow 1} = A_1 F_{1-sky}\sigma(T_{sky}^4 - T_1^4). \tag{2.18}$$

If the receiving area is gray with an absorptance $\bar{\alpha}$ equal to the emittance $\bar{\epsilon}$ the net exchange is given by

$$q_{sky \Leftrightarrow 1} = A_1 F_{1-sky}\bar{\alpha}\sigma(T_{sky}^4 - T_1^4). \tag{2.19}$$

2.2.4 Transmission of Radiation Through a Medium

When radiation passes through a transparent medium such as glass or the atmosphere, the decrease in intensity can be described by Bouger's law that assumes that the attenuation is proportional to the local intensity in the medium. If $I_\lambda(x)$ is monochromatic intensity after radiation has traveled a distance x, the law is expressed by the equation

$$-dI_\lambda(x) = I_\lambda(x)K_\lambda dx, \tag{2.20}$$

where K_λ is the monochromatic extinction coefficient assumed to be a constant of the medium. If the transparent medium is a slab of thickness L and the intensity at $x = 0$ is designated by the symbol $I_{\lambda,0}$, the monochromatic transmittance τ_λ is equal to the ratio of the intensity at $x = L$ to $I_{\lambda,0}$. An expression for $I_\lambda(L)$ can be obtained by integrating Eq. (2.20) between 0 and L, which gives

$$ln\frac{I_\lambda(L)}{I_{\lambda,0}} = -K_\lambda L \text{ or } I_\lambda(L) = I_{\lambda,0}e^{-K_\lambda L} \tag{2.21}$$

Then

$$\tau_\lambda = \frac{I_\lambda(L)}{I_{\lambda,0}} = e^{-K_\lambda L} \tag{2.22}$$

The extinction coefficient K_λ is a complex property of the medium since it combines the effects of absorption, emission and scattering by the molecules and particles that make up the medium. Fortunately, for materials such as glass and plastics with known compositions, this coefficient can be determined accurately. Transmission of radiation through such materials will be discussed further in Chapter 3. In the present chapter we are concerned about the transmission of solar radiation through the atmosphere. The atmosphere consists of the molecules of gases in it, such as N_2, O_2, CO_2, H_2O, etc., and aerosols such as dust particles, water droplets and ice crystals. The extinction processes of the atmosphere consist of (a) absorption and emission by the molecules and aerosols, (b) scattering by the molecules, and (c) scattering by aerosols.

Since the atmosphere consists of a large number of components whose concentration changes as a function of time and location, determining the extinction coefficient of the atmosphere presents a formidable challenge. A major research effort is underway by scientists trying to predict global climate change. Some early attempts for the estimation of extinction coefficient for "average atmospheric conditions" were combined with an empirical approach [76] to use the above equation for the estimation of terrestrial solar radiation resource. This approach is described later in this chapter.

2.3 SUN-EARTH GEOMETRIC RELATIONSHIP

Figure 2.6 shows the annual orbit of the earth around the sun. The distance between the earth and the sun changes throughout the year, the minimum being 1.471×10^{11} m at

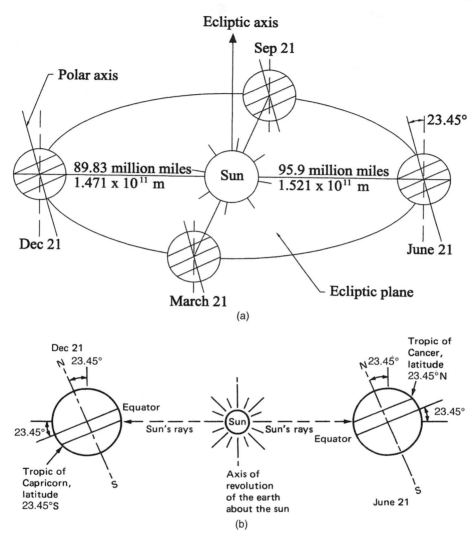

Figure 2.6. (a) Motion of the earth about the sun. (b) Location of tropics. Note that the sun is so far from the earth that all the rays of the sun may be considered as parallel to one another when they reach the earth.

winter solstice (December 21) and the maximum being 1.521×10^{11} m at summer solstice (June 21). The year-round average earth-sun distance is 1.496×10^{11} m. The amount of solar radiation intercepted by the earth, therefore, varies throughout the year, the maximum being on December 21 and the minimum on June 21.

The axis of the earth's daily rotation around itself is at an angle of 23.45° to the axis of its ecliptic orbital plane around the sun. This tilt is the major cause of the seasonal variation of the solar radiation available at any location on the earth. The angle between the earth-sun line (through their centers) and the plane through the equator is called the *solar declination*, δ_s. The declination varies between $-23.45°$ on December 21 to $+23.45°$ on June 21. Stated another way, the declination has the same numerical value as the latitude at which the sun is directly overhead at solar noon on a given day. The tropics of Cancer (23.45° N) and Capricorn (23.45° S) are at the extreme latitudes where the sun is overhead at least once a year as shown in Fig. 2.6. The Arctic and Antarctic circles are defined as those latitudes above which the sun does not rise above the horizon plane at least once per year. They are located, respectively, at 66½° N and 66½° S. Declinations north of the equator (summer in the northern hemisphere) are positive; those south, negative. The solar declination may be estimated by the relation*:

$$\delta_s = 23.45° \sin[360(284 + n)/365°], \qquad (2.23)$$

where n is the day number during a year with January 1 being $n = 1$. Approximate values of declination may also be obtained from Table 2.2 or Fig. 2.7. For most calculations, the declination may be considered constant during any given day.

For the purposes of this book, the Ptolemaic view of the sun's motion provides a simplification to the analysis that follows. It is convenient to assume the earth to be fixed and to describe the sun's apparent motion in a coordinate system fixed to the earth with its origin at the site of interest. Figure 2.8 shows an apparent path of the sun to an observer. The position of the sun can be described at any time by two angles, the altitude and azimuth angles, as shown in Fig. 2.8. The *solar altitude angle*, α, is the angle between a line collinear with the sun's rays and the horizontal plane. The *solar azimuth angle*, α_s, is the angle between a due south line and the projection of the site to sun line on the horizontal plane. The sign convention used for azimuth angle is positive west of south and negative east of south. The *solar zenith angle*, z, is the angle between the site to sun line and the vertical at the site:

$$z = 90° - \alpha. \qquad (2.24)$$

The solar altitude and azimuth angles are not fundamental angles. Hence, they must be related to the fundamental angular quantities *hour angle, latitude* and *declination*. The three angles are shown in Fig. 2.9. The solar hour angle h_s is based on the nominal time of 24 hours required for the sun to move 360° around the earth or 15° per hour. Therefore, h_s is defined as

$$h_s = 15° \cdot \text{(hours from local solar noon)} = \frac{\text{minutes from local solar noon}}{4 \text{ min/degree}}. \qquad (2.25)$$

Again, values east of due south, that is, morning values, are negative; and values west of due south are positive.

---

*A more accurate relation is $\sin \delta_s = \sin(23.45°) \sin[360(284 + n)/365]°$. Because the error is small, equation (2.23) is generally used.

Table 2.2. Summary solar ephemeris[a]

Date		Declination		Equation of Time		Date		Declination		Equation of Time	
		Deg	Min	Min	Sec			Deg	Min	Min	Sec
Jan.	1	−23	4	−3	14	Feb.	1	−17	19	−13	34
	5	22	42	5	6		5	16	10	14	2
	9	22	13	6	50		9	14	55	14	17
	13	21	37	8	27		13	13	37	14	20
	17	20	54	9	54		17	12	15	14	10
	21	20	5	11	10		21	10	50	13	50
	25	19	9	12	14		25	9	23	13	19
	29	18	9	123	5						
Mar.	1	−7	53	−12	38	Apr.	1	+4	14	−4	12
	5	6	21	11	48		5	5	46	3	1
	9	5	48	10	51		9	7	17	1	52
	13	3	14	9	49		13	8	46	−0	47
	17	1	39	8	42		17	10	12	+0	13
	21	−0	5	7	32		21	11	35	1	6
	25	+1	30	6	20		25	12	56	1	53
	29	3	4	5	7		29	14	13	2	33
May	1	+14	50	+2	50	June	1	+21	57	2	27
	5	16	2	34	17		5	22	28	1	49
	9	17	9	3	35		9	22	52	1	6
	13	18	11	3	44		13	23	10	+0	18
	17	19	9	3	44		17	23	22	−0	33
	21	20	2	3	24		21	23	27	1	25
	25	20	49	3	16		25	23	25	2	17
	29	21	30	2	51		29	23	17	3	7
July	1	+23	10	−3	31	Aug.	1	+18	14	−6	17
	5	22	52	4	16		5	17	12	5	59
	9	22	28	4	56		9	16	6	5	33
	13	21	57	5	30		13	14	55	4	57
	17	21	21	5	57		17	13	41	4	12
	21	20	38	6	15		21	12	23	3	19
	25	19	50	6	24		25	11	2	2	18
	29	18	57	6	23		29	9	39	1	10
Sep.	1	+8	35	−0	15	Oct.	1	−2	53	+10	1
	5	7	7	+1	2		5	4	26	11	17
	9	5	37	2	22		9	5	58	12	27
	13	4	6	3	45		13	7	29	13	30
	17	2	34	5	10		17	8	58	14	25
	21	1	1	6	35		21	10	25	15	10
	25	0	32	8	0		25	11	50	15	46
	29	2	6	9	22		29	13	12	16	10

Table 2.2. (*Continued*)

Date		Declination Deg	Declination Min	Equation of Time Min	Equation of Time Sec	Date		Declination Deg	Declination Min	Equation of Time Min	Equation of Time Sec
Nov.	1	−14	11	+16	21	Dec.	1	−21	41	11	16
	5	15	27	16	23		5	22	16	9	43
	9	16	38	16	12		9	22	45	8	1
	13	17	45	15	47		13	23	6	6	12
	17	18	48	15	10		17	23	20	4	47
	21	19	45	14	18		21	23	26	2	19
	25	20	36	13	15		25	23	25	+0	20
	29	21	21	11	59		29	23	17	−1	39

[a]Since each year is 365.25 days long, the precise value of declination varies from year to year. *The American Ephemeris and Nautical Almanac*, published each year by the U.S. Government Printing Office, contains precise values for each day of each year.

The latitude angle L is the angle between the line from the center of the earth to the site and the equatorial plane. The latitude may be read from an atlas and is considered positive north of the equator and negative south of the equator.

2.3.1 Solar Time and Angles

The sun angles are obtained from the local solar time, which differs from the local standard time. The relationship between the local solar time and the local standard time (LST) is

$$\text{Solar Time} = LST + ET + (l_{st} - l_{local}) \cdot 4 \text{ min/degree.} \tag{2.26}$$

ET is the equation of time, which is a correction factor that accounts for the irregularity of the speed of earth's motion around the sun; l_{st} is the standard time meridian, and l_{local} is the local longitude. ET may be estimated from Table 2.2 or calculated from the following empirical equation:

$$ET \text{ (in minutes)} = 9.87 \sin 2B - 7.53 \cos B - 1.5 \sin B \tag{2.27}$$

$$\text{where } B = 360(n - 81)/364 \text{ degrees.}$$

The solar altitude angle, α, can be found from the application of the law of cosines to the geometry of Fig. 2.9 and simplification as:

$$\sin\alpha = \sin L \sin\delta_s + \cos L \cos\delta_s \cos h_s. \tag{2.28}$$

Using a similar technique, the solar azimuth angle, a_s, can be found as:

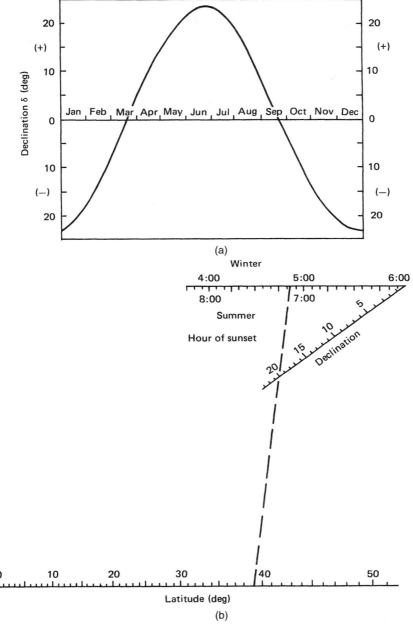

Figure 2.7. (a) Graph to determine the solar declination. (b) Sunset nomograph example. Example (b) shows determination of sunset time for summer (7:08 p.m.) and winter (4:52 p.m.) when the latitude is 39°N and the solar declination angle is 20°. From [82].

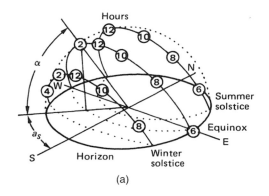

(a)

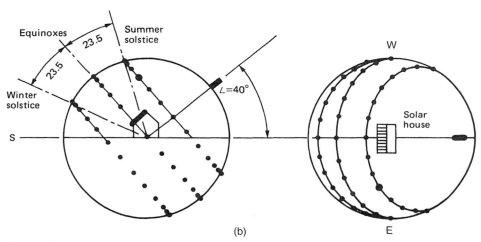

(b)

Figure 2.8. Sun paths for the summer solstice (6/21), the equinoxes (3/21 and 9/21), and the winter solstice (12/21) for a site at 40°N: (a) isometric view; (b) elevation and plan views.

$$\sin a_s = \cos \delta_s \sin h_s / \cos \alpha \qquad (2.29)$$

At local solar noon, $h_s = 0$; therefore, $\alpha = 90 - |L - \delta_s|$, and $a_s = 0$.

In calculating the solar azimuth angle from Eq. (2.29), a problem occurs whenever the absolute value of a_s is greater than 90°. A computational device usually calculates the angle as less than 90° since $\sin a_s = \sin(180 - a_s)$. The problem can be solved in the following way:

For $L > \delta_s$, the solar times when the sun is due east (t_E) or due west (t_W) can be calculated by t_E or t_W = 12:00 Noon $\mp$ ($\cos^{-1}[\tan \delta_s / \tan_L]$ degrees)/($15°$/hr) ($-$ for t_E, $+$ for t_W).

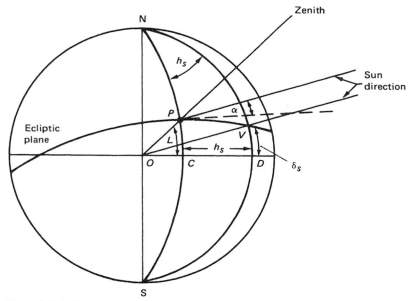

Figure 2.9. Definition of solar-hour angle h_s (CND), solar declination δ_s (VOD), and latitude L (POC); p, site of interest. Modified from Kreider, J.F. and F. Kreith, *Solar Heating and Cooling*. Washington, D.C.: Hemisphere Publ. Corp., 1982.

For solar times earlier than t_E or later than t_W the sun would be north (south in the southern hemisphere) of the east-west line and the absolute value of a_s would be greater than 90°. Therefore, the correct value of a_s is $a_s = 180° - |a_s|$.

For $L \leq \delta_s$ the sun remains north (south in the southern hemisphere) of the east-west line and the true value of a_s is greater than 90°.

Sunrise and **sunset** times can be estimated by finding the hour angle for $\alpha = 0$. Substituting $\alpha = 0$ in Eq. (2.28) gives the hour angles for sunrise (h_{sr}) and sunset (h_{ss}) as:

$$h_{ss} \text{ or } h_{sr} = \pm\cos^{-1}[-\tan L \cdot \tan\delta_s]. \tag{2.30}$$

It should be emphasized that equation (2.30) is based on the center of the sun being at the horizon. In practice, sunrise and sunset are defined as the times when the upper limb of the sun is on the horizon. Because the radius of the sun is 16′, the sunrise would occur when $\alpha = -16′$. Also, at lower solar elevations, the sun will appear on the horizon when it is actually 34′ below the horizon. Therefore, for apparent sunrise or sunset, $\alpha = -50′$.

Example 2.1(a). Find the solar altitude and azimuth angles at solar noon in Gainesville, FL on February 1. Also find the sunrise and sunset times in Gainesville on that day.

Solution. For Gainesville,

$$\text{Latitude } L = 29° + 41'\text{N or } 29.68°\text{N.}$$

$$\text{Longitude } l_{local} = 82° + 16'\text{W or } 82.27°\text{W.}$$

On February 1, day number

$$n = 32.$$

Therefore, declination

$$\delta_s = 23.45 \sin [360 (284 + 32)/365]°$$
$$= -17.5°.$$

At solar noon, $h_s = 0$. Therefore,

$$\sin \alpha = \cos L \cos \delta_s \cos h_s + \sin L \sin \delta_s$$
$$= \cos(29.68°) \cos(-17.5°) \cos(0) + \sin(29.68°) \sin(-17.5°),$$

or

$$\alpha = 42.82°$$
$$\sin a_s = \cos(-17.5°)\sin(0)/\cos(42.8°) = 0,$$

or

$$a_s = 0.$$

At solar noon, α can also be found as:

$$\alpha = 90 - |L - \delta_s|°$$
$$= 90 - |29.68 + 17.5| = 42.82°.$$
$$h_{ss} \text{ or } h_{sr} = \pm\cos^{-1}[-\tan L \cdot \tan \delta_s]$$
$$= \pm\cos^{-1}[-\tan (29.68°) \tan (-17.5°)]$$
$$= \pm79.65°.$$
$$\text{Time from solar noon} = \pm(79.65°) (4 \text{ min/degree})$$
$$= \pm319 \text{ min or } \pm(5 \text{ hr } 19 \text{ min}).$$
$$\text{Sunrise time} = 12:00 \text{ Noon} - (5 \text{ hr } 19 \text{ min})$$
$$= 6 \text{ hr } 41 \text{ min AM (Solar Time).}$$

$$\text{Sunset time} = \text{12:00 Noon} + (5 \text{ hr } 19 \text{ min})$$
$$= 5 \text{ hr } 19 \text{ min PM (Solar Time)}.$$

To convert these times to local times, we need to find ET:

$$ET = (9.87 \sin 2B - 7.53 \cos B - 1.5 \sin B) \text{ min.}$$
$$B = \frac{360}{364}(n - 81) = \frac{360}{364}(32 - 81)$$
$$= -48.46°.$$

Therefore,

$$ET = -13.67 \text{ min.}$$
$$LST = \text{Solar Time} - ET - 4(l_{st} - l_{local}).$$

Gainesville, FL is in the Eastern Standard Time (EST) zone, where $l_{st} = 75°\text{W}$. Therefore,

$$LST = \text{Solar Time} - (-13.67 \text{ min}) - 4(75 - 82.27) \text{ min}$$
$$= \text{Solar Time} + 42.75 \text{ min.}$$

Therefore,

$$\text{Sunrise Time} = \text{6:41 AM} + 43 \text{ min}$$
$$= \text{7:24 AM EST, and}$$
$$\text{Sunset Time} = \text{5:19 PM} + 43 \text{ min}$$
$$= \text{6:02 PM EST.}$$

Note: Since the sunrise and sunset times are calculated when the center of the sun is at the horizon, they differ from the apparent times. If we use $\alpha = -50'$, the apparent sunrise and sunset times would be 7:20 AM EST and 6:06 PM EST, respectively.

Example 2.1(b). Repeat the calculations of Example 2.1(a) for Canberra, Australia.

$$\text{Latitude } L = 35° - 18'\text{S or } 35.3°\text{S.}$$
$$\text{Longitude } l_{local} = 149° - 11'\text{E or } 149.18°\text{E.}$$
$$\text{Standard Meridian} = 150°\text{E.}$$

Solution. Using the values of δ_s and ET calculated in Example 2.1(a), and taking the latitude and longitude as $-35.3°$ and $-149.18°$ respectively, the solar angles α and a_s are found as:

$$\sin \alpha = \cos(-35.3°) \cos(-17.5°) \cos(0) + \sin(-35.3°) \sin(-17.5°)$$

or

$$\alpha = 72.2°.$$
$$\sin a_s = \cos(-17.5°) \sin(0)/\cos(72.2°)$$

or

$$a_s = 0°.$$

α may also be found as:

$$\alpha = 90° - |-35.3° + 17.5°|$$
$$= 72.2°.$$

The hour angles for sunrise and sunset are found as:

$$h_{ss}, h_{sr} = \pm\cos^{-1}[-\tan(-35.3°) \cdot \tan(-17.5°)]$$
$$= \pm 102.9°.$$
$$\text{Time from solar noon} = \pm(102.9°) \cdot (4 \text{ min/degree})$$
$$= \pm 412 \text{ minutes or } \pm(6 \text{ hr } 52 \text{ min}).$$
$$\text{Sunrise time} = 5{:}08 \text{ AM (Solar Time), and}$$
$$\text{Sunset time} = 6{:}52 \text{ PM (Solar Time).}$$

Taking l_{st} for Canberra as $-150°$

Therefore

$$LST = \text{Solar Time} - (-13.67 \text{ min}) - 4(-150° + 149.18°)\text{min}$$
$$= \text{Solar Time} + 17 \text{ min.}$$

Therefore,

$$\text{Sunrise Time} = 5{:}25 \text{ AM (Local Standard Time), and}$$
$$\text{Sunset Time} = 7{:}09 \text{ PM (Local Standard Time).}$$

(As explained in the previous example, the apparent sunrise and sunset times would be 5:21 AM and 7:13 PM, respectively.)

Knowledge of the solar angles is helpful in the design of passive solar buildings, especially the placement of windows for solar access and the roof overhang for shading the walls and windows at certain times of the year. The following example illustrates this point.

Example 2.2(a). Find the roof overhang L of a south facing window of height $H = 1$ m, such that the window is completely shaded at solar noon on May 1 and not shaded at all at noon on November 1. Assume that the roof extends far beyond the window on either side. Location: Gainesville, FL. Also, find the overhang if $S = 1.3$ m.

Solution. From the geometry of the figure:

$$L = S/\tan\alpha_2.$$

Also,

$$L = (S + H)/\tan\alpha_1.$$

On May 1,

$$n = 91.$$

$$\delta_s = 23.45° \sin[360 (284 + 91)/365] = 4.02°.$$

Therefore, at solar noon,

$$\alpha_1 = 90 - |29.68 - 4.02| = 64.34°.$$

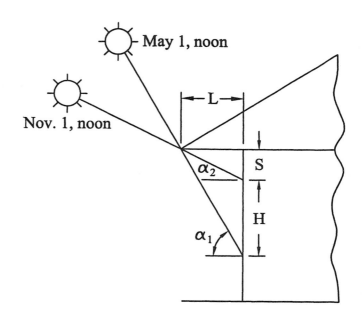

On November 1,

$$n = 305.$$

$$\delta_s = 23.45° \, Sin[360(284 + 305)/365]° = -15.4°.$$

Therefore,

$$\alpha_2 = 90 - |29.68 + 15.4|$$

$$= 44.9°.$$

$$L = \frac{H}{\tan\alpha_1 - \tan\alpha_2} = \frac{1}{\tan(64.34°) - \tan(44.9°)} = 0.92m.$$

$$S = L \tan \alpha_2 = 0.92 \times \tan(44.9°) = 0.92 \text{ m.}$$

If $S = 1.3$ m, then:

$$L = 1.3/\tan 44.9° = 1.3 \text{ m.}$$

Also,

$$L = 2.3/\tan 64.34° = 1.1 \text{ m.}$$

Therefore, $1.1 \text{ m} \le L \le 1.3$ m.

Example 2.2(b). Repeat the above example for Canberra, Australia, for a north facing window, such that the window is completely shaded at noon on November 1 and completely lit at noon on May 1.

Solution. The figure shown in Example 2.2(b) may be used if we take the solar angles α_1 and α_2 as shown in the figure to be on November 1 and May 1 respectively. From Example 2.2(a)

$$\delta_s = 4.02° \text{ on May 1, and}$$

$$= -15.4° \text{ on November 1.}$$

Therefore, at solar noon

$$\alpha_1 = 90° - |-35.3° + 15.4°|$$

$$= 70.1°$$

and

$$\alpha_2 = 90° - |-35.3° - 4.02°|$$

$$= 50.68°.$$

Following the procedure in Example 2.2(a), therefore,

$$L = \frac{1}{\tan(70.1°) - \tan(50.68°)}$$

$$= 0.65 \text{ m}$$

and

$$S = 0.65 \tan(50.68°)$$

$$= 0.79 \text{ m}.$$

If S is given as 1.3 m, then

$$L = 1.3/\tan(50.68°) = 1.07 \text{ m}.$$

Also,

$$L = 2.3/\tan(70.1°) = 0.83 \text{ m};$$

therefore, $0.83 \text{ m} \leq L \geq 1.07 \text{ m}.$

2.3.2 Sun-Path Diagram

The projection of the sun's path on the horizontal plane is called a **sun-path diagram**. Such diagrams are very useful in determining shading phenomena associated with solar collectors, windows, and shading devices. As shown earlier, the solar angles (α, a_s) depend upon the hour angle, declination, and latitude. Since only two of these variables can be plotted on a two-dimensional graph, the usual method is to prepare a different sun-path diagram for each latitude with variations of hour angle and declination shown for a full year. A typical sun-path diagram is shown in Fig. 2.10 for 30°N latitude.

Sun-path diagrams for a given latitude are used by entering them with appropriate values of declination δ_s and hour angle h_s. The point at the intersection of the corresponding δ_s and h_s lines represents the instantaneous location of the sun. The solar altitude can then be read from the concentric circles in the diagram; the azimuth, from the scale around the circumference of the diagram. A complete set of sun-path diagrams is contained in Appendix 2 (Fig. A2.1).

Example 2.3. Using Fig. 2.10, determine the solar altitude and azimuth for March 8 at 10 a.m. Compare the results to those calculated from the basic equations (Eqs. 2.28 and 2.29).

Solution. On March 8 the solar declination is $-5°$; therefore the $-5°$ sun path is used. The intersection of the 10 AM line and the $-5°$ declination line in the diagram represents the sun's location; it is marked with a heavy dot in Fig. 2.10. The sun posi-

Declination	Approx. dates
+23°27'	June 22
+20°	May 21, July 24
+15°	May 1, Aug. 12
+10°	Apr. 16, Aug. 28
+ 5°	Apr. 3, Sep. 10
0°	Mar. 21, Sep. 23
− 5°	Mar. 8, Oct. 6
−10°	Feb. 23, Oct. 20
−15°	Feb. 9, Nov. 3
−20°	Jan. 21, Nov. 22
−23°27'	Dec. 22

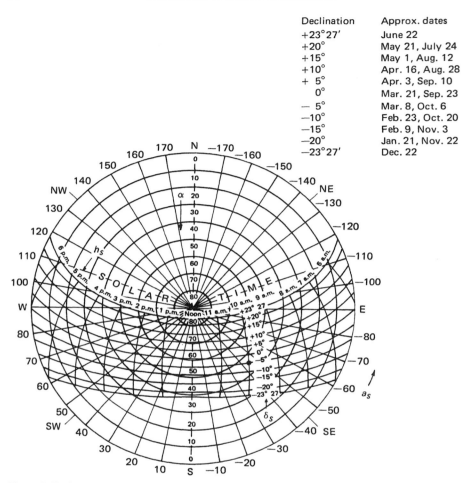

Figure 2.10. Sun-path diagram for 30°N latitude showing altitude and azimuth angles. Modified from Kreider, J.F., and F. Kreith, *Solar Heating and Cooling,* revised 1st ed. Washington, D.C.: Hemisphere Publ. Corp., 1977.

tion lies midway between the 40° and 50° altitude circles, say at 45°, and midway between the −40° and −50° azimuth radial lines, say at −45°. So $\alpha \cong 45°$ and $a_s \cong -45°$. Equations 2.28 and 2.29 give precise values for α and a_s:

$$\sin \alpha = \sin(30°)\sin(-5°) + \cos(30°)\cos(-5°)\cos(-30°)$$

$$\alpha = 44.7°$$

$$\sin a_s = \frac{\cos(-5°)\,\sin(-30°)}{\cos(44.7°)}$$

$$a_s = -44.5°$$

Therefore, the calculated values are within $\pm\ 0.5°$ (1 percent) of those read from the sun-path diagram.

2.3.3 The Shadow-Angle Protractor

The shadow-angle protractor used in shading calculations is a plot of solar-altitude angles, projected onto a given plane, versus solar-azimuth angle. The projected altitude angle is usually called the *profile angle* γ. It is defined as the angle between the normal to a surface and the projection of the sun's rays on a vertical plane normal to the same surface. The profile angle is shown in Fig. 2.11(a) with the corresponding solar-altitude angle. The profile angle, which is always used in sizing shading devices, is given by

$$\tan\gamma = \sec a \tan \alpha, \qquad (2.31)$$

where a is the solar azimuth angle with respect to the wall normal.

Figure 2.11(b) shows the shadow-angle protractor to the same scale as the sun-path diagrams in Fig. 2.10 and Appendix 2. It is used by plotting the limiting values of profile angle γ and azimuth angle a, which will start to cause shading of a particular point. The shadow-angle protractor is usually traced onto a transparent sheet so that the shadow map constructed on it can be placed over the pertinent sun-path diagram to indicate the times of day and months of the year during which shading will take place. The use of the shadow-angle protractor is best illustrated by an example.

Example 2.4. A solar building with a south-facing collector is sited to the north-northwest of an existing building. Prepare a shadow map showing what months of the year and what part of the day point C at the base of the solar collector will be shaded. Plan and elevation views are shown in Fig. 2.12. Latitude $= 40°$N.

Solution. The limiting profile angle for shading is $40°$ and the limiting azimuth angles are $-45°$ and $+10°$ as shown in Fig. 2.12. These values are plotted on the shadow-angle protractor (Fig. 2.13a). The shadow map, when superimposed on the sun-path diagram (Fig. 2.13b), shows that point C will be shaded during the following times of day for the periods shown:

Declination	Date	Time of Day
$-23°27'$	Dec 22	8:45 AM–12:40 PM
$-20°$	Jan 21, Nov 22	8:55 AM–12:35 PM
$-15°$	Feb 9, Nov 3	9:10 AM–12:30 PM

In summary, during the period from November 3 to February 9, point C will be shaded between 3 and 4 hours. It will be shown later that this represents about a 50 percent loss in collector performance for point C, which would be unacceptable for a collector to be used for heating a building in winter.

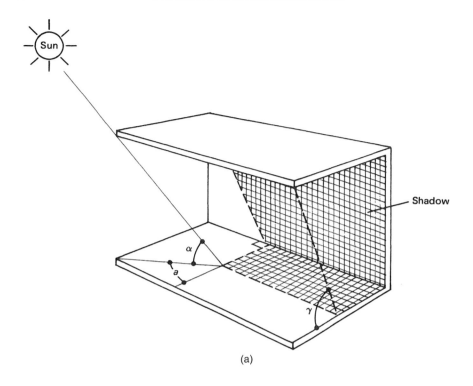

(a)

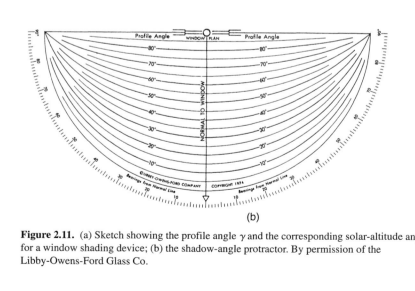

(b)

Figure 2.11. (a) Sketch showing the profile angle γ and the corresponding solar-altitude angle α for a window shading device; (b) the shadow-angle protractor. By permission of the Libby-Owens-Ford Glass Co.

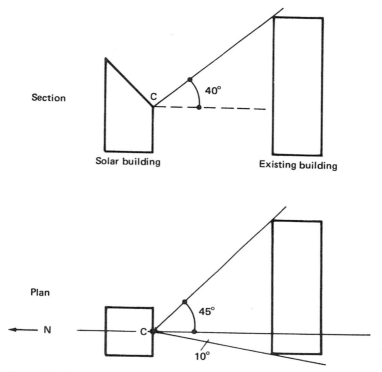

Figure 2.12. Plan and elevation view of proposed solar building and existing building, which may shade solar collector at point C.

2.4 SOLAR RADIATION

Detailed information about solar radiation availability at any location is essential for the design and economic evaluation of a solar energy system. Long-term measured data of solar radiation are available for a large number of locations in the United States and other parts of the world. Where long-term measured data are not available, various models based on available climatic data can be used to estimate the solar energy availability. Solar energy is in the form of electromagnetic radiation with the wavelengths ranging from about 0.3 μm (10^{-6}m) to over 3μm, which correspond to ultraviolet (less than 0.4 μm), visible (0.4 and 0.7 μm), and infrared (over 0.7 μm). Most of this energy is concentrated in the visible and the near-infrared wavelength range (see Fig. 2.14). The incident solar radiation, sometimes called **insolation**, is measured as irradiance, or the energy per unit time per unit area (or power per unit area). The units most often used are watts per square meter (W/m²), British thermal units per hour per square foot (Btu/hr−ft²), and Langleys per minute (calories per square centimeter per minute, cal/cm² −min).

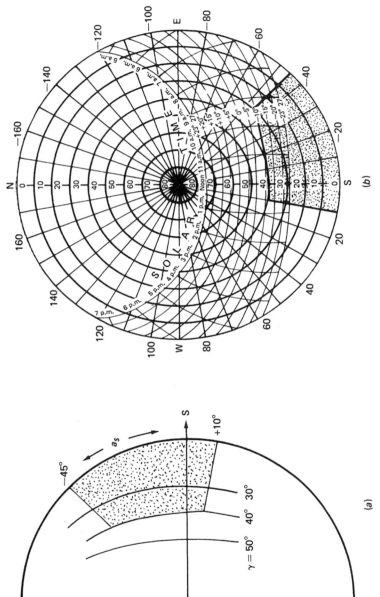

Figure 2.13. (a) Shadow map constructed for the example shown in Fig. 2.13; (b) shadow map superimposed on sun-path diagram.

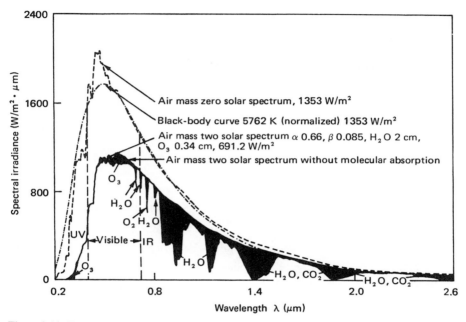

Figure 2.14. Extraterrestrial solar radiation spectral distribution. Also shown are equivalent black-body and atmosphere-attenuated spectra.

2.4.1 The Extraterrestrial Solar Radiation

The average amount of solar radiation falling on a surface normal to the rays of the sun outside the atmosphere of the earth (extraterrestrial) at mean earth-sun distance (D_o) is called the **solar constant**, I_o. Measurements by NASA indicated the value of the solar constant to be 1353 W/m² ($\pm 1.6\%$), 429 Btu/hr$-$ft² or 1.94 Cal/cm²$-$min (Langleys/min). This value was revised upward by Fröhlich et al. [16] to 1377 W/m² or 437.1 Btu/hr$-$ft² or 1.974 Langleys/min, which was the value used in compiling SOLMET data in the USA [60,61]. At present, there is no consensus on the value of the solar constant. However, considering that the difference between the two values is about 1.7% and the uncertainties in estimation of terrestrial solar radiation are of the order of 10% or higher, either value may be used. A value of 1367 W/m² is also used by many references.

The variation in seasonal solar radiation availability at the surface of the earth can be understood from the geometry of the relative movement of the earth around the sun. Since the earth's orbit is elliptical, the earth-sun distance varies during a year, the variation being $\pm 1.7\%$ from the average. Therefore, the extraterrestrial radiation, I, also varies by the inverse square law as below:

$$I = I_0(D_0/D)^2, \tag{2.32}$$

where D is the distance between the sun and the earth, and D_0 is the yearly mean earth-sun distance (1.496×10^{11}m). The $(D_0/D)^2$ factor may be approximated as [70]:

$$(D_0/D)^2 = 1.00011 + 0.034221\cos(x) + 0.00128\sin(x)$$
$$+ 0.000719\cos(2x) + 0.000077\sin(2x), \qquad (2.33)$$

where

$$x = 360(n - 1)/365°, \qquad (2.34)$$

and n = Day number (starting from January 1 as 1). The following approximate relationship may also be used without much loss of accuracy:

$$I = I_0[1 + 0.034\cos(360n/365.25)°]. \qquad (2.35)$$

Figure 2.15 also shows the relationship of the extraterrestrial solar radiation to the solar constant. For many solar energy applications, such as photovoltaics and photocatalysis, it is necessary to examine the distribution of energy within the solar spectrum. Figure 2.15 shows the spectral irradiance at the mean earth-sun distance for a solar constant of 1353 W/m² as a function of wavelength according to the standard spectrum data published by NASA in 1971. The data are also presented in Table 2.3 and their use is illustrated in the following example.

Example 2.5. Calculate the fraction of solar radiation within the visible part of the spectrum, that is, between 0.40 and 0.70 μm.

Solution. The first column in Table 2.3 gives the wavelength. The second column gives the averaged solar spectral irradiance in a band centered at the wavelength in the first column. The fourth column, $D\lambda$, gives the percentage of solar total radiation at

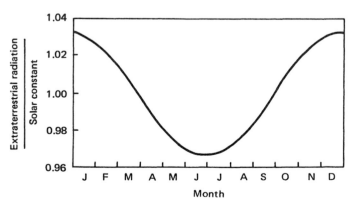

Figure 2.15. Effect of the time of year on the radio of extraterrestrial radiation to the nominal solar constant.

wavelengths shorter than the value of λ in the first column. At a value of 0.40 μm, 8.7 percent of the total radiation occurs at shorter wavelengths. At a wavelength of 0.70, 46.88 percent of the radiation occurs at shorter wavelengths. Consequently, 38 percent of the total radiation lies within the band between 0.40 and 0.70 μm, and the total energy received outside the earth's atmosphere within that spectral range is 517 W/m² (163 Btu/hr · ft²).

2.5 ESTIMATION OF TERRESTRIAL SOLAR RADIATION

As extraterrestrial solar radiation, I, passes through the atmosphere, a part of it is reflected back into space, a part is absorbed by air and water vapor, and some gets scattered by molecules of air, water vapor, aerosols and dust particles (Fig. 2.16). The part of solar radiation that reaches the surface of the earth with essentially no change in direction is called **direct** or **beam radiation**. The scattered diffuse radiation reaching the surface from the sky is called the **sky diffuse radiation**.

Although extraterrestrial radiation can be predicted with certainty,* radiation levels on the earth are subject to considerable uncertainty resulting from local climatic in-

*The effect of sunspots, which may cause up to 0.5% variation, is neglected.

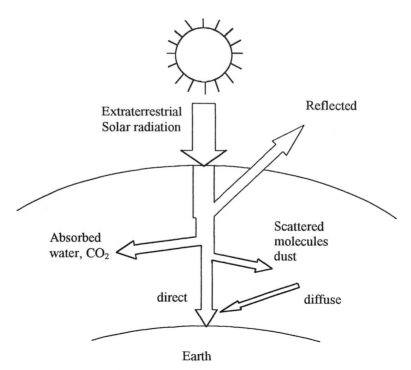

Earth

Figure 2.16. Attenuation of solar radiation as it passes through the atmosphere.

Table 2.3. Extraterrestrial solar irradiance[a,b]

λ (μm)	Eλ[c] (W/m²·μm)	Eλ[c] (Btu/hr·ft²·μm)	Dλ[d] (%)	λ (μm)	Eλ (W/m²·μm)	Eλ (Btu/hr·ft²·μm)	Dλ (%)	λ (μm)	Eλ (W/m²·μm)	Eλ (Btu/hr·ft²·μm)	Dλ (%)
0.115	0.007	0.002	1×10^{-4}	0.43	1639	520	12.47	0.90	891	283	63.37
0.14	0.03	0.010	5×10^{-4}	0.44	1810	574	13.73	1.00	748	237	69.49
0.16	0.23	0.073	6×10^{-4}	0.45	2006	636	15.14	1.2	485	154	78.40
0.18	1.25	0.397	1.6×10^{-3}	0.46	2066	655	16.65	1.4	337	107	84.33
0.20	10.7	3.39	8.1×10^{-3}	0.47	2033	645	18.17	1.6	245	77.7	88.61
0.22	57.5	18.2	0.05	0.48	2074	658	19.68	1.8	159	50.4	91.59
0.23	66.7	21.2	0.10	0.49	1950	619	21.15	2.0	103	32.7	93.49
0.24	63.0	20.0	0.14	0.50	1942	616	22.60	2.2	79	25.1	94.83
0.25	70.9	22.5	0.19	0.51	1882	597	24.01	2.4	62	19.7	95.86
0.26	130	41.2	0.27	0.52	1833	581	25.38	2.6	48	15.2	96.67
0.27	232	73.6	0.41	0.53	1842	584	26.74	2.8	39	12.4	97.31
0.28	222	70.4	0.56	0.54	1783	566	28.08	3.0	31	9.83	97.83
0.29	482	153	0.81	0.55	1725	547	29.38	3.2	22.6	7.17	98.22
0.30	514	163	1.21	0.56	1695	538	30.65	3.4	16.6	5.27	98.50
0.31	689	219	1.66	0.57	1712	543	31.91	3.6	13.5	4.28	98.72
0.32	830	263	2.22	0.58	1715	544	33.18	3.8	11.1	3.52	98.91
0.33	1059	336	2.93	0.59	1700	539	34.44	4.0	9.5	3.01	99.06
0.34	1074	341	3.72	0.60	1666	528	35.68	4.5	5.9	1.87	99.34
0.35	1093	347	4.52	0.62	1602	508	38.10	5.0	3.8	1.21	99.51
0.36	1068	339	5.32	0.64	1544	490	40.42	6.0	1.8	0.57	99.72
0.37	1181	375	6.15	0.66	1486	471	42.66	7.0	1.0	0.32	99.82
0.38	1120	355	7.00	0.68	1427	453	44.81	8.0	0.59	0.19	99.88
0.39	1098	348	7.82	0.70	1369	434	46.88	10.0	0.24	0.076	99.94
0.40	1429	453	8.73	0.72	1314	417	48.86	15.0	0.0048	0.015	99.98
0.41	1751	555	9.92	0.75	1235	392	51.69	20.0	0.0015	0.005	99.99
0.42	1747	554	11.22	0.80	1109	352	56.02	50.0	0.0004	0.0001	100.00

[a]Adapted from [75].
[b]Solar constant = 429 Btu/hr · ft² = 1353 W/m².
[c]Eλ is the solar spectral irradiance averaged over a small bandwidth centered at λ.
[d]Dλ is the percentage of the solar constant associated with wavelengths shorter than λ.

teractions. The most useful solar radiation data is based on long-term (30 years or more) measured average values at a location, which unfortunately are not available for most locations in the world. For such locations, an estimating method (theoretical model) based on some measured climatic parameter may be used. This chapter describes several ways of estimating terrestrial solar radiation; all have large uncertainties (as much as $\pm 30\%$) associated with them.

2.5.1 Atmospheric Extinction of Solar Radiation

As solar radiation I travels through the atmosphere, it is attenuated due to absorption and scattering. If K is the local extinction coefficient of the atmosphere, the beam solar radiation at the surface of the earth can be written according to Bouger's law Eq. (2.21) as:

$$I_{b,N} = Ie^{-\int Kdx},\qquad(2.36)$$

where $I_{b,N}$ is the instantaneous beam solar radiation per unit area normal to the sun's rays, and x is the length of travel through the atmosphere. If L_o is the vertical thickness of the atmosphere and

$$\int_0^{L_o} Kdx = k,\qquad(2.37)$$

the beam normal solar radiation for a solar zenith angle of z will be:

$$I_{b,N} = Ie^{-k\sec z} = Ie^{-k/\sin \alpha} = Ie^{-km},\qquad(2.38)$$

where m is a dimensionless path length of sunlight through the atmosphere, sometimes called the **air mass ratio** (Fig. 2.17). When solar altitude angle is $90°$ (sun is overhead), $m = 1$.

Threlkeld and Jordan [76] estimated values of k (also known as optical depth) for average atmospheric conditions at sea level with a moderately dusty atmosphere and the amount of precipitable water vapor equal to the average value for the United States for each month. These values are given in Table 2.4. To account for the differences in local conditions from the average sea level conditions, Eq. (2.38) is modified by a parameter called clearness number C_n, introduced by Threlkeld and Jordan [76]:

$$I_{b,N} = C_n Ie^{-k/\sin \alpha}.\qquad(2.39)$$

2.5.2 Solar Radiation on Clear Days

Total instantaneous solar radiation on a horizontal surface (see Fig. 2.18), I_h, is a sum of the beam or direct radiation, $I_{b,h}$ and the sky diffuse radiation $I_{d,h}$:

$$I_h = I_{b,h} + I_{d,h}.\qquad(2.40)$$

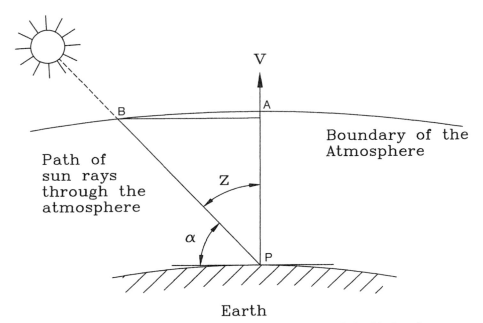

Figure 2.17. Air mass definition; air mass $m = BP/AP = \text{cosec } \alpha$, where α is the altitude angle. The atmosphere is idealized as a constant thickness layer.

According to the Threlkeld and Jordan model [76], the sky diffuse radiation on a clear day is proportional to the beam normal solar radiation and can be estimated by using an empirical sky diffuse factor C. Therefore, I_h can be estimated as:

$$I_h = I_{b,N} \cos z + CI_{b,N}$$
$$= I_{b,N} \sin \alpha + CI_{b,N}$$
$$= C_n I e^{-k/\sin \alpha} (C + \sin \alpha). \tag{2.41}$$

Values of C are given in Table 2.4.

Table 2.4. Average values of atmospheric optical depth (k) and sky diffuse factor (C) for 21st day of each month, for average atmospheric conditions at sea level for the United States

Month	1	2	3	4	5	6	7	8	9	10	11	12
k	0.142	0.144	0.156	0.180	0.196	0.205	0.207	0.201	0.177	0.160	0.149	0.142
C	0.058	0.060	0.071	0.097	0.121	0.134	0.136	0.122	0.092	0.073	0.063	0.057

Source: Threlkeld, J.L. and Jordan, R.C., *ASRAE Trans.*, 64:45 (1958) [76].

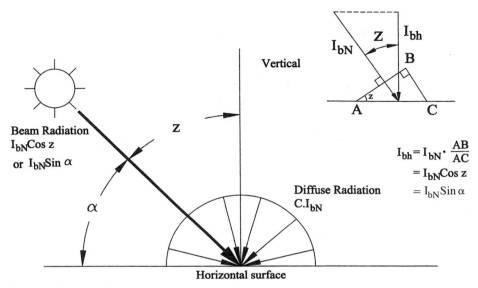

Figure 2.18. Solar radiation on a horizontal surface.

2.5.3 Solar Radiation on a Tilted Surface

Solar radiation on an arbitrary tilted surface having a tilt angle of β from the horizontal and an azimuth angle of a_w (assumed + west of south), as shown in Fig. 2.19, is the sum of components consisting of beam ($I_{b,c}$), sky diffuse ($I_{d,c}$) and ground reflected solar radiation ($I_{r,c}$):

$$I_c = I_{b,c} + I_{d,c} + I_{r,c}. \tag{2.42}$$

If i is the *angle of incidence* of the beam radiation on the tilted surface, it is simple to show that the instantaneous beam radiation on the surface per unit area is:

$$I_{b,c} = I_{b,N} \cos i. \tag{2.43}$$

From the geometry in Fig. 2.19, it can be shown that the angle of incidence i for the surface (angle between the normal to the surface and a line collinear with the sun's rays) is related to the solar angles as

$$\cos i = \cos \alpha \cos (a_s - a_w)\sin \beta + \sin \alpha \cos \beta. \tag{2.44}$$

The diffuse radiation on the surface ($I_{d,c}$) can be obtained by multiplying the sky diffuse radiation on a horizontal surface by the view factor between the sky and the surface:*

---

*The surface has been assumed infinitely large for this view factor. See Section 2.2.3.

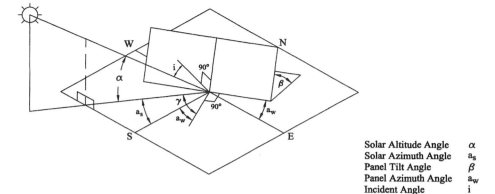

Figure 2.19. Definitions of solar angles for a tilted surface.

Solar Altitude Angle α
Solar Azimuth Angle a_s
Panel Tilt Angle β
Panel Azimuth Angle a_w
Incident Angle   i

$$I_{d,c} = I_{d,h}(1 + \cos\beta)/2$$

$$= CI_{b,N}(1 + \cos\beta)/2$$

$$= CI_{b,N}\cos^2(\beta/2). \tag{2.45}$$

The ground reflected solar radiation can be found from the total solar radiation incident on a horizontal surface and the ground reflectance ρ as:

$$I_{r,c} = I_h\rho. \tag{2.46}$$

The part of I_r intercepted by the tilted surface can be found by multiplying the ground reflected radiation by the view factor between the surface* and the ground:

$$I_{r,c} = \rho I_h(1 - \cos\beta)/2 = \rho I_h \sin^2(\beta/2) \tag{2.47}$$

$$= \rho I_{b,N}(\sin\alpha + C)\sin^2(\beta/2).$$

For ordinary ground or grass, ρ is approximately 0.2 and for snow covered ground it can be taken as approximately 0.8.

Example 2.6(a). Find the instantaneous solar radiation at 12:00 noon Eastern Standard Time on a solar collector surface ($\beta = 30°$, $a_w = +10°$) on February 1 in Gainesville, FL.

Solution. From Example 2.1, for February 1:

$$n = 32, \quad \delta_s = -17.5°, \text{ and } ET = -13.7 \text{ min}$$

*The tilted surface and the ground in front of it have been assumed to be infinitely large for this view factor.

For finding the values of solar radiation on the collector, we will need to calculate angles α, a_s, h_s, and i.

$$\text{Solar Time} = LST + ET + 4(l_{st} - l_{local})$$

$$= 12:00 - 13.7 \text{ min} + 4(75° - 82.27°)$$

$$= 11:17.2 \text{ AM},$$

$$h_s = \frac{\text{minutes from solar noon}}{4 \text{ min/deg}},$$

$$= \frac{-42.8}{4} = -10.7° \quad (-\text{before noon}).$$

From Eq. (2.28),

$$\alpha = \sin^{-1}(\sin(29.68°)\sin(-17.5°) + \cos(29.68°)\cos(-17.5°)\cos(-10.7°))$$

$$= 41.7°.$$

From Eq. (2.29),

$$a_s = \sin^{-1}(\cos(-17.5°)\sin(-10.7°)/\cos(41.7°))$$

$$= -13.7°.$$

Angle of incidence i for the solar collector is given by Eq. (2.44):

$$\cos i = \cos(41.7°)\cos(-13.7° - 10°)\sin(30°) + \sin(41.7°)\cos(30°)$$

$$i = 23.4°$$

To calculate the solar radiation using Eqs. (2.42, 2.43, 2.45, and 2.47), we need to find I_{bN} and I.

Extraterrestrial Solar Radiation

$$I = I_0[1 + 0.034 \cos(360n/365.25)°]$$

$$= 1353\text{W/m}^2[1 + 0.034 \cos(360x32/365.25)]$$

$$= 1392\text{W/m}^2,$$

$$I_{b,N} = C_n I e^{-k/\sin \alpha}$$

(Find k from Table 2.4. Assume $C_n = 1$)

$$= 1392 e^{-0.144/\sin(41.7°)}$$

$$= 1121\text{W/m}^2.$$

Beam radiation on the collector (Eq. 2.43):

$$I_{b,c} = I_{b,N} \cos i$$
$$= (1121)\cos(23.4°)$$
$$= 1029 \text{W}/\text{m}^2.$$

Sky diffuse radiation on the collector (Eq. 2.45):

$$I_{d,c} = CI_{b,N} \cos^2(\beta/2)$$
$$= (0.060)(1121)\cos^2(30/2)$$

(Find C from Table 2.4.)

$$63 \text{W}/\text{m}^2.$$

Ground reflected radiation on the collector (Eq. 2.47):

$$I_{r,c} = \rho I_{b,N}(\sin \alpha + C)\sin^2(\beta/2)$$

Assume

$$\rho = 0.2$$
$$I_{r,c} = (0.2)(1121)(\sin 41.7° + 0.060)\sin^2(15°)$$
$$= 11 \text{W}/\text{m}^2.$$

Total insolation on the collector:

$$I_c = 1029 + 63 + 11 = 1103 \text{W}/\text{m}^2.$$

Example 2.6(b). Repeat the calculations in Example 2.6(a) for a north facing solar collector ($\beta = 30°$, $a_w = 10°$) in Buenos Aires, Argentina. (latitude = $34° - 35'$S, longitude = $58° - 29'$W, Standard Meridian = $45°$W)

Solution. Following the procedure in Example 2.6(a), the solar angles and the angle of incidence are found as:

$$\text{Solar Time} = 12{:}00 - 13.7 \text{ min} + 4(45° - 58.48°)\text{min}$$
$$= 10{:}52 \text{ AM}$$
$$h_s = -68/4 = -17°$$
$$\alpha = \sin^{-1}[\sin(-34.58°)\sin(-17.5°) + \cos(-34.58°)\cos(-17.5°)\cos(-17°)]$$
$$= 67.2°$$

$$a_s = \sin^{-1}[\cos(-17.5°)\sin(-17°)/\cos(67.2°)]$$
$$= -46°$$

Therefore, the angle of incidence

$$i = \cos^{-1}[\cos(67.2°)\cos(-46° - 10°)\sin(30°) + \sin(67.2°)\cos(30°)]$$
$$= 24.95°$$

Assuming that the solar radiation model used in Example 2.2(a) applies. (Note: the model was developed for locations in the U.S.)

$$I_{b,N} = 1392\, e^{-0.144/\sin(67.2°)}$$
$$= 1191\ \text{W/m}^2$$
$$I_{b,c} = 1191\cos(24.95°)$$
$$= 1080\ \text{W/m}^2$$
$$I_{d,c} = (0.060)(1191)\cos^2(30°/2)$$
$$= 67\ \text{W/m}^2$$
$$I_{r,c} = (0.2)(1191)(\sin 67.2° + 0.060)\sin^2(15°)$$
$$= 16\ \text{W/m}^2$$

Total insolation on the collector

$$I_c = 1080 + 67 + 16$$
$$= 1163\ \text{W/m}^2$$

2.5.4 Monthly Solar Radiation Estimation Models

One of the earliest methods of estimating solar radiation on a horizontal surface was proposed by the pioneer spectroscopist Angström. It was a simple linear model relating average horizontal radiation to clear-day radiation and to the sunshine level, that is, percent of possible hours of sunshine. Since the definition of a clear day is somewhat nebulous, Page [57] refined the method and based it on extraterrestrial radiation instead of the ill-defined clear day:

$$\overline{H}_h = \overline{H}_{o,h}\left(a + b\frac{\overline{n}}{N}\right)$$

$$= \overline{H}_{o,h}\left(a + b\frac{\overline{PS}}{100}\right) \tag{2.48}$$

where $\overline{H}_h$ and $\overline{H}_{o,h}$ are the horizontal terrestrial, and horizontal extraterrestrial radiation levels averaged for a month, $\overline{PS}$ is the monthly averaged percent of possible sunshine (that is, hours of sunshine/maximum possible duration of sunshine $\times$ 100), a and b are constants for a given site, and $\overline{n}$ and $\overline{N}$ are the monthly average numbers of hours of bright sunshine and day length respectively. The ratio $\overline{n}/\overline{N}$ is also equivalent to the monthly average percent sunshine $(\overline{PS})$. $\overline{H}_{o,h}$ can be calculated by finding $H_{o,h}$ from the following equation (2.49), using Eqs. (2.35) and (2.44) and averaging $I_{o,h}$ for the number of days in each month; or data in Appendix 2 (Table A2.2) can be used.

$$H_{o,h} = \int_{t_{sr}}^{t_{ss}} I \sin \alpha dt \qquad (2.49)$$

Some typical values of a and b are given in Table 2.5 [46]. Additional values for worldwide locations are given in Appendix 2 (Table A2.4).

Example 2.7. Using the predictive method of Angström or Page, estimate the monthly solar radiation for the North Central Sahara Desert (Tamanrasset, Algeria area) at latitude $= 25°$N. Percentages of possible sunshine and extraterrestrial radiation for this site are given in the table below.

Solution. Using the climate data given, the expected monthly average horizontal radiation for the North Sahara is calculated in the following table using $a = 0.30$ and $b = 0.43$ from Table 2.5.

Month	$\overline{H}_{o,h}{}^a$		$\overline{H}_h$	
	$\overline{PS}/100$	(kJ/m² · day)	kJ/m² · day	Btu/ft² · day
Jan.	0.88	23,902	16,215	1425
Feb.	0.83	28,115	18,469	1626
Mar.	0.90	32,848	22,143	1950
Apr.	0.85	37,111	24,697	2174
May	0.80	39.356	25,345	2231
June	0.76	40,046	25,101	2210
Jul.	0.86	39,606	36,528	2336
Aug.	0.83	37,832	24,852	2188
Sep.	0.77	34,238	21,608	1902
Oct.	0.86	29,413	19,701	1735
Nov.	0.85	24,909	16,577	1460
Dec.	0.80	22,669	14,599	1285

*a*Monthly averaged, daily extraterrestrial radiation.

The high levels of radiation predicted above are typical of the North Sahel region and are higher than most U.S. locations except the Mojave Desert.

A number of researchers found Angström-Page type correlations for specific locations which are listed in Table 2.6. Some of these include additional parameters such as

Table 2.5. Coefficients *a* and *b* in the Angström-Page regression equation[+]

| Location | Climate[++] | Sunshine hours in percentage of possible | | *a* | *b* |
		Range	Avg.		
Albuquerque, NM	BS-BW	68–85	78	0.41	0.37
Atlanta, GA	Cf	45–71	59	0.38	0.26
Blue Hill, MA	Df	42–60	52	0.22	0.50
Brownsville, TX	BS	47–80	62	0.35	0.31
Buenos Aires, Arg.	Cf	47–68	59	0.26	0.50
Charleston, SC	Cf	60–75	67	0.48	0.09
Dairen, Manchuria	Dw	55–81	67	0.36	0.23
El Paso, TX	BW	78–88	84	0.54	0.20
Ely, NV	BW	61–89	77	0.54	0.18
Hamburg, Germany	Cf	11–49	36	0.22	0.57
Honolulu, HI	Af	57–77	65	0.14	0.73
Madison, WI	Df	40–72	58	0.30	0.34
Malange, Angola	Aw-BS	41–84	58	0.34	0.34
Miami, FL	Aw	56–71	65	0.42	0.22
Nice, France	Cs	49–76	61	0.17	0.63
Poona, India (monsoon)	Am	25–49	37	0.30	0.51
(dry)		65–89	81	0.41	0.34
Stanleyville, Congo	Af	34–56	48	0.28	0.39
Tamanrasset, Algeria	BW	76–88	83	0.30	0.43

[+]From Löf et al. [46] with permission.

[++] Af = tropical forest climate, constantly moist, rainfall all through the year

    Am = tropical forest climate, monsoon rain, short dry season, but total rainfall sufficient to support rain forest

    Aw = tropical forest climate, dry season in winter

    BS = steppe or semiarid climate

    BW = desert or arid climate

    Cf = mesothermal forest climate, constantly moist, rainfall all through the year

    Cs = mesothermal forest climate, dry season in winter

    Df = microthermal snow forest climate, constantly moist, rainfall all through the year

    Dw = microthermal snow forest climate, dry season in winter

relative humidity and ambient temperature. Correlations listed in the table may be used for the specific locations for which they were developed.

    Another meteorological variable that could be used for solar radiation prediction is the opaque cloud cover recorded at many weather stations around the world. This quantity is a measure of the percent of the sky dome obscured by opaque clouds. Because this parameter contains even less solar information than sunshine values, it has not been useful in predicting long-term solar radiation values. A subsequent section, however, will show that cloud cover, when used with solar altitude angle or air mass, is a useful estimator of hourly direct radiation.

Table 2.6. Angström-Page type correlations for specific locations

Authors	Measured data correlated	Correlation equations*
Iqbal [28]	Canada, 3 locations	$\dfrac{\overline{D}_h}{\overline{H}_h} = 0.791 - 0.635\left(\dfrac{\overline{n}}{\overline{N}}\right)$ $\dfrac{\overline{H}_d}{\overline{H}_h} = 0.163 + 0.478\left(\dfrac{\overline{n}}{\overline{N}}\right) - 0.655\left(\dfrac{\overline{n}}{\overline{N}}\right)^2$ $\dfrac{\overline{H}_b}{\overline{H}_{o,h}} = -0.176 + 1.45\left(\dfrac{\overline{n}}{\overline{N}}\right) - 1.12\left(\dfrac{\overline{n}}{\overline{N}}\right)^2$
Garg [17]	India, 11 locations, 20 years' data	$\dfrac{\overline{H}_h}{\overline{H}_{o,h}} = 0.3156 + 0.4520\left(\dfrac{\overline{n}}{\overline{N}}\right)^2$ $\dfrac{\overline{D}_h}{\overline{H}_{o,h}} = 0.3616 - 0.2123\left(\dfrac{\overline{n}}{\overline{N}}\right)$ $\dfrac{\overline{D}_h}{\overline{H}_h} = 0.8677 - 0.7365\left(\dfrac{\overline{n}}{\overline{N}}\right)$
Hussain [25]	India	$\dfrac{\overline{H}_h}{\overline{H}_{o,h}} = 0.394 + 0.364\left[\dfrac{\overline{n}}{\overline{N'}}\right] - 0.0035 W_{at}$ $\dfrac{\overline{D}_h}{\overline{H}_{o,h}} = 0.306 - 0.165\left[\dfrac{\overline{n}}{\overline{N'}}\right] - 0.0025 W_{at}$
Coppolino [11]	Italy	$\dfrac{\overline{H}_h}{\overline{H}_{o,h}} = 0.67\left(\dfrac{\overline{n}}{\overline{N}}\right)^{0.45} \sin(\alpha_{sn})^{0.05}$ α_{sn} = Solar Elevation at noon on the 15th of each month, degrees $0.15 \leq \dfrac{\overline{n}}{\overline{N}} \leq 0.90$
Akinoglu & Ecevit [2]	Italy	$\dfrac{\overline{H}_h}{\overline{H}_{o,h}} = 0.145 + 0.845\left(\dfrac{\overline{n}}{\overline{N}}\right) - 0.280\left(\dfrac{\overline{n}}{\overline{N}}\right)^2$
Ögelman et al. [56]	Turkey, 2 locations, 3 years' data	$\left(\dfrac{\overline{H}_h}{\overline{H}_{o,h}}\right) = 0.204 + 0.758\left(\dfrac{\overline{n}}{\overline{N}}\right) - 0.250\left\{\left[\left(\dfrac{\overline{n}}{\overline{N}}\right)^2\right]^2 + \sigma^2_{\frac{\overline{n}}{\overline{N}}}\right\}$ $\sigma^2_{\frac{\overline{n}}{\overline{N}}} = 0.035 + 0.326\left(\dfrac{\overline{n}}{\overline{N}}\right) - 0.433\left(\dfrac{\overline{n}}{\overline{N}}\right)^2$

Table 2.6. (*Continued*)

Authors	Measured data correlated	Correlation equations*
Gopinathan [19]	40 locations around the world	$\dfrac{\overline{H}_h}{\overline{H}_{o,h}} = a + b\left(\dfrac{\overline{n}}{\overline{N}}\right)$
		$a = -.309 + .539 \cos L - .0639\, h + 0.290\left(\dfrac{\overline{n}}{\overline{N}}\right)$
		$b = 1.527 - 1.027 \cos L + .0926\, h - .359\left(\dfrac{\overline{n}}{\overline{N}}\right)$

*$\overline{H}_a, \overline{H}_b, \overline{H}_{o,h}, \overline{D}_h$ are monthly averaged daily values.
N' = maximum duration for which Campbell-Stokes recorder can be active, i.e., Solar Elevation >5°.
W_{at} = relative humidity $\times$ (4.7923 + 0.3647T + 0.055T^2 + 0.0003T^3)
T = ambient temperature °C
W_{at} = gm moisture/m³
h = Elevation in km above sea level.
L = latitude

2.6 MODELS BASED ON LONG-TERM MEASURED HORIZONTAL SOLAR RADIATION

Long-term measured solar radiation data is usually available as monthly averaged total solar radiation per day on horizontal surfaces. In order to use this data for tilted surfaces, the total solar radiation on a horizontal surface must first be broken down into beam and diffuse components. A number of researchers have proposed models to do that, prominent among them being Liu and Jordan, Collares-Pereira and Rabl, and Erbs, Duffie and Klein.

2.6.1 Monthly Solar Radiation on Tilted Surfaces

In a series of papers, Liu and Jordan [41–45] have developed an essential simplification in the basically complex computational method required to calculate long-term radiation on tilted surfaces. This is called the LJ method. The fundamental problem in such calculations is the decomposition of long-term measured total horizontal radiation into its beam and diffuse components.

If the decomposition can be computed, the trigonometric analysis presented earlier can be used to calculate incident radiation on any surface in a straightforward manner. Liu and Jordan (LJ) correlated the diffuse-to-total radiation ratio $(\overline{D}_h/\overline{H}_h)$ with the **monthly clearness index** $\overline{K}_T$, which is defined as

$$\overline{K}_T = \frac{\overline{H}_h}{\overline{H}_{o,h}}, \tag{2.50}$$

where $\overline{H}_h$ is the monthly averaged terrestrial radiation per day on a horizontal surface. $\overline{H}_{o,h}$ is the corresponding extraterrestrial radiation, which can be calculated from Eq.

(2.49) by averaging each daily total for a month. The original LJ method was based upon the extraterrestrial radiation at midmonth, which is not truly an average.

The LJ correlation predicts the monthly diffuse $(\overline{D}_h)$ to monthly total $\overline{H}_h$ ratio. It can be expressed by the empirical Eq.

$$\frac{\overline{D}_h}{\overline{H}_h} = 1.390 - 4.027\,\overline{K}_T + 5.531\overline{K}_T^2 - 3.108\overline{K}_T^3. \tag{2.51}$$

Note that the LJ correlation is based upon a solar constant value of 1394 W/m² (442 Btu/hr · ft²), which was obtained from terrestrial observations, whereas the newer value, based on satellite data, is 1377 W/m² (437 Btu/hr · ft²). The values of $\overline{K}_T$ must be based on this earlier value of the solar constant to use the LJ method. Collares-Pereira and Rabl [10] conducted a study and concluded that although Liu and Jordan's approach is valid, their correlations would predict significantly smaller diffuse radiation components. They also concluded that Liu and Jordan were able to correlate their model with the measured data because they used the measured data which was not corrected for the shade ring (see solar radiation measurements). Collares-Pereira and Rabl (C-P&R) also introduced the sunset hour angle h_{ss} in their correlation to account for the seasonal variation in the diffuse component. The C-P&R correlation is:

$$\frac{\overline{D}_h}{\overline{H}_h} = 0.775 + 0.347\left(h_{ss} - \frac{\pi}{2}\right) - \left[0.505 + 0.0261\left(h_{ss} - \frac{\pi}{2}\right)\right]\cos(2K_T - 1.8), \tag{2.52}$$

where h_{ss} is the sunset hour angle in radians. The C-P&R correlation agrees well with the correlations for India [8], Israel [71] and Canada [66] and is, therefore, preferred to equation (2.51).

The monthly average beam component $\overline{B}_h$ on a horizontal surface can be readily calculated by simple subtraction since $\overline{D}_h$ is known:

$$\overline{B}_h = \overline{H}_h - \overline{D}_h. \tag{2.53}$$

It will be recalled on an instantaneous basis from Eqs. (2.41) and (2.43) and Fig. (2.19) that

$$I_{b,N} = \frac{I_{b,h}}{\sin\alpha}, \tag{2.54}$$

$$I_{b,c} = I_{b,N}\cos i, \tag{2.43}$$

where $I_{b,h}$ is the instantaneous horizontal beam radiation. Solving for $I_{b,c}$, the beam radiation on a surface,

$$I_{b,c} = I_{b,h}\left(\frac{\cos i}{\sin\alpha}\right). \tag{2.55}$$

The ratio in parentheses is usually called the beam radiation *tilt factor* R_b. It is a purely geometric quantity that converts instantaneous horizontal beam radiation to beam radiation intercepted by a tilted surface.

Equation (2.55) cannot be used directly for the long-term beam radiation $\overline{B}_h$. To be strictly correct, the instantaneous tilt factor R_b should be integrated over a month with the beam component $I_{b,h}$ used as a weighting factor to calculate the beam tilt factor. However, the LJ method is used precisely when such short-term data as $I_{b,h}$ are not available. The LJ recommendation for the monthly mean tilt factor $\overline{R}_b$ is simply to calculate the monthly average of cos i and divide it by the same average of sin α. In equation form for south-facing surfaces, this operation yields:

$$\overline{R}_b = \frac{\cos(L - \beta)\cos\delta_s \sin h_{sr} + h_{sr} \sin(L - \beta) \sin\delta_s}{\cos L \cos\delta_s \sin h_{sr} (\alpha = 0) + h_{sr}(\alpha = 0) \sin L \sin \delta_s}, \tag{2.56}$$

where the sunrise hour angle $h_{sr}(\alpha = 0)$ in radians is given by Eq (2.30) and h_{sr} is the min [$|h_s (\alpha = 0)|$, $|h_s (i = 90°)|$], respectively, and are evaluated at midmonth. Non-south-facing surfaces require numerical integration or iterative methods to determine $\overline{R}_b$. The long-term beam radiation on a tilted surface $\overline{B}_c$ is then,

$$\overline{B}_c = \overline{R}_b \overline{B}_h, \tag{2.57}$$

which is the long-term analog of Eq. (2.43). Values of $\overline{R}_b$ are tabulated in Appendix 2, Table A2.5.

Diffuse radiation intercepted by a tilted surface differs from that on a horizontal surface, because a tilted surface does not view the entire sky dome, which is the source of diffuse radiation. If the sky is assumed to be an isotropic source of diffuse radiation, the instantaneous and long-term tilt factors for diffuse radiation, R_d and $\overline{R}_d$ respectively, are equal and are simply the radiation view factor from the plane to the visible portion of a hemisphere. In equation form:

$$R_d = \overline{R}_d = \cos^2 \frac{\beta}{2} = (1 + \cos\beta)/2. \tag{2.58}$$

In some cases where solar collectors are mounted near the ground, some beam and diffuse radiation reflected from the ground can be intercepted by the collector surface. The tilt factor $\overline{R}_r$ for reflected total radiation ($\overline{D}_h + \overline{B}_h$) is then calculated to be

$$\overline{R}_r = \frac{\overline{R}}{\overline{D}_h + \overline{B}_h} = \rho \sin^2 \frac{\beta}{2} = \rho(1 - \cos\beta)/2, \tag{2.59}$$

in which ρ is the diffuse reflectance of the surface south of the collector assumed uniform and of infinite extent.

For snow, $\rho \cong 0.75$; for grass and concrete, $\rho \cong 0.2$. A more complete list of reflectances is provided in Table A2.7 of Appendix 2. The total long-term radiation intercepted by a surface $\overline{H}_c$ is then the total of beam, diffuse, and diffusely reflected components:

$$\overline{H}_c = \overline{R}_b\overline{B}_h + \overline{R}_d\overline{D}_h + \overline{R}_r(\overline{D}_h + \overline{B}_h). \tag{2.60}$$

Using Eqs. (2.61) and (2.62), we have

$$\overline{H}_c = \overline{R}_b\overline{B}_h + \overline{D}_h\cos^2\frac{\beta}{2} + (\overline{D}_h + \overline{B}_h)\rho\sin^2\frac{\beta}{2}, \tag{2.61}$$

in which $\overline{R}_b$ is calculated from Eq. (2.56).

Example 2.8. Using the $\overline{H}_h$ data calculated in Example 2.7 in place of the long-term measured data for the North Central Sahara Desert at latitude 25°N, find the monthly averaged insolation per day on a south-facing solar collector tilted at an angle of 25° from the horizontal.

Solution. The solution below is for the month of January. Values for the other months can be found by following the same method.

$$\overline{H}_h = 16,215\text{kJ/m}^2 - \text{day}.$$

From Table A2.2a:

$$\overline{H}_{o,h} = 23,902.$$

Therefore,

$$\overline{K}_T = \overline{H}_h/\overline{H}_{o,h} = 0.678.$$

δ_s and h_{sr} can be found for the middle of the month (January 16).

$$\delta_s = 23.45° \sin[360(284 + 16)365]°$$
$$= -21.1°,$$
$$h_{sr}(\alpha = 0) = -\cos^{-1}(-\tan L \tan \delta)$$
$$= -79.6° \text{ or } -1.389\text{rad}.$$
$$\text{and } h_{ss} = 1.389$$

Using CP & R correlation,

$$\frac{\overline{D_h}}{\overline{H_h}} = 0.775 + 0.347(1.389 - 1.5708)$$

$$- [0.505 + 0.0261(1.389 - 1.5708)]\cos(2 \times 0.678 - 1.8)$$

$$= 0.212$$

Therefore,

$$\overline{D_h} = 0.212 \times 16{,}215 = 3{,}438\text{kJ/m}^2 - \text{day}$$

$$\text{and } \overline{B_h} = \overline{H_h} - \overline{D_h} = 12{,}777\text{kJ/m}^2 - \text{day}$$

Insolation on a tilted surface can be found from Eq. (2.60). We need to find $\overline{R_B}$ from Eq. (2.56).

Therefore,

$$\overline{R_b} = \frac{\cos(0)\cos(-21.1°)\sin(-79.6°) - 1.389\sin(0)\sin(-21.1°)}{\cos(25°)\cos(-21.1°)\sin(-79.6°) - 1.389\sin(25°)\sin(-21.1°)}$$

$$= 147.$$

$$\overline{R_d} = \cos^2(25/2) = 0.953.$$

$$\overline{R_r} = \rho \sin^2(\beta/2) \text{ (Assume } \rho = 0.2)$$

$$= 0.2 \sin^2(12.5°)$$

$$= 0.009.$$

Therefore,

$$\overline{H_c} = (1.47)(12{,}777) + 0.953(3{,}438) + 0.009(16{,}215)$$

$$= 22{,}205 \text{ kJ/m}^2.$$

2.6.2 Circumsolar or Anisotropic Diffuse Solar Radiation

The models described in the above sections assume that the sky diffuse radiation is isotropic. However, this assumption is not true because of circumsolar radiation (brightening around the solar disk). Although the assumption of isotropic diffuse solar radiation does not introduce errors in the diffuse values on horizontal surfaces, it can result in errors of 10 to 40% in the diffuse values on tilted surfaces. A number of researchers have studied the anisotropy of the diffuse solar radiation because of circumsolar radiation. Temps and Coulson [73] introduced an anisotropic diffuse radiation algorithm for tilted surfaces for clear sky conditions. Klucher [37] refined the Temps and Coulson algorithm by adding a cloudiness function to it:

$$R_d = \frac{1}{2}(1 + \cos\beta)M_1 M_2, \tag{2.62}$$

where

$$M_1 = 1 + F \sin^3(\beta/2), \tag{2.63}$$

$$M_2 = 1 + F \cos^2 i \sin^3(z), \text{ and} \tag{2.64}$$

$$F = 1 - (D_h/H_h)^2. \tag{2.65}$$

Examining F, we find that under overcast skies ($D_h = H_h$), R_d in Eq. 2.62 reduces to the isotropic term of Liu and Jordan. The Klutcher algorithm reduces the error in diffuse radiation to about 5%.

In summary, monthly averaged, daily solar radiation on a surface is calculated by first decomposing total horizontal radiation into its beam and diffuse components using Eq. (2.51) or (2.52). Various tilt factors are then used to convert these horizontal components to components on the surface of interest.

2.6.3 Daily Solar Radiation on Tilted Surfaces

Prediction of daily horizontal total solar radiation for sites where solar data are not measured can be done using the Angström-Page model. Instead of monthly values, however, daily values are used for percent sunshine PS and extraterrestrial radiation I_{day}. The results of using this simple model would be expected to show more scatter than monthly values, however.

All U.S. National Weather Service (NWS) stations with solar capability report daily horizontal total (beam and diffuse) radiaton. Liu and Jordan have extended their monthly method described above to apply to daily data. The equation, analogous to Eq. (2.51), used to calculate the daily diffuse component $\bar{I}_{d,h}$ is [6]

$$\frac{\bar{I}_{d,h}}{\bar{I}_h} = 1.0045 + 0.04349K_T - 3.5227K_T^2 + 2.6313K_T^3, \text{ and} \tag{2.66}$$

$$K_T \leq 0.75,$$

where K_T (no overbar) is the daily clearness index analogous to the monthly $\bar{K}_T$. (In this section, overbars indicate daily radiation totals.) For values of $K_T > 0.75$, the diffuse-to-total ratio is constant at a value of 0.166. K_T is given by

$$K_T = \frac{\bar{I}_h}{\bar{I}_{o,h}}. \tag{2.67}$$

The daily extraterrestrial total radiation $\bar{I}_{o,h}$ is calculated from Eq. (2.49). Note that Eq. (2.66) is based on the early solar constant value of 1394 W/m² (442 Btu/hr · ft²).

The daily horizontal beam component $\bar{I}_{b,h}$ is given by simple subtraction:

$$\bar{I}_{b,h} = \bar{I}_h - \bar{I}_{d,h}.$$ (2.68)

The beam, diffuse, and reflected components of radiation can each be multiplied by their tilt factors R_b, R_d, and R_r to calculate the total radiation on a tilted surface

$$\bar{I}_c = R_b\bar{I}_{b,h} + R_d\bar{I}_{d,h} + R_r(\bar{I}_{b,h} + \bar{I}_{d,h}),$$ (2.69)

in which

$$R_b = \frac{\cos(L - \beta)\cos\delta_s \sin h_{sr} + h_{sr} \sin(L - \beta)\sin\delta_s}{\cos L \cos\delta_s \sin h_{sr}(\alpha = 0) + h_{sr}(\alpha = 0)\sin L \sin\delta_s},$$ (2.70)

$$R_d = \cos^2\frac{\beta}{2}, \text{ and}$$ (2.71)

$$R_r = \rho \sin^2 \frac{\beta}{2},$$ (2.72)

by analogy with the previous, monthly analysis.

If daily solar data are available, they can be used for design, the same as monthly data. Daily calculations are necessary when finer time-scale performance is required. In addition, daily data can be decomposed into hourly data, which are useful for calculations made with large, computerized solar system simulation models.

2.6.4 Hourly Solar Radiation on Tilted Surfaces

Hourly solar radiation can be predicted in several ways. Correlations between hourly total and hourly diffuse (or beam) radiation or meteorological parameters such as cloud cover or air mass may be used. Alternatively, a method proposed by Liu and Jordan based on the disaggregation of daily data into hourly data could be used [41–45]. Even if hourly NWS data are available, it is necessary to decompose these total values into beam and diffuse components depending upon the response of the solar conversion device to be used to these two fundamentally different radiation types.

Randall and Leonard [62] have correlated historical data from the NWS stations at Blue Hill, Massachusetts, and Albuquerque, New Mexico to predict *hourly beam radiation* I_b and *hourly total horizontal radiation* I_h. This method can be used to decompose NWS data into its beam and diffuse components.

The hourly beam radiation was found to be fairly well correlated by hourly *percent of possible insolation* k_t, defined as

$$k_t = \frac{I_h}{I_{o,h}},$$ (2.73)

in which $I_{o,h}$ is the hourly horizontal extraterrestrial radiation, which can be evaluated from Eq. (2.49) using one-hour integration periods. Carrying out the integration over a one hour period yields

$$I_{o,h} = I_o\left(1 + 0.034 \cos\frac{360n}{365}\right)(0.9972 \cos L \cos\delta_s \cos h_s + \sin L \sin \delta_s), \quad (2.74)$$

where the solar hour angle h_s is evaluated at the center of the hour of interest.
The direct normal-beam correlation based on k_t is [45]

$$I_{b,N} = -520 + 1800k_t(W/m^2) \quad 0.85 > k_t \geq 0.30 \quad (2.75)$$

$$I_{b,N} = 0 \quad k_t < 0.30 \quad (2.76)$$

This fairly simple correlation gives more accurate $I_{b,N}$ values than the more cumbersome Liu and Jordan procedure, at least for Blue Hill and Albuquerque. Vant-Hull and Easton [80] have also devised an accurate predictive method for beam radiation.

Randall and Leonard [62] have made a correlation of *total* horizontal hourly radiation $I_h(W/m^2)$ on the basis of *opaque cloud cover CC* and *air mass m* using data for Riverside, Los Angeles, and Santa Monica, California. Cloud cover is defined as $CC = 1.0$ for fully overcast and $CC = 0.0$ for clear skies. A polynomial fit was used:

$$I_h = \frac{I_{o,h}}{100}(83.02 - 3.847m - 4.407CC + 1.1013CC^2 - 0.1109CC^3) \quad (2.77)$$

The average predictive error for this correlation was $\pm$ 2.3 percent of the NWS data; the correlation coefficient of I_h with CC is 0.76 for the data used. Equation (2.80) was used to predict I_h for Inyokern, California, a site not used in the original correlation. Predictions of solar radiation for Inyokern were within 3.2 percent of NWS. The diffuse radiation can be calculated from Eqs. (2.75), (2.76), and (2.77):

$$I_{d,h} = I_h - I_b \sin\alpha \quad (2.78)$$

The American Society of Heating, Refrigerating, and Air Conditioning Engineers (ASHRAE) has calculated hourly clear-sky values on vertical, horizontal, and tilted surfaces. One table for a full year has been prepared for each of six values of latitude spanning the continental United States. In addition, tables of solar azimuth and altitude angle have been prepared for the same six latitudes. The solar radiation data represent the maximum that could be expected on a clear day and are therefore of limited usefulness in the design of solar systems. They can be used, however, to calculate cooling loads on buildings and the like. The ASHRAE tables are contained in Appendix 2.

2.7 MEASUREMENT OF SOLAR RADIATION

Solar radiation measurements of importance to most engineering applications, especially thermal applications, include total (integrated over all wavelengths) direct or beam and sky diffuse values of solar radiation on instantaneous, hourly, daily and monthly bases. Some applications such as photovoltaics, photochemical and daylighting require knowledge of spectral (wavelength specific) or band (over a wavelength range—e.g., ultraviolet, visible, infra-red) values of solar radiation. This section describes some of the instrumentation used to measure solar radiation and sunshine, and some sources of long-term measured data for different parts of the world. Also described briefly in this section is the method of satellite-based measurements.

2.7.1 Instruments for measuring solar radiation and sunshine

There are two basic types of instruments used to measure solar radiation, *pyranometer* and *pyrheliometer*. A pyranometer has a hemispherical view of the surroundings and therefore is used to measure total, direct and diffuse, solar radiation on a surface. A pyrheliometer, on the other hand, has a restricted view (about 5°) and is, therefore, often used to measure the direct or beam solar radiation by pointing it toward the sun. Pyranometers are also used to measure the sky diffuse radiation by using a shadow band to block the direct sun view. A detailed discussion of the instrumentation and calibration standards is given by Iqbal [27] and Zerlaut [84].

A pyranometer consists of a flat sensor/detector (described later) with an unobstructed hemispherical view, which allows it to convert and correlate the total radiation incident on the sensor to a measurable signal. The pyranometers using thermal detectors for measurements can exhibit serious errors at tilt angles from the horizontal due to free convection. These errors are minimized by enclosing the detector in double hemispherical high-transmission glass domes. The second dome minimizes the error due to infra-red radiative exchange between the sensor and the sky. A desiccator is usually provided to eliminate the effect due to condensation on the sensor or the dome. Figure 2.20 shows pictures of typical commercially available precision pyranometer.

A pyranometer can be used to measure the sky diffuse radiation by fitting a shade ring to it, as shown in Figure 2.21, in order to block the beam radiation throughout the day. The position of the shade ring is adjusted periodically as the declination changes. Since the shade ring obstructs some diffuse radiation from the pyranometer, correction factors must be applied.

Geometric correction factors (GCF) that account for the part of the sky obstructed by the shade ring can be easily calculated. However a GCF assumes isotropic sky, which results in errors because of the circumsolar anisotropy. Eppley Corp. recommends additional correction factors to account for anisotropy as: +7% for clear sky, +4% for partly cloudy condition, and +3% for cloudy sky. Mujahid and Turner [53] determined that these correction factors gave less than 3% errors on partly cloudy days but gave errors of −11% for clear sky conditions and +6% on overcast days. They suggested correction factors due to anisotropy as tabulated in Table 2.7 which reduce

(a)

(b)

Figure 2.20. Typical commercially available pyranometers with (a) Thermal detector and (b) Photovoltaic detector.

the errors to less than ±3%. It must be remembered that these correction factors are in addition to the geometric correction factors. Recently, a sun occulting disk has been employed for shading the direct sun.

Beam or direct solar radiation is usually measured with an instrument called a pyrheliometer. Basically a pyrheliometer places the detector at the base of a long tube. This geometry restricts the sky view of the detector to a small angle of about 5°. When the tube points toward the sun the detector measures the beam solar radiation and a small part of the diffuse solar radiation within the view angle. Figure 2.22 shows the geometry of a pyrheliometer sky occulting tube.

Figure 2.21. A pyranometer with a shade ring to measure sky diffuse radiation.

Table 2.7. Shading band correction factors due to anisotropy [53]

Solar Altitude Angle	k_T									
	0.	0.1	0.2	0.3	0.4	0.5	0.6	0.7	0.8	0.9
<20°	0.0	0.0	0.0	0.0	0.015	0.06	0.14	0.23	0.24	0.24
20 to 40°	0.0	0.0	0.0	0.0	0.006	0.05	0.125	0.205	0.225	0.225
40 to 60°	0.0	0.0	0.0	0.0	0.003	0.045	0.115	0.175	0.205	0.205
60°+	0.0	0.0	0.0	0.0	0.0	0.035	0.09	0.135	0.17	0.17

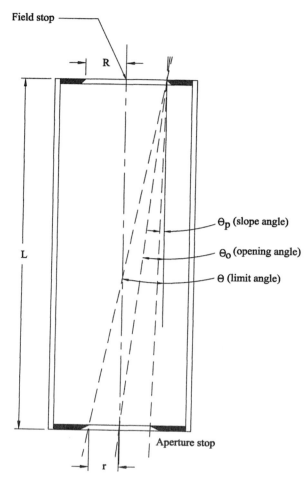

Figure 2.22. Geometry of a pyrheliometer sky occulting tube.

In this figure,

The opening half angle

$$\theta_o = \tan^{-1} R/L. \tag{2.79}$$

The slope angle

$$\theta_p = \tan^{-1}[(R - r)/L]. \tag{2.80}$$

The limit half angle

$$\theta = \tan^{-1}[(R + r)/L]. \tag{2.81}$$

The field of view is $2\theta_o$. The World Meteorological Organization (WMO) recommends the opening half angle θ_o to be 2.5° [84] and the slope angle θ_p to be 1°.

Continuous tracking of the sun is required for the accuracy of the measurements. This is obtained by employing a tracking mechanism with two motors, one for altitude and the other for azimuthal tracking. Another problem is that the view angle of a pyrheliometer is significantly greater than the angle subtended by the solar disk (about 0.5°). Therefore the measurements using a pyrheliometer include the beam and the circumsolar radiation. These measurements may present a problem in using the data for central receiver systems which use only direct beam radiation. However, this is not a significant problem for parabolic trough concentrators which in most cases have field of view of the order of 5°.

2.7.2 Detectors For Solar Radiation Instrumentation

Solar radiation detectors are of four basic types [27,84]: thermomechanical, calorimetric, thermoelectric and photoelectric. Of these, thermoelectric and photoelectric are the most common detectors in use today.

A *thermoelectric detector* uses a thermopile which consists of a series of thermocouple junctions. The thermopile generates a voltage proportional to the temperature difference between the hot and cold junctions which, in turn, is proportional to the incident solar radiation. Figure 2.23 shows different types of thermopile configurations. The Eppley black and white pyranometer uses a radial differential thermopile with the hot junction coated with 3M Velvet Black™ and the cold junction coated with a white barium sulphate paint.

Photovoltaic detectors normally use silicon solar cells measuring the short circuit current. Such detectors have the advantage of being simple in construction. Because heat transfer is not a consideration, they do not require clear domes or other convection suppressing devices. They are also insensitive to tilt as the output is not affected by natural convection. One of the principal problems with photovoltaic detectors is their spectral selectivity. Radiation with wavelengths greater than the band gap of the photovoltaic detector cannot be measured. Silicon has a bandgap of 1.07 eV corresponding to a wavelength of 1.1 μm. A significant portion of the infra-red part of solar radiation has wavelengths greater than 1.1 μm. Therefore, photovoltaic detectors are insensitive to changes in the infra-red part of solar radiation.

2.7.3 Measurement of Sunshine Duration

The time duration of bright sunshine data is available at many more locations in the world than the solar radiation. That is why a number of researchers have used this data to estimate the available solar radiation. Two instruments are widely used to measure the sunshine duration. The device used by the U.S. National Weather Service is called a *sunshine switch*. It is composed of two photovoltaic cells—one shaded, the other not. During daylight a potential difference is created between the two cells, which in turn operates the recorder. The intensity level required to activate the device is that just sufficient to cast a shadow. The other device commonly used to measure the sunshine duration is called the *Campbell-Stokes sunshine recorder*. It uses a solid, clear glass sphere as a lens to concentrate the solar beam on the opposite side of the sphere. A strip

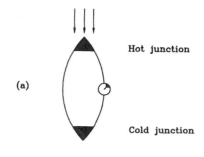

(a)

Hot junction

Cold junction

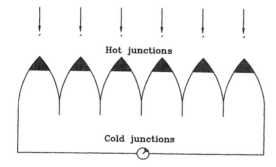

Hot junctions

Cold junctions

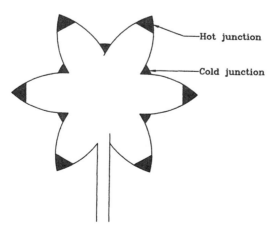

Hot junction

Cold junction

Figure 2.23. Various thermopile configurations. Source: Zerlaut [84].

of standard treated paper marked with time graduations is mounted on the opposite side of the sphere where the solar beam is concentrated. Whenever the solar radiation is above a threshold, the concentrated beam burns the paper. The length of the burned part of the strip gives the duration of bright sunshine. The problems associated with the Campbell-Stokes sunshine recorder include the uncertainties of the interpretation of burned portions of the paper, especially on partly cloudy days, and the dependence on the ambient humidity. Figure 2.24 shows a Campbell-Stokes sunshine recorder.

2.7.4 Measurement of Spectral Solar Radiation

Spectral solar radiation measurements are made with spectroradiometers. Full spectrum scanning is difficult, requires constant attention during operation and is therefore expensive. Zerlaut [84] has described a number of solar spectroradiometers. These

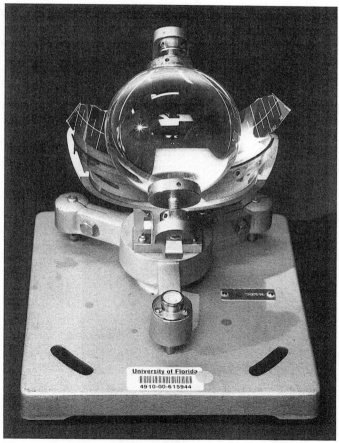

Figure 2.24. Campbell-Stokes sunshine recorder.

instruments consist basically of a monochromator, a detector-chopper assembly, an integrating sphere, and a signal conditioning/computer package. They have the capability of measuring solar radiation in the wavelength spectrum of 280–2500 nm.

2.7.5 Wide Band Spectral Measurements

Some applications of solar energy require solar radiation data in wide band wavelength ranges such as, visible, ultraviolet and infra-red rather than complete spectral data. For example, solar photocatalytic detoxification using TiO_2 as the catalyst needs data in the UV wavelength range while passive solar applications need data in the infra-red wavelength range. Instruments such as pyranometers and pyrheliometers can be adapted for wide band spectral measurements by using cut-on and cut-off filters. Eppley instruments provide standard cut-off filters at wavelengths 530 nm (orange), 630 nm (red) and 695 nm (dark red). They are provided as plain filters at the aperture of a pyrheliometer tube and as outer glass domes for pyranometers. Instrument manufacturers provide various interference filters peaking at different wavelengths in the solar spectrum.

Solar UV measurements are important in general since prolonged exposure to solar UV can cause skin cancer, fading of colors and degradation of plastic materials. Such measurements have become even more important because the photocatalytic effect based on TiO_2 as a catalyst, depends only on the solar UV wavelength range. Figure 2.25 shows an Eppley Model TUVR, total ultraviolet radiometer that measures total hemispherical UV radiation from 295 to 385 nm. This radiometer uses a selenium

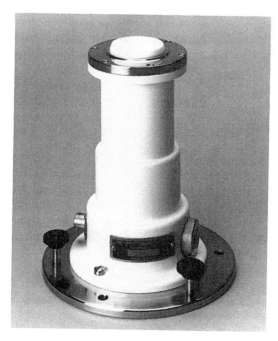

Figure 2.25. Eppley TUVR® Total UV Radiometer (Courtesy, Eppley Lab).

photoelectric cell detector, a pair of band pass filters to allow wavelengths from 295 to 385 nm to pass through, and a beveled teflon diffuser.

2.7.6 Solar Radiation Data

Measured solar radiation data is available at a number of locations throughout the world. Data for many other locations have been estimated based on measurements at similar climatic locations. Some of the available data from various locations in the world is presented in Appendix 2 (Tables A2.3a & A2.3b). This appendix also provides tables of modeled clear sky data for various latitudes (Table A2.6).

Solar radiation data for U.S.A. are available from the National Climatic Data Center (NCDC) of the National Oceanic and Atmospheric Administration (NOAA), and the National Renewable Energy Laboratory (NREL). In the mid nineteen seventies, NOAA compiled a data base of measured hourly global horizontal solar radiation for 28 locations for the period 1952–75 (called SOLMET) and of data for 222 additional sites (called ERSATZ) estimated from SOLMET data and some climatic parameters such as sunshine duration and cloudiness. NOAA also has two data sets of particular interest to engineers and designers: the typical meteorological year (TMY) and the Weather Year for Energy Calculations (WYEC) data sets. TMY data set represents typical values from 1952 to 1975 for hourly distribution of direct beam and global horizontal solar radiation. WYEC data set contains monthly values of temperature, direct beam and diffuse solar radiation and estimates of *illuminance* (for daylighting applications). Illuminance is solar radiation in the visible range to which the human eye responds. Recently, NREL compiled a National Solar Radiation Data Base (NSRDB) for 239 stations in the United States [50,55]. NSRDB is a collection of hourly values of global horizontal, direct normal, and diffuse solar radiation based on measured and estimated values for a period of 1961–1990. Since long-term measurements were available for only about 50 stations measured data makes up only about 7% of the total data in the NSRDB. A typical meteorological year data set from NSRDB is available as TMY2.

The data for other locations in the world is available from national government agencies of most countries of the world. Worldwide solar radiation data is also available from the World Radiation Data Center (WRDC) in St. Petersburg, Russia, based on worldwide measurements made through local weather service operations [81]. WRDC, operating under the auspices of the World Meteorological Organization (WMO), has been archiving data from over 500 stations and operates a worldwide web site in collaboration with NREL with an address of http://wrdc.mgo.nrel.gov. Data for some cities of the world from WRDC is given in Appendix 2 (Table A2.3). An International Solar Radiation Data Base was also developed by the University of Lowell [79].

2.8 SOLAR RADIATION MAPPING USING SATELLITE DATA

Remote sensing satellite data has been used since the early nineteen sixties to extract quantitative and qualitative cloud data. The most important application of cloud cover mapping has been the observation of storms and hurricanes, etc. Recently, however, considerable interest has been developed in using the cloud mapping data to estimate terrestrial solar radiation. Since meteorological satellites from a number of countries can now cover most of the earth, the data can be used to estimate solar radiation where no measured data exists or none is being measured.

Weather satellites are available in three main orbiting configurations—equatorial, polar, and geostationary. The equatorial satellites are low level orbiting satellites (~600 km altitude) that generally orbit the earth in a west to east direction in a sinusoidal path that crosses the equator at least twice per orbit. Polar satellites are also low orbit satellites that orbit the earth from the north to the south pole while the earth rotates underneath. Sun synchronous polar orbits have their orbits synchronized with the sun such that the same point on the earth is viewed at the same time each day. Low orbit satellites are capable of gathering high resolution spatial data. A geostationary satellite orbits in such a way that it is always over the same point on the earth's surface. Geostationary satellites have very high altitudes (approximately 36000 km) and can provide high temporal resolution images over a large portion of the earth's surface. A number of countries maintain geostationary satellites including the U.S. (GOES, longitudes 70°W and 140°W), Europe (METEOSAT, longitude 0°), India (INSAT, longitude 70°E) and Japan (GMS, longitude 140°E).

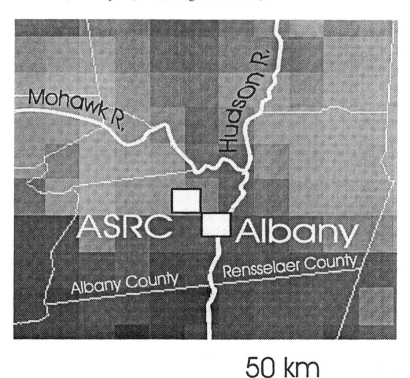

Figure 2.26. GOES-8 Intermediate resolution image close-up around Albany (Source [58]).

Various types of high resolution radiometers collect radiative data images of the earth's atmosphere below. These radiometers scan spectral measurements in the wavelength ranges of shortwave (0.2–3.0 μm), longwave (6.5–25 μm), and total irradiance (0.2–100 μm). The spatial resolution of images from the satellite is given by a *pixel* which represents the smallest area of data, generally of the order of 2 km $\times$ 2 km. However, several pixels of data are required to derive a surface value giving a surface resolution of the order of 10 km $\times$ 10 km. Figure 2.26 shows an example of an intermediate resolution GOES-8 image around Albany, New York, overlaid on a local map.

2.8.1 Estimation of Solar Resource from Satellite Data

The signal recorded by a radiometer on a satellite measures the solar radiation flux reflected back from the earth's atmosphere. The basic method behind estimation of ground solar radiation from this data is to apply the principle of energy conservation in the earth-atmosphere system [54], as shown in Fig. 2.27. From this figure we can write:

$$I_{in} = I_{out} + I_a + I_g, \tag{2.82}$$

where I_{in} represents the solar radiation incident on the atmosphere, I_{out} represents the outward radiation from the atmosphere, I_a is the radiation absorbed by the atmosphere, and I_g is the radiation absorbed by the ground.

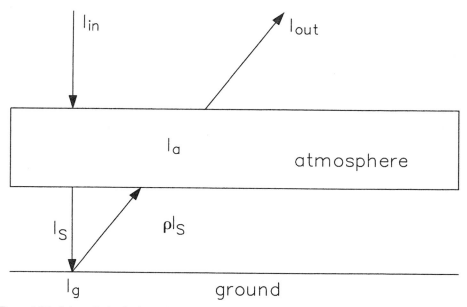

Figure 2.27. Solar radiation in the earth-atmosphere model.

I_g can be expressed in terms of the surface albedo* ρ (reflectivity) and the solar radiation I_s, incident on the earth's surface:

$$I_g = I_s(1 - \rho).$$ (2.83)

From Eqs. (2.82) and (2.83) we can obtain

$$I_s = (I_{in} - I_{out} - I_a)/(1 - \rho).$$ (2.84)

I_{out} is measured by the satellite radiometers. I_{in} depends on the sun-earth distance and the solar zenith angle and can be calculated using Eq. (2.32) as:

$$I_{in} = I_o(D_o/D)^2 \cdot \cos(z).$$ (2.85)

If we could estimate I_a, and A were known a priori, I_s could be estimated using Eq. (2.84) from the value of I_{out} measured by a satellite. However, I_a cannot be estimated easily since it depends on the atmospheric conditions such as cloud cover, dust particles and air mass, and surface albedo (reflectance) ρ varies for every point of the region under consideration. In order to deal with these factors, two types of empirical methods are under development. These are known as statistical and physical methods. These methods have been reviewed in detail by Schmetz [67], Hay [21], Noia et al. [54], Islam [29] and Pinker et al. [59].

Statistical Methods. Statistical methods are based on finding a relationship between the radiative flux measured by a satellite radiometer and the simultaneous solar radiation value measured at the earth's surface in the area under consideration. Some of the models developed on statistical approach include Hay and Hanson [22], Tarpley [72], Justus et al. [32], Cano [7], and Sorapipatana et al. [68].

Physical Methods. Physical methods are based on the analysis of radiative processes in the atmosphere as the solar radiation passes through it. Some of the models developed with this approach include Gautier [18], Moser and Raschke [51], Dedieu et al. [12], and Marullo et al. [48].

The simplest of the above models is by Hay and Hanson [22] which gives the atmospheric transmittance T as:

$$T = \frac{I_s}{I_{in}} = a - b\frac{I_{out}}{I_{in}},$$

$$\text{or } I_s = aI_{in} - bI_{out}.$$ (2.86)

*The term albedo is used mostly in the field of meteorology.

The values of regression coefficients given by Hay and Hanson [22] are: $a = 0.79, b = 0.71$. This method is simple, however, the coefficients, particularly b, vary considerably with parameters such as cloud reflectivity. More recent investigations suggest that it is necessary to determine the coefficients a and b for different locations.

It is beyond the scope of this book to discuss all the models. It suffices, however, to point out that all the models, including the Hay and Hanson model, more or less give values within 10% of the ground measured values [54]. The methods usually break down under partly cloudy conditions and under snow covered ground conditions.

PROBLEMS

2.1. Calculate the declination, the zenith angle, and the azimuth angle of the sun for New York City (latitude 40.77°N) on October 1 at 2:00 PM solar time.

2.2. A solar energy system in Gainesville, FL requires two rows of collectors facing south and tilted at a fixed 30° angle. Find the minimum distance at which the second row should be placed behind the first row for no shading at noon at winter solstice. What percentage of the second row is shaded on the same day at 9:00 AM?

2.3. Find the sunrise and sunset times for a location of your choice on September 1.

2.4. Construct a table of hourly sun angles for the 15th day of each month for a location of your choice. Also show the sunrise and sunset times for those days.

2.5. Referring to Fig. 2.19, prove equation (2.44) for the angle of incidence. (Hint: Use direction cosines of the sun-ray vector and a vector normal to the tilted surface to find the angle between them. Dot product of two unit vectors gives the cosine of the angle between them.)

2.6. Determine the following for a south facing surface at 30° slope in Gainesville, FL (Latitude = 29.68°N, Longitude = 82.27°W) on September 21 at noon solar time:
   A. Zenith Angle
   B. Angle of Incidence
   C. Beam Radiation
   D. Diffuse Radiation
   E. Reflected Radiation
   F. Total Radiation
   G. Local Time

2.7. Show that the hourly averaged, extraterrestrial radiation for a given hour is the same, to within 1 percent, as instantaneous radiation at the hour's midpoint. This is equivalent to deriving Eq. (2.74).

2.8. Prepare shadow maps for point P on the sun-path diagrams for 35°N and 40°N for the three geometries shown in a, b and c below. Determine the hours of shading that occur each month.

2.9. Repeat Problem 2.8c if the surface containing point P faces due west instead of due south for a 40°N location.

2.10. Calculate the incidence angle at noon and 9 AM on a fixed, flat-plate collector located at 40°N latitude and tilted 70° up from the horizontal. Find i for June 21 and December 21.

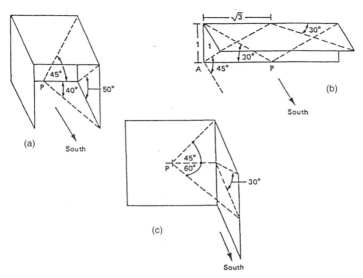

Figure for Problem 2.8.

2.11. A. If the surface in Problem 2.10 faces S 45°E, what are the incidence angles?
      B. If the collector in Problem 2.10 has a cylindrical surface, what are the inci-
         dence angles on June 21 and December 21?
2.12. Using a one-term Fourier cosine series, develop an empirical equation for solar
      declination as a function of day number counted from January 1 (see Table 2.2
      and Fig. 2.7).
2.13. Derive an equation for the lines of constant declination in a sun-path diagram,
      for example, Fig. 2.10. Check your equation by plotting a few declination lines
      on a piece of polar coordinate graph paper.
2.14. Derive Eq. (2.31) relating profile angle γ to azimuth angle α and altitude angle
      α.
2.15. Based on Eq. (2.44), what value of β would result in the annual minimum value
      of the incidence angle i? Note that this tilt angle would result in maximum col-
      lection of beam radiation on a fixed, flat, south-facing surface. *Hint:* Use a dou-
      ble integration procedure.
2.16. At what time does the sun set in Calcutta (23°N) on May 1 and December 1?
2.17. What is the true solar time in Sheridan, Wyoming (107°W) at 10:00 AM Moun-
      tain Daylight Time on June 10? What is the true solar time at 10:00 AM Moun-
      tain Standard Time on January 10?
2.18. If a compass points magnetic north in Cleveland, Ohio, how many degrees east
      or west of the true north-south line is it pointing?
2.19. Using the Angström-Page method, calculate the average horizontal insolation in
      Hamburg in May and in October with $\overline{PS}$ 40 percent and 60 percent, respec-
      tively.

2.20. Equation (2.51) is based on an early solar constant value of 1394 W/m². Derive a modified form of Eq. (2.51) based on the presently accepted value for the solar constant of 1353 W/m².

2.21. Predict the hourly beam and diffuse radiation on a horizontal surface for Denver (40°N) on September 9 at 9:30 AM on a clear day.

2.22. Derive an expression for the minimum allowable distance between east-west rows of solar collectors that will assure no shading of one row by the row immediately to the south. Use the law of sines and express the result in terms of the collector tilt and face length and the controlling value of the solar profile angle.

REFERENCES AND SUGGESTED READINGS

1. Akinoglu, B.G. 1991. A review of sunshine-based models used to estimate monthly average global solar radiation. *Renewable Energy* 1, no. 3, 479–499.
2. Akinoglu, B.G., and A. Ecevit. 1990. Construction of a quadratic model using modified angstrom coefficients to estimate global solar radiation. *Sol. Energy* 45:85.
3. Anane-Fenin, K. 1986. Estimating solar radiation in Ghana. *International Centre for Theoretical Physics* Internal Report No. IC/86/68.
4. Bird, R.E., and R.L. Hulstrom. Direct insolation models. SERI/TR-335-344.
5. Boes, E.C. et al. 1976. Distribution of direct and total solar radiation availabilities for the USA. In *Sharing the Sun,* vol. 1. Winnipeg: ISES, 238–263. See also Boes, E.C. et al. "Availability of Direct, Total and Diffuse Solar Radiation for Fixed and Tracking Collectors—the USA." *Sandia Laboratory Rept.* SAND 77-0885 1977.
6. Budde, W. *Physical Detectors of Optical Radiation.* Vol. 4 of *Optical Radiation Measurements.* New York: Academic Press.
7. Cano, D. et al. 1986. "Method for the Determination of the Global Solar Radiation from Meteorological Satellite Data." *Solar Energy* 37, no. 1, 31–39.
8. Choudhury, N.K.O. 1963. Solar radiation at New Delhi. *Solar Energy* 7, no. 44.
9. Chuah, D.G.S., and S.L. Lee. 1981. "Solar Radiation Estimates in Malaysia." *Sol. Energy* 26: 33.
10. Collares-Pereira, M., and A. Rabl. 1979. The average distribution of solar radiation-correlation between diffuse and hemispherical. *Sol. Energy* 22, 155–166.
11. Coppolino, S. 1994. "New Correlation between Clearness Index and Relative Sunshine." *Renewable Energy* 4 (4): 417–423.
12. Dedieu, G., P.Y. Deschamps, and Y.H. Kerr. 1987. Satellite estimation of solar irradiance at the surface of the earth and of surface albedo using a physical model applied to METEOSAT data. *J. Clim. Appl. Meteorol.* 26: 79–87.
13. DeJong, B. 1973. *Net Radiation Received by a Horizontal Surface on Earth.* Delft: Delft University Press.
14. Eze, A.E., and S.C. Ododo. 1988. Solar radiation prediction from sunshine in eastern Nigeria. *Energy Conversion and Management* 28: 69.
15. Fritz, S. 1949. Solar radiation in the United States. *Heat. Vent.* July 61–64.
16. Frohlich, C. et al. 1973. The third international comparison of pyrheliometers and a comparison of radiometric scales. *Solar Energy* 14: 157–166.
17. Garg, H.P., and S.N. Garg. 1985. Correlation of the monthly average daily global, diffuse and beam radiation with bright sunshine hours. *Energy Conversion and Management* 25, (4): 409–417.
18. Gautier, C. 1980. "Simple Physical Model to Estimate Solar Radiation at the Surface from GOES Satellite Data." *J. of Applied Meteorology,* 19 (8): 1005–1012.
19. Gopinathan, K. 1988. A general formula for computing the coefficients of the correlation connecting global solar radiation to sunshine duration. *Sol. Energy* 41: 499.
20. Goswami D.Y. et al. 1981. Seasonal variations of atmospheric clearness numbers for use in solar radiation modelling. AIAA-80-396R, *Journal of Energy,* 5(2): 185–187.

21. Hay, J.E. 1993. Satellite based estimates of solar irradiance at the earth's surface: I. Modelling approaches. *Renewable Energy* 3: 381–393.
22. Hay, J.E., and K.J. Hanson. 1978. A satellite-based methodology for determining solar irradiance at the ocean surface during GATE. *Bull. Am. Meteorol. Soc.* 59: 1549.
23. Hottel, H.C. 1976. A simple model for estimating the transmittance of direct solar radiation through clear atmospheres. *Sol. Energy* 18: 129–134.
24. Howell, J.R. and R. Siegel. 1992. *Thermal Radiation Heat Transfer.* New York: McGraw-Hill Book Co.
25. Hussain, M. 1994. Estimation of the global and diffuse irradiation from sunshine duration and atmospheric water vapor content, *Sol. Energy* 33, no. 2: 217–220.
26. Ibrahim, S.M.A. 1985. "Predicted and measured global solar radiation in Egypt," *Sol. Energy* 35: 185.
27. Iqbal, M. 1983. *An Introduction to Solar Radiation.* New York: Academic Press.
28. Iqbal, M. 1979. Correlation of average diffuse and beam radiation with hours of bright sunshine, *Sol. Energy* 23: 169–173.
29. Islam, M.R. 1994. Evolution of methods for solar radiation mapping using satellite data. *RERIC International Energy Journal* 16(2).
30. Jain, P.C. 1986. Irradiation estimation for Italian localities. *Solar Wind Technology,* 3: 323.
31. Jain, S., and P.G. Jain. 1985. A comparison of the angstrom type correlations and the estimation of monthly average daily global irradiation. *International Centre for Theoretical Physics,* Internal Report, No. IC85-269.
32. Justus, C., M.V. Paris and J.D. Tarpley. 1986. Satellite-measured insolation in the United States, Mexico and South America. *Remote Sensing Environ.* 20: 57–83.
33. Khogali, A. 1983. Star radiation over Sudan: Comparison of measured and predicted data. *Sol. Energy* 31: 45.
34. Khogali, A. et al. 1983. Global and diffuse star irradiance in Yemen. *Sol. Energy* 31: 55.
35. Kimura, K and D.G. Stephenson. 1969. Solar radiation on Cloudy days. *Trans. ASHRAE* 75: 227–234.
36. Klein, S.A. 1977. "A design procedure for solar heating systems," (Ph.D. diss., University of Wisconsin, Madison, 1976); see also *Sol. Energy* 19: 325.
37. Klucher, T.M. 1979. Evaluation of models to predict insolation on tilted surfaces. *Sol. Energy,* 23, no. 2: 111–114.
38. Kreith, F. 1962. *Radiation Heat Transfer for Spacecraft and Solar Power Plant Design.* Scranton, Pa.: International Textbook Co.
39. Kreith, F and M. Bohm. 1993. *Principles of Heat Transfer,* 5th ed. St. Paul, MN: West Publishing Co.
40. Leung, C.T. 1980. "The Fluctuation of Solar Irradiance in Hong Kong." *Sol. Energy* 25: 485.
41. Liu, B.Y.H., and R.C. Jordan. 1961. Daily insolation on surfaces titled toward the equator. *Trans. ASHRAE* 67: 526–541.
42. Liu, B.Y.H., and R.C. Jordan. 1961. Daily insolation on surface titled toward the equator. *Trans. ASHRAE* 3(10): 53–59.
43. Liu, B.Y.H., and R.C. Jordan. 1967. Availability of solar energy for flat-plate solar heat collectors. In *Low Temperature Engineering of Solar Energy,* chap. 1. New York: ASHRAE; see also 1977 revision.
44. Liu, B.Y.H., and R.C. Jordan. 1963. A rational procedure for predicting the long-term average performance of flat-plate solar energy collectors. *Sol. Energy* 7: 53–74.
45. Liu, B.Y.H., and R.C. Jordan 1960. The interrelationship and characteristic distribution of direct, diffuse and total solar radiation. *Sol. Energy* 4: 1–19. See also Liu, B.Y.H. "Characteristics of solar radiation and the performance of flat plate solar energy collectors" (Ph.D. dissertation, University of Minnesota, Minneapolis, 1960).
46. Löf, G.O.G. et al. 1966. World distribution of solar energy. *University of Wisconsin, Madison, Eng. Expt. Station Rept.* 21.
47. Malik, A.Q., A. Mufti, H.W. Hiser, N.T. Vezrioglu, L. Kazi. 1991. Application of geostationary satellite data in determining solar radiation over pakistan. *Renewable Energy* 1,(3–4): 455–461.
48. Marullo, S., G. Dalu, and A. Viola. 1987. "Incident short-wave radiation at the surface from METEOSAT data." *Il Nuovo Cimento* 10C: 77–90.
49. Massaquoi, J.G.M. 1988. Global Solar Radiation in Sierre Leone (West Africa). *Solar Wind Technolgoy* 5: 281.

50. Maxwell, E.L. 1998. "METSTAT—The solar radiation model used in the production of the national solar radiation data base (NSRDB)." *Sol. Energy* 62: 263–279.

51. Moeser, W and E. Raschke. 1984. Incident solar radiation over Europe estimated from METEOSTAT data. *J. of Clim. Appl. Meteorology* 23(1): 166–170.

52. Moon, P. 1940. Proposed standard solar radiation curves for engineering use. *J. Franklin Inst.* 230: 583–617.

53. Mujahid, Aziz and W.D. Turner. 1979. Diffuse sky measurement and model. ASME Pap. no. 79–WA/Sol–5.

54. Noia, M., C. Ratto and R. Festa. 1993. Solar irradiance estimation from geostationary satellite data: I. statistical models; II. physical models. *Sol. Energy* 51: 449–465.

55. NSRDB. 1992. *Volume 1: Users Manual: National Solar Radiation Data Base (1961–1990), Version 1.0.* Golden, CO: National Renewable Energy Laboratory.

56. Ögelman, H., A. Ecevit, and E. Tasdemiroglu. 1984. A new method for estimating solar radiation from bright sunshine data. *Sol. Energy* 33(6): 619–625.

57. Page, J.K. 1966. The estimation of monthly mean values of daily total short-wave radiation on vertical and inclined surfaces from sunshine records or latitudes 40°N–40°S. *Proc. U.N. Conf. New Sources Energy* 4: 378.

58. Perez, R., R. Seals, A. Zelenka, and D. Renne. 1997. The Strengths of Satellite Based Resource Assessment. *Proc. of the 1997 ASES Annual Conf.,* pp. 303–308, Washington, DC.

59. Pinker, R.T., R. Frouin, and Z. Li. 1995. A review of satellite methods to derive surface shortwave irradiance. *Remote Sens. Environ.* 51: 108–124.

60. Quinlan, F.T. ed. 1977. *SOLMET Vol. 1: Hourly Solar Radiation Surface Meteorological Observations,* Asheville, NC: NOAA.

61. Quinlan, F.T., ed. 1979. *SOLMET Vol. 2: Hourly Solar Radiation Surface Meteorological Observations,* Asheville, NC: NOAA.

62. Randall, C.M., and S.L. Leonard. 1974. Reference insolation data base: A case history, with recommendations. *Rept. Recommendations Solar Energy Data Workshop* 93–103.

63. Randall, C.M., and R. Bird. 1989. Insolation models and algorithms. In *Solar Resources,* Roland L. Hulstrom ed. Cambridge, MA: MIT Press.

64. Raphael, C., and J.E. Hay. 1984. An assessment of models which use satellite data to estimate solar irradiance at the earth's surface. *J. Clim. Appl. Meteorol.* 23: 832–844.

65. Robinson, N. 1966. *Solar Radiation.* New York: Elsevier Publ. Co.

66. Ruth, D.W. and R.E. Chant. 1976. The relationship of diffuse radiation to total radiation in Canada. *Sol. Energy* 18: 153.

67. Schmetz, J. 1989. Towards a surface radiation climatology: Retrieval of downward irradiances from satellites. *Atmosph. Res.* 23: 287–321.

68. Sorapipatana, C., R.H.B. Exell and D. Borel. 1988. A bispectral method for determining global solar radiation from meteorological satellite data. *Sol. Wind Technology* 5(3): 321–327.

69. Sparrow, E.M., and R.D. Cess. 1978. *Radiation Heat Transfer.* Belmont, CA: Wadsworth Publ. Co.

70. Spencer, J.W. 1971. Fourier series representation of the position of the sun. *Search* 2: 172.

71. Stanhill, G. 1966. Diffuse sky and cloud radiation in Israel. *Sol. Energy* 10: 66.

72. Tarpley, J.D. 1979. Estimating incident solar radiation at the surface from geostationary satellite data. *J. of Appl. Meteorol.* 18(9): 1172–1181.

73. Temps, R.C. and K.L. Coulson. 1977. Solar radiation incident upon slopes of different orientations. *Sol. Energy* 19(2): 179–184.

74. Thekaekara, M.P. 1976. Insolation data for solar energy conversion derived from satellite measurements of earth radiance. In *Sharing the Sun,* vol. 1, pp. 313–328. Winnipeg: ISES.

75. Thekaekara, M.P. 1973. Solar energy outside the Earth's atmosphere. *Sol. Energy* 14: 109–127.

76. Threlkeld, J.L., and R.C. Jordan. 1958. *Trans. ASHRAE* 64: 45.

77. Trewartha, G.T. 1954. *An Introduction to Climate.* New York: McGraw-Hill.

78. Turner, C.P. 1974. Bibliography for Solar Energy Workshop. *Solar Energy Data Workshop, NSF Rept.,* NSF/RA/N/74/062, (available from NTIS).

79. *University of Lowell Photovoltaic Program.* 1990. *International Solar Irradiation Database.* Version 1.0. Lowell, MA: University of Lowell Research Foundation.

80. Vant-Hull, L., and C.R. Easton. 1975. Solar thermal power systems based on optical transmission. *NTIS Rept.* PB253167.

81. Voeikov Main Geophysical Observatory. 1999. "Worldwide Daily Solar Radiation." Available from Worldwide Web: http://www.mgo.rssi.ru.

82. Whillier, A. 1965. Solar radiation graphs. *Sol. Energy* 9: 165–166.

83. Zabara, K. 1986. Estimation of the global solar radiation in Greece. *Sol. Wind Technology* 3: 267.

84. Zerlaut, G. 1989. Solar radiation instrumentation. In *Solar Resources,* Roland L. Hulstrom ed. Cambridge, MA: MIT Press.

METHODS OF SOLAR COLLECTION AND THERMAL CONVERSION (**3**)

Always take more notice of the corners than you do of a surface to be polished; the middle you cannot very well forget . . . but the corners and crevices are often neglected.

The French Polishers Handbook

Converting the sun's radiant energy to heat is the most common and well-developed solar conversion technology today. The temperature level and amount of this converted energy are the key parameters that must be known to match a conversion scheme to a specified task effectively.

The basic principle of solar thermal collection is that when solar radiation strikes a surface, part of it is absorbed, thereby increasing the temperature of the surface. The efficiency of that surface as a solar collector depends not only on the absorption efficiency, but also on how the thermal and reradiation losses to the surroundings are minimized and how the energy from the collector is removed for useful purposes. Various solar thermal collectors range from unglazed flat plate-type solar collectors operating at about 5–10° C above the ambient, to central receiver concentrating collectors operating at above 1000° C. Table 3.1 lists various types of solar thermal collectors and their typical temperature and concentration ranges.

This chapter analyzes in detail the thermal and optical performance of several solar-thermal collectors. They range from air- and liquid-cooled nonconcentrating, flat-plate types to compound-curvature, continuously tracking types with concentration ratios up to 3000 or more. Applications of the energy converted by solar thermal collectors are described in Chapters 6, 7 and 8. The next section describes some fundamental radiative properties of materials, knowledge of which helps in the design of solar thermal collectors.

Table 3.1. Types of solar thermal collectors and their typical temperature range

Type of Collector	Concentration Ratio	Typical Working Temperature Range (°C)
Flat plate collector	1	≤70
High efficiency flat plate collector	1	60–120
Fixed concentrator	3–5	100–150
Parabolic trough collector	10–50	150–350
Parabolic dish collector	200–500	250–700
Central receiver	500–>3000	500–>1000

3.1 RADIATIVE PROPERTIES AND CHARACTERISTICS OF MATERIALS

When radiation strikes a body, a part of it is reflected, a part is absorbed and, if the material is transparent, a part is transmitted, as shown in Fig. 3.1.

The fraction of the incident radiation reflected is defined as the reflectance ρ, the fraction absorbed as the absorptance α, and the fraction transmitted as the transmittance τ. According to the first law of thermodynamics these three components must add up to unity, or

$$\alpha + \tau + \rho = 1. \tag{3.1}$$

Opaque bodies do not transmit any radiation and $\tau = 0$.

The reflection of radiation can be ***specular*** or ***diffuse***. When the angle of incidence is equal to the angle of reflection, the reflection is called specular; when the reflected radiation is uniformly distributed into all directions it is called diffuse (see Fig. 3.2). No real surface is either specular or diffuse, but a highly polished surface approaches specular reflection, whereas a rough surface reflects diffusely.

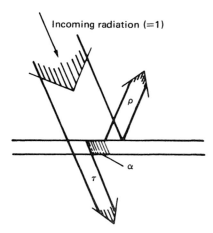

Incoming radiation (=1)

ρ

α

τ

Figure 3.1. Schematic representation of transmittance τ, absorptance α, and reflectance ρ.

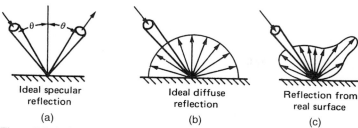

Ideal specular reflection

Ideal diffuse reflection

Reflection from real surface

(a)

(b)

(c)

Figure 3.2. Reflections from ideal specular, ideal diffuse, and real surfaces.

Another important radiative property is called emittance ϵ, which is the ratio of the radiative emissive power of a real surface to that of an ideal "black" surface as defined in Chapter 2.

All of the radiative properties of materials, α, τ, ρ and ϵ can be functions of the wavelength and direction. In fact, such dependence is used in the design of solar energy devices and systems. For example, selective absorbers are used for solar collectors and passive heating systems, and glazing materials for daylighting and solar collectors.

Referring to figure 2.5 (from Chapter 2) the monochromatic directional emittance of a surface, $\epsilon_\lambda(\theta, \phi)$, in a direction signified by an azimuth angle ϕ and a polar angle θ, is

$$\epsilon_\lambda(\theta,\phi) = \frac{I_\lambda(\theta,\phi)}{I_{b\lambda}}. \tag{3.2}$$

From the above, the total directional emittance $\epsilon(\theta,\phi)$ over all the wavelengths or the monochromatic hemispherical emittance ϵ_λ can be obtained by integration of $\epsilon_\lambda(\theta,\phi)$ over all the wavelengths or the entire hemispherical space, respectively. The overall emittance ϵ, is found by integrating the hemispherical emittance over all the wavelengths.

$$\epsilon = \frac{1}{\sigma T^4} \int_0^\infty \epsilon_\lambda E_{b\lambda} d\lambda. \tag{3.3}$$

Observe that both ϵ_λ and ϵ are properties of the surface.

The next most important surface characteristic is the absorptance. We begin by defining the monochromatic directional absorptance as the fraction of the incident radiation at wavelength λ from the direction θ,ϕ that is absorbed, or

$$\alpha_\lambda(\theta,\phi) = \frac{I_{\lambda,a}(\theta,\phi)}{I_{\lambda,i}(\theta,\phi)}, \tag{3.4}$$

where the subscripts a and i denote absorbed and incident radiation, respectively. The monochromatic directional absorptance is also a property of the surface.

More important than $\alpha_\lambda(\theta,\phi)$ is the overall directional absorptance $\alpha(\theta,\phi)$, defined as the fraction of the total radiation from the direction θ,ϕ that is absorbed, or

$$\alpha(\theta,\phi) = \frac{\int_0^\infty \alpha_\lambda(\theta,\phi)I_{\lambda,i}(\theta,\phi)\,d\lambda}{\int_0^\infty I_{\lambda,i}(\theta,\phi)\,d\lambda} = \frac{1}{I_i(\theta,\phi)}\int_0^\infty \alpha_\lambda(\theta,\phi)I_{\lambda,i}(\theta,\phi)\,d\lambda. \qquad (3.5)$$

The overall absorptance is a function of the characteristics of the incident radiation and is, therefore, unlike the monochromatic absorptance, not a property of a surface alone. It is this characteristic that makes it possible to have selective surfaces that absorb the radiation from one source at a higher rate than from another. In other words, even though according to Kirchhoff's law the monochromatic emittance at λ must equal the monochromatic absorptance

$$\alpha_\lambda(\theta,\phi) = \epsilon_\lambda(\theta,\phi), \qquad (3.6)$$

the overall emittance is not necessarily equal to the overall absorptance unless thermal equilibrium exists, and the incoming and outgoing radiation have the same spectral characteristics.

The effect of incidence angle on the absorptance is illustrated in Table 3.2 where the angular variation of the absorptance for a nonselective black surface, typical of those used on flat-plate collectors, is shown. The absorptance of this surface for diffuse radiation is approximately 0.90.

The third characteristic to be considered is the reflectance. Reflectance is particularly important for the design of focusing collectors. As mentioned previously, there are two limiting types of reflection: specular and diffuse. As illustrated in Fig. 3.2a, when a ray of incident radiation at an angle θ is reflected at the same polar angle and the azimuthal angles differ by 180°, as for a perfect mirror, the reflection is said to be specular. The reflection is said to be diffuse if the incident radiation is scattered equally in all directions, as shown in Fig. 3.2b.

Table 3.2. Angular variation of absorptance of lampblack paint[a]

Incidence angle $i(°)$	Absorptance $\alpha(i)$
0–30	0.96
30–40	0.95
40–50	0.93
50–60	0.91
60–70	0.88
70–80	0.81
80–90	0.66

[a]Adapted from Löf and Tybout [53].

3.1.1 Selective Surfaces

Two types of special surfaces of great importance in solar collector systems are selective and reflecting surfaces. Selective surfaces combine a high absorptance for solar radiation with a low emittance for the temperature range in which the surface emits radiation. This combination of surface characteristics is possible because 98 percent of the energy in incoming solar radiation is contained within wavelengths below 3 μm, whereas 99 percent of the radiation emitted by black or gray surfaces at 400K is at wavelengths longer than 3 μm. The dotted line in Fig. 3.3 illustrates the spectral reflectance of an ideal, selective semi-gray surface having a uniform reflectance of 0.05 below 3 μm, but 0.95 above 3 μm. Real surfaces do not approach this performance. Table 3.3 lists properties of some selective coatings.

3.1.2 Reflecting Surfaces

Concentrating solar collectors require the use of reflecting surfaces with high specular reflectance in the solar spectrum or refracting devices with high transmittance in the solar spectrum. Reflecting surfaces are usually highly polished metals or metal coatings on suitable substrates. With opaque substrates, the reflective coatings must always be front-surfaced, for example, chrome plating on copper or polished aluminum. If a transparent substrate is used, however, the coating may be front- or back-surfaced. In any back-surfaced reflector the radiation must pass through the substrate twice and the transmittance of the material becomes very important.

Table 3.4 presents typical values for the normal specular reflectance of new surfaces for beam solar radiation.

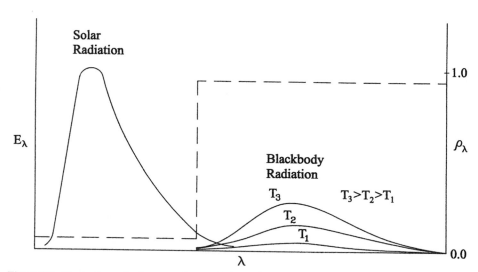

Figure 3.3. Illustrations of reflectance characteristics of an ideal selective surface. Shows radiation from ideal surfaces at different temperatures, and solar radiation.

Table 3.3. Properties of some selective plated coating systems[a]

Coating[b]	Substrate	$\bar{\alpha}_s$	$\bar{\epsilon}_i$	Durability	
				Breakdown temperature (°F)	Humidity-Degradation MIL STD 810B
Black nickel on nickel	Steel	0.95	0.07	>550	Variable
Black chrome on nickel	Steel	0.95	0.09	>800	No effect
Black chrome	Steel	0.91	0.07	>800	Completely rusted
	Copper	0.95	0.14	600	Little effect
	Galvanized steel	0.95	0.16	>800	Complete removal
Black copper	Copper	0.88	0.15	600	Complete removal
Iron oxide	Steel	0.85	0.08	800	Little effect
Manganese oxide	Aluminum	0.70	0.08		
Organic overcoat on iron oxide	Steel	0.90	0.16		Little effect
Organic overcoat on black chrome	Steel	0.94	0.20		Little effect

[a]From U.S. Dept. of Commerce, "Optical Coatings for Flat Plate Solar Collectors," NTIS No. PN-252–383, Honeywell, Inc., 1975.

[b]Black nickel coating plated over a nickel-steel substrate has the best selective properties ($\bar{\alpha}_s = 0.95$, $\bar{\epsilon}_i = 0.07$), but these degraded significantly during humidity tests. Black chrome plated on a nickel-steel substrate also had very good selective properties ($\bar{\alpha}_s = 0.95$, $\bar{\epsilon}_i = 0.09$) and also showed high resistance to humidity.

Table 3.4. Specular reflectance values for solar reflector materials

Material	ρ
Silver (unstable as front surface mirror)	0.94 ± 0.02
Gold	0.76 ± 0.03
Aluminized acrylic, second surface	0.86
Anodized aluminum	0.82 ± 0.05
Various aluminum surfaces-range	0.82–0.92
Copper	0.75
Back-silvered water-white plate glass	0.88
Aluminized type-C Mylar (from Mylar side)	0.76

3.1.3 Transparent Materials

The optical transmission behavior can be characterized by two wavelength dependent physical properties—the index of refraction n and the extinction coefficient k. The index of refraction, which determines the speed of light in the material, also determines the amount of light reflected from a single surface, while the extinction coefficient determines the amount of light absorbed in a substance in a single pass of radiation as described in Chapter 2.

Figure 3.4 defines the angles used in analyzing reflection and transmission of light. The angle i is called the **angle of incidence**. It is also equal to the angle at which a beam is specularly reflected from the surface. Angle θ_r is the **angle of refraction**, which is defined as shown in the figure. The incidence and refraction angles are related by Snell's law:

$$\frac{\sin(i)}{\sin(\theta_r)} = \frac{n'_r}{n'_i} = n_r, \tag{3.7}$$

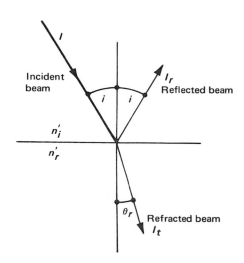

Figure 3.4. Diagram showing incident, reflected and refracted beams of light and incidence and refraction angles for a transparent medium.

Table 3.5. Refractive index for various substances in the visible range based on air.

Material	Index of Refraction
Air	1.000
Clean polycarbonate	1.59
Diamond	2.42
Glass (solar collector type)	1.50–1.52
Plexiglass[a] (polymethyl methacrylate, PMMA)	1.49
Mylar[a] (polyethylene terephthalate, PET)	1.64
Quartz	1.54
Tedlar[a] (polyvinyl fluoride, PVF)	1.45
Teflon[a] (polyfluoroethylenepropylene, FEP)	1.34
Water–liquid	1.33
Water–solid	1.31

[a]Trademark of the duPont Company, Wilmington, Delaware.

where n_i' and n_r' are the two refractive indices and n_r is the index ratio for the two substances forming the interface. Typical values of refractive indices for various materials are shown in Table 3.5. For most materials of interest in solar applications, the values range from 1.3 to 1.6, a fairly narrow range.

By having a gradual change in index of refraction, reflectance losses are reduced significantly. The reflectance of a glass-air interface common in solar collectors may be reduced by a factor of four by an etching process. If glass is immersed in a silica-

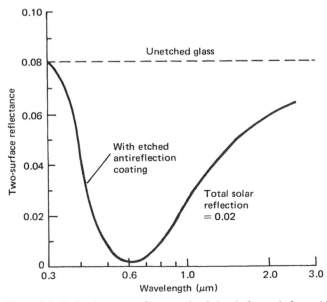

Figure 3.5. Reflection spectra for a sample of glass before and after etching.

supersaturated fluosilicic acid solution, the acid attacks the glass and leaves a porous silica surface layer. This layer has an index of refraction intermediate between glass and air. Figure 3.5 shows the spectral reflectance of a pane of glass before and after etching.

3.2 FLAT PLATE COLLECTORS

A simple flat plate collector consists of an absorber surface (usually a dark, thermally conducting surface); a trap for reradiation losses from the absorber surface (such as glass which transmits shorter wavelength solar radiation but blocks the longer wavelength radiation from the absorber); a heat transfer medium such as air, water, etc.; and some thermal insulation behind the absorber surface. Flat plate collectors are used typically for temperature requirements up to 75°C although higher temperatures can be obtained from high efficiency collectors. These collectors are of two basic types based on the heat transfer fluid:

Liquid type: where heat transfer fluid may be water, mixture of water and antifreeze, oil, etc.;

Air type: where heat transfer medium is air (used mainly for drying and space heating requirements).

3.2.1 Liquid-Type Collectors

Figure 3.6 shows a typical liquid-type flat plate collector. In general it consists of:

1. *Glazing*: one or more covers of transparent material like glass, plastics, etc. Glazing may be left out for some low-temperature applications.

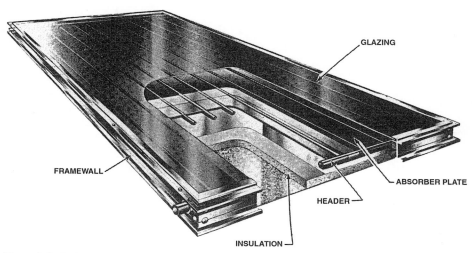

Figure 3.6. Typical liquid-type flat plate collector (courtesy of Morning Star Corporation, Orange Park, Florida).

2. *Absorber*: a plate with tubes or passages attached to it for the passage of a working fluid. The absorber plate is usually painted flat black or electroplated with a selective absorber.
3. *Headers* or manifolds: to facilitate the flow of heat transfer fluid.
4. *Insulation*: to minimize heat loss from the back and the sides.
5. *Container*: box or casing.

3.2.2 Air-Type Collectors

Air types of collectors are more commonly used for agricultural drying and space heating applications. Their basic advantages are low sensitivity to leakage and no need for an additional heat exchanger for drying and space heating applications. However, because of the low heat capacity of the air and the low convection heat transfer coefficient between the absorber and the air, a larger heat transfer area and higher flow rates are needed. Figure 3.7 shows some common configurations of air heating collectors. Common absorber materials include corrugated aluminum or galvanized steel sheets, black metallic screens, or simply any black painted surface.

Unglazed, transpired solar air collectors offer a low-cost opportunity for some applications such as preheating of ventilation air and agricultural drying and curing [49]. Such collectors consist of perforated absorber sheets that are exposed to the sun and through which air is drawn. The perforated absorber sheets are attached to the vertical walls, which are exposed to the sun. Kutcher and Christensen [50] have given a de-

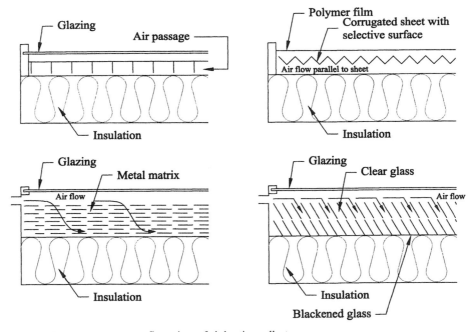

Figure 3.7. Some common configurations of air heating collectors.

tailed thermal analysis of unglazed transpired solar collectors. (See Chapter 5, Section 5.4.5 for additional details.)

The most important components, whose properties determine the efficiency of solar thermal collectors, are glazings and absorbers.

3.2.3 Glazings

The purpose of a glazing or transparent cover is to transmit the shorter wavelength solar radiation but block the longer wavelength reradiation from the absorber plate, and to reduce the heat loss by convection from the top of the absorber plate. Consequently, an understanding of the process and laws that govern the transmission of radiation through a transparent medium is important. Section 3.1 describes in brief the transmission of radiation through materials.

Glass is the most common glazing material. Figure 3.8 shows transmittance of glass as a function of wavelength. Transparent plastics, such as polycarbonates and acrylics are also used as glazings for flat plate collectors. The main disadvantage of plastics is that their transmittance in the longer wavelength is also high, therefore, they are not as good a trap as glass. Other disadvantages include deterioration over a period of time due to ultraviolet solar radiation. Their main advantage is resistance to breakage. Although glass can break easily, this disadvantage can be minimized by using tempered glass.

In order to minimize the upward heat loss from the collector, more than one transparent glazing may be used. However, with the increase in the number of cover plates, transmittance is decreased. Figure 3.9 shows the effect of number of glass cover plates on transmittance.

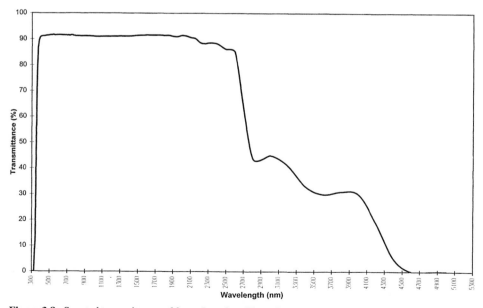

Figure 3.8. Spectral transmittance of 3 mm Low Iron Float glass.

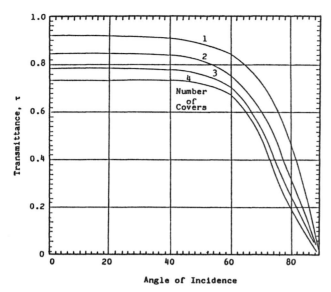

Figure 3.9. Transmittance of multiple glass covers having an index of refraction 1.526".

Absorbers. The purpose of the absorber is to absorb as much of the incident solar radiation as possible, re-emit as little as possible, and allow efficient transfer of heat to a working fluid. The most common forms of absorber plates in use are shown in Figure 3.10. The materials used for absorber plates include copper, aluminum, stainless steel, galvanized steel, plastics and rubbers. Copper seems to be the most common material used for absorber plates and tubes because of its high thermal conductivity and high corrosion resistance. However, copper is quite expensive. For low-temperature applications (up to about 50°C or 120°F) a plastic material called ethylene propylene polymer (trade names EPDM, HCP, etc.) can be used to provide inexpensive absorber ma-

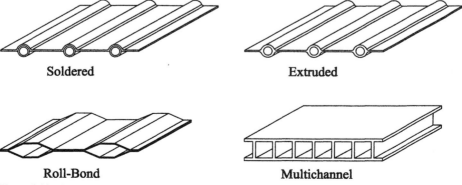

Figure 3.10. Common types of absorber plates.

Manifold

HCP® Absorber Panel

Substrate

Frame

Back Brace

Figure 3.11. Cutaway view of a typical collector made from ethylene propylene polymer (Courtesy of Sealed Air Corporation, Saddlebrook, NJ).

terial. To compensate for the low thermal conductivity, a large surface area is provided for heat transfer. Figure 3.11 shows a typical collector made from such material.

In order to increase the absorption of solar radiation and to reduce the emission from the absorber, the metallic absorber surfaces are painted or coated with flat black paint or some selective coating. A selective coating has high absorptivity in the solar wavelength range (0.3 to 3.0 μm). Absorptivities and emissivities of some common selective surfaces are given in Table 3.3.

A simple and inexpensive collector consists of a black painted corrugated metal absorber on which water flows down open, rather than enclosed in tubes. This type of collector is called a *trickle collector* and is usually built on-site. Although such a collector is simple and inexpensive, it has the disadvantages of condensation on the glazing and a higher pumping power requirement.

3.2.4 Energy balance for a flat-plate collector. The thermal performance of any type of solar thermal collector can be evaluated by an energy balance that determines the portion of the incoming radiation delivered as useful energy to the working fluid. For a flat-plate collector of an area A_c this energy balance on the absorber plate is

$$I_c A_c \tau_s \alpha_s = q_u + q_{loss} + \frac{de_c}{dt}, \tag{3.8}$$

where I_c = solar irradiation on a collector surface,
τ_s = effective solar transmittance of the collector cover(s),
α_s = solar absorptance of the collector-absorber plate surface,
q_u = rate of heat transfer from the collector-absorber plate to the working fluid,
q_{loss} = rate of heat transfer (or heat loss) from the collector-absorber plate to the surroundings, and
de_c/dt = rate of internal energy storage in the collector.

The instantaneous efficiency of a collector η_c is simply the ratio of the useful energy delivered to the total incoming solar energy, or

$$\eta_c = \frac{q_u}{A_c I_c}. \tag{3.9}$$

In practice, the efficiency must be measured over a finite time period. In a standard performance test, this period is on the order of 15 or 20 min, whereas for design, the performance over a day or over some longer period t is important. Then we have for the average efficiency

$$\eta_c = \frac{\int_0^t q_u dt}{\int_0^t A_c I_c dt}, \tag{3.10}$$

where t is the time period over which the performance is averaged.

A detailed and precise analysis of the efficiency of a solar collector is complicated by the nonlinear behavior of radiation heat transfer. However, a simple linearized analysis is usually sufficiently accurate in practice. In addition, the simplified analytical procedure is very important because it illustrates the parameters of significance for a solar collector and how these parameters interact. For a proper analysis and interpretation of these test results an understanding of the thermal analysis is imperative, although for design and economic evaluation the results of standardized performance tests are generally used.

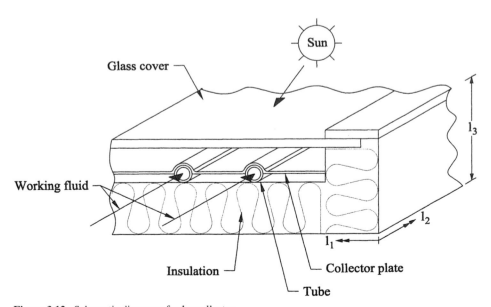

Figure 3.12. Schematic diagram of solar collector.

Collector heat-loss conductance. In order to obtain an understanding of the parameters determining the thermal-efficiency of a solar collector, it is important to develop the concept of **collector heat-loss conductance**. Once the collector heat-loss conductance U_c is known, and when the collector plate is at an average temperature T_c, the collector heat loss can be written in the simple form

$$q_{loss} = U_c A_c (T_c - T_a) \tag{3.11}$$

The simplicity of this relation is somewhat misleading because the collector heat-loss conductance cannot be specified without a detailed analysis of all the heat losses. Figure 3.12 shows a schematic diagram of a single-glazed collector, while Fig. 3.13(a) shows the thermal circuit with all the elements that must be analyzed before they can be combined into a single conductance element shown in Figure 3.13(b). The analysis below shows an example of how this combination is accomplished.

In order to construct a model suitable for a thermal analysis of a flat-plate collector, the following simplifying assumptions will be made:

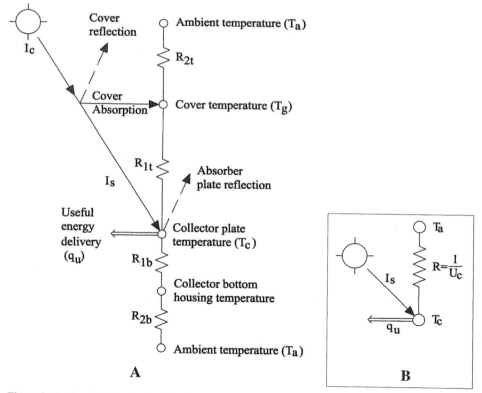

Figure 3.13. Thermal circuits for a flat-plate collector shown in Fig. 3.12: (a) detailed circuit; (b) approximate, equivalent circuit to (a). In both circuits, the absorber energy is equal to $\alpha_s I_s$, where $I_s = \tau_s I_c$. Collector assumed to be at uniform temperature T_c.

1. The collector is thermally in steady state.
2. The temperature drop between the top and bottom of the absorber plate is negligible.
3. Heat flow is one-dimensional through the cover as well as through the back insulation.
4. The headers connecting the tubes cover only a small area of the collector and provide uniform flow to the tubes.
5. The sky can be treated as though it were a black-body source for infrared radiation at an equivalent sky temperature.
6. The irradiation on the collector plate is uniform.

For a quantitative analysis let the plate temperature be T_c and assume solar energy is absorbed at the rate $I_s \alpha_s$. Part of this energy is then transferred as heat to the working fluid, and if the collector is in the steady state, the other part is lost as heat to the ambient air if $T_c > T_a$. Some of the heat loss occurs through the bottom of the collector. It passes first through the back to the environment. Since the collector is in steady state, according to Eq. (3.8),

$$q_u = I_c A_c \tau_s \alpha_s - q_{loss}, \tag{3.12}$$

where q_{loss} can be determined using the equivalent thermal circuit as shown in figure 3.13.

$$q_{loss} = U_c A_c (T_c - T_a) = \frac{A_c(T_c - T_a)}{R}. \tag{3.13}$$

There are three parallel paths to heat loss from the hot collector absorber plate at T_c to the ambient at T_a: the top, bottom and edges. Because the edge losses are quite small compared to the top and the bottom losses, they are quite often neglected. However, they can be estimated easily if the insulation around the edges is of the same thickness as the back. The edge loss can be accounted for by simply adding the areas of the back (A_c) and the edges (A_e) for back heat loss.

Therefore, the overall heat loss coefficient is:

$$U_c A_c = \frac{A_c}{R} = \frac{A_c}{R_{1t} + R_{2t}} + \frac{A_c + A_e}{R_{1b} + R_{2b}}. \tag{3.14}$$

The thermal resistances can be found easily from the definition. For example,

$$R_{1b} = \frac{l_i}{k_i} \text{ and } R_{2b} = \frac{1}{h_{c,bottom}}, \tag{3.15}$$

where k_i and l_i are, respectively, the thermal conductivity and thickness of the insulation, and $h_{c,bottom}$ is the convective heat transfer coefficient between the collector and the air below the collector. In a well-insulated collector, R_{2b} is much smaller than R_{1b} and usually neglected. Referring to figure 3.12

$$A_e = 2(l_1 + l_2)\,l_3 \tag{3.16}$$

Since the heat loss from the top is by convection and radiation, it is more complicated than the bottom heat loss. Convection and radiation provide two parallel paths for heat loss from the absorber plate at T_c to the glass cover at T_g, and from the glass cover to the ambient. That is, the series resistance of R_{1t} and R_{2t} consists of:

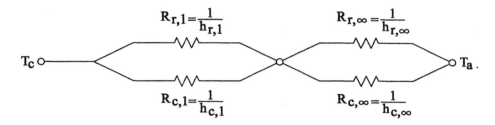

Therefore,

$$\frac{1}{R_{1t}} = \frac{1}{R_{r,1}} + \frac{1}{R_{c,1}} = h_{r,1} + h_{c,1} \text{ and} \tag{3.17}$$

$$\frac{1}{R_{2t}} = \frac{1}{R_{r,\infty}} + \frac{1}{R_{c,\infty}} = h_{r,\infty} + h_{c,\infty} \tag{3.18}$$

Since thermal radiative heat transfer is proportional to the fourth power of the temperature, R_r and h_r are found as follows:

Radiative heat transfer from the plate to the glass cover

$$q_{r_{c \to g}} = \sigma A_c \frac{(T_c^4 - T_g^4)}{(1/\epsilon_{p,i} + 1/\epsilon_{g,i} - 1)} = h_{r,1} A_c (T_c - T_g), \tag{3.19}$$

where

$$\epsilon_{p,i} = \text{infrared emittance of the plate}$$

$$\epsilon_{g,i} = \text{infrared emittance of the glass cover.}$$

Therefore,

$$h_{r,1} = \frac{\sigma(T_c + T_g)(T_c^2 + T_g^2)}{(1/\epsilon_{p,i} + 1/\epsilon_{g,i} - 1)}. \tag{3.20}$$

Similarly, from the radiative heat transfer between the glass plate (at T_g) and the sky (at T_{sky}) we can find that:

$$q_{r_{g \to sky}} = \epsilon_{g,i} \sigma A_c (T_g^4 - T_{sky}^4) = h_{r,\infty} A_c (T_g - T_a), \tag{3.21}$$

or

$$h_{r,\infty} = \epsilon_{g,i}\sigma(T_g^4 - T_{sky}^4)/(T_g - T_a). \tag{3.22}$$

Evaluation of the collector heat-loss conductance defined by Eq. (3.14) requires iterative solution of Eqs. (3.19) and (3.21), because the unit radiation conductances are functions of the cover and plate temperatures, which are not known a priori. A simplified procedure for calculating U_c for collectors with all covers of the same material, which is often sufficiently accurate and more convenient to use, has been suggested by Hottel and Woertz [36] and Klein [41]. It is also suitable for application to collectors with selective surfaces. For this approach the collector top loss in watts is written in the form [1]:

$$q_{toploss} = \frac{(T_c - T_a)A_c}{N/(C/T_c)[(T_c - T_a)/(N + f)]^{0.33} + 1/h_{c,\infty}}$$

$$+ \frac{\sigma(T_c^4 - T_a^4)A_c}{1/[\epsilon_{p,i} + 0.05N(1 - \epsilon_{p,i})] + (2N + f - 1)/\epsilon_{g,i} - N}, \tag{3.23}$$

where $f = (1 - 0.04h_{c,\infty} + 0.005h_{c,\infty}^2)(1 + 0.091N)$,
$\quad C = 250[1 - 0.0044(\beta - 90)]$,
$\quad N$ = number of covers,
$\quad h_{c,\infty} = 5.7 + 3.8V$, and
$\quad \epsilon_{g,i}$ = infrared emittance of the covers,
$\quad V$ = wind speed in m/sec.

The values of $q_{top\,loss}$ calculated from Eq. (3.23) agreed closely with the values obtained from Eq. (3.22) for 972 different observations encompassing the following conditions:

$320 < T_c < 420$ K
$260 < T_a < 310$ K
$0.1 < \epsilon_{p,i} < 0.95$
$0 \le V \le 10$ m/sec
$1 \le N \le 3$
$0 \le \beta \le 90$

The standard deviation of the differences in $U_c = q_{top\,loss}/A_c(T_c - T_a)$ was 0.14 W/m² · K for these comparisons.

3.2.5 Thermal Analysis of Flat-Plate Collector-Absorber Plate

In order to determine the efficiency of a solar collector, the rate of heat transfer to the working fluid must be calculated. If transient effects are neglected [35,41,79], the rate of heat transfer to the fluid flowing through a collector depends on the temperature of the collector surface from which heat is transferred by convection to the fluid, the temperature of the fluid, and the heat-transfer coefficient between the collector and the fluid. To analyze the rate of heat transfer consider first the condition at a cross section of the collector with flow ducts of rectangular cross sections as shown in Fig. 3.14. Solar radiant energy impinges on the upper face of the collector plate. A part of the total solar radiation falls on the upper surface of the flow channels, while another part is incident on the plates connecting any two adjacent flow channels. The latter is conducted in a transverse direction toward the flow channels. The temperature is a maximum at any midpoint between adjacent channels, and the collector plate acts as a fin attached to the walls of the flow channel. The thermal performance of a fin can be expressed in terms of its efficiency. The fin efficiency η_f is defined as the ratio of the rate of heat flow through the real fin to the rate of heat flow through a fin of infinite thermal conductivity, that is, a fin at a uniform temperature. We shall now derive a relation to evaluate this efficiency for a flat-plate solar collector.

If U_c is the overall unit conductance from the collector-plate surface to the ambient air, the rate of heat loss from a given segment of the collector plate at x,y in Fig. 3.14 is

$$q(x,y) = U_c[T_c(x,y) - T_a]dxdy, \qquad (3.24)$$

where T_c = local collector-plate temperature $(T_c > T_a)$
  T_a = ambient air temperature
  U_c = overall unit conductance between the plate and the ambient air

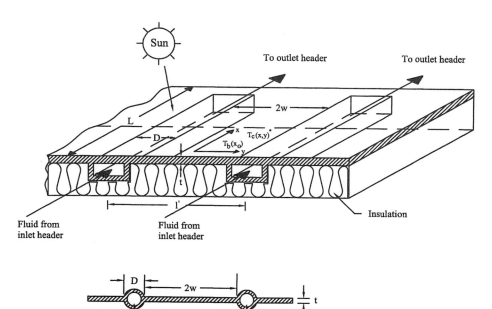

Figure 3.14. Sketch showing coordinates and dimensions for collector plate and fluid ducts.

U_c includes the effects of radiation and free convection between the plates, the radiative and convective transfer between the top of the cover and the environment, and conduction through the insulation. Its quantitative evaluation has been previously considered.

If conduction in the x direction is negligible, a heat balance at a given distance x_0 for a cross section of the flat-plate collector per unit length in the x direction can be written in the form

$$\alpha_s I_s \, dy - U_c(T_c - T_a) dy + \left(-kt \frac{dT_c}{dy} \bigg|_{y,x_0} \right) - \left(-kt \frac{dT_c}{dy} \bigg|_{y+dy,x_0} \right) = 0. \quad (3.25)$$

If the plate thickness t is uniform and the thermal conductivity of the plate is independent of temperature, the last term in Eq. (3.25) is

$$\frac{dT_c}{dy} \bigg|_{y+dy,x_0} = \frac{dT_c}{dy} \bigg|_{y,x_0} + \left(\frac{d^2 T_c}{dy^2} \right)_{y,x_0} dy,$$

and Eq. (3.25) can be cast into the form of a second-order differential equation:

$$\frac{d^2 T_c}{dy^2} = \frac{U_c}{kt} \left[T_c - \left(T_a + \frac{\alpha_s I_s}{U_c} \right) \right] \quad (3.26)$$

The boundary conditions for the system described above at a fixed x_0, are:

1. At the center between any two ducts the heat flow is 0, or at $y = 0$, $dT_c = 0$.
2. At the duct the plate temperature is $T_b(x_0)$, or at $y = w = (l' - D)/2$, $T_c = T_b(x_0)$, where $T_b(x_0)$ is the fin-base temperature.

If we let $m^2 = U_c/kt$ and $\phi = T_c - (T_a + \alpha_s I_s/U_c)$, Eq. (3.26) becomes

$$\frac{d^2 \phi}{dy^2} = m^2 \phi, \quad (3.27)$$

subject to the boundary conditions

$$\frac{d\phi}{dy} = 0 \text{ at } y = 0, \text{ and}$$

$$\phi = T_b(x_0) - \left(T_a + \frac{\alpha_s I_s}{U_c} \right) \text{ at } y = w.$$

The general solution of Eq. (3.27) is

$$\phi = C_1 \sinh my + C_2 \cosh my. \quad (3.28)$$

The constants C_1 and C_2 can be determined by substituting the two boundary conditions and solving the two resulting equations for C_1 and C_2. This gives

$$\frac{T_c - (T_a + \alpha_s I_s/U_c)}{T_b(x_0) - (T_a + \alpha_s I_s/U_c)} = \frac{\cosh my}{\cosh mw}. \tag{3.29}$$

From the preceding equation the rate of heat transfer to the conduit from the portion of the plate between two conduits can be determined by evaluating the temperature gradient at the base of the fin, or

$$q_{fin} = -kt\frac{dT_c}{dy}\bigg|_{y=w} = \frac{1}{m}\{\alpha_s I_s - U_c[T_b(x_0) - T_a]\tanh mw\}. \tag{3.30}$$

Since the conduit is connected to fins on both sides, the total rate of heat transfer is

$$q_{total}(x_0) = 2w\{\alpha_s I_s - U_c[T_b(x_0) - T_a]\}\frac{\tanh mw}{mw}. \tag{3.31}$$

If the entire fin were at the temperature $T_b(x_0)$, a situation corresponding physically to a plate of infinitely large thermal conductivity, the rate of heat transfer would be a maximum, $q_{total,max}$. As mentioned previously, the ratio of the rate of heat transfer with a real fin to the maximum rate obtainable is the fin efficiency η_f. With this definition, Eq. (3.31) can be written in the form

$$q_{total}(x_0) = 2w\eta_f\{\alpha_s I_s - U_c[T_b(x_0) - T_a]\}, \tag{3.32}$$

where $\eta_f \equiv \tanh mw/mw$.

The fin efficiency η_f is plotted as a function of the dimensionless parameter $w(U_c/kt)^{1/2}$ in Fig. 3.15. When the fin efficiency approaches unity, the maximum portion of the radiant energy impinging on the fin becomes available for heating the fluid.

In addition to the heat transferred through the fin, the energy impinging on the portion of the plate above the flow passage is also useful. The rate of useful energy from this region available to heat the working fluid is

$$q_{duct}(x_0) = D\{\alpha_s I_s - U_c[T_b(x_0) - T_a]\}. \tag{3.33}$$

Thus, the useful energy per unit length in the flow direction becomes

$$q_u(x_0) = (D + 2w\eta)\{\alpha_s I_s - U_c[T_b(x_0) - T_a]\}. \tag{3.34}$$

The energy $q_u(x_0)$ must be transferred as heat to the working fluid. If the thermal resistance of the metal wall of the flow duct is negligibly small and there is no contact resistance between the duct and the plate, the rate of heat transfer to the fluid is

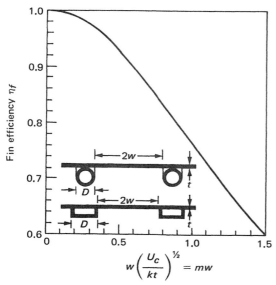

Figure 3.15. Fin efficiency for tube and sheet flat-plate solar collectors.

$$q_u(x_0) = P\bar{h}_{c,i}[T_b(x_0) - T_f(x_0)], \tag{3.35}$$

where P is the perimeter of the flow duct, which is $2(D + d)$ for a rectangular duct. Contact resistance may become important in poorly manufactured collectors in which the flow duct is clamped or glued to the collector plate. Collectors manufactured by such methods are usually not satisfactory.

3.2.6 Collector Efficiency Factor

To obtain a relation for the useful energy delivered by a collector in terms of known physical parameters, the fluid temperature, and the ambient temperature, the collector temperature must be eliminated from Eqs. (3.34) and (3.35). Solving for $T_b(x_0)$ in Eq. (3.35) and substituting this relation in Eq. (3.34) gives

$$q_u(x_0) = l'F'\{\alpha_s I_s - U_c[T_f(x_0) - T_a]\}, \tag{3.36}$$

where F' is called the collector efficiency factor [9], and $l' = (2w+D)$. F' is given by

$$F' = \frac{1/U_c}{l'[1/(U_c(D + 2w\eta_f)) + 1/(\bar{h}_{c,i}P)]}. \tag{3.37}$$

For a flow duct of circular cross section of diameter D, F' can be written as

Table 3.6. Typical values for the parameters that determine the collector efficiency factor F' for a flat-plate collector in Eq. (3.37).

		U_c
2 glass covers	4 W/m^2 · K	0.685 Btu/hr · ft^2 · °F
1 glass cover	8 W/m^2 · K	1.37 B Btu/hr · ft^2 · °F
		kt
copper plate, 1 mm thick	0.38 W/K	0.72 Btu/hr · °F
steel plate, 1 mm thick	0.045 W/K	0.0866 Btu/hr · °F
		$\bar{h}_{c,i}$
water in laminar flow forced convection	300 W/m^2 · K	52 Btu/hr · ft^2 · °F
water in turbulent flow forced convection	1500 W/m^2 · K	254 Btu/hr · ft^2 · °F
air in turbulent forced convection	100 W/m^2 · K	17.6 Btu/hr · ft^2 · °F

$$F' = \frac{1/U_c}{l'[1/(U_c(D + 2w\eta_f)) + 1/(\bar{h}_{c,i}\pi D)]}$$

Physically, the denominator in Eq. (3.37) is the thermal resistance between the fluid and the environment, whereas the numerator is the thermal resistance between the collector surface and the ambient air. The collector-plate efficiency factor F' depends on U_c, $\bar{h}_{c,i}$, and η_f. It is only slightly dependent on temperature and can, for all practical purposes, be treated as a design parameter. Typical values for the factors determining the value of F' are given in Table 3.6.

The collector efficiency factor increases with increasing plate thickness and plate thermal conductivity, but it decreases with increasing distance between flow channels. Also, increasing the heat-transfer coefficient between the walls of the flow channel and the working fluid increases F', but an increase in the overall conductance U_c will cause F' to decrease.

3.2.7 Collector Heat-Removal Factor

Equation (3.36) yields the rate of heat transfer to the working fluid at a given point x along the plate for specified collector and fluid temperatures. However, in a real collector the fluid temperature increases in the direction of flow as heat is transferred to it. An energy balance for a section of flow duct dx can be written in the form

$$\dot{m}c_p(T_f|_{x+dx} - T_f|_x) = q_u(x)dx. \tag{3.38}$$

Substituting Eq. (3.36) for $q_u(x)$ and $T_f(x) + (dT_f(x)/dx)\,dx$ for $T_f|_{x+dx}$ in Eq. (3.38) gives the differential equation

$$\dot{m}c_p\frac{dT_f(x)}{dx} = l'F'\{\alpha_s I_s - U_c[T_f(x) - T_a]\}. \tag{3.39}$$

Separating the variables gives, after some rearranging,

$$\frac{dT_f(x)}{T_f(x) - T_a - \alpha_s I_s/U_c} = \frac{l'F'U_c}{\dot{m}c_p}dx. \tag{3.40}$$

Equation (3.40) can be integrated and solved for the outlet temperature of the fluid $T_{f,out}$ for a duct length L, and for the fluid inlet temperature $T_{f,in}$ if we assume that F' and U_c are constant, or

$$\frac{T_{f,out} - T_a - \alpha_s I_s/U_c}{T_{f,in} - T_a - \alpha_s I_s/U_c} = \exp\left(-\frac{U_c l'F'L}{\dot{m}c_p}\right). \tag{3.41}$$

To compare the performance of a real collector with the thermodynamic optimum, it is convenient to define the heat-removal factor F_R as the ratio between the actual rate of heat transfer to the working fluid and the rate of heat transfer at the minimum temperature difference between the absorber and the environment.

The thermodynamic limit corresponds to the condition of the working fluid remaining at the inlet temperature throughout the collector. This can be approached when the fluid velocity is very high. From its definition F_R can be expressed as

$$F_R = \frac{Gc_p(T_{f,out} - T_{f,in})}{\alpha_s I_s - U_c(T_{f,in} - T_a)}, \tag{3.42}$$

where G is the flow rate per unit surface area of collector $\dot{m}/A_c$. By regrouping the right-hand side of Eq. (3.42) and combining with Eq. (3.41), it can easily be verified that

$$F_R = \frac{Gc_p}{U_c}\left[1 - \frac{\alpha_s I_s/U_c - (T_{f,out} - T_a)}{\alpha_s I_s/U_c - (T_{f,in} - T_a)}\right]$$

$$= \frac{Gc_p}{U_c}\left[1 - \exp\left(-\frac{U_c F'}{Gc_p}\right)\right]. \tag{3.43}$$

Inspection of the above relation shows that F_R increases with increasing flow rate and approaches as an upper limit F', the collector efficiency factor. Since the numerator of the right-hand side of Eq. (3.42) is q_u, the rate of useful heat transfer can now be expressed in terms of the fluid inlet temperature, or

$$q_u = A_c F_R[\alpha_s I_s - U_c(T_{f,in} - T_a)]. \tag{3.44}$$

If a glazing above the absorber plate has transmittance τ_s then

$$q_u = A_c F_R[\tau_s \alpha_s I_c - U_c(T_{f,in} - T_a)],\qquad(3.45)$$

and instantaneous efficiency η_c is

$$\eta_c = \frac{q_u}{I_c A_c} = F_R[\tau_s \alpha_s - U_c(T_{f,in} - T_a)/I_c]\qquad(3.46)$$

Equation (3.46) is also known as the **Hottel-Whillier-Bliss** equation. This is a convenient form for design, because the fluid inlet temperature to the collector is usually known or can be specified.

Example 3.1. Calculate the averaged hourly and daily efficiency of a water solar collector on January 15, in Boulder, CO. The collector is tilted at an angle of 60° and has an overall conductance of 8.0 W/m² · K on the upper surface. It is made of copper tubes, with a 1-cm ID, 0.05 cm thick, which are connected by a 0.05-cm-thick plate at a center-to-center distance of 15 cm. The heat-transfer coefficient for the water in the tubes is 1500 W/m² · K, the cover transmittance is 0.9, and the solar absorptance of the copper surface is 0.9. The collector is 1 m wide and 2 m long, the water inlet temperature is 330 K, and the water flow rate is 0.02 kg/sec. The horizontal insolation (total) I_h and the environmental temperature are tabulated below. Assume the diffuse radiation accounts for 25 percent of the total insolation.

Solution. The total radiation received by the collector is calculated from Eq. (2.42) and neglecting the ground reflected radiation:

$$I_c = I_{d,c} + I_{b,c} = 0.25 I_h \cos^2\left(\frac{60}{2}\right) + (1 - 0.25)I_h R_b.$$

Time (hr)	I_h (W/m²)	T_{amb} (K)
7–8	12	270
8–9	80	280
9–10	192	283
10–11	320	286
11–12	460	290
12–13	474	290
13–14	395	288
14–15	287	288
15–16	141	284
16–17	32	280

The tilt factor R_b is obtained from its definition in Chapter 2 [see Eq. (2.58)]:

$$R_b = \frac{\cos i}{\sin \alpha} = \frac{\sin(L - \beta)\sin\delta_s + \cos(L - \beta)\cos\delta_s\cos h_s}{\sin L \sin\delta_s + \cos L \cos\delta_s\cos h_s}$$

where $L = 40°$ $\delta_s = -21.1$ on January 15 (from Fig. 2.8), and $\beta = 60°$. The hour angle h_s equals $15°$ for each hour away from noon.

The fin efficiency is obtained from Eq. (3.32):

$$\eta_f = \frac{\tanh m(l' - D)/2}{m(l' - D)/2}$$

where

$$m = \left(\frac{U_c}{kt}\right)^{1/2} = \left(\frac{8}{390 \times 5 \times 10^{-4}}\right)^{1/2} = 6.4, \text{ and}$$

$$\eta_f = \frac{\tanh 6.4(0.15 - 0.01)/2}{6.4(0.15 - 0.01)/2} = 0.938.$$

The collector efficiency factor F' is, from Eq. (3.37),

$$F' = \frac{1/U_c}{l'[1/(U_c(D + 2w\eta_f)) + 1/(h_{c,i}\pi D)]}$$

$$= \frac{1/8.0}{0.15[1/8.0(0.01 + 0.14 \times 0.938) + 1/1500\pi \times 0.01]} = 0.92.$$

Then we obtain the heat-removal factor from Eq. (3.43):

$$F_R = \frac{Gc_p}{U_c}\left[1 - \exp\left(-\frac{U_cF'}{Gc_p}\right)\right].$$

Time (hr)	I_h (W/m²)	R_b	$I_{d,c}$ (W/m²)	$I_{b,c}$ (W/m²)	I_c (W/m²)	q_u (W)	T_{amb} (K)	η_c
7–8	12	10.9	1	98	99	0	270	0
8–9	80	3.22	5	193	198	0	280	0
9–10	192	2.44	12	351	363	0	283	0
10–11	320	2.18	20	523	543	148	286	0.137
11–12	460	2.08	29	718	747	482	290	0.322
12–13	474	2.08	30	739	769	512	290	0.333
13–14	395	2.18	25	646	671	351	288	0.261
14–15	287	2.49	18	525	543	175	288	0.162
15–16	141	3.22	9	341	350	0	284	0
16–17	32	10.9	2	261	263	0	280	0

$$F_R = \frac{0.01 \times 4184}{8.0}\left[1 - \exp\left(-\frac{8.0 \times 0.922}{0.01 \times 4184}\right)\right] = 0.845.$$

From Eq. (3.45), the useful heat delivery rate is

$$q_u = 2 \times 0.845[I_c \times 0.81 - 8.0(T_{f,in} - T_{amb})].$$

The efficiency of the collector is $\eta_c = q_u / A_c I_c$ and the hourly averages are calculated in the table above.

Thus, $\Sigma I_c = 4546$ W/m^2 and $\Sigma q_u = 1668$W. The daily average efficiency is obtained by summing the useful energy for those hours during which the collector delivers heat and dividing by the total insolation between sunrise and sunset. This yields

$$\overline{\eta}_{c,day} = \frac{\Sigma q_u}{\Sigma A_c I_c} = \frac{1668}{2 \times 4546} = 0.183 \text{ or } 18.3 \text{ percent}$$

3.2.8 Transient Effects

The preceding analysis assumed that steady-state conditions exist during the operation of the collector. Under actual operating conditions the rate of insolation will vary and the ambient temperature and the external wind conditions may change. To determine the effect of changes in these parameters on the performance of a collector it is necessary to make a transient analysis that takes the thermal capacity of the collector into account.

As shown in [42], the effect of collector thermal capacitance is the sum of two contributions: the **collector storage** effect, resulting from the heat required to bring the collector up to its final operating temperature, and the **transient** effect, resulting from fluctuations in the meteorological conditions. Both effects result in a net loss of energy delivered compared with the predictions from the zero capacity analsysis. This loss is particularly important on a cold morning when all of the solar energy absorbed by the collector is used to heat the hardware and the working fluid, thus delaying the delivery of useful energy for some time after the sun has come up.

Transient thermal analyses can be made with a high degree of precision [47], but the analytical predictions are no more accurate than the weather data and the overall collector conductance. For most engineering applications, a simpler approach is therefore satisfactory [19]. For this approach, it will be assumed that the absorber plate, the ducts, the back insulation, and the working fluid are at the same temperature. If back losses are neglected, an energy balance on the collector plate and the working fluid for a single-glazed collector delivering no useful energy can be written in the form

$$(\overline{mc})_p \frac{d\overline{T}_p(t)}{dt} = A_c I_s \alpha_s + A_c U_p[\overline{T}_g(t) - \overline{T}_p(t)], \tag{3.47}$$

where $(\overline{mc})_p$ is the sum of the thermal capacities of the plate, the fluid, and the insolation, I_s is the insolation on the absorber plate, and U_p is the conductance between the absorber plate at $\overline{T}_p$ and its cover at $\overline{T}_g$. Similarly, a heat balance on the collector cover gives

$$(mc)_g \frac{d\overline{T}_g(t)}{dt} = A_c U_p[\overline{T}_p(t) - \overline{T}_g(t)] - A_c U_\infty[\overline{T}_g(t) - T_a], \tag{3.48}$$

where $U_\infty = (h_{c,\infty} + h_{r,\infty})$ [see Eq. (3.18)], and
$(mc)_g$ = thermal capacity of the cover plate.

Equations (3.47) and (3.48) can be solved simultaneously and the transient heat loss can then be determined by integrating the instantaneous loss over the time during which transient effects are pronounced. A considerable simplification in the solution is possible if one assumes that at any time the collector heat loss and the cover heat loss are equal, as in a quasi-steady state, so that

$$U_\infty A_c[\overline{T}_g(t) - T_a] = U_c A_c[\overline{T}_p(t) - T_a] \tag{3.49}$$

Then, for a given air temperature, differentiation of Eq. (3.49) gives

$$\frac{d\overline{T}_g(t)}{dt} = \frac{U_c}{U_\infty} \frac{d\overline{T}_p(t)}{dt}. \tag{3.50}$$

Adding Eqs. (3.48), (3.49) and (3.50) gives a single differential equation for the plate temperature

$$\left[(\overline{mc})_p + \frac{U_c}{U_\infty}(mc)_g \right] \frac{d\overline{T}_p(t)}{dt} = [\alpha_s I_s - U_c(\overline{T}_p(t) - T_a)]A_c \tag{3.51}$$

Equation (3.51) can be solved directly for given values of I_s and T_a. The solution to Eq. (3.51) then gives the plate temperature as a function of time, for an initial plate temperature $T_{p,0}$, in the form

$$\overline{T}_p(t) - T_a = \frac{\alpha_s I_s}{U_c} - \left[\frac{\alpha_s I_s}{U_c} - (T_{p,0} - T_a) \right] \exp\left[-\frac{U_c A_c t}{(\overline{mc})_p + (U_c/U_\infty)_c(mc)_g} \right]. \tag{3.52}$$

Collectors with more than one cover can be treated similarly, as shown in [12].

For a transient analysis, the plate temperature $\overline{T}_p$ can be evaluated at the end of a specified time period if the initial value of $\overline{T}_p$ and the values of α_s, I_s, U_c and T_a during the specified time are known. Repeated applications of Eq. (3.52) provide an approximate method of evaluating the transient effects. An estimate of the net decrease in useful energy delivered can be obtained by multiplying the effective heat capacity of the collector, given by $(\overline{mc})_p + (U_c/U_\infty)(mc)_g$, by the temperature rise necessary to bring the collector to its operating temperature. Note that the parameter $[(\overline{mc})_p + (U_c/U_\infty)(mc)_g]/U_cA_c$ is the **time constant** of the collector (18,47) and small values of this parameter will reduce losses resulting from transient effects.

Example 3.2. Calculate the temperature rise between 8 and 10 AM of a 1 m × 2 m single-glazed water collector with a 0.3-cm-thick glass cover if the heat capacities of the plate, water, and back insulation, are 5, 3, and 2 kJ/K, respectively. Assume that the unit surface conductance from the cover to ambient air is 18 W/m² · K and the unit surface conductance between the collector and the ambient air is U_c = 6 W/m² · K. Assume that the collector is initially at the ambient temperature. The absorbed insolation α_sI_s during the first hour averages 90 W/m² and between 9 and 10 AM, 180 W/m². The air temperature between 8 and 9 AM is 273 K and that between 9 and 10 AM, 278 K.

Solution. The thermal capacitance of the glass cover is $(mc)_g = (\rho Vc_p)_g = (2500$ kg/m³) (1 m × 2 m × 0.003 m) (1 kJ/kg · K) = 15 kJ/K. The combined collector, water and insulation thermal capacity is equal to

$$(\overline{mc})_p + \frac{U_c}{U_\infty}(mc)_g = 5 + 3 + 2 + 0.3 \times 15 = 15.5 \text{kJ/K}.$$

From Eq. (3.52) the temperature rise of the collector given by

$$T_p - T_a = \frac{\alpha_sI_s}{U_c} - \left[\frac{\alpha_sI_s}{U_c} - (T_{p,0} - T_a)\right]\exp\left[-\frac{U_cA_ct}{(\overline{mc})_p + (U_c/U_\infty)(mc)_g}\right].$$

At 8 AM, $T_{p,0} = T_a$, therefore the temperature rise, between 8 and 9 AM, is

$$= \frac{90}{6}\left[1 - \exp\left(-\frac{2 \times 6 \times 3600}{15,500}\right)\right] = 15 \times 0.944 = 14.2 \text{K}.$$

Thus, at 9 AM the collector temperature will be 287.2 K. Between 9 and 10 AM the collector temperature will rise as shown below:

$$= 278 + \frac{180}{6} - (30 - 9.2)0.056 = 306.3 \text{K}(91°\text{F})$$

Thus, at 10 AM the collector temperature has achieved a value sufficient to deliver useful energy at a temperature level of 306 K.

3.2.9 Air-Cooled Flat-Plate Collector Thermal Analysis

The basic air-cooled flat-plate collector shown in Fig. 3.16 differs fundamentally from the liquid-based collectors described in preceding sections because of the relatively poor heat-transfer properties of air. For example, in turbulent flow in a given conduit

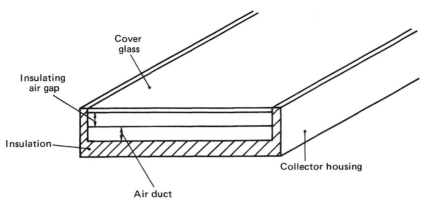

Figure 3.16. Schematic diagram of a basic air-heating flat-plate collector with a single glass (or plastic) cover.

for a fixed value of Reynolds number, the convection heat-transfer coefficient for water is about 50 times greater than that for air. As a result, it is essential to provide the largest heat-transfer area possible to remove heat from the absorber surface of an air-heating collector.

The most common way to achieve adequate heat transfer in air collectors is to flow air over the entire rear surface of the absorber as shown. The heat-transfer analysis of such a collector does not involve the fin effect or the tube-to-plate bond conductance problem, which arises in liquid collectors. The heat-transfer process is essentially that of an unsymmetrically heated duct of large aspect ratio (typically 20–40).

Malik and Buelow [54,55] surveyed the fluid mechanics and heat-transfer phenomena in air collectors. They concluded that a suitable expression for the Nusselt number for a smooth air-heating collector is

$$Nu_{sm} = \frac{0.0192 Re^{3/4} Pr}{1 + 1.22 Re^{-1/8}(Pr - 2)} \tag{3.53}$$

If the surface is hydrodynamically rough, they recommended multiplying the smooth-surface Nusselt number by the ratio of the rough-surface friction factor f to the smooth-surface friction factor f_{sm} from the Blasius equation

$$f_{sm} = 0.079 Re^{-1/4} \tag{3.54}$$

The convective coefficient h_c in the Nusselt number is based on a unit absorber area but the hydraulic diameter D_H in the Nusselt and Reynolds numbers is based on the ***entire duct perimeter***.

Air-Collector Efficiency Factor

The collector efficiency factor F' is defined as the ratio of energy collection rate to the collection rate if the absorber plate were at the local fluid temperature. F' is particularly simple to calculate, since no fin analysis or bond-conductance term is present for air collectors. For the collector shown in Fig. 3.16, F' is given by

$$F' = \frac{\bar{h}_c}{\bar{h}_c + U_c} \tag{3.55}$$

where U_c is calculated from Klein's equation, Eq. (3.23), and the duct convection coefficient $\bar{h}_c$ is calculated from Eq. (3.53).

Air-Collector Heat-Removal Factor

The collector heat-removal factor is a convenient parameter, since it permits useful energy gain to be calculated by knowledge of only the easily determined fluid inlet temperature as shown in Eq. (3.43). The heat-removal factor F_R for a typical air collector such as the one shown in Fig. 3.16 is given by

$$F_R = \frac{\dot{m}_a c_{p,a}}{U_c A_c}\left[1 - \exp\left(-\frac{F' U_c A_c}{\dot{m}_a c_{p,a}}\right)\right], \tag{3.56}$$

where F' is given by Eq. (3.55), $\dot{m}_a$ is the air flow rate (kg/sec), and $c_{p,a}$ is its specific heat. The heat-removal factor for air collectors is usually significantly less than that for liquid collectors because $\bar{h}_c$ for air is much smaller than for a liquid such as water.

Example 3.3. Calculate the collector-plate efficiency factor F' and heat-removal factor F_R for a smooth, 1 m-wide, 5 m-long air collector with the following design. The flow rate per unit collector area is 0.7 m³/min-m_c^2 (2.1 ft³/min-ft_c^2). The air duct height is 1.5 cm (0.6 in), the air density is 1.1 kg/m³ (0.07 lb/ft³), the specific heat is 1 kJ/kg-K (0.24 Btu/lb-°F), and the viscosity is 1.79×10^{-5} kg/m-sec (1.2×10^{-5} lb/ft-sec). The collector heat-loss coefficient U_c is 18 kJ/hr-m²-K (5 W/m²-K; 0.88 Btu/hr-ft²-°F).

Solution. The first step is to determine the duct heat-transfer coefficient h_c from Eq. (3.53). The Reynolds number is defined as

$$\text{Re} = \frac{\rho \bar{V} D_H}{\mu},$$

in which the average velocity $\bar{V}$ is the volume flow rate divided by the flow area.

$$\bar{V} = \frac{0.7 \times 1 \times 5}{1 \times 0.015} = 233 \text{ m/min} = 3.89 \text{ m/sec},$$

and the hydraulic diameter D_H is

$$D_H = \frac{4(0.015 \times 1)}{1 + 1 + 0.015 + 0.015} = 0.0296\text{m, and}$$

$$\text{Re} = \frac{(1.1)(3.89)(0.0296)}{1.79 \times 10^{-5}} = 7066.$$

From Eq. (3.53) the Nusselt number is

$$Nu = \frac{0.0192(7066)^{3/4}(0.72)}{1 + 1.22(7066)^{-1/8}(0.72 - 2.0)} = 22.0.$$

The heat-transfer coefficient is

$$h_c = Nu\frac{k}{D_H} = \frac{Nu c_p \mu}{\text{Pr} D_H}.$$

The Prandtl number for air is ~0.72. Therefore,

$$h_c = \frac{(22.0)(1.0)(1.79 \times 10^{-5})}{(0.72)(0.0296)} = 0.0185\text{kJ/m}^2\cdot\text{sec}\cdot\text{K}$$

$$= 66.5 \text{ kJ/m}^2\cdot\text{hr}\cdot°\text{C} \ (3.26 \text{ Btu/hr}\cdot\text{ft}\cdot°\text{F}).$$

The plate efficiency is then, from Eq. (3.55),

$$F' = \frac{66.5}{66.5 + 18} = 0.787,$$

and the heat-removal factor F_R can be calculated from Eq. (3.56). The mass flow rate per unit area is

$$\frac{\dot{m}}{A_c} = \rho(q/A_c) = \frac{1.1 \times 0.7}{60} = -0.0128 \text{ kg/sec}\cdot\text{m}_c^2.$$

Then

$$F_R = \frac{(0.0128)(1)}{(18/3600)}\left\{1 - \exp\left[-\frac{(0.787)(18/3600)}{(0.0128)(1)}\right]\right\} = 0.677.$$

That is, the particular collector in question can collect 67.7 percent of the heat it could collect if its surface were at the air-inlet temperature. F_R varies weakly with the fluid temperature through the temperature effect upon U_c [see Eq. (3.23)].

3.3 Tubular Solar Energy Collectors

Two general methods exist for significantly improving the performance of solar collectors above the minimum flat-plate collector level. The first method increases solar flux incident on the receiver. It will be described in the next section on concentrators. The second method involves the reduction of parasitic heat loss from the receiver surface. Tubular collectors, with their inherently high compressive strength and resistance to implosion, afford the only practical means for completely eliminating convection losses by surrounding the receiver with a vacuum on the order of 10^{-4} mm Hg. The analysis of evacuated tubular collectors is the principal topic of this section.

Tubular collectors have a second application. They may be used to achieve a small level of concentration (1.5 to 2.0) by forming a mirror from part of the internal concave surface of a glass tube. This reflector can focus radiation on a receiver inside this tube. Since such a receiver is fully illuminated, it has no parasitic "back" losses. The performance of a nonevacuated tubular collector may be improved slightly by filling the envelope with high-molecular-weight noble gases. External concentrators of radiation may also be coupled to an evacuated receiver for improvement of performance over the simple evacuated tube. Collectors of this type are described briefly below.

3.3.1 Evacuated-Tube Collectors

Evacuated-tube devices have been proposed as efficient solar energy collectors since the early twentieth century. In 1909, Emmett [24] proposed several evacuated-tube concepts for solar energy collection, two of which are being sold commercially today. Speyer [74] also proposed a tubular evacuated flat-plate design for high-temperature operation. With the recent advances in vacuum technology, evacuated-tube collectors can be reliably mass-produced. Their high-temperature effectiveness is essential for the efficient operation of solar air-conditioning systems and process heat systems.

Figure 3.17 shows schematic cross sections of several glass evacuated-tube collector concepts. The simplest design is basically a small flat-plate collector housed in an evacuated cylinder (Fig. 3.17a). If the receiver is metal, a glass-to-metal seal is required to maintain a vacuum. In addition, a thermal short may occur from inlet to outlet tube unless special precautions are taken.

Mildly concentrating, tubular collectors can be made using the design of Fig. 3.17c. Either a single flow-through receiver with fins or a double U tube as shown can be used.

One of Emmett's designs is shown by Fig. 3.17d. It consists of an evacuated vacuum bottle much like an unsilvered, wide-mouth Dewar flask into which a metal heat exchanger is inserted. The outer surface of the inner glass tube is the absorber. The heat generated is transferred through the inner glass tube to the metal slip-in heat exchanger. Since this heat transfer is through a glass-to-metal interface that has only intermittent point contacts, significant axial temperature gradients can develop, thereby stressing the glass tube. In addition, a large temperature difference can exist between the inner and outer glass tubes. At the collector ends where the tubes are joined, a large temperature gradient and consequent thermal stress can exist.

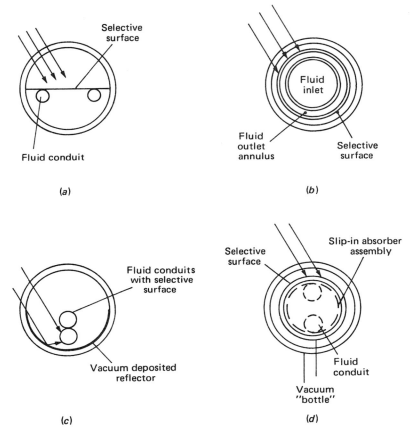

Figure 3.17. Evacuated-tube solar energy collectors: (a) flat plate; (b) concentric tubular; (c) concentrating; (d) vacuum bottle with slip-in heat exchanger contacting rear surface of receiver.

The level of evacuation required for suppression of convection and conduction can be calculated from basic heat-transfer theory. As the tubular collector is evacuated, reduction of heat loss first occurs because of the reduction of the Rayleigh number. The effect is proportional to the square root of density. When the Rayleigh number is further reduced below the lower threshold for convection, the heat-transfer mechanism is by conduction only. For most gases, the thermal conductivity is independent of pressure if the mean free path is less than the heat-transfer path length.

For very low pressure, the conduction heat transfer in a narrow gap [20] is given by

$$q_k = \frac{k\Delta T}{g + 2p} \tag{3.57}$$

where g is the gap width and p is the mean free path. For air, the mean free path at atmospheric pressure is about 70 μm. If 99 percent of the air is removed from a tubular collector, the mean free path increases to 7 mm, and conduction heat transfer is affected very little. However, if the pressure is reduced to 10^{-3} torr, the mean free path is 7 cm, which is substantially greater than the heat-transfer path length, and conduction heat transfer is effectively suppressed. The relative reduction in heat transfer as a function of mean free path can be derived from Eq. (3.57):

$$\frac{q_{vac}}{q_k} = \frac{1}{1 + 2p/g} \tag{3.58}$$

where q_k is the conduction heat transfer if convection is suppressed and q_{vac} is the conduction heat transfer under a vacuum. Achieving a vacuum level of 10^{-3}–10^{-4} torr for a reasonably long period of time is within the grasp of modern vacuum technology.

3.3.2 Thermal Analysis of a Tubular Collector

The heat loss from a tubular collector occurs primarily through the mechanism of radiation from the absorber surface. The rate of heat loss per unit absorber area q_L then can be expressed as

$$q_L = U_c(T_r - T_a) \tag{3.59}$$

Total thermal resistance $1/U_c$ is the sum of three resistances:

R_1 = radiative exchange from absorber tube to cover tube,
R_2 = conduction through glass tube, and
R_3 = convection and radiation to environment.

The overall resistance is then

$$\frac{1}{U_c} = R_1 + R_2 + R_3. \tag{3.60}$$

The conductances R_i^{-1} are given by

$$R_1^{-1} = \frac{1}{1/\epsilon_r + 1/\epsilon_e - 1} \sigma(T_r + T_{ei})(T_r^2 + T_{ei}^2), \tag{3.61}$$

$$R_2^{-1} = \frac{2k}{D_r \ln(D_{eo}/D_{ei})}, \text{ and} \tag{3.62}$$

$$R_3^{-1} = [h_c + \sigma\epsilon_e(T_{eo} + T_a)(T_{eo}^2 + T_a^2)]\frac{D_{eo}}{D_r}, \tag{3.63}$$

where the subscript e denotes envelope (tube) properties, subscripts i and o denote inner and outer surfaces of the envelope (tube), T_r is the receiver (absorber) temperature, and

h_c is the external convection coefficient for the envelope. Test data have shown that the loss coefficient U_c is between 0.5 and 1.0 W/m² · °C, thus confirming the analysis.

The energy delivery rate q_u on an aperture area basis can be written as

$$q_u = \tau_e \alpha_r I_{\text{eff}} \frac{A_t}{A_c} - U_c(T_r - T_a)\frac{A_r}{A_c} \tag{3.64}$$

where I_{eff} is the effective solar radiation both directly intercepted and intercepted after reflection from the back reflector, A_t is the projected area of a tube (its diameter) and A_r is the receiver or absorber area. The receiver-to-collector aperture area ratio is $\pi D_r/d$, where d is the center to center distance between the tubes. Therefore

$$q_u = \frac{D_r}{d}[\tau_e \alpha_r I_{\text{eff}} - \pi U_c(T_r - T_a)]. \tag{3.65}$$

Beekley and Mather [7] have shown that a tube spacing one envelope diameter D_{eo} apart maximizes daily energy gain. A specularly reflecting cylindrical back surface improves performance by 10 percent or more.

3.4 Experimental testing of collectors. The performance of solar thermal systems for heating and cooling depends largely on the performance of solar collectors. Therefore, experimental measurement of thermal performance of solar collectors by standard methods is important and necessary. The experimentally determined performance data is needed for design purposes and for determining the commercial value of the collectors. The thermal performance of a solar collector is determined by establishing an efficiency curve from the measured instantaneous efficiencies for a combination of values of incident solar radiation, ambient temperature and inlet fluid temperature. An instantaneous efficiency of a collector under steady state conditions can be established by measuring the mass flow rate of the heat transfer fluid, its temperature rise across the collector ($T_{f,out} - T_{f,in}$) and the incident solar radiation intensity (I_c) as:

$$\eta_c = \frac{q_u}{A_c I_c} = \frac{\dot{m}C_p(T_{f,out} - T_{f,in})}{A_c I_c}. \tag{3.66}$$

The efficiency, η_c, of a collector under steady state can also be written according to the Hottel-Whillier-Bliss equation (Eq. 3.46) as:

$$\eta_c = F_R \tau_s \alpha_s - F_R U_c \frac{(T_{f,in} - T_a)}{I_c} \tag{3.46}$$

Equation 3.46 suggests that for constant values of F_R and U_c, if η_c is plotted with respect to $(T_{f,in} - T_a)/I_c$, a linear curve will result, with a "y" intercept of $F_R \tau_s \alpha_s$ and a slope of $-F_R U_c$. Figure 3.18 shows a typical thermal performance curve for a flat plate collector. Since τ_s and α_s can be measured independently, a thermal performance curve of a flat

COLLECTOR TILT ANGLE, 45°
INLET TEMPERATURE, 32° to 60°C
FLOW RATE, 0.0136kg/(s.m²)
SOLAR FLUX, 599 to 1009 W/m²
Slope= -FRUC
Intercept =FRτsαs
COLLECTOR EFFICIENCY, %
80
60
40
20
0
0
0.02
0.04
0.06
0.08
0.1
tf,i-ta / I , °C.m² / W

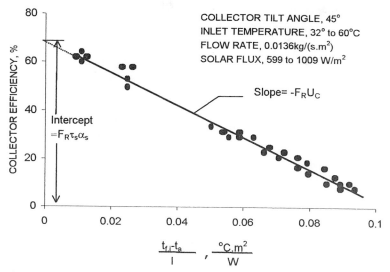

Figure 3.18. Thermal efficiency curve for a double-glazed flat-plate liquid type of solar collector.

plate collector allows us to establish the value of F_R and U_c also. Figure 3.19 shows typical performance curves of various glazed and unglazed flat plate solar collectors.

In Eq. (3.46) the product $\tau_s \alpha_s$ will change with the angle of incidence. Since flat plate collectors are normally fixed, the angle of incidence changes throughout the day. A relationship can be written between the actual or effective $(\tau_s \alpha_s)$ and $(\tau_s \alpha_s)_n$ for normal incidence as:

$$(\tau_s \alpha_s) = (\tau_s \alpha_s)_n K_{\tau\alpha} \tag{3.67}$$

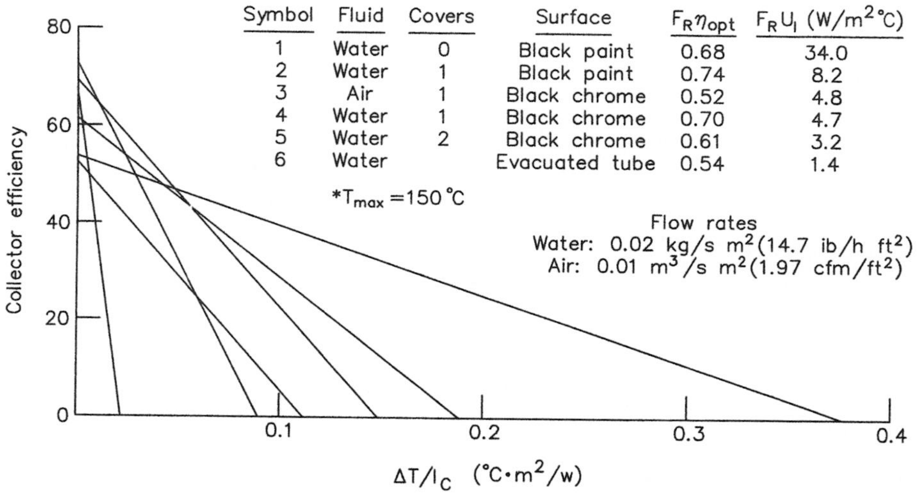

Symbol Fluid Covers Surface FRηopt FRUl (W/m²°C)
1 Water 0 Black paint 0.68 34.0
2 Water 1 Black paint 0.74 8.2
3 Air 1 Black chrome 0.52 4.8
4 Water 1 Black chrome 0.70 4.7
5 Water 2 Black chrome 0.61 3.2
6 Water Evacuated tube 0.54 1.4
*Tmax =150 °C
Flow rates
Water: 0.02 kg/s m²(14.7 ib/h ft²)
Air: 0.01 m³/s m²(1.97 cfm/ft²)
Collector efficiency
80
60
40
20
0
0.1
0.2
0.3
0.4
ΔT/IC (°C•m²/w)

Figure 3.19. Typical performance curves for various flat-plate solar collectors.

where $(\tau_s\alpha_s)_n$ is the value of the product for normal angle of incidence, and $K_{\tau\alpha}$ is called the incidence angle modifier. Therefore, the thermal performance of a flat plate collector may be written as:

$$\eta_c = F_R\left[K_{\tau\alpha}(\tau_s\alpha_s)_n - U_c\frac{(T_{f,in} - T_a)}{I_c}\right]. \tag{3.68}$$

Thermal performance of collectors is usually found experimentally for normal angles of incidence in which case $K_{\tau\alpha} = 1.0$. The incidence angle modifier is then measured separately. It has been established that $K_{\tau\alpha}$ is of the form:

$$K_{\tau\alpha} = 1 - b\left(\frac{1}{\cos i} - 1\right), \tag{3.69}$$

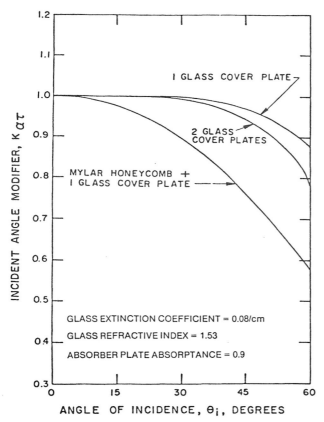

Figure 3.20. Incident angle modifier for three flat-plate solar collectors. (Reprinted by permission of the American Society of Heating, Refrigerating and Air-Conditioning Engineers, Inc., Atlanta, from ASHRAE Standard 93-77, "Methods of Testing to Determine the Thermal Performance of Solar Collectors.")

where b is a constant and i is the angle of incidence. Figure 3.20 shows a typical curve for $K_{\tau\alpha}$.

3.4.1 Testing Standards for Solar Thermal Collectors

Standard testing procedures adopted by regulating agencies in various countries establish ways of comparing the thermal performance of various collectors under the same conditions. In the USA, the thermal performance standards established by the American Society of Heating, Refrigeration and Air-Conditioning Engineers (ASHRAE) have been accepted for thermal performance testings of solar thermal collectors. Similar standards have been used in a number of countries. The standards established by ASHRAE include:

ASHRAE Standard 93–77, "Methods of Testing to Determine the Thermal Performance of Solar Collectors"; and

ASHRAE Standard 96–80, "Methods of Testing to Determine the Thermal Performance of Unglazed Solar Collectors."

ASHRAE Standard 93–77 specifies the procedures for determining the time constant, thermal performance and the incidence angle modifier of solar thermal collectors using a liquid or air as a working fluid. Figure 3.21 shows a schematic of a standard testing configuration for thermal performance testing. The tests are conducted under quasi-steady state conditions.

Time constant. In order to avoid the transient effects, performance of a collector is measured and integrated over at least the time constant of the collector. The time constant is measured by operating the collector under steady or quasi-steady conditions with a solar flux of at least 790 W/m² and then abruptly cutting the incident flux to zero, while continuing to measure the fluid inlet ($T_{f,i}$) and exit ($T_{f,e}$) temperatures. The fluid inlet temperature is maintained at $\pm$ 1°C of the ambient temperature. The time constant is then determined as the time required to achieve:

$$\frac{T_{f,e} - T_{f,i}}{T_{f,e,initial} - T_{f,i}} = 0.30. \tag{3.70}$$

Thermal performance. For determining the thermal performance of a collector, ASHRAE Standard 93–77 specifies test conditions to get at least four data points for the efficiency curve. The test conditions include:

1. Near normal incidence ($i \le 5°$) angle tests close to solar noon time;
2. At least four tests for each $T_{f,i}$, two before and two after solar noon; and
3. At least four different values of $T_{f,i}$ to obtain different values of $\Delta T/I_c$, preferably to obtain ΔT at 10, 30, 50 and 70% of stagnation temperature rise under the given conditions of solar intensity and ambient conditions.

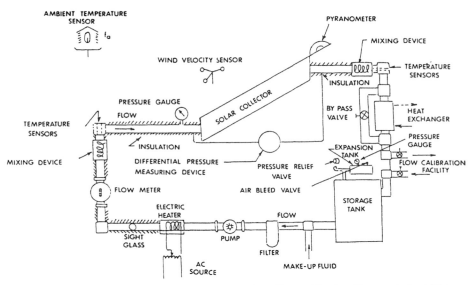

Figure 3.21. A testing configuration for a liquid-type solar collector. (Reprinted by permission of the American Society of Heating, Refrigerating and Air-Conditioning Engineers, Inc., Atlanta, from ASHRAE Standard 93-77, "Methods of Testing to Determine the Thermal Performance of Solar Collectors.")

Incidence angle modifier. ASHRAE 93–77 specifies that a curve for incidence angle modifier be established by determining the efficiencies of a collector for average angles of incidence of 0°, 30°, 45° and 60° while maintaining $T_{f,i}$ at $\pm$ 1°C of the ambient temperature. Since $(T_{f,i} - T_a) \simeq 0$ the incident angle modifier according to Eq. (3.68) is:

$$K_{\tau\alpha} = \frac{\eta_c}{F_R(\tau_s\alpha_s)_n}. \tag{3.71}$$

The denominator is the y intercept of the efficiency curve.

3.5 CONCENTRATING SOLAR COLLECTORS

Concentration of solar radiation is achieved by reflecting or refracting the flux incident on an aperture area A_a onto a smaller receiver/absorber area A_r. An optical concentration ratio, CR_o, is defined as the ratio of the solar flux I_r on the receiver to the flux, I_a, on the aperture, or

$$CR_o = I_r/I_a, \tag{3.72}$$

while a geometric concentration ratio CR is based on the areas, or

$$CR = A_a/A_r. \tag{3.73}$$

CR_o gives a true concentration ratio because it accounts for the optical losses from the reflecting and refracting elements. However, since it has no relationship to the receiver area it does not give an insight into the thermal losses which are proportional to the receiver area. In the analyses in this book only geometric concentration ratio CR will be used.

Concentrators are inherently more efficient at a given temperature than are flat-plate collectors, since the area from which heat is lost is smaller than the aperture area. In the flat-plate device both areas are equal in size. A simple energy balance illustrates this principle. The useful energy delivered by a collector q_u is given by

$$q_u = \eta_o I_c A_a - U_c(T_c - T_a)A_r, \tag{3.74}$$

in which η_o is the optical efficiency and other terms are as defined previously. The instantaneous collector efficiency is given by

$$\eta_c = \frac{q_u}{I_c A_a}, \tag{3.75}$$

from which, using Eq. (3.74),

$$\eta_c = \eta_o - \frac{U_c(T_c - T_a)}{I_c} \frac{1}{CR} \tag{3.76}$$

For the flat plate $CR \cong 1$ and for concentrators $CR > 1$. As a result, the loss term (second term) in Eq. (3.76) is smaller for a concentrator and the efficiency is higher. This analysis is necessarily simplified and does not reflect the reduction in optical efficiency that frequently, but not always, occurs because of the use of imperfect mirrors or lenses in concentrators. The evaluation of U_c in Eq. (3.76) in closed form is quite difficult for high-temperature concentrators, because radiation heat loss is usually quite important and introduces nonlinearities ($\propto T^4$). One disadvantage of concentrators is that they can collect only a small fraction of the diffuse energy incident at their aperture. This property is an important criterion in defining the geographic limits to the successful use of concentrators and is described shortly.

3.5.1 The Thermodynamic Limits to Concentration

A simple criterion is developed below for the upper limit of concentration [59] of a solar collector. Figure 3.22 is a schematic diagram of any concentrating device in which the source, aperture and receiver are shown. The source represents a diffuse source or a diffuse-like source that could be formed by a moving point source, that is, the sun. The evaluation of the maximum achievable concentration ratio CR_{max} uses the concept of radiation exchange factors described in Chapter 2.

The factor $\Im_{12}$ is defined as the fraction of radiation emitted from surface 1 that reaches surface 2 by whatever means—direct exchange, reflection, or refraction. It is

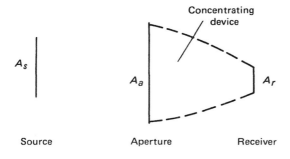

Figure 3.22. Generalized schematic diagram of any two-dimensional solar energy concentrating device showing source of radiation A_s, aperture A_a, and receiver A_r areas.

shown in Chapter 2 that reciprocity relations exist for area exchange factors; for this analysis the second law of thermodynamics requires, in addition, that

$$A_s \Im_{sa} = A_a \Im_{as}, \tag{3.77}$$

and

$$A_s \Im_{sr} = A_r \Im_{rs}, \tag{3.78}$$

where a denotes aperture, r denotes receiver, and s denotes the source. Here we use $\Im$ symbols for exchange factor between nonblack surfaces and F symbols for black surfaces. By means of these expressions, the concentration ratio CR can be expressed as

$$CR \equiv \frac{A_a}{A_r} = \frac{\Im_{sa} \Im_{rs}}{\Im_{as} \Im_{sr}} \tag{3.79}$$

For the best concentrator possible, all radiation entering the aperture A_a reaches the receiver A_r, that is,

$$\Im_{sa} = \Im_{sr}. \tag{3.80}$$

In addition, if the source is representable as a black body,

$$\Im_{as} = F_{as}, \tag{3.81}$$

where F_{as} is the radiation shape factor between two black surfaces. Using Eqs. (3.79) and (3.81), we have

$$CR = \frac{\Im_{rs}}{F_{as}}. \tag{3.82}$$

Because $\Im_{rs} \leq 1$ by the second law,

$$CR \leq CR_{max} = \frac{1}{F_{as}}. \tag{3.83}$$

Equation (3.83) states the simple and powerful result that the maximum concentration permitted by the second law is simply the reciprocal of the radiation shape factor F_{as}.

The shape factor F_{as} for a solar concentrator in two dimensions can be calculated from the diagram in Fig. 3.23. This sketch represents a trough-like or single-curvature concentrator formed from mirrors or a linear (cylindrical) lens. It is illuminated by a line source of light of length $2r$ representing a portion of the sun's virtual trajectory. By reciprocity, we have

$$F_{as} = F_{sa}\frac{A_s}{A_a}. \tag{3.84}$$

If an angle $2\theta_{max}$ is defined as the maximum angle within which light is to be collected, we have

$$F_{as} = \sin\theta_{max} \tag{3.85}$$

from Hottel's crossed-string method [34] for $L \ll r$. The angle θ_{max} is called the **acceptance half-angle**. From Eq. (3.83) the maximum concentration is then

$$CR_{max,2D} = \frac{1}{\sin\theta_{max}} \tag{3.86}$$

The term acceptance half-angle denotes coverage of one-half of the angular zone within which radiation is accepted (that is, "seen") by the receiver of a concentrator. Radiation is said to be accepted over an acceptance angle $2\theta_{max}$ because radiation incident within this angle reaches the absorber after passing through the aperture. Practical acceptance angles range from a minimum subtending the sun's disk (about ½°) to 180°, a value characterizing a flat-plate collector accepting radiation from a full hemisphere.

Double-curvature or dish-type concentrators have an upper limit of concentration that can be evaluated by extending the method used above to three dimensions. The result of such a calculation for a compound curvature device is given by

$$CR_{max,3D} = \frac{1}{\sin^2\theta_{max}}. \tag{3.87}$$

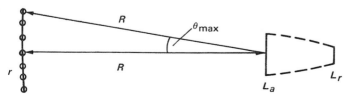

Source

Figure 3.23. Schematic diagram of two-dimensional sun-concentrator geometry used to calculate shape factor F_{as} and showing acceptance half-angle θ_{max} and several positions of the sun.

In the above analyses, if the concentrator has an index of refraction $n > 1$, the maximum concentration will be [84]:

$$CR_{max,2D} = \frac{n}{\sin\theta_{max}}, \text{ and} \tag{3.88}$$

$$CR_{max,3D} = \frac{n^2}{\sin^2\theta_{max}}. \tag{3.89}$$

The second law prescribes not only the geometric limits of concentration as shown above, but also the operating temperature limits of a concentrator. The radiation emitted by the sun and absorbed by the receiver of a concentrator q_{abs} is

$$q_{abs} = \tau\alpha_s A_s F_{sa}\sigma T_s^4, \tag{3.90}$$

where T_s is the effective temperature of the sun and τ is the overall transmittance function for the concentrator, including the effects of any lenses, mirrors, or glass covers. If the acceptance half-angle θ_{max} is selected to just accept the sun's disk of angular measure $\theta_s(\theta_s \sim \frac{1}{4}°)$, we have, by reciprocity [Eq. (3.84)], for a concentrator with compound curvature,

$$q_{abs} = \tau\alpha_s A_a \sin^2\theta_s\sigma T_s^4. \tag{3.91}$$

If convection and conduction could be eliminated, all heat loss q_L is by radiation, and

$$q_L = \epsilon_{ir}A_r\sigma T_r^4, \tag{3.92}$$

where ϵ_{ir} is the infrared emittance of the receiver surface. Radiation inputs to the receiver from a glass cover or the environment can be ignored for this upper limit analysis.

An energy balance on the receiver is then

$$q_{abs} = q_L + \eta_c q_{abs}, \tag{3.93}$$

where η_c is the fraction of energy absorbed at the receiver that is delivered to the working fluid. Substituting for Eqs. (3.91) and (3.92) in Eq. (3.93) we have

$$(1 - \eta_c)\tau\alpha_s A_a \sin^2\theta_s\sigma T_s^4 = \epsilon_{ir} A_r\sigma T_r^4, \tag{3.94}$$

Because $CR = A_a/A_r$ and $CR_{max} = 1/\sin^2\theta_s$,

$$T_r = T_s\left[(1 - \eta_c)\tau\frac{\alpha_s}{\epsilon_{ir}}\frac{CR}{CR_{max}}\right]^{1/4}. \tag{3.95}$$

In the limit as $\eta_c \to 0$ (no energy delivery) and $\tau \to 1$ (perfect optics), we have

$$\lim T_r \to T_s \left(\frac{CR}{CR_{max}} \right)^{1/4}. \tag{3.96}$$

Since $\epsilon_{ir} \to \alpha_s$ as $T_r \to T_s$, Eq. (3.96) shows that

$$T_r \le T_s, \tag{3.97}$$

as expected for an optically and thermally idealized concentrator. Equation (3.97) is equivalent to the Clausius statement of the second law for a solar concentrator.

3.5.2 Optical Limits to Concentrations

Equations (3.86) to (3.89) define the upper limit of concentration that may be achieved for a given concentration viewing angle. Of interest are the upper and lower limits of concentration defined by practical viewing angle limits—the maximum CR limited only by the size of the sun's disk and achieved by continuous tracking; and the minimum CR, based on a specific number of hours of collection with no tracking.

The upper limit of concentration for two- and three-dimensional concentrators is on the order of

$$CR_{max,2D} = \frac{1}{\sin 1/4°} \simeq 216 \text{ in air},$$

$$= \frac{1 \cdot 5}{\sin 1/4°} \simeq 324 \text{ in glass } (n = 1 \cdot 5), \text{ and}$$

$$CR_{max,3D} = \frac{1}{\sin^2 1/4°} \simeq 46{,}000 \text{ in air},$$

$$= \frac{(1 \cdot 5)^2}{\sin^2 1/4°} = 103{,}500 \text{ in glass}.$$

In practice, these levels of concentration are not achievable because of the effects of tracking errors and imperfections in the reflecting- or refracting-element surface, as described later. Gleckman et al. [28] are reported to have achieved the highest concentration of 56,000.

3.5.3 Acceptance of Diffuse Radiation

Diffuse or scattered radiation is not associated with a specific direction as is beam radiation. It is expected, therefore, that some portion of the diffuse component will fall beyond the acceptance angle of a concentrator and not be collectable. The minimum amount of diffuse radiation that is collectable can be estimated by assuming that the diffuse component is isotropic at the aperture. The exchange factor reciprocity relation shows that

$$A_a \Im_{as} = A_r \Im_{rs}. \tag{3.98}$$

For most practical concentrating devices that can accept a significant fraction of diffuse radiation, $\Im_{rs} = 1$. That is, all radiation leaving the receiver reaches the aperture and the environment eventually. Then

$$\Im_{as} = \frac{A_r}{A_a} = \frac{1}{Cr} \tag{3.99}$$

for any concentrator. Equation (3.99) indicates that at least $1/CR$ of the incident diffuse radiation reaches the receiver. In actual practice, the collectable diffuse portion will be greater than $1/CR$, since diffuse radiation is usually concentrated near the solar disk except during the cloudiest of days (see Chapter 2).

3.5.4 Ray Tracing Diagrams

Ray tracing diagrams are helpful in understanding the distribution of concentrated flux on the receiver and to design the receiver-absorber configurations. They are drawn by tracing a beam of parallel rays specularly reflected from the reflector surface. Figure 3.24 shows simple examples of ray tracing diagrams for a circular reflector and a parabolic reflector. From this figure it is easy to understand that a receiver for a linear parabolic trough could be designed as a small diameter tube, while a larger area rectangular surface would be required for a linear circular trough. Figure 3.25 demonstrates that the size of the receiver can be reduced, and therefore concentration increased, by using a smaller rim angle. A three-dimensional concentrator can be visualized by rotating each reflector profile about its axis. If a parabolic or circular reflecting concentrator tracks the sun in such a way that the axis of symmetry is always parallel to the solar beam, the reflected flux will be incident on the receiver. However, if the tracking is not

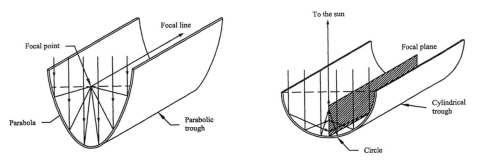

Figure 3.24. Examples of Ray Tracing Diagrams for (a) parabolic reflector, and (b) circular reflector.

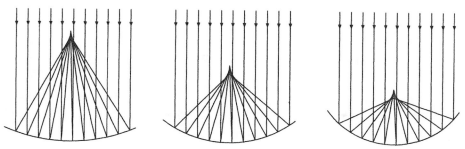

Figure 3.25. Focusing of parallel rays of light using circular mirrors with different rim angles.

perfect or if the concentrator is not designed to track, the beam may make an angle with the axis of symmetry. In which case the incident beam will not focus on the receiver, as demonstrated in Fig. 3.26, unless the receiver design is modified. As seen from this figure, tracking errors will penalize a parabolic concentrator more severely than a circular concentrator. Therefore, a parabolic concentrator requires more accurate sun tracking for best utilization.

In order to formulate a ray-tracing procedure suitable for numerical computation, laws of reflection and refraction are used in vector form as explained in detail in Welford and Winston [84]. Figure 3.27 shows the incident and reflected rays as unit vectors R_{inc} and R_{ref}, and a unit vector along the normal pointing into the reflecting surface is shown as n. Then, according to the laws of reflection,

$$R_{ref} = R_{inc} - 2(n \cdot R_{inc})n \tag{3.100}$$

Thus, to trace a ray we first find the incidence point at the reflecting surface, then a unit normal to the surface and finally the reflected ray using Eq. (3.100). For example, if the reflecting surface is an ideal paraboloid with a focal length F, it can be expressed in an x, y, z coordinate system as

$$z = \frac{x^2 + y^2}{4F}, \tag{3.101}$$

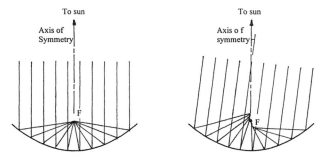

Figure 3.26. Concentration by parabolic reflector for a beam (a) parallel to the axis of symmetry, and (b) at an angle to the axis.

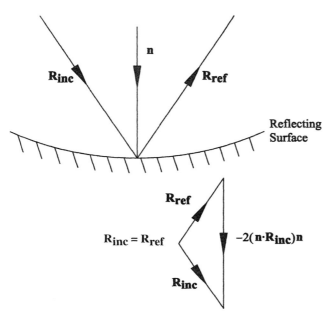

Figure 3.27. Vector formulation of reflection. Adapted from [84].

or

$$f(x,y,z) = 4Fz - x^2 - y^2. \tag{3.102}$$

The normal vector or gradient to this surface is defined as

$$\nabla f = \frac{df}{dx}\boldsymbol{i} + \frac{df}{dy}\boldsymbol{j} + \frac{df}{dz}\boldsymbol{k}. \tag{3.103}$$

Upon evaluation of the derivative ∇f from Eq. (3.103) a normal to the paraboloid at x, y, z is

$$\nabla f = -2x\boldsymbol{i} - 2y\boldsymbol{j} + 4F\boldsymbol{k}, \tag{3.104}$$

and a unit normal is defined as:

$$\boldsymbol{n} = \frac{\nabla f}{|\nabla f|}. \tag{3.105}$$

After finding $\boldsymbol{n}$ at the incidence point, $\boldsymbol{R}_{ref}$ can be found from Eq. (3.100). For a refracting surface the vector relationship between the incident and the refracted rays can be expressed as (Fig. 3.28)

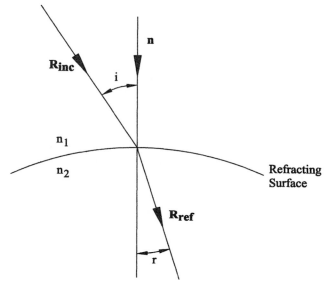

Figure 3.28. Vector formulation of refraction. Adapted from [84].

$$n_2 \boldsymbol{R}_{\text{ref}} \times \boldsymbol{n} = n_1 \boldsymbol{R}_{\text{inc}} \times \boldsymbol{n} \qquad (3.106)$$

The procedure to follow is similar to that for reflection. Welford [85] gives details of the application to lens systems.

3.5.5 Concentrator Types

Concentrators can be classified according to:

1. Amount of tracking required to maintain the sun within the acceptance angle, and
2. Type of tracking—single or double axis

As shown in the preceding sections, tracking requirements depend on the acceptance half-angle θ_{max}: the larger the θ_{max}, the less frequently and less accurately the tracker must operate. Two tracking levels may be identified:

1. Intermittent tilt change or completely fixed; and
2. Continuously tracking reflector, refractor, or receiver. If oriented east-west, it requires an approximate $\pm 30°$/day motion; if north-south, a $\sim 15°$/hr motion.

Both must accommodate to a $\pm 23\frac{1}{2}°$/yr declination excursion.

The least complex concentrators are those not requiring continuous accurate tracking of the sun. These are necessarily of large acceptance angle, moderate concentration ratio, and usually single-curvature design. Because the smallest diurnal, angular excursion of the sun is in a north-south plane, the fixed or intermittently turned concentrators must be oriented with the axis of rotation perpendicular to this plane, that is, in an east-west direction, in order to capitalize on the large acceptance angles.

3.5.6 Fixed Concentrators

A simple fixed-concentrator concept is based on using flat reflectors to boost the performance of a flat-plate collector as shown in Fig. 3.29.

Figure 3.30 shows the boost to summer solar radiation pickup that can be achieved by the use of tilted reflectors. By proper tilt, a seasonal peak of collected energy can be matched to a seasonal peak of energy demand as shown. The curves in Fig. 3.30 are calculated from optical analyses. Other variations of this concept are shown in Fig. 3.31. The concentration ratio of these concentrators can be as high as 3. For higher concentrations, spherical or parabolic reflecting or refracting surfaces are used.

3.6 Parabolic Trough Concentrator (PTC)

The most common commercially available solar concentrator is the Parabolic Trough Concentrator (PTC). Figure 3.32 shows a photograph of a commercial PTC and Fig. 3.33 shows sketches that describe it. PTC collectors usually track the sun with one degree of freedom using one of three orientations: east-west, north-south or polar. Sun tracking can be done for each by programming computers that control the tracking motors. The east-west and north-south configurations are the simplest to assemble into large arrays, but have higher incidence angle cosine losses than the polar mount. Also, the polar mount intercepts more solar radiation per unit area. The absorber of a PTC is usually tubular, enclosed in a glass tube to reduce radiative and convective losses. The convective losses can be minimized by creating a vacuum in the annular space between the absorber and the glass cover.

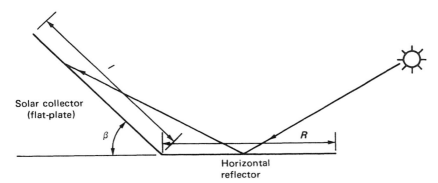

Figure 3.29. Horizontal reflecting surface used to boost flat-plate collector performance.

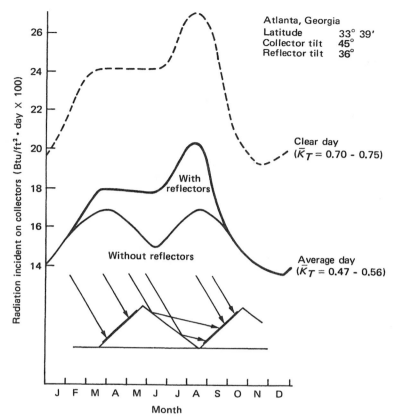

Figure 3.30. Effect of tilted reflector booster for flat-plate collector array in Atlanta, Georgia, for both clear-sky and average conditions (1 Btu/ft^2 = 11.4 kJ/m^2). From [85].

3.6.1 Optical analysis of the PTC. The *optical efficiency* of a collector is the ratio of the solar radiation absorbed in the absorber to that intercepted by the collector apera-ture directly facing the sun. The following major collector optical performance effects can be combined into the optical efficiency denoted by η_0:

1. mirror reflectance ρ_m (not present in refracting devices);
2. absorber or receiver cover system transmittance τ_r;
3. absorber absorptance for solar radiation α_r;
4. mirror surface slope errors—parameterized by the slope error ψ_1;
5. solar image spread—parameterized by angular deviation from a beam from the center of the sun ψ_2;
6. mirror tracking error—parameterized by aperature misalignment ψ_3;
7. shading of the mirror by the absorber cover tube and supports—parameterized by f_t, the unobstructed fraction of the aperature; and
8. off-normal incidence effects.

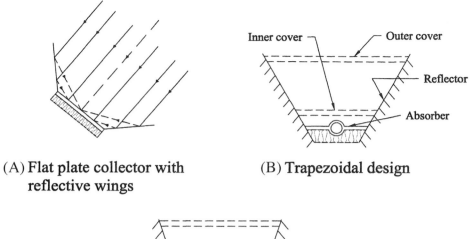

(A) Flat plate collector with   (B) Trapezoidal design
reflective wings

(C) Two-facet design

Figure 3.31. Examples of flat reflectors used to concentrate sunlight [3].

Figure 3.32. Commercial parabolic trough concentrator (PTC).

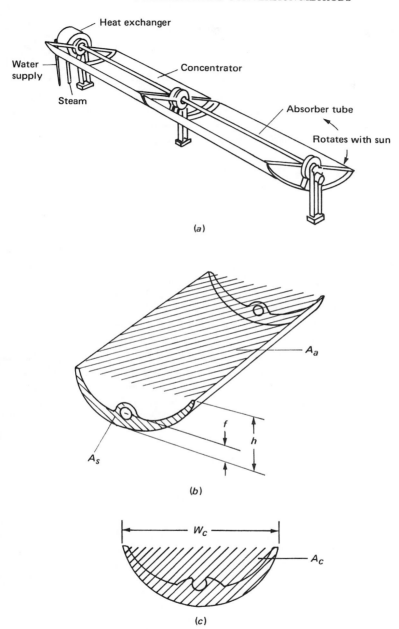

Figure 3.33. (a) Parabolic trough concentration isometric drawing; sketches (b) and (c) identify important parameters for optical design. From [65].

One expression for the optical efficiency of a PTC is

$$\eta_0 = (\rho_m \tau_r \alpha_r f_t)[\delta(\psi_1, \psi_2) F(\psi_3)][(1 - A \tan i)\cos i], \qquad (3.107)$$

where $\delta(\psi_1, \psi_2)$ is the fraction of rays reflected from a real mirror surface which are intercepted by the absorber for perfect tracking. $F(\psi_3)$ is the fraction of rays intercepted by the absorber for perfect optics and a point sun for a mirror tracking error ψ_3. A is a geometric factor accounting for off-normal incidence effects including blockages, shadows, and loss of radiation reflected from the mirror to beyond the end of the receiver.

For the PTC in Fig. 3.33, A is given by [65]

$$A = [W_c(f + h) + A_s - A_c]/A_a. \qquad (3.108)$$

A is usually of the order of 0.2–0.3.

Tracking errors, denoted by $F(\psi_3)$ in Eq. (3.107), have been measured on a prototype PTC with CR = 15 by Ramsey [65] and are shown in Fig. 3.34. The figure shows that a 1° error in tracking reduces the optical capture by 20 percent. An alternative method of considering tracking errors is to combine the ψ_3 effect into δ in Eq. (3.107). This method is described below.

The order of magnitude of η_0 can be estimated from the foregoing. For a well-designed and fabricated PTC, the following values of the various optical constants in Eq. (3.107) should apply: $\rho_m \sim 0.85$, $\tau_r \sim 0.9$, $\alpha_r \sim 0.95$, $f_t \sim 0.95$, $\delta(\psi_1, \psi_2) \sim 0.95$, $F(\psi_3) \sim 0.95$. At normal incidence, therefore, the optical efficiency is about 62 percent. If an etched glass absorber cover were used, the optical efficiency could be raised to about 66–67 percent. Other optical properties are unlikely to improve substantially.

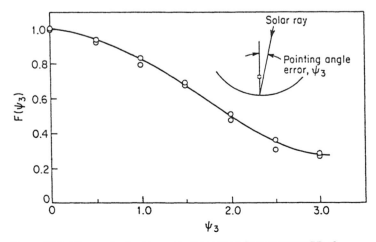

Figure 3.34. The angular tracking error factor $F(\psi_3)$ for a prototype PR of concentration ratio 15. From [65].

Optical intercept factor δ. The values of all optical properties other than $\delta(\psi_1,\psi_2)$ and $F(\psi_3)$ can be obtained from standard handbooks [46], whereas the angular factors must be measured on a collector assembly. During the design process it may be useful to be able to calculate ψ_i effects to determine tolerances on trackers and mirror surfaces. A method of calculating $\delta(\psi_1,\psi_2)$ and, by extension, $F(\psi_3)$ is given below [17].

Figure 3.35 shows the relationship of a mirror surface segment of a PTC to the tubular absorber. The nominally reflected ray is that ray from the center of the sun which is reflected from the theoretical mirror contour to the center of the absorber. Rays at an angular distance ψ_2 from the center of the sun will intercept the absorber at an angle ψ_2 away from the center of the absorber. Likewise, rays reflected from a mirror surface element whose slope is ψ_1 degrees in error from the theoretical surface slope will intercept the absorber at an angle $2\psi_1$ from the nominal ray.

From Fig. 3.35 it is clear that y, the distance from the receiver center to the ray intersection with a diameter, is

$$y = r\sin(2\psi_1 + \psi_2) \approx r(2\psi_1 + \psi_2). \tag{3.109}$$

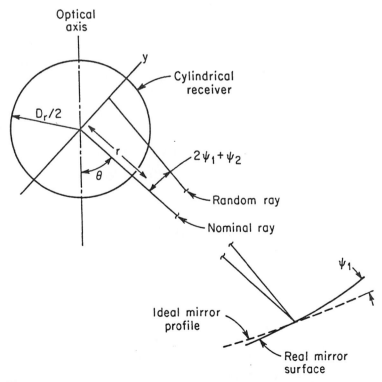

Figure 3.35. PTC collector section showing a mirror segment whose slope is ψ_1 degrees away from the theoretical slope. Also shown is a ray from the sun at a distance ψ_2 from the center of the sun and reflected from the real mirror surface.

The distance r from the mirror to the absorber tube center for a parabola is

$$r = [L_a(1 + \cos\phi)]/[2\sin\phi(1 + \cos\theta)], \qquad (3.110)$$

where L_a is the aperture width, ϕ is the trough rim half-angle, and θ is shown in Fig. 3.35. Using Eq. (3.110), y can be expressed as

$$y = [L_a(1 + \cos\phi)(2\psi_1 + \psi_2)]/[2\sin\phi(1 + \cos\theta)] \qquad (3.111)$$

Random rays are seen to intercept the absorber (and cover if present) at off-normal incidence. Hence, reflection losses are larger than for the nominal ray owing to increased reflectance losses. In the analysis which follows this effect is not included. However, if unetched glass and selective surfaces without antireflectants are used, the effect may need to be considered.

If the variables ψ_1 and ψ_2 are normally distributed with variances σ_{ψ_1} and σ_{ψ_2} and mean zero, the variance σ_y^2 of the ray intercept point is from Eq. (3.111)

$$\sigma_y^2 = \frac{L_a^2(1 + \cos\phi)^2(4\sigma_{\psi_1}^2 + \sigma_{\psi_2}^2)}{(4\sin^2\phi)(2\phi)} \int_{-\phi}^{\phi} \frac{d\theta}{(1 + \cos\theta)^2} \qquad (3.112)$$

carrying out the integration,

$$\sigma_y^2 = [L_a^2(4\sigma_{\psi_1}^2 + \sigma_{\psi_2}^2)(2 + \cos\phi)]/(12\phi \sin\phi). \qquad (3.113)$$

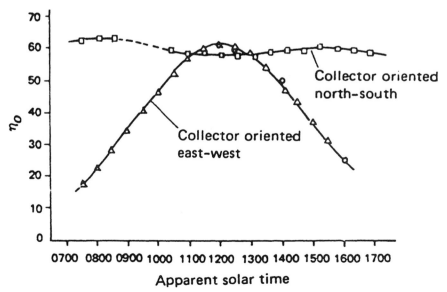

Figure 3.36. Measured optical efficiency of a PTC in north-south and east-west orientations [65]. Adapted from Kreider [43].

The assumption of a Gaussian sun with variance $\sigma_{\psi_2}^2$ is not physically correct because the sun has a nonnormal flux distribution. However, the Gaussian model may be used if $\sigma_{\psi_1}^2$ is of the same order as $\sigma_{\psi_2}^2$ and other errors described shortly. If other optical errors are small, however, the Gaussian sun approximation should not be used.

The intercept factor δ is the amount of energy contained in the normal probability flux distribution of variance σ_y^2 which is intercepted by the receiver tube of diameter D_r. That is,

$$\delta(\psi_1,\psi_2) = \frac{1}{\sqrt{2\pi}} \int_{-d_r/2}^{d_r/2} (e^{-Z^2/2})dZ, \tag{3.114}$$

where the change of variables $Z = y/\sigma_y$ and $d_r = D_r/\sigma_y$ has been made. The right-hand side of Eq. (3.114) is the normal probability integral included in many handbooks.

The method used above can be extended to include other optical effects including tracking and location errors ψ_3 and ψ_4, respectively. The effective variance Σ^2 to be used in Eq. (3.113) would be

$$\Sigma^2 = 4\sigma_{\psi_1}^2 + \sigma_{\psi_2}^2 + 4\sigma_{\psi_3}^2 + \sigma_{\psi_4}^2 \tag{3.115}$$

if the absorber is not rigidly attached to the reflector assembly. If the absorber is attached rigidly to the mirror, the constant 4 in the third term is replaced with unity.

Example 3.4. Find the maximum concentration ratio and receiver diameter for a PTC to intercept 95 percent of the incident radiation striking a perfectly tracking mirror with negligible dispersion. The mirror rim half-angle is 60°, and mirror surface accuracy measurements showed that $\sigma_{\psi_1} = 0.2°$. The aperture is 0.5 m wide.

Solution. From Eq. (3.113), σ_y^2 is

$$\sigma_y^2 = \{(0.5)^2(\pi/180)^2[4 \times 0.2^2 + 0.125^2](2 + \cos 60°)\}/[12(\pi/180)(60°)\sin 60°]$$

$$= 3.07 \times 10^{-6}\ m^2;$$

$$\sigma_y = 1.175\ mm.$$

Since the distribution of the sun's rays is more rectangular than normal, the effective standard deviation σ_{ψ_2} for the sun has been taken as ⅛° above.

The optical intercept factor is to be 95 percent; therefore, d_r from the probability integral tables is 1.96. Hence,

$$D_r = d_r\sigma_y = 3.43\ mm.$$

The concentration CR is

$$CR = A_c/A_r = L_a/(\pi D_r) = 500/(\pi \times 3.43)$$

$$= 46.4.$$

The example is finished at this point.

The optical intercept factor for a PTC with a *flat* receiver can be computed from Eq. (3.114) and the value of σ_y^2 to be used is [16]

$$
\sigma_y^2 = \frac{L_a^2(4\sigma_{\psi_1}^2 + \sigma_{\psi_2}^2)}{4\phi\tan^2(\phi/2)} \left[\frac{-1}{3\sin^3\phi\cos\phi} + \frac{2}{3\sin^3\phi} + \frac{2}{\sin\phi} - \frac{\cos\phi}{3\sin^3\phi} - \frac{2\cos\phi}{\sin\phi} \right.
$$
$$
\left. + \frac{4\sin\phi}{3\cos\phi} - \ln(\tan((\pi/4) + (\phi/2))) + \ln(\tan((\pi/4) - (\phi/2))) \right] \quad (3.116)
$$

3.6.2 Off-Normal Incidence Effects

The preceding results apply for normal incidence of beam radiation onto the aperture plane. A major off-normal incidence angle effect is the increased shading and blocking of the reflector plus end effects as represented by the last term in brackets in Eq. (3.116). Measurements at off-normal incidence have shown this term accurately represents shading, blocking, and end effects of a PTC [65]. Measured collector optical efficiency data are shown in Fig. 3.36. As expected, the east-west orientation shows the greatest incidence angle sensitivity. At normal incidence an optical efficiency in the range 0.60–0.65 was achieved as expected from the order of magnitude estimate above.

At off-normal incidence the incidence angle on the receiver cover and absorber surface increases, thereby increasing the reflective losses from both. This effect must be considered in the prediction of long-term performance.

3.6.3 Thermal performance of PTC collector. Thermal performance of a PTC collector can be analyzed similarly to that of a flat plate collector. Therefore, the instantaneous efficiency η_c can be written as:

$$
\eta_c = \frac{q_u}{I_c A_a} = \frac{q_{absorbed} - q_L}{I_c A_a}
$$
$$
= \eta_o - U_c(T_r - T_a)A_r/I_c A_a
$$
$$
= \eta_o - U_c \frac{(T_r - T_a)}{I_c} \cdot \frac{A_r}{A_a} \quad (3.117)
$$

where T_r and A_r are the temperature and the area of the receiver and A_a is the aperture area of the collector. A major difference between the thermal performance of a PTC and a flat plate collector is that the losses are from the receiver which is much smaller than the aperture area, however, the radiative losses are high because T_r is high. If the receiver is surrounded by an evacuated glass tube the heat losses due to convection are negligible as compared to the radiative losses, therefore,

$$
q_L = \epsilon_r \sigma(T_r^4 - T_c^4)A_r \quad (3.118)
$$

where ϵ_r is the emissivity of the receiver.

Therefore, $U_c = \epsilon_r \sigma(T_r^2 + T_a^2)(T_r + T_a)$. If conduction loss from the end seals of the receiver tube is included

$$q_L = A_{cond.} \frac{T_r - T_a}{R_k} + \epsilon_r \sigma(T_r^4 - T_a)A_r \qquad (3.119)$$

where R_k is the composite thermal resistance of the absorber end supports and seals and A_k is the effective areas for conduction heat transfer.

3.7 Compound-curvature solar concentrators.

The collectors used for high-temperature solar processes are of the double-curvature type and require a tracking device with two degrees of freedom. Concentration ratios above 50 are generally used. Examples of concentrator designs include: spherical mirror, CR = 50–150; paraboloidal mirror, CR = 500–3000; Fresnel lens, CR = 100–1000; and Fresnel mirror, CR = 1000–3000. These concentrator types will be discussed briefly in this section.

3.7.1 A. Paraboloidal concentrators.

The surface produced by rotating a parabola about its optical axis is called a paraboloid. The ideal optics of such a reflector are the same, in cross section, as those of the parabolic trough described earlier in this chapter. However, owing to the compound curvature, the focus occurs ideally at a point instead of along a line. Figure 3.37 shows a commercial paraboloid collector.

The optical efficiency η_o (defined relative to the direct-normal solar flux) of a paraboloid is the product of six terms:

$$\eta_o = \rho_m \tau_r \alpha_r f_t \delta(\psi_1, \psi_2) F(\psi_3), \qquad (3.120)$$

where ρ_m is the mirror reflectance, τ_r the receiver cover (if any) transmittance, α_r the receiver absorptance, f_t the fraction of the aperture not shaded by supports and absorber, $\delta(\psi_1, \psi_2)$ the intercept factor depending on mirror slope errors ψ_1 and solar beam spread ψ_2, and $F(\psi_3)$ the tracking error factor where ψ_3 is the angle between the solar direction and the aperture normal. The optical intercept factor for a paraboloid is given by [16]

$$\delta(\psi_1, \psi_2) = 1 - \exp[-(\pi D_r^2/4)/\sigma_y^2], \qquad (3.121)$$

where D_r is the receiver diameter and σ_y^2 is the beam spread variance at the receiver. For a spherical segment receiver, σ_y^2 is given by [17]

$$\sigma_y^2 = 2A_a(4\sigma_{\psi_1}^2 + \sigma_{\psi_2}^2)(2 + \cos\phi)/3\phi \sin\phi, \qquad (3.122)$$

where A_a is the aperture area, and ϕ is the paraboloid rim half-angle. The beam variances σ_{ψ_1} and σ_{ψ_2} are described earlier in this chapter.

Figure 3.37. Commercial paraboloidal solar concentrator. The receiver assembly has been removed from the focal zone for this photograph.

For a flat receiver with paraboloid optics, σ_y^2 is given by [17]

$$\sigma_y^2 = 2A_a(4\sigma_{\psi_1}^2 + \sigma_{\psi_2}^2)/\sin^2 \phi. \tag{3.123}$$

The concentration ratio of paraboloids can be determined easily from basic geometry. The aperture area is πR^2 where R is the aperture radius and the area for a spherical receiver is where $4\pi R_r^2$ where R_r is the spherical receiver radius. If the receiver is only a spherical segment, not a complete sphere, the receiver area to be used below is ΩR_r^2, where Ω is the segment included solid angle, instead of $4\pi R_r^2$. The receiver radius for perfect optics is sized to collect all rays with the acceptance half-angle θ_{max}. Hence,

$$R_r = (R/\sin\phi)\sin\theta_{max}. \tag{3.124}$$

Therefore, the concentration CR is

$$CR = (\pi R^2)/[4\pi \times (R/\sin\phi)^2 \sin^2 \theta_{max}],$$

or

$$CR = \sin^2\phi/(4\sin^2\theta_{max}). \tag{3.125}$$

The small effect of absorber shading of the mirror is ignored. Its inclusion would reduce CR given by Eq. (3.125). For $\theta_{max} = \frac{1}{4}°$ (the sun's half-angle) $CR \sim 13,000$ if $\phi = 90°$. For $\theta_{max} = \frac{1}{2}°$, $CR \sim 3300$ and for $\theta_{max} = 1°$, $CR \sim 800$. Note that concentrations for this configuration are one-fourth the thermodynamic limit, which is $(\sin \theta_{max})^{-2}$ for compound-curvature collectors. See Eq. (3.87).

For a flat absorber the concentration for perfect optics is

$$CR = \sin^2 \phi \cos^2(\phi + \theta_{max})/\sin^2 \theta_{max} \qquad (3.126)$$

where the small effect of mirror shading by the absorber is again ignored.

Example 3.4. Calculate the concentration ratio and diameter of a flat receiver for a 10-m diameter paraboloid concentrator which is designed to accept 90 percent of the incident beam radiation. The rim half-angle is 90° and the mirror surface variance $\sigma_{\psi_1}^2$ is expected to be $(0.25°)^2$. Assume that the standard deviation for the sun's disk is $0.125°$.

Solution. From Eq. (3.121)

$$\pi D_r^2/4\sigma_y^2 = -\ln(1 - 0.90) = 2.303,$$

and from Eq. (3.123)

$$\sigma_y^2 = 2A_a(4\sigma_{\psi_1}^2 + \sigma_{\psi_2}^2)/\sin^2 \phi$$

$$= 2 \times (\pi/4 \times 10^2)[4 \times (0.25)^2 + (0.125)^2](\pi/180)^2/\sin^2 90°$$

$$= 0.0127 \text{ m}^2 \ (\sigma_y = 11.3 \text{ cm}).$$

Solving the first equation for D_r,

$$D_r = [2.303 \times 4 \times \sigma_y^2/\pi]^{1/2} = 0.193 \text{ m}.$$

The concentration ratio is

$$CR = A_a/A_r = \frac{\pi \times 10^2/4}{\pi(0.193)^2/4} = 2685.$$

Note that the effect of tracking error ψ_3 can be included in the intercept factor $\delta(\psi_1,\psi_2)$ by defining an appropriate $\sigma_{\psi_3}^2$ and adding it to the slope and solar image variances in Eqs. (3.122) and (3.123).

The thermal losses from a paraboloid are quite small and primarily radiative. Since the absorber area is so small, it is generally not worthwhile to use any type of convection-suppressing cover. For example, consider the thermal losses from a planar absorber of $CR = 1500$ paraboloid with a receiver surface temperature of 600°C. For a cavity absorber ($\epsilon_{ir} \sim 1.0$) the radiation heat loss is about 20 W/m² aperture if the ambient temperature is 60°C. For a typical convection coefficient in light winds of 25 W/m² °C, the convection loss is about 10 W/m² aperture. If the insolation is 900 W/m²

(a)

Figure 3.38. Examples of commercially developed multifaceted and stretched membrane paraboloidal concentrators: (a) multifaceted mirror.

(*continues*)

and the optical efficiency 65 percent, the total heat loss at 600°C represents less than 6 percent of the absorbed flux. Stated another way, the collector loss coefficient U_c at 600°C is only 0.064 W/m² °C. The performance of high-concentration paraboloids is, therefore, much more sensitive to optical properties than to thermal losses.

Recent developments in paraboloidal concentrators include multifaceted mirror concentrators, single faceted stretched membrane mirror concentrators and multifaceted stretched membrane mirror concentrators. Figure 3.38 shows examples of such commercially developed paraboloidal concentrators.

3.7.2 Spherical concentrators.

A second type of compound-curvature collector uses spherical geometry instead of parabolic geometry. Figure 3.39c shows ray traces in a plane of symmetry for normal incidence (two-axis tracking) on a spherical concentrator. Spherical aberration is seen to be present and causes the reflected flux to be along a line instead of at a point as is the case for other compound-curvature mirrors. It is also clear that rays intercepted near the pivot end of the absorber intercept the absorber at very large incidence angles. Therefore, an absorber envelope with very low reflectance or no envelope at all is used to avoid severe penalties to the optical efficiency.

(b)

Figure 3.38. (*continued*) (b) Stretched single membrane (Schlaich Bergermann & Partner [Germany]).

(*continues*)

(c)

Figure 3.38. (*continued*) (c) Multifaceted stretched membrane (SAIC, Colorado, USA).

The concentration ratio of a spherical concentrator can be calculated from the definition of CR:

$$CR = A_a/A_r = \pi R_o^2 \sin^2 \phi / \pi D_r L_r. \tag{3.127}$$

Taking the receiver length $L_r = R_o/2$

$$CR = (2R_o/D_r) \sin^2 \phi, \tag{3.128}$$

If the absorber is a cylinder and is to intercept all singly reflected rays within the acceptance half-angle θ_{max},

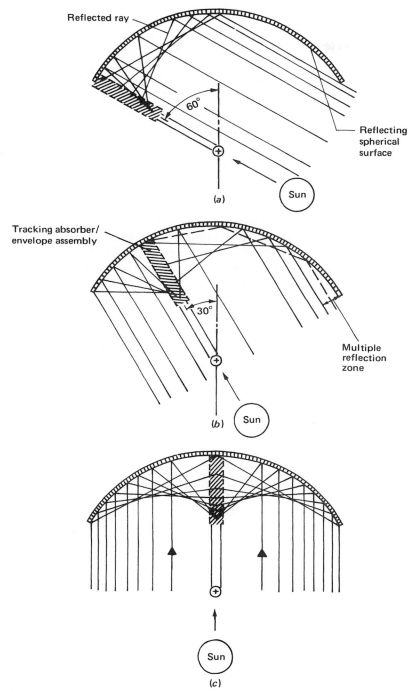

Figure 3.39. Ray traces for a fixed spherical concentrator for three incidence angles: (a) normal incidence for two-degree-of-freedom tracking of the mirror; (b) and (c) off-normal incidence for a fixed mirror and a moving absorber.

$$D_r = 2R_o \sin \theta_{max}, \tag{3.129}$$

and

$$CR = \sin^2 \phi / \sin \theta_{max}, \tag{3.130}$$

if the acceptance half-angle is ¼° for $\phi = 90°$, $CR = 229$; for $\theta_{max} = ½°$, $CR = 115$ for perfect optics. It is obvious that the spherical concentrator cannot achieve concentration ratios approaching those of paraboloids owing to spherical aberration and hence this concentrator is much more sensitive to operating temperature than the point-focus type.

The efficiency of this concentrator is lower than that of other compound-curvature devices, because of smaller CR values, therefore the unit cost of solar heat will be higher. One method of reducing the unit cost is to use the concentrator with a *fixed mirror*, thereby eliminating the expensive mirror tracking and structural components. This configuration is known as a Stationary Reflector Tracking Absorber (SRTA) concentrator. The collector is kept "in focus" by tracking only the absorber, aiming it directly at the sun's center. However, the effective aperture in this fixed mirror mode is reduced by the cosine of the incidence angle. Hence, the saving in cost by eliminating the reflector tracker is partly offset by reduced energy capture. If a value of rim half-angle $\phi = 60°$ is used as shown in Fig. 3.39 and if the fixed aperture faces the sun directly at noon, then the day-long cosine penalty will reduce the captured flux by about 17%. Of course, if the aperture does not directly face the sun at noon, the cosine loss will be larger.

3.7.3 Compound parabolic concentrator.
A non-imaging concentrator concept called the Compound Parabolic Concentrator (CPC) was developed by Winston [86–88] and Rabl [57–59] and described by Baranov and Melnikov [4]. The CPC can approach the thermodynamic limit of concentration discussed earlier, i.e.,

$$CR_{CPC} = CR_{max,2D} = 1/\sin\theta_{max}$$

Figure 3.40 shows a schematic cross section of the original CPC concept. It is seen to be formed from two distinct parabolic segments, the foci of which are located at the opposing receiver surface end points. The axes of the parabolic segments are oriented away from the CPC axis by the acceptance angle θ_{max}. The slope of the reflector surfaces at the aperture is parallel to the CPC optical axis. Figure 3.41 is a photograph of a CPC collector.

Different types of planar and tubular receivers have been proposed for CPCs. Of most interest in this book is the tubular type receiver shown in Fig. 3.42 since high pressure heat transfer fluid can flow through it.

3.7.4 Optical analysis of CPC collector.
A number of the optical losses present in tracking collectors described above are negligible in CPC collectors because of their

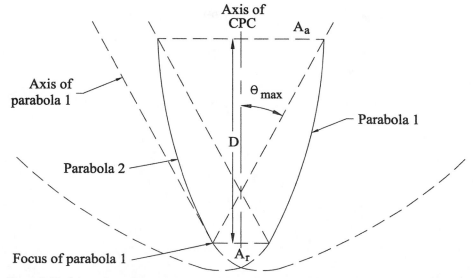

Figure 3.40. Schematic cross section of a CPC showing parabolic segments, aperture, and receiver.

broad acceptance band and ability to use imprecise optical elements. Of the eight optical loss mechanisms listed in section 3.3.7, at least four are reduced or nonexistent in CPCs.

The three optical parameters—mirror reflectance, cover transmittance, and absorber absorptance—are first-order effects for CPCs. Mirror reflectance effects are slightly different for a CPC than for other line-focus reflectors since the CPC does not form a sharp image of the sun at the absorber. Figure 3.43 shows that some solar radiation incident near the aperture edges is reflected more than once on its way to the absorber. The effect of multiple reflectance can be simply accounted for in optical efficiency calculations, to lowest order, by using $\rho_m^{\bar{n}}$ where $\bar{n}$ is the average number of reflections of all incident rays over the aperture. Some effect of incident angle on $\bar{n}$ has been noted but Rabl [60] recommends use of a constant $\bar{n}$ for engineering purposes.

Figure 3.44 can be used to determine the average number of reflections $\bar{n}$ for various concentration ratios CR for the basic CPC in Fig. 3.40. For tubular receiver CPCs (Fig. 3.42) $\bar{n}$ should be increased by about 0.5 over the Fig. 3.44 values. The dashed line in Fig. 3.44 represents $\bar{n}$ for a fully developed CPC. Since the upper half of the reflector for most CPCs is nearly parallel to the optical axis, it affords little concentration effect. In practice, the upper portion (about 40–60 percent of the mirror) is normally eliminated to reduce CPC size and cost. Figure 3.45 shows the concentration achievable for various truncation amounts. This CR value and the corresponding value of acceptance angle then determine a point on the $\bar{n}$ - CR map (Fig. 3.44), and $\bar{n}$ can be evaluated for any truncation. For careful designs a ray trace procedure must be used to find $\bar{n}$ and its dependence on incidence angle.

A final consideration in the optical analysis of the CPC family is the effect of partial acceptance of diffuse radiation. Equation (3.99) showed that at least (CR)$^{-1}$ of the

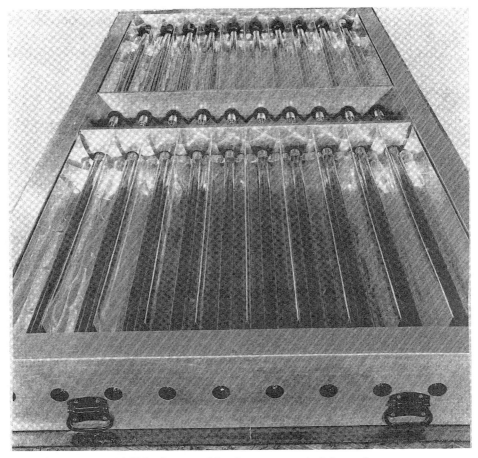

Figure 3.41. Commercial CPC collector module using an evacuated tubular receiver.
(Courtesy of Energy Design Corp., Memphis, Tennessee.)

incident diffuse flux reached the absorber. Since CPC collectors operate in the concentration range of 2 to 10 to capitalize on the corresponding reduced tracking requirement, from one-half to one-tenth of the incident diffuse radiation is accepted. This property of CPCs is conveniently included in the CPC optical efficiency by defining the intercept factor δ used for PTC analysis somewhat differently. If δ is defined as the fraction of total radiation accepted by a CPC, it can be expressed as

$$\delta \equiv [I_{b,c} + I_{d,c}CR]/I_{tot,c} \tag{3.131}$$

where subscripts b, d, and tot refer to beam, diffuse, and total flux incident on the collector aperture.

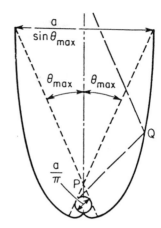

Figure 3.42. CPC collector concept with tubular receiver.

The optical efficiency of a CPC can then be written as*

$$\eta_o = \rho_m^{\bar{n}} \tau_r \alpha_r \delta. \tag{3.132}$$

Although this expression has the same appearance as Eq. (3.107), it is to be emphasized that δ depends on the characteristics of local solar flux and is not a purely geometric factor such as $\delta(\psi_1, \psi_2)$. For the tubular CPC, $\bar{n} \sim 1.2$, $\rho_m \sim 0.85$, $\alpha_r \sim 0.95$, $\delta \sim 0.95$, and $\tau_r \sim 0.90$. Therefore, $\eta_o \sim 0.67$. If an etched glass cover is used, $\eta_o \sim 0.71$. These values of optical efficiency are 7–8 percent greater than those for a PTC device. Improved optical efficiency partially offsets the lower concentration and associated heat loss effect usually imposed on CPCs ($CR < 10$) in order to benefit from their reduced tracking requirements.

3.7.5 Thermal Performance of the CPC Collector

For a tubular receiver CPC the same heat-loss analysis can be used as for a PTC collector with a tubular receiver. Some uncertainty in the convection heat loss from the tube to the environment exists, since the tube is in a partial enclosure and protected from the environment.

The thermal efficiency of a CPC (based on total beam and diffuse collector plane flux) is given by

$$\eta_c = \rho_m^{\bar{n}} \tau_r \alpha_r \delta - U_c(T_r - T_a)/(I_c CR), \tag{3.133}$$

---

*For the tubular receiver CPC a small gap of width g is required between the mirror cusp and the absorber pipe to accommodate the absorber envelope. Some otherwise collectable rays escape the receiver at this point, reducing the optical efficiency by the factor $(1 - g/p_r)$, where p_r is the absorber perimeter. Other design details are given in Ref. [61].

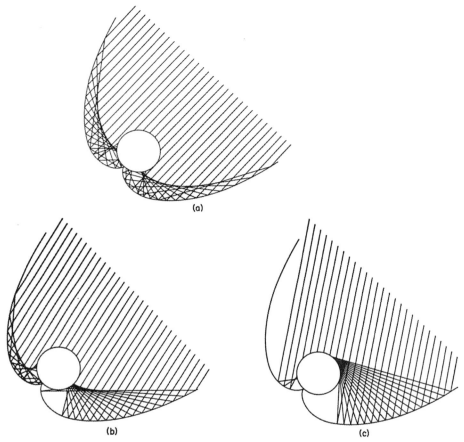

Figure 3.43. Ray trace diagrams of the tubular CPC collector at three values of incidence angle: (a) normal incidence; (b) intermediate; and (c) the limit of acceptance. (Courtesy of W. McIntire, Argonne National Laboratory.)

where U_c is based on the aperture area.

By analogy with the PTC analysis developed earlier for an evacuated receiver,

$$U_c \sim [\epsilon_r \sigma(T_r^2 + T_a^2)(T_r + T_a)] + (A_k/A_a)(1/R_k)(CR), \qquad (3.134)$$

where R_k is the conduction heat transfer resistance for all conduction paths of total effective area A_k. Hsieh [38] and Hsieh and Mei [39] presented a detailed thermal analysis of a CPC with an evacuated tubular selective coated receiver and the entire collector covered with a transparent cover. They proposed the following empirical equations for the heat loss coefficient U_c.

For $29°C \le (T_r - T_a) \le (137 + 0.0283T_a - 0.0000616T_a^2)$,

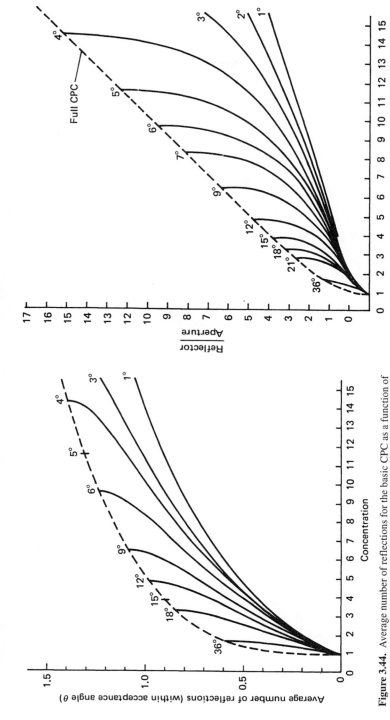

Figure 3.44. Average number of reflections for the basic CPC as a function of CR. The dashed line represents a full CPC. The number reflections for various truncations can be determined by using Fig. 3.45 along with this figure. Values of acceptance angle are shown on the upper curve. [From [60], © 1976, Pergamon Press, Ltd.]

Figure 3.45. Effect of truncation on the basic CPC concentration ratio for various values of acceptance angle. [From [60], © 1976, Pergamon Press, Ltd.]

$$U_c = (0.18 + 16.95\epsilon_r)[0.212 + 0.00255T_a + (0.00186 + 0.000012T_a)(T_r - T_a)],$$

$$(3.135a)$$

and for $(137 + 0.0283T_a - 0.0000616T_a^2) \leq (T_r - T_a) \leq 260°C,$

$$U_c = (0.168 + 17.16\epsilon_r)[0.086 + 0.00225T_a + (0.00278 + 0.000014T_a)(T_r - T_a)],$$

$$(3.135b)$$

where U_c = collector loss coefficient, W/K·m² of absorber area,
$\quad\quad T_r$ = receiver (absorber) temperature, °C,
$\quad\quad T_a$ = ambient temperature, °C, and
$\quad\quad \epsilon_r$ = emissivity of absorber surface.

These empirical equations are valid under the following conditions: $\epsilon_r = 0.05$ to 0.2, wind velocity $V = 0$ to 10 m/s, concentration $C = 1.5$ to 6; correspondingly, average number of reflections $\bar{n} = 0.55$ to 1.05, ambient temperature $T_a = -10$ to 35°C, total radiation $I_t = 720$ to 1200 W/m² aperture area, diffuse to beam radiation ratio $I_d/I_b = 0.1$ to 0.5.

Central receiver collector. A central receiver collector consists of a large field of mirrors on the ground that track the sun in such a way that the reflected radiation is concentrated on a receiver/absorber on top of a tower. The mirrors are called *heliostats*. Central receivers can achieve temperatures of the order of 1000°C. Therefore, a central receiver concentrator is suitable for thermal electric power production in the range of 10–1000 MW. The concept of a central receiver solar thermal power (SRSTP) has been known for a long time. Francia [25] built a pilot model of a solar power tower plant in 1967 at the University of Genoa, Italy. Since the early 1970s, when the power tower concept was first proposed by Hildebrandt and Lorin Vant-Hull [31], the central receiver technology has been actively developed in the US. Since then, the technology has been pursued in Germany, Spain, Switzerland, France, Russia, Italy and Japan [14].

A central receiver collector consists of a heliostat field, a receiver/absorber and the tracking controls for the heliostats. The following sections briefly describe the heliostats and the receiver.

3.8.1 Heliostats

Although several heliostats designs are possible, the one that has been used at all of the pilot plants consists of silvered glass mirrors on a steel support structure. Figure 3.46 shows the back side of a heliostat used at the Solar One central receiver power plant. It consists of several mirror panels supported on a steel structure so that it forms a slightly concave mirror surface. The heliostat focal length is approximately equal to the distance of the heliostat from the receiver.

Recent research and development of heliostats has been concentrated on using stretched membrane reflectors in order to reduce the cost. A stretched membrane helio-

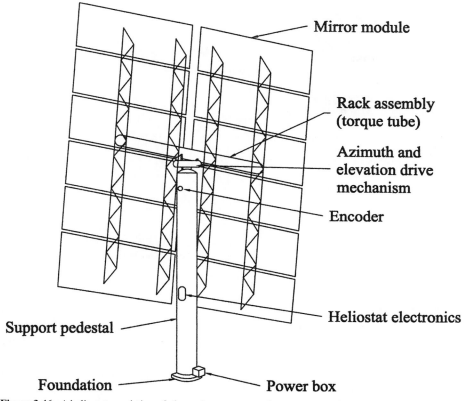

Mirror module

Rack assembly
(torque tube)

Azimuth and
elevation drive
mechanism

Encoder

Heliostat electronics

Support pedestal

Foundation

Power box

Figure 3.46. A heliostat consisting of glass mirrors supported on a structure (back view).

stat consists of two polymer membranes stretched over a metal ring support structure. The front polymer is laminated with a silvered polymer reflector. The space between the two membranes is evacuated. The vacuum pressure is adjusted to achieve the desired focal length.

Figure 3.47 shows a stretched membrane heliostat developed by Science Applications International Corp. (SAIC). These heliostats have the potential to reduce the cost of the central receiver systems, however, the problems include uncertainties regarding their durability and lifetimes.

3.8.2 Receiver

Receivers are designed to accept and absorb very high solar flux and to transfer the heat to a fluid at a very high temperature. Two designs have emerged—external and cavity types.

External receivers. External receivers consist of several panels of vertical tubes connected at the top and the bottom by welded headers. The panels are connected with

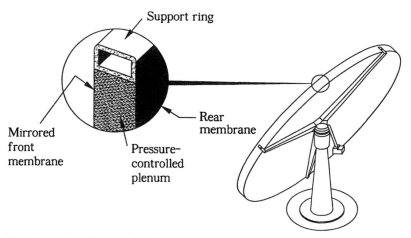

Figure 3.47. Stretched-membrane heliostat consists of two membranes, a mirrored front one and supporting rear one, which are stretched across a support structure. Curvature of the heliostat is adjusted by changing the vacuum pressure of the space between the membranes. Adapted from Ref. [13].

each other to approximate a vertical cylinder. The Solar One central receiver facility used this type of receiver absorber as shown in Fig. 3.48. It consists of 24 panels, each containing 70 tubes of 12.7 mm diameter. The tubes are made of Incoloy 800 and coated with high absorptance black paint. The overall dimensions of the receiver are 7 m diameter and 13.7 m height.

The area of the receiver is kept to a minimum to reduce the heat loss depending on the radiative flux and the heat transfer fluid.

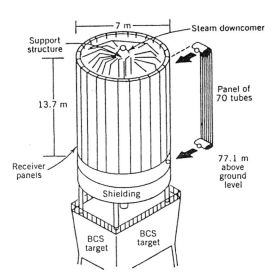

Figure 3.48. The receiver of the Solar One central receiver facility at Barstow, CA. This is an external-type receiver.

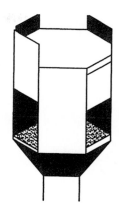

Figure 3.49. A cavity-type receiver design incorporating four apertures. It would operate in the 510–565°C (950–1050°F) temperature range with steam, molten salt, or sodium [5].

Cavity receivers. A cavity receiver directs the concentrated flux inside a small insulated cavity in order to reduce the radiative and convective heat losses. Figure 3.49 shows an example of a cavity receiver. Typical designs have an aperture area about one-third to one-half of the internal absorbing surface area. The acceptance angles range from 60–120 degrees [5].

Heat Flux Consideration

Heat flux on the receiver is limited by the temperature limitation of the heat transfer surface as well as the fluids and other considerations such as temperature gradients and temperature cycling. Table 3.7 lists the peak flux design values based on Ref. [5].

3.8.3 Design of Mirror Field

A detailed design of the mirror field is beyond the scope of this book. However, some macroscale features are described here.

A heliostat is oriented in such a way that a solar ray incident on the heliostat is reflected toward the receiver as shown in Fig. 3.50. In vector notation, the angle i be-

Table 3.7. Typical receiver peak flux design values*

Heat-Transfer Fluid	Configuration	Peak Flux (M W/m²)
Liquid sodium	In tubes	1.5
Liquid sodium	In heat pipes - (transferring to air)	1.2
Molten nitrate salt	In tubes	0.7
Liquid water	In tubes	0.7
Steam vapor	In tubes	0.5
Air	In metal tubes	0.22

*Source: Battleson [5].

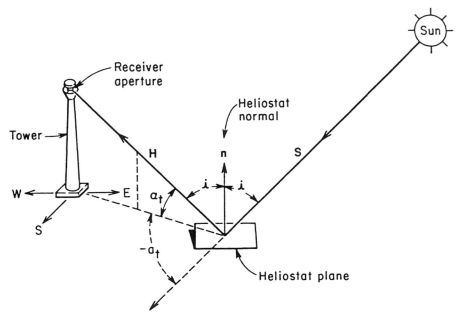

Figure 3.50. Sun-heliostat-tower geometry for calculating tracking requirements for CRSTP heliostats. Tower unit vector H altitude and azimuth angles are shown.

tween a heliostat to receiver vector H and the heliostat normal n is given by the scalar product

$$\cos i = n \cdot H / |n||H|. \qquad (3.136)$$

This equation along with Eq. (2.44) gives the incidence angle between a solar ray S and the heliostat normal n. The second equation relating the two heliostat rotation angles (surface tilt and azimuth angles) is the requirement that the incident ray S, n and H all be coplanar; i.e.,

$$(S \times n) \cdot H = 0. \qquad (3.137)$$

Unit vectors S, n and H are shown in Fig. 3.50.

Riaz [66] has solved Eqs. (3.136), (3.137) and (2.44) to provide the components of n, H and S:

$$n = \begin{bmatrix} \cos\alpha_n \, \sin a_n \\ -\cos\alpha_n \, \cos a_n \\ \sin\alpha_n \end{bmatrix}, \qquad (3.138)$$

$$\mathbf{H} = \begin{bmatrix} \cos\alpha_t \ \sin a_t \\ -\cos\alpha_t \ \cos a_t \\ \sin\alpha_t \end{bmatrix}, \tag{3.139}$$

$$\mathbf{S} = \begin{bmatrix} \cos\alpha \ \sin a_s \\ -\cos\alpha \ \cos a_s \\ \sin\alpha \end{bmatrix} \tag{3.140}$$

where α and a_s are the solar altitude and azimuth angles, α_t and a_t are the fixed tower-vector altitude and azimuth angles shown in Fig. 3.50, and α_n and a_n are the heliostat-normal altitude and azimuth angles measured in the same way as α and a_s. The Cartesian coordinate system in which all vectors are defined is oriented in such a way that (x,y,z) correspond to east, north and vertical.

A heliostat positioned to reflect flux onto the central receiver can generally cast a shadow on other heliostats "behind" it or can interrupt ("block") light reflected from another heliostat. It is not always cost effective to completely eliminate shadowing and blocking in winter when the sun is low in the sky. If this were done, the optical performance in summer would be penalized excessively, and an unnecessarily sparse field would result. An economic analysis is necessary to resolve this question. If Φ is the ratio of mirror-to-ground area, Vant-Hull and Hildebrandt [81] have shown that

$$\Phi = 1.06 - 0.23 \tan \phi, \tag{3.141}$$

where ϕ is the Fresnel mirror (i.e., heliostat field) rim half-angle. A typical value of Φ is 40–50 percent.

Depending upon the instantaneous position of the sun, a given heliostat can direct to the receiver either more or less energy than strikes an equivalent horizontal surface. P is defined as the ratio of redirected flux for perfect optics and reflectance ($\rho_m = 1.0$), to horizontal flux. It has been shown that P is given by [37]

$$P = 0.78 + 1.5(1 - \alpha/90)^2, \tag{3.142}$$

where α is the solar altitude in degrees. P in Eq. (3.142) is for an entire array optimized for low sun angles in winter at 2 PM. Obviously, other expressions apply for other configurations. For optimum summer performance, the high solar altitude dictates a dense field symmetrically located about the receiver tower. For winter optimization, the tower is in the southern half of the field (for sites north of the equator). Mirrors placed far south of the towers are relatively ineffectual because low sun angles and corresponding large incidence angles cause drastic foreshortening of the mirrors. In addition, long shadows result in a rather sparse field. Likewise, afternoon or morning demand peaks cause a western or eastern shift of the tower from the center of the field. Of course, the local ground cover ratio Φ_{local} (heliostat area-to-ground area) will vary throughout the field, depending on the daily and seasonal peaking time. For winter optimized fields it will range from 0.2 to 0.5; for summer fields from 0.4 to 0.6 [79]. A method for field design is presented in [79].

The optical efficiency η_o of a CRSTP heliostat field is then given by

$$\eta_o = \Phi P \rho_m \tau_r \alpha_r f_t \delta, \tag{3.143}$$

where the intercept factor δ includes solar beam spread, mirror surface errors, and tracking inaccuracies, and f_t represents the fraction of receiver area not shaded by supports. Depending upon the tower and heliostat design f_t may range from 0.94 to nearly 1.0. Typical values of optical efficiency are about $\eta_o \sim 0.67 \Phi P$. (Here, in accordance with the convention of CRSTP analysts, η_o is based on horizontal beam radiation and total ground area.)

In order to minimize shading of heliostats and blocking of the reflected light an optimized layout of the heliostat field was developed by Lipps and Vant-Hull [51]. The layout, as shown in Fig. 3.51, is called a radial stagger layout. The radial spacing R and azimuthal spacing A as defined in Fig. 3.51 are given by Dellin [15] for high reflectance heliostats in a large field as:

$$R = h(1.44\cot\alpha_t - 1.094 + 3.068\alpha_t - 1.1256\alpha_t^2), \tag{3.144}$$

and

$$A = w\left(1.749 + 0.6396\alpha_t + \frac{0.2873}{\alpha_t - 0.04902}\right) \tag{3.145}$$

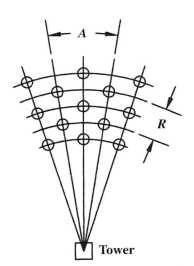

Tower

Figure 3.51. The radial stagger heliostat layout pattern developed by the University of Houston.

where h and w are the height and width of the heliostat and α_t is the tower altitude angle measured from the center of the front heliostat.

3.9 Fresnel reflectors and lenses. If the smooth optical surface of a reflector or a lens can be broken into segments to achieve essentially the same concentration, the resulting concentrator is called the Fresnel Concentrator. Figure 3.52 shows examples of a Fresnel mirror and a Fresnel lens. Use of Fresnel reflectors for large collectors reduces the wind load and simplifies manufacture. A Fresnel lens can achieve a concentration close to a corresponding plano convex lens with far less material and lower manufacturing costs. However, a disadvantage is that the facet edges get rounded in the manufacturing process which makes the edges ineffective.

Off-normal incidence effects are present in a Fresnel device owing to the change in focal length with incidence angle.

Figure 3.53 shows the focal point for normal and nonnormal incidence. Note that the effect of the shortened focal length is to cause the sun's image to appear wider at the nominal focal plane during off-normal periods. Figure 3.54 shows the decrease in focal length which occurs for various off-normal conditions [12,62]. It is seen that a $\pm 60°$ excursion, which would be encountered for 8 h of collection at the equinoxes in an east-west axis alignment, would diminish the focal length by two-thirds. Hence, the Fresnel line-focus device with $CR > 10$ is restricted to a north-south orientation ($\pm 35°$ excursion over a year) if the majority of daylight hours are to be collection hours.

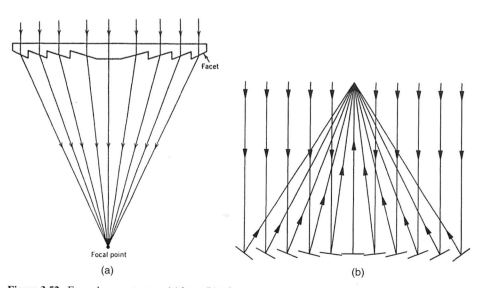

Figure 3.52. Fresnel concentrators; (a) lens; (b) mirror.

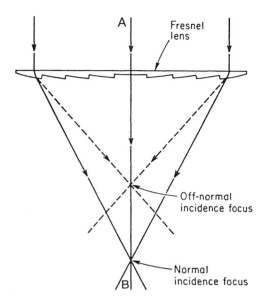

Figure 3.53. Line-Focus Fresnel lens concentrator section showing nominal focus for normal incidence and shortened focal length at off-normal incidence.

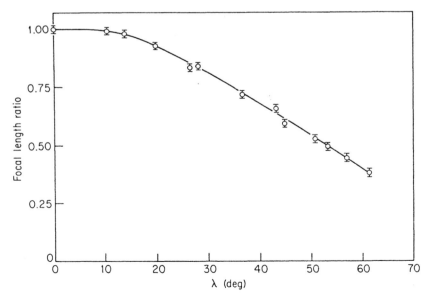

Figure 3.54. Effect of off-normal incidence angle λ (measured in plane *AB* out of the plane of Fig. 3.53) on apparent focal length of a line-focus Fresnel lens. The focal length ratio is the focal length at $\lambda \neq 0$ divided by the focal length at $\lambda = 0$ (i.e., normal incidence). From [12].

SOLAR CONCENTRATOR SUMMARY

Table 3.8 contains a summary of all important optical properties of the common reflecting solar concentrators:

Concentration ratio CR
Average number of reflections $\bar{n}$
Capture parameter δ
Beam spread variance σ_y^2

The CR standard of comparison is the thermodynamic limit derived in the first section describing concentrators. It is seen that all concentrators except the CPC family fall short of the ideal limit by a factor of 3 or more. For any meaningful concentration and rim half-angles $\phi > 45°$, the SRTA is seen to fall short of the ideal limit by the greatest amount because of its focal aberration.

The optical parameter δ and its independent variable σ_y have been developed using analyses analogous to that used for the PTC collector. Duff and Lameiro [17] have also derived expression for σ_y for reflector systems with flat mirror segments instead of curved ones. They are slightly different in form from those in Table 3.8, but are developed by using the same analysis method.

Thermal properties of concentrators cannot be summarized conveniently since they are complex functions of surface properties, geometry, and conduction-convection suppression techniques. Each collector concept must be analyzed using the approach described in detail for the PTC.

Table 3.9 summarizes the equations most often used for calculation of the incidence angles of beam radiation on the aperture of all common types of solar collectors. These equations are used to calculate daylong energy delivery of both concentrators and flat-plate collectors. A complete discussion of incidence angles is contained in Chapter 2.

In order to apply the method of section 2.61 to find monthly averaged daily solar radiation on a tracking solar concentrator, the tilt factors in Eq. (2.64) must be continuously changed due to tracking. Collares-Pareira and Rabl have given expressions for various modes of tracking that can be used to calculate $\bar{H}_c$ as

$$\bar{H}_c = [\bar{r}_T - \bar{r}_d(\overline{D}_h/\overline{H}_h)]\overline{H}_h. \tag{3.146}$$

Equations for $\bar{r}_T$ and $\bar{r}_d$ are given in Table 3.10. The collection start (or stop) time, h_{coll}, used in Table 3.10 represents the time measured from solar noon at which the collector starts (or stops) operating.

Table 3.8. Comparison of design parameters of solar concentrators*

Collector parameter	Single curvature			Compound curvature		
	Nonimaging CPC family	Parabolic mirror		Paraboloidal mirror		Spherical mirror cylindrical receiver
		Cylindrical receiver	Flat receiver	Spherical receiver	Flat receiver	
CR/CR_{max} [†] for perfect optics	1.0	$\dfrac{\sin\phi}{\pi}$ [‡]	$\sin\phi\cos(\phi+\theta_{max})^{\dagger} - \sin\theta_{max}$	$\dfrac{\sin^2\phi}{4}$ [‡]	$\sin^2\phi\cos^2(\phi+\theta_{max})^{\ddagger} - \sin^2\theta_{max}$	$2\sin^2\phi\sin\theta_{max}$
$\bar{n}$	0.7–1.2		1.0		1.0	1.0
δ	$\dfrac{1}{1+\sigma_\theta\cot\theta_{max}}$	$\dfrac{1}{\sqrt{2\pi}\sigma_y}\displaystyle\int_{-L_c/2}^{L_c/2}\exp\left[-\dfrac{1}{2}\left(\dfrac{y}{\sigma_y}\right)^2\right]dy$		$1 - e^{-(L_c/\sigma_r^2)}$		(No closed-form expression)
σ_y^2	—	$\dfrac{A_a^2\sigma_\theta^2(2+\cos\phi)}{12\,\phi\sin\phi}$	#	$\dfrac{2A_a\sigma_\theta^2(2+\cos\phi)}{3\,\phi\sin\phi}$	$\dfrac{2A_a\sigma_\theta^2}{\sin^2\phi}$	(No closed-form expression)
L_c	$\dfrac{D_r}{2}$	$\dfrac{D_r}{2}$	$\dfrac{D_r}{2}$	$\dfrac{\pi D_r^2}{4}$	$\dfrac{\pi D_r^2}{4}$	$\dfrac{D_r}{2}$
A_a	Aperture width	Aperture width		Aperture area		

*Adapted from [17].

†Symbols in table:

CR_{max} = thermodynamic limit to concentrations; see Eqs. (3.88) and (3.89),

L_c = characteristic dimension of receiver,

D_r = receiver diameter or width,

θ_{max} = acceptance angle,

ϕ = mirror rim half-angle,

$\sigma_\theta^2 = 4\sigma_{\psi_t}^2 + \sigma_{\psi_s}^2$,

δ = optical capture factor, and

$\bar{n}$ = average number of reflections.

Other parameters defined in the text.

‡For Fresnel mirror arrangements or power towers, multiply expressions given by the ground cover fraction. It is usually on the order of $\frac{1}{2}$.

§Approximate expression for radiation incident near the limits of acceptance only; otherwise $\delta = 1.0$.

$$\left\{ \frac{A_a^2(4\sigma_{\psi_t}^2 + \sigma_{\psi_s}^2)}{4\phi \tan^2(\phi/2)} \left[\frac{-1}{3\sin^2\phi\cos\phi} + \frac{2}{3\sin^3\phi} + \frac{2}{\sin\phi} - \frac{\cos\phi}{3\sin\phi} - \frac{2\phi}{\sin\phi} + \frac{4\sin\phi}{3\cos\phi} - \ln\tan\left(\frac{\pi}{4} + \frac{\phi}{2}\right) + \ln\tan\left(\frac{\pi}{4} - \frac{\phi}{2}\right) \right]^{1/2} \right\}.$$

Duff [18] has given an alternative expression for σ_y^2 in which a different integration procedure was used.

Table 3.9. Incidence angle factors for various orientations and motions of solar collectors[a]

Orientation of collector	Incidence factor $\cos i$
Fixed, horizontal, plane surface.	$\sin L \sin \delta_s + \cos \delta_s \cos h_s \cos L$
Fixed plane surface tilted so that it is normal to the solar beam at noon on the equinoxes.	$\cos \delta_s \cos h_s$
Rotation of a plane surface about a horizontal east-west axis with a single daily adjustment permitted so that its surface normal coincides with the solar beam at noon every day of the year.	$\sin^2 \delta_s + \cos^2 \delta_s \cos h_s$
Rotation of a plane surface about a horizontal east-west axis with continuous adjustment to obtain maximum energy incidence.	$\sqrt{1 - \cos^2 \delta_s \sin^2 h_s}$
Rotation of a plane surface about a horizontal north-south axis with continuous adjustment to obtain maximum energy incidence.	$[(\sin L \sin \delta_s + \cos L \cos \delta_s \cos h_s)^2 + \cos^2 \delta_s \sin^2 h_s]^{1/2}$
Rotation of a plane surface about an axis parallel to the earth's axis with continuous adjustment to obtain maximum energy incidence.	$\cos \delta_s$
Rotation about two perpendicular axes with continuous adjustment to allow the surface normal to coincide with the solar beam at all times.	1

[a]The incidence factor denotes the cosine of the angle between the surface normal and the solar beam.

Table 3.10. Parameters r_T and r_d used to calculate monthly solar flux incident on various collector types*

Collector Type	$r_T^{†,‡,§}$	$r_d^{\#}$
Fixed aperture concentrators which do not view the foreground	$[\cos(L - \beta)/(d\cos L)]\{-ah_{coll}\cos h_{sr}(i=90) + [a - b\cos h_{sr}(i=90)]\sin h_{coll} + (b/2)(\sin h_{coll}\cos h_{coll} + h_{coll})\}$	$(\sin h_{coll}/d)\{[\cos(L + \beta)/\cos L] - [1/(CR)]\} + (h_{coll}/d)\{[\cos h_{sr}(\alpha = 0)/(CR)] - [\cos(L - \beta)/\cos h_{sr}(i = 90)]\}$
East-west axis tracking‖	$\dfrac{1}{d}\displaystyle\int_0^{h_{coll}}\{[(a + b\cos x)/\cos L] \times \sqrt{\cos^2 x + \tan^2\delta_s}\}dx$	$\dfrac{1}{d}\displaystyle\int_0^{h_{coll}}\{(1/\cos L)\sqrt{\cos^2 x + \tan^2\delta_s} - [1/(CR)][\cos x - \cos h_{sr}(\alpha = 0)]\}dx$
Polar tracking	$(ah_{coll} + b\sin h_{coll})/(d\cos L)$	$(h_{coll}/d)\{(1/\cos L) + [\cos h_{sr}(\alpha = 0)/(CR)]\} - \sin h_{coll}/[d(CR)]$
Two-axis tracking	$(ah_{coll} + b\sin h_{coll})/(d\cos\delta_s\cos L)$	$(h_{coll}/d)\{1/(\cos\delta_s\cos L) + [\cos h_{sr}(\alpha = 0)/(CR)]\} - \sin h_{coll}/[d(CR)]$

*From (11); the collection hour angle value h_{coll} not used as the argument of trigonometric functions is expressed in radians. Note that the total interval 2 h_{coll} is assumed to be centered about solar noon.

†$a = 0.409 + 0.5016 \sin[h_{sr}(\alpha = 0) - 60°]$.

‡$b = 0.6609 - 0.4767 \sin[h_{sr}(\alpha = 0) - 60°]$.

§$d = \sin h_{sr}(\alpha = 0) - h_{sr}(\alpha = 0)\cos h_{sr}(\alpha = 0)$.

#CR is the collector concentration ratio.

‖Elliptic integral tables to evaluate terms of the form of $\int_0^h \sqrt{\cos^2 x + \tan^2\delta_s}\,dx$ contained in r_T and r_d are given in the Appendix. Use the identity $\cos\delta_s = \sin(90° - \delta_s)$ and multiply the intergral by $\cos\delta_s$, a constant. For computer implementation a numerical method can be used. For hand calculations, use Weddle's rule or Cote's formula.

PROBLEMS

3.1. Calculate the heat-removal factor for a collector having an overall heat-loss co-efficient of 6 W/m²K and constructed of aluminum fins and tubes. Tube-to-tube centered distance is 15 cm; fin thickness is 0.05 cm; tube diameter is 1.2 cm; fluid tube heat-transfer coefficient is 1200 W/m²K. The cover transmittance to solar radiation is 0.9 and is independent of direction. The solar absorptance of the absorber plate is 0.9, the collector is 1 m wide and 3 m long, and the water flow rate is 0.02 kg/sec. The water temperature is 330 K.

3.2. Calculate the efficiency of the collector described in Problem 3.1 on March 1 at a latitude of 40°N between 11 and 12 AM. Assume that the total horizontal insolation is 450 W/m², the ambient temperature is 280 K, and the collector is facing south.

3.3. Calculate the plate temperature in Example 3.2 at 10 AM, if the insolation during the first three hours is 0, 150, and 270 W/m² and the air temperature is 285 K.

3.4. Calculate the overall heat-transfer coefficient, neglecting edge losses, for a collector with a double glass cover, with the following specifications:

Plate-to-cover spacing	3 cm
Plate emittance	0.9
Ambient temperature	275 K
Wind speed	3 m/sec
Glass-to-glass spacing	3 cm
Glass emittance	0.88
Back insulation thickness	5 cm
Back insulation thermal conductivity	0.04 W/mK
Mean plate temperature	340 K
Collector tilt	45°

3.5. The graph in Fig. 3.18 gives the results of an ASHRAE standard performance test for a single-glazed flat-plate collector. If the transmittance for the glass is 0.90 and the absorptance of the surface of the collector plate is 0.92, determine:
The collector heat-removal factor F_R;
The overall heat-loss conductance of the collector U_c in W/m²K;
The rate at which the collector can deliver useful energy in W/m²K when the insolation incident on the collector per unit area is 600 W/m²K, the ambient temperature is 5°C, and inlet water is at 15°C; and
The maximum flow rate through the collector, with cold water entering at a temperature of 15°C, that will give an outlet temperature of at least 60°C if this collector is to be used to supply heat to a hot-water tank.

3.6. Calculate the overall heat-loss coefficient for a solar collector with a single glass cover having the following specifications:

Spacing between plate and glass cover	5 cm,
Plate emittance	0.2,
Plate absorptance at equilibrium temperature	0.2,
Ambient air temperature	283 K,
Wind speed	3 m/sec,

Back insulation thickness   3 cm,
Conductivity of back insulation material   0.04 W/mK,
Mean plate temperature   340 K, and
Collector tilt   45°

Note that the solution to this problem requires trial and error. Start by assuming a glass temperature and then determine whether the heat gained by the glass equals the heat loss from the glass at that temperature. If the heat gain is larger than the heat loss, repeat the calculations with a slightly higher temperature. After you have calculated the heat-loss coefficient, compare your answer with the one obtained from Eq. (3.23).

3.7. Standard tests on a commercially available flat plate collector gave a thermal efficiency of:

$$\eta = 0.7512 - 0.138 \, (T_{f,in} - T_a)/I_c$$
$$K_{\tau\alpha} = 1 - .15 \, (1/\cos(i) - 1)$$

where $(T_{f,in} - T_a)/I_c$ is in °C − m²/W

Find the useful energy collected from this collector each hour and for the whole day in your city on September 15.

Assume that all the energy collected is transferred to water storage with no losses. Calculate the temperature of the storage for each hour of the day. Assume a reasonable ambient temperature profile for your city. Given:

Collector area = 6 m²

Collector Tilt = 30° (South facing in northern hemisphere, North facing in southern hemisphere).

Storage Volume = 0.3 m³ (water)

Initial Storage Temp. = 30°C

3.8. What is the second law efficiency of a flat plate collector operating at 70°C if the environmental temperature is 10°C and the first law efficiency is 50 percent? Compare with a single-curvature concentrator operating at 200°C and with a double-curvature concentrator operating at 2500°C, all with first law efficiencies of 50 percent.

3.9. Show that the plate efficiency F' for an air-cooled flat-plate collector is given by Eq. (3.55).

3.10. The heat-removal factor F_R permits solar collector delivery to be written as a function of collector fluid *inlet* temperature T_f in Eq. (3.46). Derive the expression for a factor analogous to F_R, relating collector energy delivery to fluid *outlet* temperature.

3.11. In nearly all practical situations the argument of the exponential term in Eq. (3.43) for F_R is quite small. Use this fact along with a Taylor's series expansion to derive an alternate equation for F_R. Determine the range where the alternate equation and Eq. (3.43) agree to within 1 percent.

3.12. The stagnation temperature $T_{c,max}$ of a solar collector corresponds to the temperature at the zero efficiency point, i.e., the no net energy delivery point. Using Eq. (3.74), calculate the stagnation temperature of a flat-plate collector with 75 percent optical efficiency and a U_c value of 4.5 W/m²K if the insolation is 900 W/m² and the ambient temperature is 20°C.

3.13. Calculate the stagnation temperature of the collector in Problem 3.12 if it were used with an optical booster of concentration ratio 2.0 having an optical efficiency of 70 percent.

3.14. What is the operating temperature for an evacuated-tube collector operating at 50 percent efficiency if the insolation is 800 W/m²? Use data from Fig. 3.19.

3.15. A method of reducing heat loss from a flat-plate collector is to pull a partial vacuum in the dead air spaces between cover plates. What vacuum level is required to completely convection in a single cover flat-plate collector operating at 85°C and tilted at 45° if the cover plates are at 40°C and are spaced 2 cm apart? See Eq. (3.74).

3.16. Derive an expression for the heat-loss conductance U_c for a flat-plate collector in which convection and conductance are completely eliminated in the air layers by use of a hard vacuum.

3.17. The effect of air flow rate in an air-cooled flat-plate collector appears in the heat-removal factor F_R [Eq. (3.56)]. Calculate the effect of doubling the flow rate on the heat-removal factor for the collector analyzed in Example 3.3. What percentage increase in energy delivery would be achieved by doubling the fan size?

3.18. Calculate the optical efficiency on November 13 of an evacuated-tube collector array at noon and 2 PM if the direct normal insolation is 600 W/m² and the diffuse insolation is 100 W/m². The effective optical transfer function $\tau_e \alpha_r$ is 70 percent, and the tubes are spaced one diameter apart in front of a white painted surface with reflectance ρ of 60 percent.

3.19. What are the maximum concentration ratios for trough concentrators with acceptance angles of 10°, 25°, and 36°?

3.20. What is the maximum achievable temperature of a double-curvature concentrator with a concentration ratio of 5000, a nonselective surface, and an 80 percent transmittance function?

3.21. Calculate the depth of a full CPC collector if the aperture is 1 m and $CR = 5$. Repeat for an aperture of 1 cm.

3.22. What is the reflectance loss in a 50 percent truncated CPC collector using silver mirrors if it has a 36° acceptance half-angle? What is the loss for an anodized aluminum reflector?

3.23. Compare the average number of reflections for a full and 50 percent truncated CPC if the acceptance half-angle is 9°.

3.24. How much concentration effect is lost by truncating a 7° half-angle CPC by ¼, ⅓, and ½?

3.25. Explain how Figs. 3.44 and 3.45 could be used to prepare a map showing the effect of truncation on average number of reflectors of a CPC collector. Sketch qualitatively what such a map would look like.

3.26. A parabolic trough 1 m wide and 10 m long with no end support plates is 30 cm deep and has a focal length of 20.83 cm. Calculate its optical efficiency at a 30° incidence angle for perfect optics and tracking if its reflector reflectance is 80 percent, $\tau\alpha$ product is 75 percent, and if the receiver supports shade 5 percent of the aperture.

3.27. Calculate the maximum concentration ratio for a parabolic trough collector with a 50 cm aperture, which captures 92 percent of the incident radiation if the rim half-angle is 50° and the surface is accurate to 2 mrad in slope. What is the receiver diameter?

REFERENCES AND SUGGESTED READINGS

1. Agarwal, V.K., and D.C. Larson. 1981. Calculation of the top loss coefficient of a flat plate collector. *Solar Energy* 27:69–71.

2. American Society of Heating, Refrigerating and Air Conditioning Engineers, "ASHRAE Standard 93-77, Methods of testing to determine the thermal performance of solar collectors," (Atlanta, Georgia: ASHRAE).

3. Bannerot, R.B., and J.R. Howell. 1977. The effect of nondirect insolation on the radiative performance of trapezoidal grooves used as solar energy collectors. *Solar Energy* 19, No. 5:539.

4. Baranov, V.K. and G.K. Melnikov. 1966. Study of the illumination characteristics of hollow focons, *Sov. J. Opt. Technol.* 33: 408–411. Also 1975. Parabolocylindric reflecting unit and its properties. *Geliotekhnika* 11: 45–52.

5. Battleson, K.W. 1981. Solar power tower design guide: solar thermal central receiver power systems, a source of electricity and/or process heat. *Sandia National Lab Rept.* SAND 81–8005.

6. Baum, V.A. 1955. *Proc. World Symp. Solar Energy,* Phoenix, 289–298.

7. Beekley, D.C. and G.R. Mather. 1975. *Analysis and experimental tests of a high performance, evacuated tube collector.* Toledo, Ohio: Owens-Illinois.

8. Blake, F. 1975. Solar thermal power. *Proc. Energy Technol. Update, Colorado Decision Makers.* Colorado School of Mines, Golden.

9. Bliss, R.W. 1959. The derivations of several plate efficiency factors useful in the design of flat-plate solar-heat collectors. *Sol. Energy* 3: 55.

10. Clausing, A.M. 1976. The performance of a stationary reflector/tracking absorber solar concentrator. In *Sharing the Sun,* vol. 2. Winnipeg: ISES, 304–326. See also, by the same author, *ERDA Rept.* SAND 76–8039, 1976.

11. Collares-Pereira, and A. Rabl. 1979. Simple procedure for predicting long term average performance of nonconcentrating and of concentrating solar collectors. *Sol. Energy* 23:235–254.

12. Collares-Pereira, A. Rabl, and R. Winston. 1977. Lens mirror combinations with maximal concentration. *Enrico Fermi Institute Rept.,* No. EFI 77–20.

13. Colorado State University and Westinghouse Electric Corporation. 1974. *Solar thermal electric power systems, Final Report, 3 vols. Colorado State University Rept.,* NSF/RANN/SE/GI-37815/FR/74/3.

14. DeLaquil, III, P., et al. 1993. Solar thermal electric technology. In *Renewable energy–sources for fuels and electricity,* T. Johansson, et al., Eds., 213–296. Washington, D.C.: Island Press.

15. Dellin, T.A., M.J. Fish, and C.L. Yang. 1981. A user's manual for DELSOL2: a computer code for calculating the optical performance and optimal system design for solar central receiver plants." *Sandia National Lab Rept.* SAND 81–8237.

16. deWinter, F. 1975. Solar energy and the flat plate collector. *ASHRAE Rept.* S-101.

17. Duff, W.S., and G.F. Lameiro. 1974. A performance comparison method for solar concentrators. ASME Paper 74-WA/Sol-4, New York.

18. Duff. W.S. 1976. Optical and thermal performance of three line focus collectors. ASME Paper 76-WA/HT-15.

19. Duffie, J.A., and W.A. Beckman. 1980. *Solar energy thermal processes.* New York: John Wiley & Sons.

20. Dushman, S.. 1962. In *Scientific foundations of vacuum technology,* 2nd ed., J.M. Lafferty ed. New York: John Wiley & Sons.

21. Duwez, P., et al. 1958. The operation and use of a lens-type solar furnace. *Trans. Int. Conf. Use Solar Energy–Sci. Basis:* 213–221.

22. Eckert, E.R.G., et al. 1973. Research applied to solar-thermal power systems, *University of Minnesota Rept.* NSF/RANN/SE/GF-34871/PR/73/2.

23. Edenburn, M.W. 1976. Performance analysis of a cylindrical parabolic focusing collector and comparison with experimental results. *Sol. Energy*, 18: 437–444.

24. Emmett, W.L.R. 1911. *Apparatus for Utilizing Solar Heat*, U.S. Patent 980, 505.

25. Francia, G. 1968. Pilot plants of solar steam generating stations. *Sol. Energy* 12: 51–64.

26. Fraser, M.D. 1976. Survey of the applications of solar thermal energy to industrial process heat. *Proc. Solar Industrial Process Heat Workshop*, University of Maryland. See also: 1977. InterTechnology Corporation. "Analysis of the economic potential of solar thermal energy to provide industrial process heat." *InterTechnology Rept.* 00028–1 for U.S. ERDA Contract EY-76-C-02-2829, 53.

27. Garg, H.P. 1987. *Advances in solar energy technology*, Vol. 1, *Collection and storage systems*. Dordrecht, Holland: D. Reidel Publishing Co.

28. Gleckman, P., J. O'Gallagher, and R. Winston. 1989. "Concentration of sunlight to solar surface levels using non-imaging optics," *Nature* 339: 198–200.

29. Goodman, N.B., A. Rabl, and R. Winston. 1976. Optical and thermal design considerations for ideal light collectors. In *Sharing the Sun*, vol. 2. Winnipeg: ISES, 336–350. See also: Rabl, A., N.B. Goodman, and R. Winston. 1977. Optical and thermal design considerations for ideal light collectors. *Sol. Energy* 19.

30. Grimmer, D.P., and K.C. Herr. 1976. Solar process heat from concentrating flat-plate collectors. In *Sharing the Sun*, vol. 2. Winnipeg: ISES, 351–373.

31. Hildebrandt, A.F., and L.L. Vant-Hull. Tower top focus solar energy collector. *Mech. Eng.* 23, no. 9 (September 1974):23–27.

32. Hill, J.E., and E.R. Streed. 1976. A method of testing for rating solar collectors based on thermal performance. *Sol. Energy* 18: 421–431.

33. Hill, J.E., E.R. Streed, G.E. Kelly, J.C. Geist, and T. Kusuda. 1976. Development of proposed standards for testing solar collectors and thermal storage devices, *NBS Tech. Note* 899.

34. Hottel, H.C., and A.F. Sarofim. 1967. *Radiative transfer.* New York: McGraw-Hill Book Co.

35. Hottel, H.C., and A. Whillier. 1958. Evaluation of flat-plate collector performance. *Trans. Conf. Use Sol. Energy*, vol. 2, part 1, p. 74.

36. Hottel, H.C., and B.B. Woertz. 1942. Performance of flat-plate solar-heat collectors. *Trans. Am. Soc. Mech. Eng.*, vol. 64, p. 91.

37. Howell, J.R., R.B. Bannerot, and G.C. Vliet. 1982. *Solar-thermal energy systems—Analysis and design.* New York: McGraw-Hill.

38. Hsieh, C.K. 1981. Thermal analysis of CPC collectors. *Sol. Energy* 27: 19.

39. Hsieh, C.K., and F.M. Mei. 1983. Empirical equations for calculation of CPC collector loss coefficients. *Sol. Energy* 30: 487.

40. Kauer, E., R. Kersten, and F. Madrdjuri. 1975. Photothermal conversion. *Acta Electron* 18: 297–304.

41. Klein, S.A. 1975. Calculation of flat-plate collector loss coefficients. *Sol. Energy* 17: 79–80.

42. Klein, S.A., J.A. Duffie, and W.A. Beckman. 1974. Transient considerations of flat-plate solar collectors. *J. Eng. Power* 96A: 109–114.

43. Kreider, J.F. 1979. *Medium and high temperature solar processes.* New York: Academic Press.

44. Kreider, J.F. 1975. Thermal performance analysis of the stationary reflector/tracking absorber (SRTA) solar concentrator. *J. Heat Transfer* 97: 451–456.

45. Kreider, J.F., and F. Kreith. 1975. *Solar heating and cooling.* Washington, D.C.: Hemisphere Publ. Corp.

46. Kreider, J.F. 1980. Ed: *Handbook of solar energy.* New York: McGraw-Hill.

47. Kreith, F., and M.S. Bohn. 1997. *Principles of heat transfer.* 5th ed. Boston: PWS Publishers.

48. Krieth, F. 1975. Evaluation of focusing solar energy collectors. *ASTM Stand News* 3: 30–36.

49. Kutcher, C.F. 1996. "Transpired solar collector systems: A major advance in solar heating." Paper presented at the World Engineering Congress, Nov. 6–8, 1996, Atlanta, GA.

50. Kutcher, C.F., and C.B. Christensen. 1992. "Unglazed transpired solar collectors. Advances in solar energy," K. Böer, ed., vol. 7, pp. 283–307.

51. Lipps, F.W., and L.L. Vant-Hull. 1978. A cellwise method for the optimization of large central receiver systems. *Sol. Energy* 30(6): 505.

52. Löf, G.O.G., D.A. Fester and J.A. Duffie. 1962. Energy balances on a parabolic cylinder solar collector. *J. Eng. Power* 84A: 24–32.

53. Löf, G.O.G., and R.A. Tybout. 1972. Model for optimizing solar heating design. ASME Paper 72-WA/SOL-8.

54. Malik, M.A.S., and F.H. Buelow. 1976. Hydrodynamic and Heat Transfer Characteristics of a Heat Air Duct. In *Heliotechnique and development,* (COMPLES 1975), Vol. 2, 3–30. Cambridge, Massachusetts: Development Analysis Associates.

55. Malik, M.A.S., and F.H. Buelow. 1976. Heat transfer in a solar heated air duct–a simplified analysis. In *Heliotechnique and Development* (COMPLES 1975), Vol. 2, 31–37. Cambridge, Massachusetts: Development Analysis Associates.

56. McDaniels, D.K., et al. 1975. "Enhanced solar energy collectors using reflector-solar thermal collector combinations." *Sol. Energy* 17: 277–283.

57. Rabl, A. 1976. Optical and thermal properties of compound parabolic concentrators. *Sol. Energy* 18: 497. See also A. Rabl. 1977. "Radiation through specular passages." *Int. J. Heat Mass Trans.* 20: 323.

58. Rabl, A. 1976. "Ideal two dimensional concentrators for cylindrical absorbers." *Appl. Opt.* vol. 15: 871.

59. Rabl, A. 1975. Comparison of solar concentrators. *Sol. Energy* 18: 93–111. See also A. Rabl, and R. Winston. Ideal concentrators for finite sources and restricted exit angles. *Appl. Opt.* 15: 2886.

60. Rabl, A. 1976. *Sol. Energy* 18: 497.

61. Rabl, A. et al. 1979. *Sol. Energy* 22: 373.

62. Rabl, A. 1978. *Proc. Sol. Thermal Conc. Coll. Tech. Symp.* B. Gupta, ed. Denver, Colorado. 1–42.

63. Ramsey, J.W., J.T. Borzoni, and T.H. Hollands. 1975. Development of flat plate solar collectors for heating and cooling of buildings. *NASA Rept.* CR-134804.

64. Ramsey, J.W., et al. 1976. "Experimental Evaluation of a Cylindrical Parabolic Solar Collector." ASME Paper 76-WA/HT-13. See also [64].

65. Ramsey, J.W., et al. 1977. Experimental evaluation of a cylindrical parabolic solar collector. *J. Heat Transfer* 99: 163.

66. Riaz, M. 1975. ASME Paper No. 75-WA/Sol-1, New York.

67. Seitel, S.C. 1975. Collector performance enhancement with flat reflectors. *Sol. Energy* 17: 291–295.

68. Simon, F.F. 1975. Flat plate solar collector performance evaluation with a solar simulator as a basis for collector selection and performance prediction. *NASA Rept.* TM X-71792. See also 1976. *Sol. Energy* 18: 451–466.

69. Sparrow, E.M., et al. 1974. Research applied to solar-thermal power systems. *University of Minnesota Rept.* NSF/RANN/SF/GI-34871/PR/73/4.

70. Sparrow, E.M., et al. 1974. Research applied to solar-thermal power systems. *University of Minnesota Rept.* NSF/RANN/SF/GI-34871/PR/74/2.

71. Sparrow, E.M., et al. 1975. Research applied to solar-thermal power systems. *University of Minnesota Rept.* NSF/RANN/SE/GI-37871/PR/74/4.

72. Sparrow, E.M., et al. 1975. Research applied to solar-thermal power systems. *University of Minnesota Rept.* NSF-RANN-75-219.

73. Sparrow, E.M., and R.D. Cess. 1977. *Radiation heat transfer*, Washington, D.C.: Hemisphere Publ. Corp.

74. Speyer, F. 1965. Solar energy collection with evacuated tubes. *J. Eng. Power* 87, 270.

75. Steward, W.G. 1973. A concentrating solar energy system employing a stationary spherical mirror and a movable collector. *Proc. Solar Heating & Cooling Buildings Workshop* 24–25. See also Steward, W.G., et al. 1976. Experimental evaluation of a stationary spherical reflector tracking absorber solar energy collector. ASME Paper 76-WA/HT-10.

76. Steward, W.G. 1973. *Proc. Solar Heating & Cooling Buildings Workshop*, NTIS, 24.

77. Tabor, H. 1958. Radiation, convection, and conduction coefficients in solar collectors. *Bull. Res. Coun. Isr.* 6C: 155.

78. Tabor, H.Z. 1958. Stationary mirror systems for solar collectors. *Sol. Energy* 2: 27. See also Tabor, H.Z. 1966. Mirror boosters for solar collectors. *Sol. Energy* 10: 111.

79. Vant-Hull, L.L. 1976. *Proc. SPIE Solar Energy Utilization Conf. II*, 85: 104. See also Vant-Hull, L.L. 1976. 85: 111.

80. Vant-Hull, L.L., "Concentrator Optics," Chapter 3, in Solar Power Plants - Fundamentals, Technology, Systems, Economics, Winter, C.J., et al., Eds., Springer-Verlag, Berlin, Germany, 1991.

81. Vant-Hull, L.L., and A.F. Hildebrandt, *Proc. Solar Thermal Conversion Workshop, January 1973*, NTIS Rept. No. PB 239277, 1974.

82. Vant-Hull, L.L., and A.F. Hildebrandt, A Solar Thermal Power System Based on Optical Transmission, *Proc. Solar Thermal Conversion Workshop, NSF Rept.* NSF/RANN/GI- 32488, 1974; see also, same title, *Sol. Energy*, vol. 18, pp. 31–39, 1976.

83. Weinstein, A., et al., Lessons Learned from Atlanta (Towns) Solar Experiments, in "Sharing the Sun," vol. 3, pp. 153–167, ISES, Winnipeg, 1976; see also, same title, *Sol. Energy*, vol. 19, pp. 421–427, 1977.

84. Welford, W.T., and R. Winston, "High Collection Nonimaging Optics," Academic Press, San Diego, California, 1989.

85. Welford, W.T., "Aberrations of the Symmetrical Optical System," Academic Press, New York, 1974.

86. Westinghouse Electric Corporation. 1974. Solar heating and cooling experiment for a school in Atlanta. *NTIS Rept.* PB240611.

87. Winston, R. 1974. "Solar concentrators of a novel design." *Sol. Energy,* 16: 89–95. Includes references of early CPL work from 1966.

88. Winston, R. 1975. Radiant Energy Collection, U.S. Patent 3,923,381.

89. Winston, R., and H. Hinterberger. 1975. Principles of cylindrical concentrators for solar energy. *Sol. Energy,* 17: 255–258. See also 1970. *J. Opt. Soc. Am.,* 60: 265–270.

THERMAL ENERGY STORAGE
AND TRANSPORT

Energy storage becomes necessary whenever there is a mismatch between the energy available and the demand. Storage is especially important in solar energy applications because of the seasonal, diurnal, and intermittent nature of solar energy. Nature provides storage of solar energy in a number of ways, such as plant matter (also known as biomass), ocean thermal energy, and hydro-potential at high elevation by evaporation from water bodies and subsequent condensation. In fact, even fossil fuels are a stored form of solar energy because they are produced from biomass. Natural solar energy storage provides a longer term buffer between supply and demand. In this chapter, we will examine thermal energy storage. Electrochemical storage in batteries and hydrogen energy storage will be described in Chapters 9 and 11 respectively.

4.1 THERMAL ENERGY STORAGE (TES)

Thermal energy storage in various solid and liquid media is used for solar water heating, space heating, and cooling as well as for high temperature applications such as solar thermal power. The following example illustrates the effect of thermal storage.

Example 4.1. On a clear day the energy delivered by a solar system is approximately a sinusoidal function of time with a maximum near noon and a minimum at sunrise and sunset. If the energy demand on a solar-thermal system is a constant L_o, calculate the temperature history of storage for a day for the block system in Figure 4.1.

Solution. Let the useful energy delivery be given by

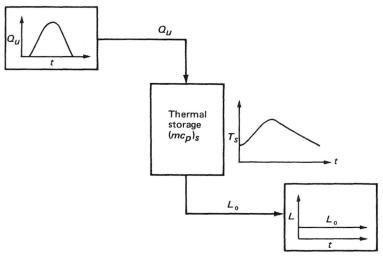

Figure 4.1. Energy flow in Example 4.1 showing solar input to storage and load withdrawals from storage.

$$Q_u = Q_{uo} \sin\frac{\pi t}{\tau},\tag{4.1}$$

where Q_{uo} is the peak delivery rate, τ is the length of a day, and t is measured from sunrise for convenience.

The energy balance on the storage tank, which acts as a summing junction for energy flows, is

$$(mc_p)_s\frac{dT_s}{dt} = Q_u - L_o,\tag{4.2}$$

where $(mc_p)_s$ is the heat capacitance of storage in kJ/K, and T_s is the storage temperature. Equation (4.2) ignores thermal losses from storage which can be made very small by proper insulation.

Equation (4.2) can be solved for $T_s(t)$ after substituting Eq. (4.1) for Q_s.

$$\frac{dT_s}{dt} = \frac{1}{(mc_p)_s}\left(Q_{uo}\sin\frac{\pi t}{\tau} - L_o\right).\tag{4.3}$$

Integrating gives

$$T_s = T_{so} - \frac{1}{(mc_p)_s}\left(\frac{\tau}{\pi}Q_{uo}\cos\frac{\pi t}{\tau}\bigg|_o^t + L_o t\right),\tag{4.4}$$

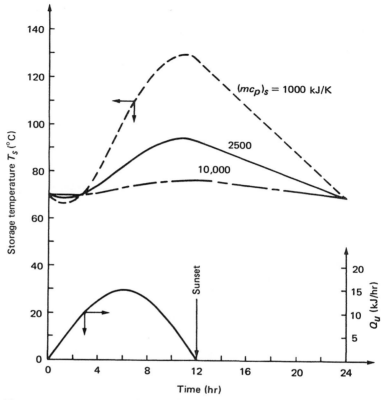

Figure 4.2. Storage temperature history for Example 4.1 showing the effect of storage size. Useful energy delivery to storage shown in lower curve Q_u.

$$T_s = T_{so} - \frac{1}{(mc_p)_s}\left(L_o t - \frac{2\tau Q_{uo}}{\pi} \sin^2 \frac{\pi t}{2\tau}\right), \qquad (4.5)$$

where $T_{so} \equiv T_s (t = 0)$, and $t \leq \tau$.

Equation (4.5) illustrates that the larger the storage capacitance $(mc_p)_s$, the smaller the fluctuations of storage temperature represented by the second term of the equation (see Fig. 4.2).

Important parameters in a storage system include the ***duration of storage, energy density*** (or specific energy), and the ***charging and discharging*** (storage and retrieval) characteristics. The energy density is a critical factor for the size of a storage system. The rate of charging and discharging depends on thermophysical properties such as thermal conductivity and design of the storage system. The following sections describe the types of TES, materials, and storage system design.

4.2 TYPES OF THERMAL ENERGY STORAGE

Thermal energy can be stored as sensible heat, latent heat, or as the heat of chemical reaction (thermochemical).

4.2.1 Sensible-Heat Storage

Sensible heat, Q, is stored in a material of mass m and specific heat c_p by raising the temperature of the storage material from T_1 to T_2 and is expressed by Eq. (4.6):

$$Q = \int_{T_1}^{T_2} mc_p dT,$$

$$= \int_{T_1}^{T_2} \rho V c_p dt, \tag{4.6}$$

where ρ and V are density and volume of the storage material, respectively. For moderate temperature changes, such as for solar space and water heating systems, the density and specific heat may be considered constants. Therefore, $Q = \rho V c_p \Delta T$. The most common sensible heat storage materials are water, organic oils, rocks, ceramics, and molten salts. Some of these materials, along with their physical properties, are listed in Table 4.1. Water has the highest specific heat value of 4190 J/kg · C.

The most common medium for storing sensible heat for use with low- and medium-temperature solar systems is water. Water is cheap and abundant and has a number of particularly desirable properties. Table 4.2 lists advantages and disadvantages of aqueous storage of thermal energy.

Water is the standard storage medium for solar-heating and -cooling systems for buildings today. For these systems, useful energy can be stored below the boiling point of water.

4.2.2 Latent Heat Storage

Thermal energy can be stored as latent heat in a material that undergoes phase transformation at a temperature that is useful for the application. If a material with phase change temperature T_m is heated from T_1 to T_2 such that $T_1 < T_m < T_2$ the thermal energy Q stored in a mass m of the material is given by Eq. (4.7):

$$Q = \int_{T_1}^{T_m} mc_p dT + m\lambda + \int_{T_m}^{T_2} mc_p dT, \tag{4.7}$$

where λ = heat of phase transformation.

Four types of phase transformations useful for latent heat storage are: solid $\rightleftarrows$ liquid, liquid $\rightleftarrows$ vapor, solid $\rightleftarrows$ vapor, and solid $\rightleftarrows$ solid. Since phase transformation is an isothermal process, thermal energy is stored and retrieved at a fixed temperature known as the transition temperature. Some common phase change materials (PCMs)

Table 4.1. Physical properties of some sensible heat storage materials

Storage Medium	Temperature Range, °C	Density (ρ), kg/m³	Specific Heat (C), J/kg K	Energy Density (ρC) kWh/m³ K	Thermal Conductivity (W/m K)
Water	0–100	1000	4190	1.16	0.63 at 38°C
Water (10 bar)	0–180	881	4190	1.03	—
50% ethylene glycol–50% water	0–100	1075	3480	0.98	—
Dowtherm A® (Dow Chemical, Co.)	12–260	867	2200	0.53	0.122 at 260°C
Therminol 66® (Monsanto Co.)	–9–343	750	2100	0.44	0.106 at 343°C
Draw salt (50NaNO$_3$–50KNO$_3$)[a]	220–540	1733	1550	0.75	0.57
Molten salt (53KNO$_3$/ 40NaNO$_2$/7NaNO$_3$)[a]	142–540	1680	1560	0.72	0.61
Liquid Sodium	100–760	750	1260	0.26	67.5
Cast iron	m.p. (1150–1300)	7200	540	1.08	42.0
Taconite	—	3200	800	0.71	—
Aluminum	m.p. 660	2700	920	0.69	200
Fireclay	—	2100–2600	1000	0.65	1.0–1.5
Rock	—	1600	880	0.39	—

[a] Composition in percent by weight.
Note: m.p. = melting point.

Table 4.2. Advantages and disadvantages of water as a thermal storage medium

Advantages	Disadvantages
Abundant	High vapor pressure
Low cost	Difficult to stratify
Nontoxic	Low-surface tension—leaks easily
Not combustible	Corrosive medium
Excellent transport properties	Freezing and consequent destructive expansion
High specific heat	Nonisothermal energy delivery
High density	
Good combined storage medium and working fluid	
Well-known corrosion control methodology	

used for thermal storage are paraffin waxes, nonparaffins, inorganic salts (both anhydrous and hydrated), and eutectics of organic and/or inorganic compounds. Table 4.3 lists some PCMs with their physical properties.

Most common PCMs used for solar energy storage undergo solid $\rightleftarrows$ liquid transformation. For such materials, the thermal energy stored may be written from Eq. (4.7) as, approximately,

$$Q = m[\bar{c}_{p_s}(T_m - T_1) + \lambda + \bar{c}_{p_l}(T_2 - T_m)], \qquad (4.8)$$

where $\bar{c}_{p_s}$ and $\bar{c}_{p_l}$ are the average specific heats in the solid and liquid phases, respectively.

4.2.3 Thermochemical Energy Storage

Thermochemical energy can be stored as heat of reaction in reversible chemical reactions. In this mode of storage, the reaction in the forward direction is endothermic (storage of heat), while the reverse action is exothermic (release of heat). For example,

$$A + \Delta H \rightleftarrows B + C. \qquad (4.9)$$

The amount of heat Q stored in a chemical reaction depends on the heat of reaction and the extent of conversion as given by Equation (4.10):

$$Q = a_r m \Delta H, \qquad (4.10)$$

where a_r = fraction reacted, ΔH = heat of reaction per unit mass, and m = mass.

Chemical reaction is generally a highly energetic process. Therefore, a large amount of heat can be stored in a small quantity of a material. Another advantage of thermochemical storage is that the products of reaction can be stored at room temperature and need not be insulated. For sensible and latent heat storage materials, insulation is very important. Examples of reactions include decomposition of metal hy-

Table 4.3. Physical properties of latent heat storage materials or PCMs

Storage Medium	Melting Point °C	Latent Heat, kJ/kg	Specific Heat (kJ/kg °C)		Density (Kg/m³)		Energy Density (kWhr/m³K)	Thermal Conductivity (W/m K)
			Solid	Liquid	Solid	Liquid		
$LiClO_3 \cdot 3H_2O$	8.1	253	—	—	1720	1530	108	—
$Na_2SO_4 \cdot 1OH_2O$ (Glauber's Salt)	32.4	251	1.76	3.32	1460	1330	92.7	2.25
$Na_2S_2O_3 \cdot 5H_2O$	48	200	1.47	2.39	1730	1665	92.5	0.57
$NaCH_3COO \cdot 3H_2O$	58	180	1.90	2.50	1450	1280	64	0.5
$Ba(OH)_2 \cdot 8H_2O$	78	301	0.67	1.26	2070	1937	162	0.653ℓ
$Mg(NO_3) \cdot 6H_2O$	90	163	1.56	3.68	1636	1550	70	0.611
$LiNO_3$	252	530	2.02	2.041	2310	1776	261	1.35
$LiCO_3/K_2CO_3$ (35:65)[a]	505	345	1.34	1.76	2265	1960	188	—
$LiCO_3/K_2CO_3/$ Na_2CO_3 (32:35:33)[a]	397	277	1.68	1.63	2300	2140	165	—
n-Tetradecane	5.5	228	—	—	825	771	48	0.150
n-Octadecane	28	244	2.16	—	814	774	52.5	0.150
HDPE (cross-linked)	126	180	2.88	2.51	960	900	45	0.361
Steric acid	70	203	—	2.35	941	347	48	0.172ℓ

[a]Composition in percent by weight.

Note: ℓ = liquid.

drides, oxides, peroxides, ammoniated salts, carbonates, sulfur trioxide, etc. Some useful chemical reactions are reported in Table 4.4.

4.3 DESIGN OF STORAGE SYSTEM

Design of a storage system involves the selection of a storage material and design of containment and heat exchangers for charging and discharging.

4.3.1 Selection of Storage Material

Selection of the storage material is the most important part of the design of a TES system. The selection depends on a number of factors. Below are some of those factors.

A. Solar collection system. The solar collection system determines the temperature at which the storage material will be charged and the maximum rate of charge. Thermophysical properties of the storage material at this temperature are important in determining the suitability of the material. For example, flat plate liquid type collectors may use water as the storage material, while air type flat plate collectors for space heating may use a rock or pebble bed as the storage medium. If the storage material can be used as the heat exchanger fluid in the collector, it avoids the need of a collector-to-storage heat exchanger. This criteria favors liquid storage materials. Water and glycol-water mixtures are the most common storage materials for flat plate and moderately concentrating collector systems. For parabolic trough concentrators high temperature oils are more appropriate. For higher concentration and higher temperature collectors such as central receiver tower, molten salts may be used.

Molten nitrate salt (50 wt% $NaNO_3$/50 wt% KNO_3), also known as Draw salt, which has a melting point of 222°C, has been used as a storage and a heat-transfer fluid in an experiment in Albuquerque, NM. This was the first commercial demonstration of generating power from storage [8]. Solar Two, a 10-MW solar thermal power demonstration project in Barstow, CA, is also designed to use this molten salt to store solar energy [4]. Another molten nitrate salt is 40 wt% $NaNO_2$/7 wt% $NaNO_3$/53 wt% KNO_3, known as HTS (heat-transfer salt) with a melting point of 142°C. This salt has been widely used in the chemical industry.

B. Application. The application determines the temperature at which the storage will be discharged and maximum rate of discharge. For hot water applications and moderate temperature industrial process heat, water would be an obvious choice for heat storage. A PCM may be used if space considerations are very important.

For applications in heating and cooling of buildings, the containment of PCM can become an integral part of the building. It may be part of the ceiling, wall, or floor of the building and may serve a structural or a nonstructural function. Tubes, trays, rods, panels, balls, canisters, and tiles containing PCMs have been studied in the 1970s and 1980s for space-heating applications [23]. The PCMs used were mostly salt hydrates such as $Na_2SO_4 \cdot 10H_2O$ (Glauber's salt), $Na_2S_2O_3 \cdot 5H_2O$ (Hypo), $NaCH_3COO \cdot$

Table 4.4. Properties of thermochemical storage media

Reaction	Condition of Reaction		Component (Phase)	Pressure, kPa	Temperature, °C	Density, kg/m³	Volumetric Storage Density, kWh/m³
	Pressure, kPa	Temperature, °C					
$MgCO_3(s) + 1200\ kJ/kg =$ $MgO(s) + CO_2(g)$	100	427–327	$MgCO_3(s)$ $CO_2(\ell)$	100 7400	20 31	1500 465	187
$Ca(OH)_2(s) + 1415\ kJ/kg =$ $CaO(s) + H_2O(g)$	100	572–402	$Ca(OH)_2(s)$ $H_2O(\ell)$	100	20	1115	345
$SO_3(g) + 1235\ kJ/kg =$ $SO_2(g) + \frac{1}{2}O_2(g)$	100	520–960	$SO_3(\ell)$ $SO_2(\ell)$ $O_2(g)$	100 630 10000	45 40 20	1900 1320 130	280

Note: s = solid; ℓ = liquid; g = gas

$3H_2O$, $Na_2HPO_4 \cdot 12H_2O$, $Ba(OH)_2 \cdot 8H_2O$, $MgCl_2 \cdot 6H_2O$, and $Mg(NO_3)_2 \cdot 6H_2O$. Paraffin mixtures have been used for thermal storage in wall boards.

C. Additional considerations. Other considerations include space requirements, cost, long term cycling stability, corrosivity, complexity of containment system, and complexity of heat exchanger design for maximum rates of charge and discharge.

4.3.2 Design of Containment

Containment design is especially important for liquid and phase change storage materials. Water can be stored in any tank able to withstand the expected pressure. Some commonly used tanks for water storage include steel tanks with glass, epoxy or stone lining, fiberglass reinforced polymer, concrete with plastic liner, and wooden tanks.

Containment of PCMs can be problematic, since phase change is accompanied by a large change in volume and other thermophysical properties. PCMs may be bulk stored in tanks or be micro- or macroencapsulated. Bulk storage requires special attention to the design of heat exchangers for charging and discharging. *Microencapsulation* involves very small particles of a PCM dispersed in a single phase matrix. It is usually considered a way of introducing high TES in building materials. Some examples include PCMs encapsulated in concrete, floor tiles, and wallboard. The U.S. Department of Energy sponsored the development of a composite wallboard by mixing up to 35 percent paraffin waxes in gypsum [28]. The waxes contain n-octadecane as the main constituent with a melting point of 23°C and a heat of fusion of 184 kJ/kg.

Table 4.5. Thermal conductivity of potential containment materials

Materials	Thermal conductivity[b]	
	(W/m K)	(Btu in/hr ft² °F)
Plastics[a]		
ABS	0.17–0.33	1.2–2.3
Acrylic	0.19–0.43	1.3–3.0
Polypropylene	0.12–0.17	0.8–1.2
Polyethylene (high density)	0.43–0.52	3.0–3.6
Polyethylene (medium density)	0.30–0.42	2.1–2.9
Polyethylene (low density)	0.30	2.1
Polyvinyl chloride	0.13	0.9
Metals[c]		
Aluminum	200	1500
Copper	390	2700
Steel	48	330

[a]*Plastics, a desk-top data bank*, 5th Ed. Book A. San Diego, CA: The International Plastics Selector, Inc., 1980.

[b]As measured by ASTM C-177.

[c]*Handbook of Chemistry and Physics*, 40th Ed.

Most of the successful developments to date have been made in ***macroencapsulation*** of PCMs. Capsules may be made out of plastics for low temperature applications and metals for higher temperatures. They may be shaped as spherical balls, cylindrical rods, rectangular panels or flexible pouches. Some of the materials used for macroencapsulation are listed in Table 4.5. Figure 4.3 shows some examples of commercially developed, macroencapsulated, PCMs.

4.3.3　Heat Exchanger Design

For liquid storage media, heat exchangers of the shell and tube type or submerged coil type are generally used for charging and discharging. These types of heat exchangers are described later in this chapter.

For solid storage media or macroencapsulated PCMs, normally a packed bed type of storage configuration is designed. In this case, the heat is transferred to or from a heat transfer fluid by flowing the heat transfer fluid through the voids in the bed. Figure 4.4 shows examples of packed bed and tube bank types of storage systems. Table 4.6 gives average void fractions for packed beds [6]. Clark [6] has reviewed recent research on heat transfer and pressure drop for flow through such storage systems.

For a bed of randomly packed spheres, Clark recommends a correlation by Beasly and Clark [2] for the heat transfer coefficient h for air flow through the bed, valid for Reynolds Numbers of 10–10,000,

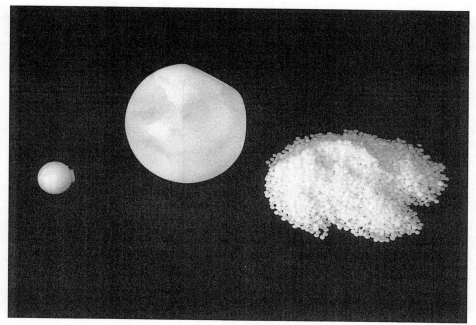

Figure 4.3. Examples of commercially developed PCM capsules.

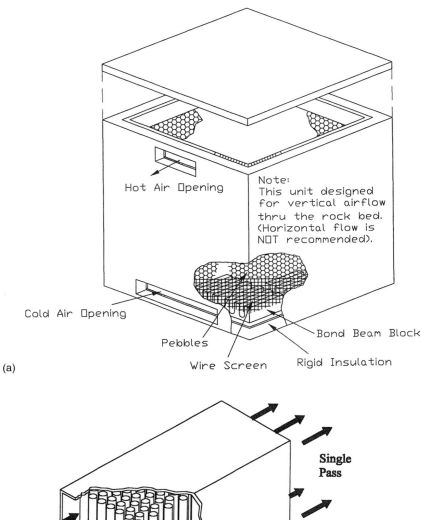

(a)

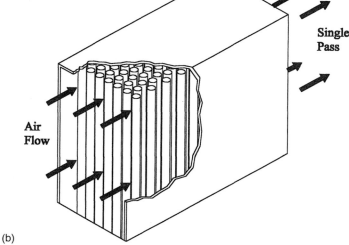

(b)

Figure 4.4. Storage systems using: (a) pebble bed storage unit (Courtesy Solaron Corporation, Englewood, CO); and (b) PCM encapsulated in tubes.

Table 4.6. Representative average void fractions, $\bar{\xi}$, for packed beds. Adapted from Clark [6].

Particle Type	Packing	Average void fraction ($\bar{\xi}$)
Sphere	Rhombohedral	0.26
Sphere	Tetragonal spheroidal	0.30
Sphere	Orthorhombic	0.40
Sphere	Cubic	0.48
Sphere	Random	0.36–0.43
Crushed rock	Granular	0.44–0.45
Sphere	Very loose, random	0.46–0.47
Sphere	Poured, random	0.37–0.39
Sphere	Close, random	0.36–0.38

$$\frac{h}{c_p G_o} = \frac{2 \cdot 0}{Re_o \, Pr} + \frac{2 \cdot 031}{Re_o^{1/2} \, Pr^{1/2}} \tag{4.11}$$

where G_o is the superficial mass velocity defined as the mass flow rate of the heat transfer fluid ($\dot{m}$) through the bed per unit face area (A_o), and Re_o and Pr are Reynolds and Prandtl numbers, respectively, as defined below:

$$G_o = \frac{\dot{m}}{A_o}, \; Re_o = \frac{D_s G_o}{\mu}, \; Pr = \frac{hc_p}{k}, \tag{4.12}$$

where D_s is the diameter of the storage particle, and μ, c_p and k are the viscosity, specific heat and thermal conductivity, respectively, of the heat transfer fluid (air).

For flow of air across tube banks of 10 rows deep and Reynolds numbers from 2,000 to 40,000, McAdams [21] recommends a mean Nusselt number, Nu, as:

$$Nu = \frac{hD_o}{k_f} = A \left(\frac{D_o G_{max}}{\mu_f} \right)^{0.6} Pr_f^{1/3}, \tag{4.13}$$

where $A = 0.33$ for Staggered tubes, and 0.26 for In-line tubes. The subscript f indicates that the thermal property is to be determined at the film temperature or the average of the tube and the air temperatures. G_{max} is defined as

$$G_{max} = \frac{\dot{m}}{A_{min}}, \tag{4.14}$$

or the mass flow rate of air divided by the minimum free flow area. To find the pressure drop in packed beds, the following equation developed by Ergun [10] may be used:

$$\Delta p = f \left(\frac{L}{D_s} \right) \frac{\rho V^2}{g_c} \left(\frac{1 - \bar{\xi}}{\bar{\xi}^3} \right), \tag{4.15}$$

where $\bar{\xi}$ is the average void fraction (Table 4.6) and the friction factor f is given by

$$f = \frac{A(1 - \bar{\xi})}{Re_o} + B. \tag{4.16}$$

McDonell et al. [22] recommended the values of the constants A and B as:

$A = 180$, $B = 1.80$ for smooth particles, and
$A = 180$, $B = 4.0$ for rough particles.

Alternatively, the pressure drop for a packed bed may be found from Figs. 4.5 and 4.6, where the cross-hatched area is the recommended design range [6]. For flow across a tube bank

$$\Delta p = 4f\left(\frac{L}{D_e}\right) \frac{G^2_{max}/\rho_m}{2g_c} \tag{4.17}$$

where ρ_m is the mean fluid density and D_e is the equivalent diameter.

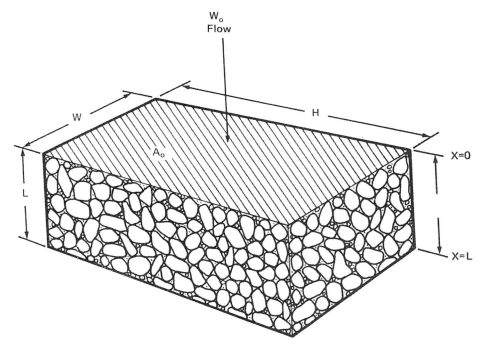

Figure 4.5. A schematic of a packed bed storage. Adapted from [6].

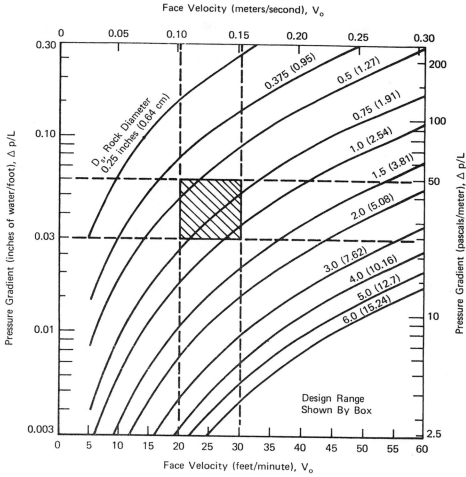

Figure 4.6. Rockbed performance Map (p = 0.15 to 0.30 in. water; length = 5 ft). Adapted from [7].

4.4 ENERGY TRANSPORT SUBSYSTEMS

To transport solar heat from a solar collector to storage, and then to an end use, an energy transport subsystem is used. It consists of pipes, pumps, expansion tanks, valves of various types, and heat exchangers. Heat exchangers are treated in the next section and pipes and pumps in this section. Valves and expansion tanks are standard items and are not described in detail.

4.4.1 Piping Systems

In most solar-thermal processes considered in this book, standard circular pipes of steel, copper, aluminum or special alloys are used to transport heat in the form of internal energy in the pumped fluid. Two parasitic losses occur in pipes–pressure drop and heat loss. Both losses are treated briefly herein.

4.4.2 Pressure Drop

Friction at the pipe walls, bends and valves, etc., results in flow resistance and pressure drop. To overcome the pressure drop, pumps are required. The sizing of pumps is a routine engineering exercise and is done by calculating two quantities, the friction factor at a given flow rate and the pipe length. The friction factor f can be read from Fig. 4.7 as a function of the Reynolds number and roughness ratio shown in Fig. 4.8. The Darcy friction factor is defined as

$$f = \Delta p / \{4(L/D)(\tfrac{1}{2}\rho \overline{V^2})\}, \tag{4.18}$$

where Δp is the pressure drop, D is the diameter, L is the pipe length including the equivalent length of smooth pipe for all fittings, expansions, and contractions, and $\overline{V}$ is the space averaged velocity.

Churchill [5] finds that the curves in Fig. (4.7) can all be represented by one equation for the laminar, transition, and turbulent regimes:

$$f = 8\{(8/Re)^{12} + (A + B)^{-3/2}\}^{1/12}, \tag{4.19}$$

where

$$A = \{-2.457 ln[(7/Re)^{0.9} + 0.27\epsilon/D]\}^{16}, \; B = (37,530/Re)^{16}. \tag{4.20}$$

and ϵ is the average height of surface irregularities.

Pump motor horsepower P_p can be calculated using Eq. (4.19) to be

$$P_p = f(L/D)(\tfrac{1}{2}\rho A \overline{V}^3)/\eta_p \eta_m, \tag{4.21}$$

where A is the flow area, η_p is the pump efficiency, and η_m is the motor efficiency, and the average velocity $\overline{V}$ is related to the volumetric flow as $\dot{Q} = A\overline{V}$.

For quick pipe sizing Kent [18] gives some guidelines which can eliminate the need to use the detailed equations above at the design development level of plant engineering. The typical diameter of liquid pipes is

$$D = 2.607(\dot{m}/1000\rho)^{0.434}, \tag{4.22}$$

where D is in inches, $\dot{m}$ is in lb/h, and ρ is in lb/ft³. If the liquid pipe is a vent or drain, D from Eq. (4.22) is increased by 35 percent. For gases

$$D = 1.065(\dot{m}/1000)^{0.408}/\rho^{0.343} \tag{4.23}$$

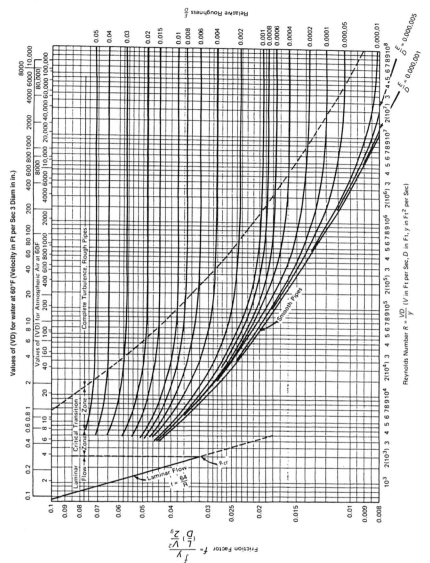

Figure 4.7. Friction factor for pipe flow as a function of Reynolds number and roughness ratio ϵ/D. From [25].

189

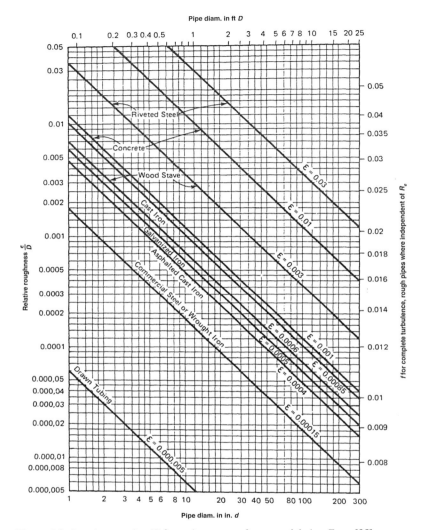

Figure 4.8. Roughness ratio ϵ/D for various types of commercial pipe. From [25].

with the same 35 percent rule applying.

If allowable pressure drop from previous experience is used as the design criterion,

$$D = 1.706\left\{\frac{[\mu^{0.16}\,(\dot{m}/1000)^{1.84}]}{(\rho(\Delta P/100)}\right\}^{0.207}, \tag{4.24}$$

where μ is viscosity in cp and $\Delta P/100$ is the pressure drop in psi/100 ft. Equation (4.24) applies for both liquids and gases.

Although the fluid circuits in most parts of solar-thermal systems are quite simple, the collector field is frequently connected in a complex series–parallel arrangement as dictated by tradeoffs between piping costs, pump size, and fluid temperature rise. The flow in series–parallel arrays can be calculated using iterative techniques such as the Hardy-Cross method. However, for preliminary design, a closed form solution has been developed for flow through arrays of solar collectors connected in parallel. For $j + 1$ identical collectors with the same flow rate connected in parallel with constant diameter D headers or manifolds, the array pressure drop (inlet header inlet port to outlet header outlet port) is given by

$$\Delta P = (K_f \dot{m}^2/6)[\, j(2j + 1)(j + 1)] + \Delta p_{coll}, \tag{4.25}$$

where Δp_{coll} is the collector pressure drop, $\dot{m}$ the mass flow rate per collector, and

$$K_f = 8\bar{f}L/(\pi^2 \rho D^5). \tag{4.26}$$

The average friction factor, $\bar{f}$, in the header is evaluated at one-half the total array flow, $(j + 1)\dot{m}/2$, and the collectors are all connected into the manifold at a uniform distance L apart. It is noted that identical flow $\dot{m}$ through each header does not occur automatically, but can be accomplished to within a few percent by requiring 90 percent or more of the total array pressure drop to occur across the collector, i.e.,

$$\Delta p_{coll} \geq 0.9\, \Delta P. \tag{4.27}$$

In the design of piping systems the absolute pressure must not be permitted to drop below the boiling point to avoid vapor lock, boiling, or pump cavitation. The critical design point is where the fluid is hottest and the pressure lowest–often at a circulating pump inlet port. Some common fluids used for high temperature solar systems are shown in Table 4.7. The fluid manufacturer should be consulted for precise viscosity values to be used for system design. Typical values are in the range 0.2 to 5.0 cp. Most are quite temperature sensitive.

Heat loss. Parasitic heat losses from storage tanks, pipes and pumps can be calculated from equations given in any heat transfer text. For cylindrical insulation of thermal conductivity k, length L, inner diameter D, and thickness t for pipes and tanks

$$q_{cyl} = \left[\frac{2\pi kL}{\ln((D + 2t)/D)} \right](T_f - T_a), \tag{4.28}$$

where T_f is the fluid temperature and T_a the ambient temperature. For spherical storage tanks the heat loss is

$$q_{sph} = [\pi kD(D + 2t)/t](T_f - T_a). \tag{4.29}$$

In the above it is assumed that the principal resistance to heat loss occurs in the insulation.

Table 4.7. Properties of common heat transfer liquids[a]

Material	Density (kg/m²)	Specific heat (kJ/°C kg)	Boiling point[f] (°C)
Water	~1000	~4.2	100
Therminol 66™ [b]	820 (250°C)	2.4 (250°C)	~330
Dowtherm A™ [c]	859 (250°C)	2.2 (250°C)	257
Hi Tec™ [d]	1890 (250°C)	1.5 (250°C)	—[d]
Caloria HT-43™ [e]	694 (250°C)	3.0 (250°C)	—
Diethylene glycol	1020 (150°C)	2.8 (150°C)	240
Tetraethylene glycol	1110 (66°C)	2.5 (100°C)	280

[a]Data collected from various manufacturers brochures.
[b]Trademark of Monsanto Chemical Co.
[c]Trademark of Dow Chemical Co; Dowtherm is a mixture of diphenyl and diphenyl oxide.
[d]Trademark of E.I. duPont; solidifies below approximately 190°C; does not boil or decompose below 500°C; consists of 40% $NaNO_2$, 7% $NaNO_3$, 53% KNO_3.
[e]Trademark of Exxon Corp.
[f]At 1 atm.

Although heat losses can be made arbitrarily small, the law of diminishing returns applies and a point is reached where added insulation costs more than the value of the extra heat retained. It can be shown that the optimum distribution of a given volume of insulation over all the components of a solar-thermal system occurs when the surface heat flux is the same everywhere; note that this does not imply equal insulation thickness t.

For low temperature applications such as heating, cooling, and domestic hot water, foam rubber or polyurethane foam insulation is normally used. For higher temperature applications the principal insulations include calcium silicate, mineral fiber, expanded silica (perlite), and fiberglass. Table 4.8 lists the properties of industrial grade insulations useful in elevated temperature solar systems. Calcium silicate is a mixture of lime and silica reinforced with fibers and molded into shape. It has good compressive strength. Mineral fiber consists of rock and slag fibers bonded together and is useful up to 1200°F (650°C). The compressive strength is less than calcium silicate, but it is available in both rigid and flexible, shaped segments.

Perlite is expanded volcanic rock consisting of small air cells enclosed by a mineral structure. Additional binders are added to decrease moisture migration and to reduce shrinkage. Cellular glass is available in flexible bats or rigid boards and shaped sections. It has very low moisture absorption.

For very high temperatures, the selection of materials is quite limited. In addition to mineral fiber and calcium silicate noted above, ceramic fibers (to 1400°C), castable ceramic insulation (to 1600°C), Al_2O_3 or ZrO_2 fibers (to 1600°C), and carbon fibers (to 2000°C) are available. It is to be noted that these temperatures correspond to the refractory range of metals where service conditions are very severe.

An excellent summary of industrial insulation practice and design methods and economics analysis is contained in [13].

Table 4.8. Properties of pipe insulation for elevated temperatures[a,b]

Insulation Type	Temperature range (°F)	Conductivity k [Btu/(h)(ft²)(°F/in.)]	Density (lb/ft³)	Applications
Urethane foam	−300–300	0.11–0.14	1.6–3.0	Hot and cold piping
Cellular glass blocks	−350–500	0.20–0.75	7.0–9.5	Tanks and piping
Fiberglass blanket for wrapping	−120–550	0.15–0.54	0.60–3.0	Piping and pipe fittings
Fiberglass preformed shapes	−60–450	0.22–0.38	0.60–3.0	Hot and cold piping
Fiberglass mats	150–700	0.21–0.38	0.60–3.0	Piping and pipe fittings
Elastomeric preformed shapes and tape	−40–220	0.25–0.27	4.5–6.0	Piping and pipe fittings
Fiberglass with vapor barrier jacket	20–150	0.20–0.31	0.65–2.0	Refrigerant lines, dual-temperature lines, chilled-water lines, fuel-oil piping
Fiberglass without vapor barrier jacket	to 500	0.20–0.31	1.5–3.0	Hot piping
Cellular glass blocks and boards	70–900	0.20–0.75	7.0–9.5	Hot piping
Urethane foam blocks and boards	200–300	0.11–0.14	1.5–4.0	Hot piping
Mineral—fiber preformed shapes	to 1200	0.24–0.63	8.0–10.0	Hot piping
Mineral—fiber blankets	to 1400	0.26–5.60	8.0	Hot piping
Fiberglass field applied jacket for exposed lines	500–800	0.21–0.55	2.4–6.0	Hot piping
Mineral—wool blocks	850–1800	0.36–0.90	11.0–18.0	Hot piping
Calcium silicate blocks	1200–1800	0.33–0.72	10.0–14.0	Hot piping

[a]From [13].
[b]1 Btu/(h)(ft²)(°F/in.) = 1/12 Btu/h ft °F = 0.144 W/m °C.

4.4.2 Heat Exchangers

Heat exchangers are devices in which two fluid streams exchange thermal energy: one stream is heated while the other one is cooled. There are a number of arrangements used to transfer heat from one fluid to another. The simplest arrangement is the double-pipe heat exchanger shown in Figure 4.9. It will be discussed first.

In this system fluid A flows inside a tube of inner radius r_i and other radius r_o. Fluid B flows in the annulus formed between the outer surface of the inner tube, A_o, and the inner surface of the outer tube. An overall heat-transfer coefficient U can be calculated from the thermal circuit shown in Figure 4.9. It may be based on any convenient area of the exchanger, but usually the outside area $A_o = 2\pi r_o L$ of the inner tube is most convenient [16]. Then,

$$U_o A_o = \frac{1}{1/h_i A_i + (\ln r_o/r_i)/2\pi kL + 1/h_o A_o}$$ (4.30)

and at any cross section the local rate of heat transfer across the tube is

$$dq = U_o \, dA_o(T_A - T_B) = U_o 2\pi r_o \, dx(T_A - T_B)$$ (4.31)

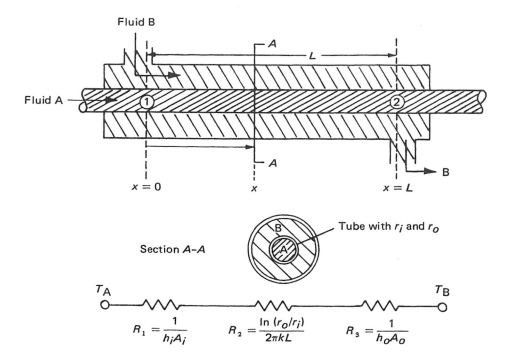

Figure 4.9. Schematic diagram of double-pipe heat exchanger.

Simplifications in the evaluation of U_o are possible when one or two of the thermal resistances dominate. For example, the thermal resistance of the tube wall, R_2 in Figure 4.9, is often small compared to the convective resistances and sometimes one of the two convective resistances is negligible compared to the other.

The flow arrangement shown in Figure 4.9, where both fluids enter from the same end, is called parallel flow, and Figure 4.10a shows the temperature distribution for both fluid streams. If one of the two fluids were to enter at the other end and flow in the opposite direction, the flow arrangement would be *counterflow*. The temperature distribution for the latter case is shown in Figure 4.10b. If a counterflow arrangement is made very long, it approaches the thermodynamically most efficient possible heat-transfer condition.

There are two basic methods for calculating the rate of heat transfer in a heat exchanger. One method employs a mean temperature difference between the two fluids in the exchanger ΔT_{mean}, and then determines the rate of heat transfer from the relation (16)

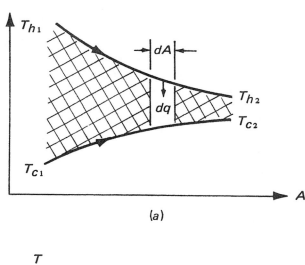

(a)

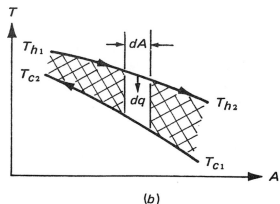

Figure 4.10. (a) Parallel flow; (b) counterflow.

(b)

$$q = UA \, \Delta T_{mean} \tag{4.32}$$

This mean temperature difference is called the logarithmic mean temperature difference (LMTD). It can be evaluated directly only when the inlet and outlet temperatures of both fluid streams are specified. The use of the LMTD approach is explained in [16]. The other method is called the effectiveness-NTU method (ϵ-NTU). The ϵ-NTU method offers many advantages over the LMTD approach and will be discussed below.

First, we define the exchanger effectiveness ϵ

$$\epsilon = \frac{\text{actual rate of heat transfer}}{\text{maximum possible rate of heat transfer}} \tag{4.33}$$

The actual rate of heat transfer can be determined by calculating either the rate of internal energy loss of the hot fluid or the rate of internal energy gain of the cold fluid, that is,

$$q = \dot{m}_h c_{p,h}(T_{h,\text{in}} - T_{h,\text{out}}) = \dot{m}_c c_{p,c}(T_{c,\text{out}} - T_{c,\text{in}}) \tag{4.34}$$

where the subscripts h and c denote the hot and cold fluid, respectively. The maximum rate of heat transfer possible for specified inlet fluid temperatures is attained when one of the two fluids undergoes the maximum temperature difference in the exchanger. This maximum equals the difference in the entering temperatures for the hot and cold fluid. Which of the two fluids can undergo this maximum temperature change depends on the relative value of the product ($\dot{m}c_p$), the mass flow rate times the specific heat of the fluid at constant pressure, called the heat capacity rate $\dot{C}$. Since a thermodynamic energy balance requires that the energy given up by the one fluid must be received by the other if there are no external heat losses, only the fluid with the smaller value of $\dot{C}$ can undergo the maximum temperature change. Thus, the fluid that may undergo this maximum temperature change is the one that has the minimum value of the heat capacity rate and the maximum rate of heat transfer is

$$q_{\text{max}} = C_{\text{min}}(T_{h,\text{in}} - T_{C,\text{in}}) \tag{4.35}$$

where $\dot{C}_{\text{min}} = (\dot{m}c_p)_{\text{min}}$

The fluid with the minimum $\dot{m}c_p$ or $\dot{C}$ value can be either the hot or the cold fluid. Using subscript h to designate the effectiveness when the hot fluid has the minimum $\dot{C}$ and the subscript c when the cold fluid has the minimum $\dot{C}$ value, we get for a parallel-flow arrangement

$$\epsilon_h = \frac{\dot{C}_h(T_{h1} - T_{h2})}{\dot{C}_h(T_{h1} - T_{c1})} = \frac{T_{h1} - T_{h2}}{T_{h1} - T_{c1}} \tag{4.36}$$

$$\epsilon_c = \frac{\dot{C}_c(T_{c2} - T_{c1})}{\dot{C}_c(T_{h1} - T_{c1})} = \frac{T_{c2} - T_{c1}}{T_{h1} - T_{c2}} \tag{4.37}$$

where the subscripts 1 and 2 refer to the left- and right-hand side of the heat exchanger as shown in Figure 4.9. Similarly, for a counterflow arrangement

$$\epsilon_h = \frac{\dot{C}_h(T_{h1} - T_{h2})}{\dot{C}_h(T_{h1} - T_{c2})} = \frac{T_{h1} - T_{h2}}{T_{h1} - T_{c2}} \tag{4.38}$$

$$\epsilon_c = \frac{\dot{C}_c(T_{c1} - T_{c2})}{\dot{C}_c(T_{h1} - T_{c2})} = \frac{T_{c1} - T_{c2}}{T_{h1} - T_{c2}} \tag{4.39}$$

Kays and London [17] have calculated effectiveness values for many types of heat exchangers. They found that ϵ_{hx} can be expressed for a given exchanger type as a function of two variables, the number of transfer units NTU,

$$\text{NTU} \equiv UA_{hx}/(\dot{m}c_p)_{min}, \tag{4.40}$$

and the capacitance ratio C,

$$C \equiv (\dot{m}c_p)_{min}/(\dot{m}c_p)_{max} = C_{min}/C_{max}. \tag{4.41}$$

For boiling or condensation $C = 0$.

Figures 4.9a–4.9d show effectiveness values for the common heat exchanger types and Table 4.9 contains equations for calculating $\epsilon_{hx}(\text{NTU},C)$. For parallel flow and counterflow devices in which boiling or condensation occurs in one stream

Table 4.9. Heat-exchanger effectiveness expressions[a]

Flow geometry	Relation
Double pipe Parallel flow	$\epsilon = \dfrac{1 - \exp[-N(1 + C)]}{1 + C}$
Counterflow	$\epsilon = \dfrac{1 - \exp[-N(1 - C)]}{1 - C\exp[-N(1 - C)]}$
Cross flow Both fluids unmixed	$\epsilon = 1 - \exp\{(C/n)[\exp(-NCn) - 1]\}$ where $n = N^{-0.22}$
Both fluids mixed	$\epsilon = \left[\dfrac{1}{1 - \exp(-N)} + \dfrac{C}{1 - \exp(-NC)} - \dfrac{1}{N}\right]^{-1}$
C_{max} mixed, C_{min} unmixed	$\epsilon = (1/C)\{1 - \exp[C(e^{-N} - 1)]\}$
C_{max} unmixed, C_{min} mixed	$\epsilon = 1 - \exp\{(1/C)[\exp(-NC) - 1]\}$
Shell and tube One shell pass, 2,4,6 tube passes	$\epsilon = 2\left\{1 + C + (1 + C^2)^{1/2}\dfrac{1 + \exp[1 + C^2)^{1/2}]}{1 - \exp[-N(1 + C^2)^{1/2}]}\right\}^{-1}$

[a] $N = \text{NTU} \equiv UA/C_{min}$, $C = C_{min}/C_{max}$. From [16].

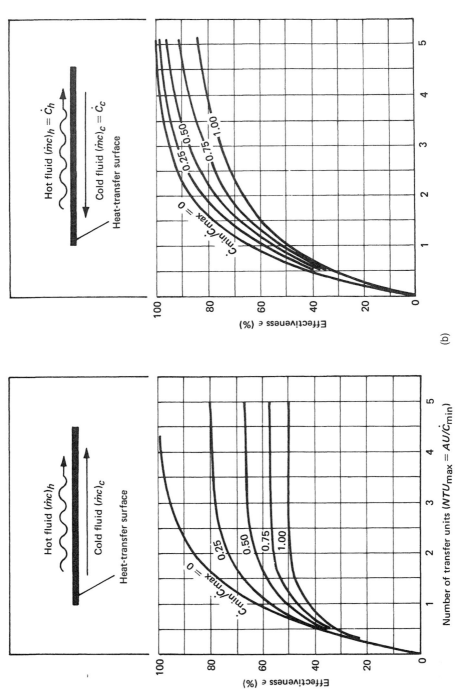

Figure 4.11. Heat exchanger effectiveness as a function of NTU and $\dot{C}_r$; (a) parallel flow, (b) counterflow.

(continues)

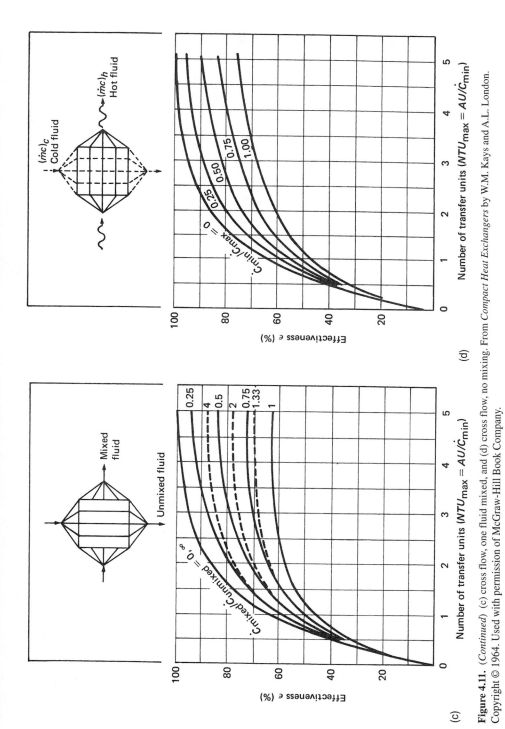

Figure 4.11. (*Continued*) (c) cross flow, one fluid mixed, and (d) cross flow, no mixing. From *Compact Heat Exchangers* by W.M. Kays and A.L. London. Copyright © 1964. Used with permission of McGraw-Hill Book Company.

$\epsilon_{hx} = 1 - e^{-\text{NTU}}$. For the special case of $(\dot{m}c_p)_{\min} = (\dot{m}c_p)_{\max}$ a special effectiveness equation applies for counterflow heat exchangers

$$\epsilon_{hx} = \text{NTU}/(\text{NTU} + 1) \tag{4.42}$$

Figure 4.12 shows that the counterflow exchanger has the highest heat transfer effectiveness and is therefore the preferred design for solar systems.

One of the most common uses of a heat exchanger is to isolate a fluid used in solar collector field from the process fluid which conveys the heat energy to its end use as shown in Fig. 4.13. If the required process inlet temperature is $T_{p,i}$, the collector outlet fluid must be at a temperature $T_{c,o}$ above $T_{p,i}$ because of the presence of the heat exchanger. Since collector efficiency decreases with increasing operating temperature, less energy is collected than if no heat exchanger were present. A heat exchanger penalty factor F_{hx}, which can be applied to the linear collector model [Eq. (3.46)] developed earlier, has been calculated by deWinter [9]. Equation (3.46) can be rewritten in terms of the heat exchanger inlet temperature (process return temperature) $T_{p,o}$ as

$$\eta_c = F_{hx}F_R[\eta_o - U_c(T_{p,o} - T_a)/I_c]. \tag{4.43}$$

The heat exchanger factor is given by

$$F_{hx} = \{1 + [F_RU_cA_c/(\dot{m}c_p)_c][(\dot{m}c_p)_c/\epsilon_{hx}(\dot{m}c_p)_{\min} - 1]\}^{-1}, \tag{4.44}$$

where the subscript c denotes collector properties, and ϵ_{hx} is the heat exchanger effectiveness. The collector loss coefficient U_c is evaluated at the mean collector operating temperature $\overline{T}_c$.

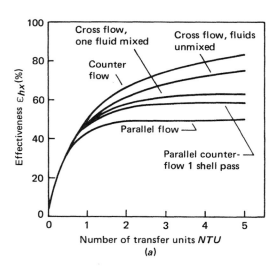

Figure 4.12. Heat-exchanger effectiveness as a function of NTU; effect of flow type for $(\dot{m}c_p)_{\min} = (\dot{m}c_p)_{\max}$. From *Compact Heat Exchangers* by W.M. Kays and A.L. London, Copyright © 1964 by McGraw-Hill Book Co. Used with permission of McGraw-Hill Book Co.

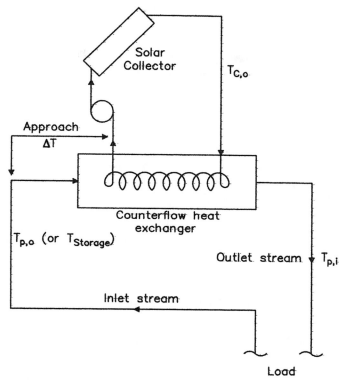

Figure 4.13. Use of a heat exchanger to isolate a collector from the thermal process.

Figure 4.14 shows F_{hx} for a range of operating conditions. Unless the collector flow rate is very low, $F_{hx} > 0.95$ for most high-performance collectors. For small $[F_R U_c A_c/(\dot{m}c_p)_c]$, F_{hx} is approximated closely by

$$F_{hx} = 1 - F_R U_c A_c \{[(\dot{m}c_p)_c - \epsilon_{hx}(\dot{m}c_p)_{\min}]/[\epsilon_{hx}(\dot{m}c_p)_{\min}(\dot{m}c_p)_c]\}. \qquad (4.45)$$

Example 4.2. Calculate the efficiency at which the solar collector in Fig. 4.15 operates in order to deliver energy to a working fluid at 65°C for several values of the approach temperature difference. The approach temperature is the difference between the incoming cool fluid and the exiting warm fluid at the heat exchanger (see Fig. 4.15). The temperature rise through the heat exchanger is 10°C and is equal to the fluid temperature rise through the solar collector.

The collector has a $\tau\alpha$ product of 0.80 and loss coefficient U_c of 5 W/m² · K. If the solar radiation normal to the collector surface is 500 W/m² and the ambient temperature T_a is 20°C, calculate the collector efficiency for the following five values of approach ΔT:

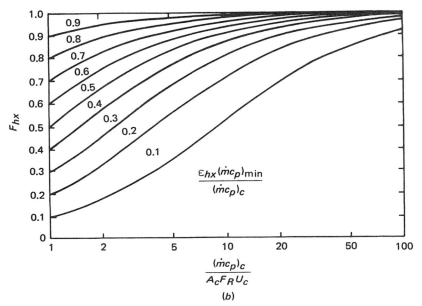

Figure 4.14. Heat exchanger factor F_{hx} calculated from Eq. (4.45).

Solution. Fluid stream temperatures are shown at several points for each case in Fig. 4.15. The collector temperature T_c can be taken as the average of inlet and outlet fluid temperatures for purposes of the example as shown in the figure.

Collector efficiency can be calculated from the following equation (see Chapter 3):

$$\eta = \tau\alpha - U_c\left(\frac{\Delta T}{I_c}\right), \qquad (4.46)$$

where $\Delta T = \overline{T}_c - T_a$.

For the collector specified here

$$\eta = \left(0.8 - 5\frac{\Delta T}{I_c}\right) \times 100\%. \qquad (4.47)$$

The collector efficiency can be calculated in tabular form as shown in Table 4.10.

Note that the heat-exchanger design has a very strong effect on collector efficiency; for relatively small changes of approach temperature difference, efficiency, and therefore energy delivery, change significantly.

Example 4.3. Calculate the energy-delivery penalty in a solar system caused by a required heat exchanger for values of effectiveness from 0.2 to 0.99. Compare the economic value of this energy penalty to the heat-exchanger cost if the annual amortized

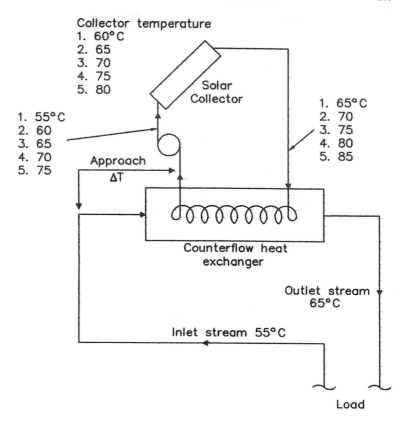

Collector temperature
1. 60°C
2. 65
3. 70
4. 75
5. 80

Solar Collector

1. 65°C
2. 70
3. 75
4. 80
5. 85

1. 55°C
2. 60
3. 65
4. 70
5. 75

Approach
ΔT

Counterflow heat exchanger

Outlet stream 65°C

Inlet stream 55°C

Load

Case	Approach ΔT (°C)
1	0 (thermodynamic limit)
2	5
3	10
4	15
5	20

Figure 4.15. Fluid stream and solar collector temperatures for Example 4.2 (heat-exchanger problem).

Table 4.10. Collector efficiency calculation for Example 4.2

Case	$\bar{T}_c$ (°C)	ΔT (°C)	I_c W/m²	$\Delta T/I_c$ (K · m²/W)	η (%)
1	60	40	500	0.08	40
2	65	45	500	0.09	35
3	70	50	500	0.10	30
4	75	55	500	0.11	25
5	80	60	500	0.12	20

cost (see the sections on economics in Chapter 11 of a heat exchanger per year is $6/m$^2_{hx}$. The economic value of the solar energy collection penalty is the value of non-solar energy required to make up for the penalty resulting from the heat exchanger. The value of nonsolar, makeup energy is $10/GJ for this example. System specifications are listed below. Water is the working fluid.

$$A_c = 100 \text{ m}^2,$$

$$(\dot{m}c_p)_{min} = (\dot{m}c_p)_{max} = 50 \text{ kg/hr·m}^2_c \times 4182 \text{ J/kg·K}.$$

Energy delivery with no heat exchanger $= 200 \text{ MJ/m}^2_c \cdot \text{yr}$

$$F_R U_c = 5\text{W/m}^2_c\cdot°\text{C}(= 5 \times 3600\text{J/hr·m}^2_c\cdot°\text{C}),$$

$$U_{hx} = 1400\text{W/m}^2_c\cdot°\text{C}.$$

Solution. The heat-exchanger penalty factor F_{hx}, with the above values substituted in Eq. (4.44) is

$$F_{hx} = \frac{1}{1 + 0.0861(1/\epsilon_{hx} - 1)}.$$

Table 4.12 summarizes the thermal penalty from heat-exchanger use. Also shown is the heat-exchanger area required to provide a given effectiveness. The heat-exchanger area A_{hx} is evaluated by the use of Eq. (4.41) and values of NTU required to provide a given effectiveness for counterflow heat exchangers from Table 4.11. The annual energy delivery Q_u in Table 4.12 is given by

$$Q_u = F_{hx}A_c(200)\text{MJ/m}^2_c\cdot\text{yr}.$$

The solar energy penalty, which must be made up by other fuels, is

$$Q_a = (1 - F_{hx})A_c(200)\text{MJ/m}^2_c\cdot\text{yr}.$$

The economic value of the heat-exchanger energy penalty is

Table 4.11. Counterflow heat-exchanger performance[a]

	ϵ_{hx} for indicated capacity rate ratios $(mc_p)_{min}/(mc_p)_{max}$							
NTU	0	0.25	0.50	0.70	0.75	0.80	0.90	1.00
0	0	0	0	0	0	0	0	0
0.25	0.221	0.216	0.210	0.206	0.205	0.204	0.202	0.200
0.50	0.393	0.378	0.362	0.350	0.348	0.345	0.339	0.333
0.75	0.528	0.502	0.477	0.457	0.452	0.447	0.438	0.429
1.00	0.632	0.598	0.565	0.538	0.532	0.525	0.513	0.500
1.25	0.713	0.675	0.635	0.603	0.595	0.587	0.571	0.556
1.50	0.777	0.735	0.691	0.655	0.645	0.636	0.618	0.600
1.75	0.826	0.784	0.737	0.697	0.687	0.677	0.657	0.636
2.00	0.865	0.823	0.775	0.733	0.722	0.711	0.689	0.667
2.50	0.918	0.880	0.833	0.788	0.777	0.764	0.740	0.714
3.00	0.950	0.919	0.875	0.829	0.817	0.804	0.778	0.750
3.50	0.970	0.945	0.905	0.861	0.848	0.835	0.807	0.778
4.00	0.982	0.962	0.928	0.886	0.873	0.860	0.831	0.800
4.50	0.989	0.974	0.944	0.905	0.893	0.880	0.850	0.818
5.00	0.993	0.982	0.957	0.921	0.909	0.896	0.866	0.833
5.50	0.996	0.998	0.968	0.933	0.922	0.909	0.880	0.846
6.00			0.975	0.944		0.921	0.892	0.857
6.50			0.980	0.953		0.930	0.902	0.867
7.00			0.985	0.960		0.939	0.910	0.875
7.50			0.988	0.966		0.946	0.918	0.882
8.00			0.991	0.971		0.952	0.925	0.889
8.50			0.993	0.975		0.957	0.931	0.895
9.00			0.994	0.979		0.962	0.936	0.900
9.50			0.996	0.982		0.966	0.941	0.905
10.00			0.997	0.985		0.970	0.945	0.909
∞	1.000	1.000	1.000	1.000	1.000	1.000	1.000	1.000

[a]From *Compact Heat Exchangers* by W.M. Kays and A.L. London. Copyright © 1964 by McGraw-Hill Book Co. Used with permission of McGraw-Hill Book Co.

$$C_a = Q_a c_a,$$

where c_a is the cost of auxiliary energy in dollars per gigajoule. The cost of heat-exchanger area C_{hx} is

$$C_{hx} = A_{hx} c_{hx},$$

where c_{hx} is the amortized cost per year of heat exchanger per unit area in dollars per square meter per year. See Table 4.13.

The results of this example show that the extra exchanger area to provide extra effectiveness ϵ_{hx} rises rapidly with ϵ_{hx}. Since the heat-exchanger cost to provide the last

Table 4.12. Thermal performance summary for Example 4.3

Effectiveness ϵ_{hx}	Exchanger penalty F_{hx}	Energy collection Q_u (GJ/yr)	NTU[a]	Exchanger area A_{hx} (m_{hx}^2)
0.2	0.744	14.88	0.25	1.04
0.333	0.853	17.06	0.50	2.07
0.5	0.921	18.42	1.00	4.15
0.6	0.946	18.92	1.50	6.22
0.7	0.964	19.28	2.33	9.75
0.8	0.979	19.58	4.00	16.60
0.9	0.991	19.82	9.00	37.34
0.99	0.999	19.98	99.00	411.00
1.0[b]	1.000[b]	20.00[b]	—[c]	0.00[b]

[a]Table 4.11 or Eq. (4.43).
[b]Baseline.
[c]No exchanger baseline.

few percent of solar energy delivery is high, the value of this energy is relatively small when compared with the additional exchanger cost. It is, therefore, not cost-effective to provide this energy. Likewise, small heat exchangers have a larger energy penalty, which has a correspondingly larger dollar value in replacement fuel. It is therefore cost-effective to add heat-exchanger area in order to recover a portion but not all of the solar energy required.

Figure 4.16 is a plot of the total extra cost required to provide a total demand of 20 GJ/yr. This extra cost is required by the design stipulation that a heat-exchanger be used. It is a cost above the basic solar and backup system costs. The least cost heat-exchanger configuration is for an effectiveness value of about 0.46. It is to be noted that this example considers only one component of the solar system. A complete system optimization requires making simultaneous trade-offs of many system compo-

Table 4.13. Economic summary for Example 4.3.

Effectiveness ϵ_{hx}	Extra nonsolar Q_a (GJ/yr)	Economic value of energy penalty C_a ($/yr)	Cost of exchanger area C_{hx} ($/yr)	Total cost[a] ($/yr)
0.2	5.12	51.20	6.24	57.44
0.333	2.94	29.40	12.42	41.82
0.5	1.58	15.80	24.90	40.70
0.6	1.08	10.80	37.32	48.12
0.7	0.72	7.20	58.50	65.70
0.8	0.42	4.20	99.60	103.80
0.9	0.18	1.80	224.04	225.85
0.99	0.02	0.20	2466.00	2466.20

[a]$C_a + C_{hx}$.

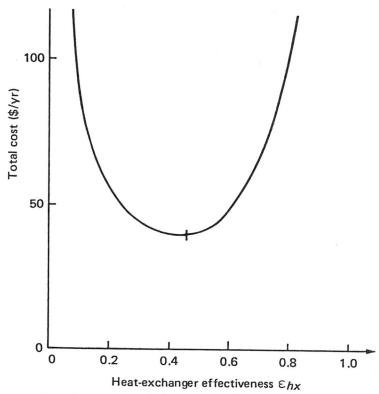

Figure 4.16. Total cost of heat exchanger and nonsolar energy penalty for Example 4.3. Least-cost configuration has a heat-exchanger effectiveness of 0.46.

nents. This methodology is described later in this book. Note that the direct application of the F_{hx} factor to ***annual*** energy delivery instead of ***instantaneous*** delivery is correct only for the case of negligible loss from storage and no heat rejection by storage boiling or other means.

PROBLEMS

4.1. Compare the energy storage capability of sodium sulfate decahydrate (Glauber's salt) in a range from 30 to 60°C with that of water and rock in the same range. Also compare the volumes of storage for the three media.

4.2. A 100-m² liquid solar collector (flat-plate type) is located 40 m from the building it serves. If the liquid pressure drop through the collector is 7 kPa, what is the total pressure drop in the collector fluid loop? What pipe size would you recommend for the collector loop? Why?

4.3. A cylindrical storage tank for the collector in Problem 4.2 is located in the basement of the dwelling it serves. If the basement temperature is 16°C and the storage tank liquid is 70°C, what is the heat-loss ratio from the tank? The tank is a cylinder with diameter equal to its height; it is insulated with 15-cm-thick fiberglass batt.

4.4. How large should a water storage tank be if it is to supply the total daily heat load (1.5GJ) of a building if storage tank temperature may vary by 20°C? Neglect parasitic transmission and conversion losses. If a 4-m³ carbon steel tank costs $1,000.00, what will this tank cost?

4.5. What is the pressure drop in 50 m of 1-in pipe for fluid flowing at 0.4 m³/min?

4.6. If the fluid velocity in a pipe is to be kept below 4 ft/sec, what pipe size should be used to flow 500 gal/min? What is the pressure drop in 500 ft of this pipe?

4.7. An array of twenty 40-ft² solar panels is connected in parallel. What is the pressure drop through the manifold if water is used at the rate 0.03 gal/min · ft_c^2 and the collectors are spaced 5 ft apart. Pipe with 1-in diameter is used.

4.8. A heat exchanger is used to isolate a collector loop from the storage loop in a solar-cooling system. Collector and storage fluid capacitance rates are 300 kJ/m_c^2 · hr · °C and the collector heat loss conductance is 5 W/m_c^2 K with a heat removal factor $F_R = 0.9$. What is the energy-delivery penalty from heat-exchanger use for heat-exchanger effectiveness levels of 0.25, 0.5, 0.9?

4.9. Water flowing at a rate of 70 kg/min is to be heated from 340 to 350 K in a counterflow double pipe heat exchanger by a 50 percent glycol solution that enters the heat exchanger at 350 K and exits at 345 K. If the overall heat-transfer coefficient is 300 W/m²·K, calculate the required heat exchanger area.

4.10. Repeat Problem 4.9, but assume that instead of a counterflow double pipe heat exchanger a shell-and-tube heat exchanger will be used with the water making one shell pass and the glycol solution making two passes.

4.11. A shell-and-tube heat exchanger with one shell pass and two tube passes, having an area of 4.5 m², is to be used to heat high-pressure water initially at 290 K with hot air initially at 400 K. If the exit water temperature is not to exceed 350 K, the air flow rate is 0.5 kg/sec, and the overall heat-transfer coefficient is 300 W/m²·K, calculate the water flow rate.

4.12. A cross-flow fin-and-tube heat exchanger uses hot water to heat air from 290 to 300 K. Water enters the heat exchanger at 340 K and exits at 310 K. If the total heat-transfer rate is to be 300 kW and the average heat-transfer coefficient between the water and the air is 50 W/m²·K, calculate the area of the heat exchanger required and the mass rate of air flow through the exchanger.

4.13. A small steam condenser is to be designed to condense 0.8 kg/m steam at 80 kN/m² with cooling water at 290 K. If the exit temperature of the water is not to exceed 320K, calculate the area required for a shell-and-tube heat exchanger with the steam making one shell pass and the water making two tube passes. The overall heat-transfer coefficient is 3000 W/m²·K.

REFERENCES AND SUGGESTED READINGS

1. ASHRAE. 1995. Thermal storage. In *ASHRAE Handbook, HVAC Application,* 40.14. Atlanta, Georgia: American Society of Heating, Refrigerating and Air-Conditioning Engineers.

2. Beasley, D.E., and J.A. Clark. 1984. Transient response of a packed bed for thermal energy storage. *Int. J. of Heat & Mass Transfer* 27 (9): 1659–1669.

3. Beckman, G., and P.V. Gilli. 1984. *Topics in energy–Thermal energy storage.* New York: Springer-Verlag.

4. Chavez, J.M., et al. 1995. The solar two power tower project: A 10-MW power plant. In *Proc. of the 1995 IECEC,* vol. 2, 469–475. New York: ASME.

5. Churchill, S.W. 1977. Friction-factor equation spans all fluid-flow regimens. *Chem. Eng.* 84(24), 91.

6. Clark, J.A. 1986. Thermal design principles. In *Solar Heat Storage: Latent Heat Materials,* G.A. Lane, ed. 185–223. CRC Press.

7. Cole, R.L., et al. 1980. *Design and installation manual for thermal energy storage,* ANL-79-15, 2nd ed. Argonne, IL: Argonne National Lab.

8. Delameter, W.R., and N.E. Bergen. 1986. Review of molten salt electric experiment: Solar central receiver project, SAND 86-8249. Albuquerque: Sandia National Laboratory.

9. deWinter F. 1975. Heat exchanger penalties in double-loop solar water heating systems. *Sol. Energy* 17: 335.

10. Ergun, S. 1952. Fluid flow through packed columns. *Chem. Eng. Prog.* 48, 89.

11. Garg, H.P., S.C. Mullick, and A.K. Bhargava. 1985. *Solar thermal energy storage.* Boston: D. Reidel.

12. Glendenning, I. 1981. Advanced mechanical energy storage. In *Energy storage and transportation,* G. Beghe, ed., 50–52. Boston: D. Reidel.

13. Harrison, M., and C. Pelanne. 1977. "Cost-effective thermal insulation." *Chem. Eng.* 84: 62.

14. Jensen, J. 1980. *Energy storage.* Boston: Newnes-Butterworth.

15. Jotshi, C.K., and D.Y. Goswami. 1998. Energy storage. In *Mechanical engineering handbook,* F. Kreith, ed. 8–104. Boca Raton FL: CRC Press.

16. Karlekar, B.V., and R.M. Desmond. 1977. *Engineering heat transfer.* St. Paul, Minnesota: West Publ.

17. Kays, W.M., and A.L. London. 1980. *Compact heat exchangers.* New York: McGraw-Hill.

18. Kent, G.R. 1978. "Preliminary pipeline sizing." *Chem. Eng.* 85, 199.

19. Kreider, J.F. 1979. *Medium and high temperature processes.* New York: Academic Press.

20. Makansi, J. 1994. Energy storage reinforces competitive business practices. *Power* 138(9):63.

21. McAdams, W.H. 1954. *Heat transmission,* 3rd ed. New York: McGraw-Hill.

22. McDonnell, I.F., M.S. El-Sayad, K. Mow, and F.A.L. Dullien. 1979. Flow through porous media–the Ergun equation revisited. *Ind. Eng. Chem. Fundam.* 18: 199.

23. Moses, P.J., and G.A. Lane. 1986. Encapsulation of PCMs. In *Solar heat storage: Latent heat materials,* vol. II, 93–152, Boca Raton, FL: CRC Press.

24. Perry, R.H., et al. 1984. *Chemical engineer's handbook.* New York: McGraw-Hill.

25. Potter, P.J. 1959. *Power plant theory and design.* New York: Ronald Press.

26. Rabl, A. 1977. "A note on the optics of glass tubes." *Sol. Energy* 19: 215.

27. Sharma, S.K., and C.K. Jotshi. 1979. Dicussion on storage subsystems. In *Proc. of the First National Workshop on Solar Energy Storage,* 301–308, Chandigarh, India: Panjab University, March 16–18.

28. Swet, C.J. 1987. "Storage of thermal energy–status and prospects." In *Progress in solar engineering,* D.Y. Goswami, ed. Washington, D.C.: Hemisphere Publ.

29. Tomlinson, J.J., and L.F. Kannberg. 1990. Thermal energy storage. *Mech. Eng.* 9, 68–72.

SOLAR-HEATING SYSTEMS

For the well being and health . . . the homesteads should be airy in summer and sunny in winter. A homestead promising these qualities would be longer than it is deep and the main front would face south.

Aristotle

The use of solar energy for heat production dates from antiquity. Historically, methods used for collecting and transferring solar heat were passive methods, that is, without active means such as pumps, fans and heat exchangers. *Passive* solar heating methods utilize natural means such as radiation, natural convection, thermosyphon flow and thermal properties of materials for collection and transfer of heat. *Active* solar heating methods, on the other hand, use pumps and fans to enhance the rate of fluid flow and heat transfer. Active methods for water heating, space heating and industrial process heat have been developed mainly in the last four decades. This chapter describes in detail the function and design of active systems for heating buildings and service water. Passive solar heating, cooling and daylighting is covered in Chapter 7. Other applications, such as low-temperature solar heat for agriculture, agricultural drying, or aquaculture, can be analyzed with the principles set forth in this chapter and Chapters 3 and 4.

Energy for heating buildings and hot water consumes about one-fourth of the annual energy production in the United States. In many areas of the United States and the world, solar heating can compete economically with other types of fuel for heating, without even considering the environmental benefits.

5.1 CALCULATIONS OF HEATING AND HOT WATER LOADS IN BUILDINGS

Energy requirements for space heating or service water heating can be calculated from basic conservation of energy principles. For example, the heat required to maintain the interior of a building at a specific temperature is the total of all heat transmission losses from the structure and heat required to warm and humidify the air exchange with the environment by infiltration and ventilation.

Comfort in buildings has long been a subject of investigation by the American Society of Heating, Refrigerating and Air-Conditioning Engineers (ASHRAE). ASHRAE has developed extensive heat load calculation procedures embodied in the *ASHRAE Handbook of Fundamentals* [2]. The most frequently used load calculation procedures

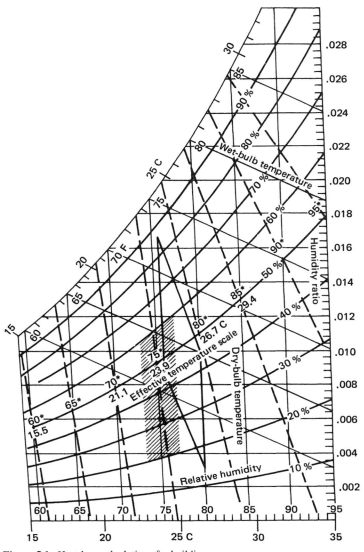

Figure 5.1. Heat loss calculations for buildings.

will be summarized in this section; the reader is referred to the ASHRAE handbook for details.

Figure 5.1 shows the combinations of temperature and humidity that are required for human comfort. The shaded area is the standard U.S. comfort level for sedentary persons. Many European countries have human comfort levels from 3 to 7°C below U.S. levels. If activity of a continuous nature is anticipated, the comfort zone lies to the left of the shaded area; if extra clothing is worn, the comfort zone is displaced similarly.

5.1.1 Calculation of Heat Loss

It is outside the scope of this book to describe the details of the heat load calculations for buildings. However, the method is described in brief in this section. For details one should refer to the *ASHRAE Handbook of Fundamentals* [2] or some textbook on heating and air-conditioning. Table 5.1 lists the components of heat loss calculations of a building.

Complete tables of thermal properties of building materials are given in Appendix 5. Appendix 5 also contains web addresses that list the average wind and design temperature data for many cities in the U.S. and the rest of the world.

Transmission heat losses through attics, unheated basements, and the like are buffered by the thermal resistance of the unheated space. For example, the temperature of an unheated attic lies between that of the heated space and that of the environment. As a result, the ceiling of a room below an attic is exposed to a smaller temperature

Table 5.1. Heating load calculations for buildings

Heating load component	Equations (5.1–5.4)	Descriptions/References
Walls, Roof, Ceilings, Glass	$q = U \cdot A (T_i - T_o)$	T_i, T_o are inside and outside air temperature, respectively. U values of composite section are calculated from the thermal properties of components given in Appendix 5.
Basement floors and walls below ground level	$q_f = U_f^* A_f$	U_f^* has special units of W/m². Values of U_f^* for various ground water temperatures are given in Appendix 5.
Concrete floors on ground	$q_{fe} = F_e P_e (T_i - T_o)$	P_e is the perimeter of the slab. F_e values are given in Appendix 5.
Infiltration and ventilation air	$q_{sensible} = Q \rho_a C p_a (T_i - T_o)$ or $= 1200*Q(T_i - T_o)$ Watts	Q is volume of air flow in m³/sec. ρ_a and Cp_a are density and specific heat of air.
	$q_{latent} = Q \rho_a h_{fg} \Delta W$ or $= 2808*Q \Delta W$ Watts	h_{fg} is the latent heat of water at room temperature. ΔW is humidity ratio difference between inside and outside air.

*Assuming ρ_a = 1.2 Kg/m³; h_{fg} = 2340 J/kg, Cp_a = 1000 J/kg °C

difference and consequent lower heat loss than the same ceiling without the attic would be. The effective conductance of thermal buffer spaces can easily be calculated by forming an energy balance on such spaces.

The example below is an illustration of the heat-loss calculation method described in this section.

Example 5.1. Calculate the heat load on a house for which the wall area is 200 m², the floor area is 600 m², the roof area is 690 m², and the window area totals 100 m². Inside wall height is 3 m. The construction of the wall and the roof is shown in Fig. 5.2.

Solution. The thermal resistance of the wall shown in Fig. 5.2 can be found by the electrical resistance analogy as:

$$R_{wa} = R_{\text{outside air}} + R_{\text{wood siding}} + R_{\text{sheathing}} + R_{\text{comb}} + R_{\text{wall board}} + R_{\text{inside air}}$$

Combined thermal resistance for the studs and insulation (R_{comb}) is found as:

$$\frac{1}{R_{\text{comb}}} = \left(\frac{A_{\text{stud}}}{R_{\text{stud}}} + \frac{A_{\text{insulation}}}{R_{\text{insulation}}}\right) \frac{1}{A_{\text{stud}} + A_{\text{insulation}}}$$

Assuming that the studs occupy 15 percent of the wall area

$$\frac{1}{R_{\text{comb}}} = \frac{0.15}{0.77} + \frac{0.85}{1.94}$$

or

$$R_{\text{comb}} = 1.58 \frac{\text{m}^2{}^\circ\text{C}}{\text{W}}.$$

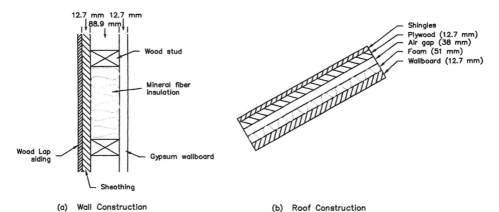

(a) Wall Construction   (b) Roof Construction

Figure 5.2. Cross-sections of the wall and the roof for Example 5.1.

Therefore the wall thermal resistance, R_{wa}, can be found as:

Element	Thermal Resistance (m² °C/W)
Outside Air (6·7 m/s wind)	0.030
Wood Bevel Lap Siding	0.14
12.7 mm Sheathing	0.23
88.9 mm Combined Wood Stud and Mineral Fiber Insulation	1.58
12.7 mm Gypsum Wallboard	0.079
Inside Air (still)	0.12
	$R_{wa} = 2.179$

Therefore,

$$U_{wa} = \frac{1}{R_{wa}} = \frac{1}{2.179} = 0.46 \text{ W/m}^2\text{°C.}$$

The heat loss through the windows depends on whether they are single- or double-glazed. In this example, single-glazed windows are installed, and a U factor equal to 4.7 W/m²-°C is used. (If double-glazed windows were installed, the U factor would be 2.4 W/m²-°C.)

The roof is constructed of 12.7 mm gypsum wall board, 51 mm foam insulation board, 38 mm still air, 12.7 mm plywood, and asphalt shingles (wooden beams and roofing paper are neglected for the simplified calculations here). Therefore,

$$U_{rf} = \frac{1}{\underset{\substack{\text{Outside} \\ \text{Air}}}{0.030} + \underset{\text{Shingles}}{0.077} + \underset{\text{Plywood}}{0.11} + \underset{\text{Air Gap}}{0.17} + \underset{\text{Foam}}{2.53} + \underset{\text{Wallboard}}{0.079} + \underset{\text{Inside Air}}{0.1}} = 0.32 \text{ W/m}^2\text{°C.}$$

If the respective areas and U factors are known, the rate of heat loss per hour for the walls, windows, and roof can be calculated.

Walls: $q_{wa} = (200 \text{ m}^2) \times 0.46 \text{ W/m}^2\text{°C} = 92 \text{ W/°C}$
Windows: $q_{wi} = (100 \text{ m}^2) \times 4.7 \text{ W/m}^2\text{°C} = 470 \text{ W/°C}$
Roof: $q_{rf} = (690 \text{ m}^2) \times 0.32 \text{ W/m}^2 \text{°C} = 220 \text{ W/°C}$
                       Total $q_{tr} = 782 \text{ W/°C}$

If double-glazed windows were used, the heat loss would be reduced to 552 W/°C.

The infiltration and ventilation rate Q for this building is assumed to be 0.5 ACH (Air Changes per hour). The sensible and latent heat loads of the infiltration air may be calculated using the equations given in Tabl⸱ ⸱.1. Therefore,

$$Q = 0.5 \times (600 \text{ m}^2 \times 3 \text{ m}) = 900 \text{ m}^3\text{/hr} = 0.25 \text{ m}^3\text{/s,}$$
$$\text{(volume)}$$

$$q_{sensible} = 0.25 \text{ m}^3/\text{s} \times (1.2 \text{ kg/m}^3)(1000 \text{ J/kg } °\text{C})$$
$$= 300 \text{ W/}°\text{C}.$$

In residential buildings, humidification of the infiltration air is rarely done. Neglecting the latent heat, the total rate of heat loss q_{tot} is the sum of $q_{sensible}$ and q_{tr}:

$$q_{tot} = (782 + 300) = 1082 \text{ W/}°\text{C}.$$

This calculation is simplified for purposes of illustration. Heat losses through the slab surface and edges have been neglected, for example.

More refined methods of calculating energy requirements on buildings do not use the steady-state assumption used above [23]. The thermal inertia of buildings may be expressly used as a load-leveling device. If so, the steady-state assumption is not met and the energy capacitance of the structure must be considered for accurate results. Many adobe structures in the U.S. Southwest are built intentionally to use daytime sun absorbed by 1-ft-thick walls for nighttime heating for example.

5.1.2 Internal Heat Sources in Buildings

Heat supplied to a building to offset energy losses is derived from both the heating system and internal heat sources. Table 5.2 lists the common sources of internal heat generation for residences. Commercial buildings such as hospitals, computer facilities, or supermarkets will have large internal gains specific to their function. Internal heat gains tend to offset heat losses from a building but will add to the cooling load of an

Table 5.2. Some common internal sensible heat gains that tend to offset the heating requirements of buildings[a]

Type	Magnitude (W or J/s)
Incandescent lights	total W
Fluorescent lights	total W
Electric motors	746 × (hp/efficiency)
Natural gas stove	8.28 × m³/hr
Appliances	total W
A dog	50–90
People	
Sitting	70
Walking	75
Dancing	90
Working hard	170
Sunlight	Solar heat gain x fenestration transmittance x shading factor[b]

[a]For more data see [2].

[b]Shading factor is the amount of a window not in a shadow expressed as a decimal between 1.0 and 0.0.

air-conditioning system. The magnitude of the reduction in heating system operation will be described in the next section.

5.1.3 The Degree-day Method

The preceding analysis of heat loss from buildings expresses the loss on a per unit temperature difference basis (except for unexposed floor slabs). In order to calculate the peak load and total annual load for a building, appropriate design temperatures must be defined for each. The outdoor design temperature is usually defined statistically, such that the actual outdoor temperature will exceed the design temperature 97.5 or 99 percent of the time over a long period. The design temperature difference (ΔT) is then the interior building temperature minus the outdoor design temperature. The design ΔT is used for rating non-solar heating systems, but is not useful for selection of solar systems, since solar systems rarely provide 100 percent of the energy demand of a building at peak conditions.

A more useful index of heating energy demand is the total annual energy requirement for a building. This quantity is somewhat more difficult to calculate than the peak load. It requires a knowledge of day-to-day variations in ambient temperature during the heating season and the corresponding building heat load for each day. Building heat loads vary with ambient temperatures as shown in Fig. 5.3. The environmental

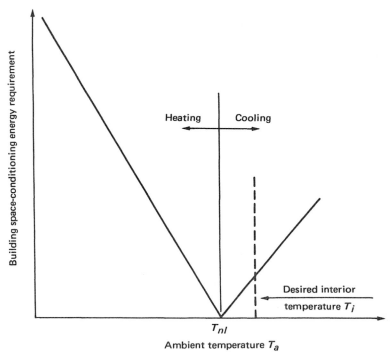

Figure 5.3. Building load profile versus ambient temperature showing no-load temperature T_{nl} and desired interior temperature T_i.

temperature T_{nl}, above which no heat need be supplied to the building, is a few degrees below the required interior temperature T_i because of internal heat-generation effects.

The no-load temperature at which internal source generation q_i just balances transmission and infiltration losses can be determined from the energy balance

$$q_i = \overline{UA}(T_i - T_{nl}), \tag{5.5}$$

where $\overline{UA}$ is the overall loss coefficient for the building (W/°C). Then

$$T_{nl} = T_i - \frac{q_i}{\overline{UA}} \tag{5.6}$$

The total annual heat load on the building, Q_T, can be expressed as

$$Q_T = \int_{365 \text{ days}} \overline{UA}(T_{nl} - T_a)^+ \, dt, \tag{5.7}$$

in which all arguments of the integral are functions of time. The superscript $+$ indicates that only positive values are considered. In practice, it is difficult to evaluate this integral; therefore, three simplifying assumptions are made:

1. $\overline{UA}$ is independent of time.
2. T_{nl} is independent of time.
3. The integral can be expressed by the sum.

Thus,

$$\overline{UA} \sum_{n=1}^{365} (T_{nl} - \overline{T}_a)_n^+ \tag{5.8}$$

where n is the day number, and the daily average temperature $\overline{T}_a$ can be approximated by $\frac{1}{2}(T_{a,\max} + T_{a,\min})$, in which $T_{a,\max}$ and $T_{a,\min}$ are the daily maximum and minimum temperatures, respectively.

The quantity $(T_{nl} - \overline{T}_a)^+$ is called the **degree-day unit**. For example, if the average ambient temperature for a day is 5°C and the no-load temperature is 20°C, 15 degree C-days are said to exist for that day. However, if the ambient temperature is 20°C or higher, 0 degree-days exist, indicating 0 demand for heating that day. Degree-day totals for monthly ($\Sigma_{month} (T_{nl} - \overline{T}_a)^+$) and annual periods can be used directly in Eq. (5.8) to calculate the monthly and annual heating energy requirements.

In the past, a single value of temperature has been used throughout the United States as a universal degree-day base, 65.0°F or 18.3°C.* This practice is now out-

---

*The degree-day base in SI units is defined as 19.0°C, not 18.3°C, which corresponds to 65.0°F. Therefore, precise conversion between the two systems is not possible by a simple multiplication by 5/9.

dated, since many homeowners and commercial building operators have lowered their thermostat settings in response to increased heating fuel costs, thereby lowering T_{nl}. Likewise, warehouses and factories operate well below the 19°C level. Therefore, a more generalized database of degree-days to several bases (values of T_{nl}) has been created by the U.S. National Weather Service (NWS).

Thom [32–34], in a series of papers, developed a statistically rigorous method of calculating degree-days to any base from values of ambient temperature and monthly standard deviations of ambient temperature. The details of this method are too lengthy to present here. However, the NWS has used this method to prepare tabulations of degree F-days for many U.S. locations. In addition, maps of degree F-days to the standard base are available on a monthly basis [35].

A variable base degree-day method, which recognizes that T_{nl} may vary not only with location but also from building to building, is more accepted now [3, 9, 12, 25]. Kreider and Rabl [25] have described this method in detail. Erbs et al. [12] developed a model to estimate variable base degree-days that needs as input only $\overline{T}_a$ for each month. Balcomb et al. [3] have listed monthly degree-days to bases of 50°F, 55°F, 60°F, 65°F, and 70°F for 209 U.S. and Canadian cities.

Example 5.2. A building located in Denver, CO, has a heat-loss coefficient $\overline{UA}$ of 1000 kJ/hr · °C and internal heat sources of 4440 kJ/hr. If the interior temperature is 20°C (68°F), what are the monthly and annual heating energy requirements? A gas furnace with 65 percent efficiency is used to heat the building.

Solution. In order to determine the monthly degree-day totals, the no-load temperature (degree-day basis) must be evaluated from Eq. (5.6).

$$T_{nl} = 20 - \frac{4440}{1000} = 15.6° \ C(60° \ F)$$

The monthly degree C-days for Denver are taken from the U.S. National Weather Service and given in Table 5.3. The energy demand is calculated as

$$\text{Energy Demand} = \overline{UA} \times 24\frac{\text{hr}}{\text{day}} \times \text{Degree C} - \text{days.} \qquad (5.9)$$

The monthly energy demand is given in Table 5.3.

The annual energy demand of 62.9 GJ is delivered by a 65 percent efficient device. Therefore,

$$\text{Average Annual Purchased Energy} = \frac{62.9}{0.65}\text{GJ} = 96.8 \text{ GJ.}$$

Table 5.3. Monthly and annual energy demands for Example 5.2.

Month	Degree C-days	Energy demand[a] (GJ)
Jan.	518	12.4
Feb.	423	10.2
Mar.	396	9.5
Apr.	214	5.2
May	68	1.6
June	14	0.3
July	0	0
Aug.	0	0
Sep.	26	0.6
Oct.	148	3.6
Nov.	343	8.2
Dec.	472	11.3
Total	2622	62.9

[a]Energy demand equals $\overline{UA}$ × degree C-days × 24 hrs/day.

5.1.4 Service Hot-Water Load Calculation

Service hot-water loads can be calculated precisely with the knowledge of only a few variables. The data required for calculation of hot-water demand are

Water source temperature (T_s)
Water delivery temperature (T_d)
Volumetric demand rate (Q)

The energy requirement for service water heating q_{hw} is given by

$$q_{hw}(t) = \rho_w Q(t) c_{pw}[T_d - T_s(t)], \qquad (5.10)$$

where ρ_w is the water density and c_{pw} is its specific heat. The demand rate, $Q(t)$, varies in general with time of day and time of year; likewise, the source temperature varies seasonally. Source temperature data are not compiled in a single reference; local water authorities are the source of such temperature data.

Few generalized data exist with which to predict the demand rate Q. Table 5.4 indicates some typical usage rates for several common building types. Process water heating rates are peculiar to each process and can be ascertained by reference to process specifications.

Example 5.3. Calculate the monthly energy required to heat water for a family of four in Nashville, TN. Monthly source temperatures for Nashville are shown in Table 5.5, and the water delivery temperature is 60°C (140°F).

Table 5.4. Approximate service hot-water demand rates

Usage type	Demand per person	
	liters/day	gal/day
Retail store	2.8	0.75
Elementary school	5.7	1.5
Multifamily residence	76.0	20.0
Single-family residence	76.0	20.0
Office building	11.0	3.0

Solution. For a family of four, the demand rate Q may be found using a demand recommended from Table 5.4:

$$Q = 4 \times 76 \text{ liters/day} = 0.30 \text{ m}^3/\text{day}$$

The density of water can be taken as 1000 kg/m³ and the specific heat as 4.18 kJ/kg · °C.

Monthly demands are given by

$$q_m = (Q \times \text{days/month})(\rho_w c_{pw})[T_d - T_s(t)]$$
$$= (0.30 \times \text{days/month})(1000 \times 4.18)[60 - T_s(t)].$$

The monthly energy demands calculated from the equation above with these data are tabulated in Table 5.5.

Table 5.5. Water heating energy demands for Example 5.3.

Month	Days/month	Demand (m³/month)	Source temperature C°	Energy requirement (GJ/month)
Jan.	31	9.3	8	2.0
Feb.	28	8.4	8	1.8
Mar.	31	9.3	12	1.9
Apr.	30	9.0	19	1.5
May	31	9.3	17	1.7
June	30	9.0	21	1.5
July	31	9.3	22	1.5
Aug.	31	9.3	24	1.4
Sep.	30	9.0	24	1.4
Oct.	31	9.3	22	1.5
Nov.	30	9.0	14	1.7
Dec.	31	9.3	12	1.9

5.2 SOLAR WATER HEATING SYSTEMS

Solar water-heating systems represent the most common application of solar energy at the present time. Small systems are used for domestic hot water applications while larger systems are used in industrial process heat applications. There are basically two types of water-heating systems: ***natural circulation*** or passive solar system (thermosyphon) and ***forced circulation*** or active solar system. Natural circulation solar water heaters are simple in design and low cost. Their application is usually limited to nonfreezing climates, although they may also be designed with heat exchangers for mild freezing climates. Forced circulation water heaters are used in freezing climates and for commercial and industrial process heat.

5.2.1 Natural Circulation Systems

The natural tendency of a less dense fluid to rise above a denser fluid can be used in a simple solar water heater to cause fluid motion through a collector [10]. The density difference is created within the solar collector where heat is added to the liquid. In the system shown in Fig. 5.4, as water gets heated in the collector, it rises to the tank, and the cooler water from the tank moves to the bottom of the collector, setting up a natural circulation loop. It is also called a ***thermosyphon loop.*** Since this water heater does not use a pump, it is a passive water heater. For the thermosyphon to work, the storage tank must be located higher than the collector.

The flow pressure drop in the fluid loop (ΔP_{FLOW}) must equal the bouyant force "pressure difference" ($\Delta P_{BOUYANT}$) caused by the differing densities in the hot and cold legs of the fluid loop:

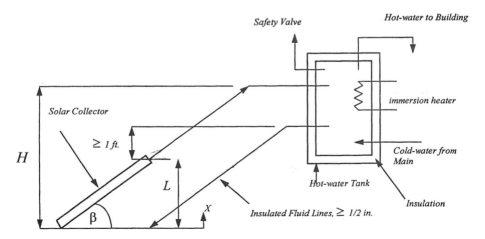

Figure 5.4. Schematic diagram of thermosyphon loop used in a natural circulation, service water-heating system. The flow pressure drop in the fluid loop must equal the bouyant force "pressure" $\left[\int_{o}^{L} g\rho(x)dx - \rho_{stor}gL \right]$ where $\rho(x)$ is the local collector fluid density and ρ_{stor} is the tank fluid density, assumed uniform.

$$\Delta P_{FLOW} = \Delta P_{BOUYANT}$$

$$= \rho_{stor}gH - \left[\int_0^L \rho(x)gdx + \rho_{out}g(H - L)\right], \qquad (5.11)$$

where H is the height of the legs and L the height of the collector (see Fig. 5.4), $\rho(x)$ is the local collector fluid density, ρ_{stor} is the tank fluid density, and ρ_{out} is the collector outlet fluid density, the latter two densities assumed uniform. The flow pressure term ΔP_{FLOW}, is related to the flow loop system head loss, which is in turn directly connected to friction and fitting losses and the loop flow rate:

$$\Delta P_{FLOW} = \oint_{LOOP} \rho d(h_L), \qquad (5.12)$$

where $h_L = KV^2$, with K being the sum of the component loss velocity factors (see any fluid mechanics text) and V the flow velocity.

Since the driving force in a thermosyphon system is only a small density difference and not a pump, larger-than-normal plumbing fixtures must be used to reduce pipe friction losses [30]. In general, one pipe size larger than normal would be used with a pump system is satisfactory. Figure 5.5 shows some passive water heaters.

Since the hot-water system loads vary little during a year, the angle of tilt is that equal to the latitude, that is, $\beta = L$. The temperature difference between the collector inlet water and the collector outlet water is usually 8–11°C during the middle of a sunny day [10]. After sunset, a thermosyphon system can reverse its flow direction and lose heat to the environment during the night. To avoid reverse flow, the top header of the absorber should be at least 30 cm below the cold leg fitting on the storage tank, as shown, otherwise a check valve would be needed.

To provide heat during long cloudy periods, an electrical immersion heater can be used as a backup for the solar system. The immersion heater is located near the top of the tank to enhance stratification and so that the heated fluid is at the required delivery temperature. Tank stratification is desirable in a thermosyphon to maintain flow rates as high as possible. Insulation must be applied over the entire tank surface to control heat loss.

Several features inherent in the thermosyphon design limit its utility. If it is to be operated in a freezing climate, a nonfreezing fluid must be used, which in turn requires a heat exchanger between collector and potable water storage. (If potable water is not required, the collector can be drained during cold periods instead.) Heat exchangers of either the shell-and-tube type or the immersion-coil type require higher flow rates for efficient operation than a thermosyphon can provide. Therefore, the thermosyphon is usually limited to nonfreezing climates. For mild freeze climates, a heat exchanger coil welded to the outer surface of the tank and filled with an antifreeze may work well.

Example 5.4. Determine the "pressure difference" available for a thermosyphon system with 1 meter high collector and 2 meter high legs. The water temperature input to the collector is 25°C and the collector output temperature is 35°C. If the overall system loss velocity factor (K) is 15.6, estimate the system flow velocity.

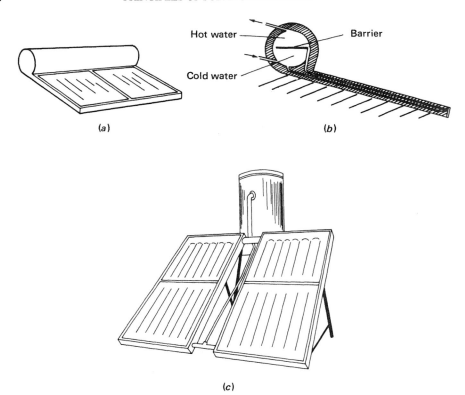

Figure 5.5. Passive solar water heaters; (a) compact model using combined collector and storage, (b) section view of the compact model, and (c) tank and collector assembly.

Solution. Equation (5.11) is used to calculate the pressure difference, with the water densities being found from the steam tables (see Appendix 3 and Tables 3.8 and 3.9).

$$\rho_{stor}(25°C) = 997.009 \text{ kg}/m^3$$

$$\rho_{out}(35°C) = 994.036 \text{ kg}/m^3$$

$\rho_{coll.ave.}(30°C) = 996.016$ kg/m³ (note: average collector temperature used in 'integral') and with H = 2 and L = 1 m:

$$\Delta P_{BOUYANT} = (997.009)9.81(2) - [(996.016)9.81(1) + (994.036)9.81(1)]$$

$$= 38.9 \text{ N}/m^2 \text{ (Pa)}.$$

The system flow velocity is estimated from the system K given, the pressure difference calculated above, taking the average density of the water around the loop (at 30°C), and substituting into Eq. (5.12):

$$\Delta P_{\text{BOUYANT}} = (\rho_{\text{loop.ave.}})(h_L)_{\text{loop}} = (\rho_{\text{loop.ave.}}) \, KV^2$$
$$V^2 = 38.9/(996.016)(15.6)$$
$$V = 0.05 \text{ m/s}$$

5.2.2 Forced-Circulation Systems

If a thermosyphon system cannot be used for climatic, structural, or architectural reasons, a forced-circulation system is required.

Figure 5.6 shows three configurations of forced circulation systems: (1) open loop, (2) closed loop, and (3) closed loop with drainback. In an open loop system (Fig. 5.6a) the solar loop is at atmospheric pressure, therefore, the collectors are empty when they are not providing useful heat. A disadvantage of this system is the high pumping power required to pump the water to the collectors every time the collectors become hot. This disadvantage is overcome in the pressurized closed loop system (Fig. 5.6b) since the pump has to overcome only the resistance of the pipes. In this system, the solar loop remains filled with water under pressure.

In order to accommodate the thermal expansion of water from heating, a small (about 2 gallon capacity) expansion tank and a pressure relief valve are provided in the solar loop. Because water always stays in the collectors of this system, antifreeze (propylene glycol or ethylene glycol) is required for locations where freezing conditions can occur. During stagnation conditions (in summer), the temperature in the col-

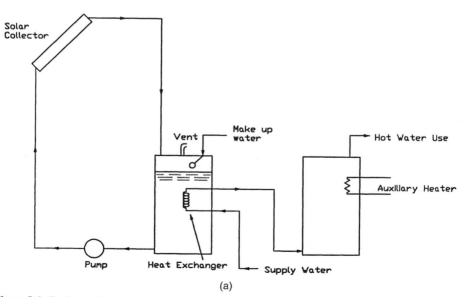

Figure 5.6. Typical configurations of solar water-heating systems: (a) open loop system.

(*continues*)

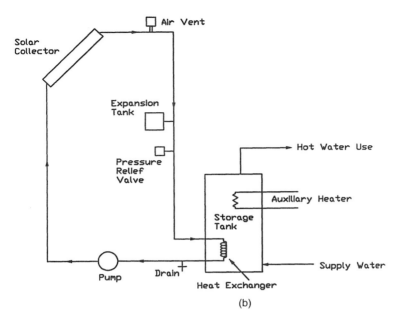

(b)

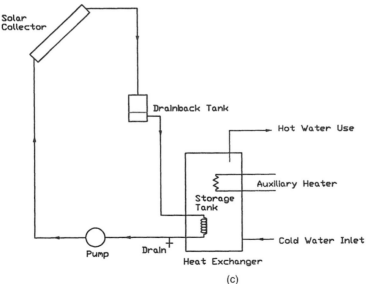

(c)

Figure 5.6. (*Continued*) (b) closed loop system; and (c) closed loop drainback system. (Adapted from Goswami, D.Y. *Alternative Energy in Agriculture*, Vol. 1. Boca Raton, FL: CRC Press, (1986).

lector can become very high, causing the pressure in the loop to increase. This can cause leaks in the loop unless some fluid is allowed to escape through a pressure-release valve. Whether the result of leaks or of draining, air enters the loop causing the pumps to run dry. This disadvantage can be overcome in a closed loop drainback system which is not pressurized (Fig. 5.6c). In this system, when the pump shuts off, the water in the collectors drains back into a small holding tank while the air in the holding tank goes up to fill the collectors. The holding tank can be located where freezing does not occur, but still at a high level to reduce pumping power. In all three configurations, a differential controller measures the temperature differential between the solar collector and the storage, and turns the circulation pump on when the differential is more than a set limit (usually 5°C) and turns it off when the differential goes below a set limit (usually 2°C). Alternatively, a photovoltaic (PV) panel and a DC pump may be used. The PV panel will turn on the pump only when solar radiation is above a minimum level. Therefore, the differential controller and the temperature sensors may be eliminated.

5.2.3 Industrial Process Heat Systems

For temperatures of up to about 100°C, required for many industrial process heat applications, forced ciculation water-heating systems described above can be used. The systems, however, will require a large collector area, storage and pumps, etc. For higher temperatures, evacuated tube collectors or concentrating collectors must be used. Industrial process heat systems are described in more detail in Chapter 8.

5.3 LIQUID-BASED SOLAR-HEATING SYSTEMS FOR BUILDINGS

The earliest active solar space-heating systems were constructed from enlarged water-heating components. Experiments beginning in 1938 at the Massachusetts Institute of Technology (MIT) showed that solar heating with liquid working fluids could be done without any major technical problems. The early MIT work formed the basis of many of the design techniques used today. Other experiments after World War II provided additional fundamental information on collector designs and storage operation for liquid-based heating systems. Fig. 5.7 shows a modern solar-heated, multifamily building in Colorado.

Solar space-heating systems can be classified as active or passive depending on the method utilized for heat transfer. A system that uses pumps and/or blowers for fluid flow in order to transfer heat is called an active system. On the other hand, a system that utilizes natural phenomena for heat transfer is called a passive system. Examples of passive solar space-heating systems include direct gain, attached greenhouse, and storage wall (also called Trombe wall). Passive solar heating systems are described in Chapter 7. In this section, configurations, design methods, and control strategies for active solar-heating systems are described.

Figure 5.7. Multi-unit residence located in Boulder, CO, heated by an active solar-heating system. Courtesy of Joint Venture, Inc., Boulder, CO.

5.3.1 Physical Configurations of Active Solar Heating Systems

Figure 5.8 is a schematic diagram of a typical space-heating system. The system consists of three fluid loops–collector, storage, and load. In addition, most space-heating systems are integrated with a domestic water-heating system to improve the year long solar load factor.

Since space heating is a relatively low-temperature use of solar energy, a thermodynamic match of collector to task indicates that an efficient flat-plate collector or low-concentration solar collector is the thermal device of choice.

The collector fluid loop contains fluid manifolds, the collectors, the collector pump and heat exchanger, an expansion tank, and other subsidiary components. A collector heat-exchanger and antifreeze in the collector loop are normally used in all solar space-heating systems, since the existence of a significant heating demand implies the existence of some subfreezing weather.

The storage loop contains the storage tank and pump as well as the tube side of the collector heat exchanger. To capitalize on whatever stratification may exist in the storage tank, fluid entering the collector heat exchanger is generally removed from the bottom of storage. This strategy ensures that the lowest temperature fluid available in the collector loop is introduced at the collector inlet for high efficiency.

The energy delivery-to-load loop contains the load device, baseboard heaters or fin-and-tube coils, and the backup system with a flow control (mode selector) valve.

5.3.2 Solar Collector Orientation

The best solar collector orientation is such that the average solar incidence angle is smallest during the heating season. For tracking collectors this objective is automati-

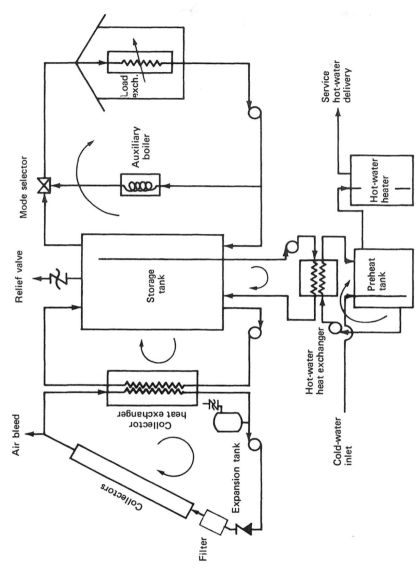

Figure 5.8. Typical solar-thermal system for space heating and hot-water heating showing fluid transport loops and pumps.

cally realized. For fixed collectors in the northern hemisphere the best orientation is due south (due north in the southern hemisphere), tilted up from the horizon at an angle of about 15° greater than the local latitude.

Although due south is the optimum azimuthal orientation for collectors in the northern hemisphere, variations of 20° east or west have little effect on annual energy delivery [24]. Off-south orientations greater than 20° may be required in some cases because of obstacles in the path of the sun. These effects may be analyzed using sun-path diagrams and shadow-angle protractors as described in Chapter 2.

5.3.3 Fluid Flow Rates

For the maximum energy collection in a solar-collector, it is necessary that it operates as closely as possible to the lowest available temperature, which is the collector inlet temperature. Very high fluid flow rates are needed to maintain a collector-absorber surface nearly isothermal at the inlet temperature. Although high flows maximize energy collection, practical and economic constraints put an upper limit on useful flow rates. Very high flows require large pumps and excessive power consumption and lead to fluid conduit erosion.

Figure 5.9 shows the effect of mass flow rate on annual energy delivery from a solar system. It is seen that the law of diminishing returns applies and that flows beyond about 50 kg/hr $m_c^2 \cdot$ (~10 lb/hr ft_c^2) have little marginal benefit for collectors with loss

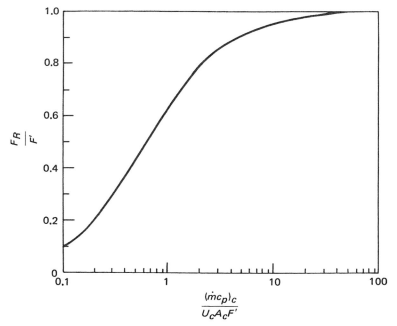

Figure 5.9. Effect of fluid flow rate on collector performance as measured by the heat-removal factor F_R; F' is the plate efficiency factor (see Chapter 4).

coefficients on the order of 6 W/m_c^2 °C ($\sim$1 Btu/hr ft_c^2 °F). In practice, liquid flows in the range of 50–75 kg/hr m_c^2 (10–15 lb/hr ft_c^2) of water equivalent are the best compromise among collector heat-transfer coefficient, fluid pressure drop and energy delivery. However, an infinitely large flow rate will deliver the most energy if pumping power is ignored for a nonstratified storage. If storage stratification is desired, lower flow rates must be used, since high flow destroys stratification.

In freezing climates, an antifreeze working fluid is recommended for collectors. Attempts to drain collectors fully for freeze protection have usually been unsuccessful unless collector fluid conduits are very large and smooth and unless all piping is sloped to assure drainage. The potential damage risk from incomplete draining will usually dictate that the additional investment in antifreeze be made.

5.3.4 Thermal Storage

Thermal storage tanks must be insulated to control heat loss, as described in Chapter 4. If a storage tank is located within a structure, any losses from the tank tend to offset the active heating demands of the building. However, such storage loss is uncontrolled and may cause overheating in seasons with low heat loads. Some solar-heating systems have used a ventilatable structure surrounding storage. This enclosure may be vented to the building interior in winter and to the environment in summer.

Safety concerns may cause storage to be located external to a building in some cases. Large volumes of hot water could be released to a building interior if a storage tank were to fail. Potential personal injury or property damage, which could result from an accident, must be assessed in siting storage tanks. Tank burial would seem to be the safest approach in some cases. Buried storage tanks must be sealed from ground moisture, insulated with waterproof insulation, and galvanically protected.

The amount of thermal storage used in a solar-heating system is limited by the law of diminishing returns. Although larger storage results in larger annual energy delivery, the increase at the margin is small and hence not cost-effective. Seasonal storage is, therefore, usually uneconomic, although it can be realized in a technical sense. Experience has shown that liquid storage amounts of 50–75 kg H_2O/m_c^2 (10–15 lb/ft_c^2) are the best compromise between storage tank cost and useful energy delivery. Klein [20] has calculated the effect of storage size on annual energy delivery. His results, shown in Fig. 5.10, exhibit the expected diminishing returns to scale.

Since solar energy heating systems operate at temperatures relatively close to the temperatures of the spaces to be heated, storage must be capable of delivering and receiving thermal energy at relatively small temperature differences. The designer must consider the magnitude of these driving forces in sizing heat exchangers, pumps, and air-blowers. The designer must also consider the nonrecoverable heat losses from storage—even though storage temperatures are relatively low, surface areas of storage units may be large and heat losses therefore appreciable.

Some investigators have proposed heating the storage medium with conventional fuels to maintain its temperature at useful levels during sunless periods. This approach has two major flaws:

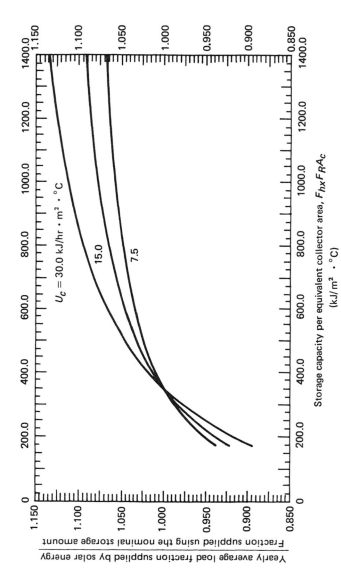

Figure 5.10. Effect of liquid storage capacity on liquid-based solar-heating system energy delivery. From [20].

1. If storage is heated with conventional fuels, it cannot be heated with solar energy when available; therefore, some collected solar energy cannot be used.
2. If storage is partially heated with conventional fuels, the collector inlet temperatures will be higher and efficiency lower than it would be if storage were not boosted. Therefore, the useful return on the solar system investment would be diminished.

In conclusion, it should be emphasized that storage heating with conventional fuels is uneconomical in any practical solar-thermal systems designed to date.

5.3.5 Other Mechanical Components

Other mechanical components in solar heating systems include pumps, heat exchangers, air bleed valves, pressure release valves and expansion tanks. Heat exchangers in solar systems are selected based on economic criteria described in Chapter 4. The best trade-off of energy delivery increase with increasing heat-exchanger size usually results from use of an exchanger with effectiveness in the range of 0.6–0.8. Counterflow heat exchangers are required for this level of effectiveness. A detailed example of heat-exchanger selection is contained in Chapter 4.

Achievement of the required effectiveness level may dictate fairly high flow rates in the storage tank side of the collector heat exchanger. Flows up to twice that in the collector side can improve exchanger performance significantly in many cases. Since the storage side loop is physically short and has a small pressure drop, increased flow in this loop increases pump energy requirements by a negligible amount. Typical solar heat-exchangers sizes range from 0.05 to 0.10 m^2 of heat-exchanger surface per square meter of net collector area.

In hydronic heating systems, it is essential that all air be pumped from the system. To facilitate this process, air bleed valves located at the high points in a system are used. These are opened during system fill and later if air should collect. Air bleeds are required at points of low velocity in piping systems where air may collect because the local fluid velocity is too low for entrainment.

5.3.6 Controls in Liquid Systems

Control strategies and hardware used in current solar system designs are quite simple and are similar in several respects to those used in conventional systems. The single fundamental difference lies in the requirement for differential temperature measurement instead of simple temperature sensing. In the space heating system shown in Fig. 5.8, two temperature signals determine which of three modes is used. The signals used are the collector-storage differential and room temperature. The collector-storage difference is sensed by two thermistors or thermocouples, the difference being determined by a solid-state comparator, which is a part of the control device. Room temperature is sensed by a conventional dual-contact thermostat.

The control system operates as follows. If the first room thermostat contact closes, the mode selector valve and distribution pump are activated in an attempt to deliver the thermal demand from solar-thermal storage. If room temperature continues to drop, indicating inadequate solar availability, the mode selector diverts flow through the backup system instead of the solar system, and the backup is activated until the load is satisfied.

The collector-storage control subsystem operates independently of the heating subsystem described above. If collector temperature, usually sensed by a thermistor thermally bonded to the absorber plate, exceeds the temperature in the bottom of the storage tank by 5–10°C (9–18°F), the collector pump and heat-exchanger pump (if present) are activated and continue to run until the collector and storage temperature are within about 1–2°C (2–4°F) of each other. At this point it is no longer worthwhile to attempt to collect energy and the pumps are turned off. The collector-storage subsystem also has a high temperature cutout that turns the collector loop pump off when the storage temperature exceeds a set limit.

5.3.7 Load Devices in Liquid Solar-Heating Systems

A heating load device transfers heat from the solar storage to the air in the space. Therefore, a liquid-to-air heat exchanger is sized based on the energy demand of a building. Several generic types of load devices are in common use.

1. Forced-air systems—tube-and-fin coil located in the main distribution duct of a building or zone of a building (see Fig. 5.11).
2. Baseboard convection systems—tube-and-fin coils located near the floor on external walls. These operate by natural convection from the convectors to the room air.
3. Heated floors or ceilings—water coils. These transfer heat to large thermal masses that in turn radiate or convect into the space. This heating method is usually called radiant heating.

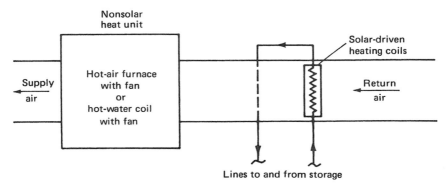

Figure 5.11. Forced-air heating system load device location upstream of nonsolar heat exchanger or furnace.

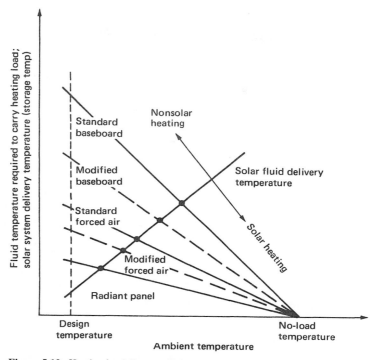

Figure 5.12. Heating load diagram for baseboard, forced-air and radiant systems. Modified baseboard and forced-air systems are oversized in order to carry heating demands at lower temperature. Balance points are indicated by large dots at intersections.

Each load device requires fluid at a different temperature in order to operate under design load conditions as shown in Fig. 5.12. Since baseboard heaters are small in heat-transfer area and rely on the relatively ineffectual mechanism of natural convection, they require the highest fluid temperature. Forced-air systems involve the more efficient forced-convection heat-transfer mode and, hence, are operable at lower fluid temperatures (see Fig. 5.12). Radiant heating can use very large heat-transfer areas and is, therefore, operable at relatively low fluid temperatures.

In Fig. 5.12 the intersection of the solar fluid temperature line and the load line for a specific configuration is called the ***balance point***. At ambient temperatures below the balance point solar energy cannot provide the entire demand, and some backup is required; above the balance point solar capacity is sufficient to carry the entire load. Note that the load lines are specific to a given building. The solar fluid temperature line is not fixed for a building but depends on solar collector and storage size as well as local solar radiation levels. The line shown in Fig. 5.12 is, therefore, an average line. The instantaneous solar line changes continuously in response to load and climatic forcing functions as described in Chapter 4.

It is possible to modify load devices to lower the balance point, as shown in Fig. 5.12. For example, a forced-air tube-and-fin exchanger can be enlarged by adding one

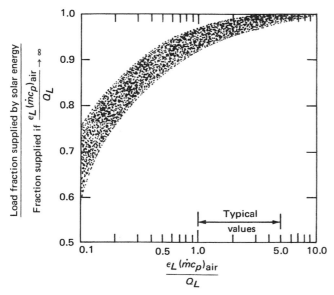

Figure 5.13. Effect of load device capacity for forced-air heating systems. Q_L is the building heat load expressed in units consistent with $(\dot{m}c_p)_{air}$. From [20].

or more additional rows of tubes. This increased heat-transfer area will permit the same energy delivery at a lower fluid temperature. Figure 5.13 depicts the effect of forced-air load device size (heat-transfer effectiveness ϵ_L) on annual energy delivery. The law of diminishing returns is evident as increasing effectiveness returns progressively less energy. The effectiveness of a cross-flow heat exchanger of the type used in forced-air systems is calculated in Chapter 4 and shown in Fig. 4.9.

5.4 SOLAR AIR-HEATING SYSTEMS

Air has been used as the working fluid in solar-heating systems since World War II. Although demonstrated in fewer buildings than liquid systems, air systems have several advantages that can lead to their use in smaller installations in single- and multifamily residences. In addition, air systems are well suited to crop drying and air preheating in certain processes.

Table 5.6 lists some of the advantages and disadvantages of air and liquid systems. Several of the disadvantages of an air system follow from the poor heat-transfer properties of air. Large space requirements for ducts preclude the use of air systems in large buildings because of space limitations. This is usually not a disadvantage in air systems used in residences. However, the thermal performance and costs for both systems are nearly identical if the air system does not have leaks.

In all air-based solar-heating systems a particle-bed storage device is used. Although water offers a higher storage energy density than most solids, the difficulty of

Table 5.6. Advantages and disadvantages of air and liquid space-heating systems.

Air Systems	
Advantages	Disadvantages
No freezing problem.	Space heating only.
No internal corrosion problem with dry air.	Large space requirements for ducts.
Leaks of smaller consequence.	Larger storage volume required for rocks.
No heat exchanger between collector-storage and storage-building loops.	Cannot store heat and heat building at the same time—major problem in low load seasons.
No boiling or pressure problems.	Low (ρc_p) product for air.
Easy for do-it-yourselfers.	
Simple and reliable.	

Liquid Systems	
Advantages	Disadvantages
Higher transport energy density.	Freezing problems.
Better heat-transfer properties.	Leakage problems.
Water storage has higher energy density.	Corrosion problem—water chemistry needs monitoring.
Suitable for space heating and cooling.	Heat exchanger required for collector-storage and storage-building loops.
Small fluid conduits.	Boiling and fluid expansion provisions required.

economically exchanging heat from an air stream to a liquid storage fluid stream precludes its use. Storage controls and operating parameters are described in detail in a subsequent section.

Flat-plate collectors are usually required for air-heating systems to provide sufficient area to transfer heat effectively from the absorber plate to the air stream. A flat-plate collector provides a good thermodynamic match for temperature demands in space-heating applications. Air-heating flat-plate collectors are described in Chapter 3.

5.4.1 Heating System Physical Configuration

In many ways liquid-based and air-based solar systems are similar in operation. Fig. 5.14 is a schematic diagram of a typical air-heating system. Similar flow regimes and collector orientations are used in both types of systems. The air system is simpler, however, since a collector heat exchanger and associated pumps, pipes and expansion tanks are not present. The use of pebble-bed storage is advantageous because of the stratification that results. Stratification ensures a low inlet temperature to the collector; the collector inlet temperature is approximately the temperature of the cool zone of storage, which, in turn, is the building return air temperature ($\sim$15–20°C). Cool collector inlet temperatures are essential to efficient operation of an air system. Since stratification cannot be achieved with isothermal phase-change storage media, they are

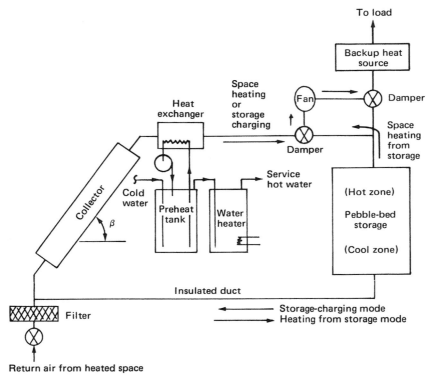

Figure 5.14. Typical air-based solar space-heating system showing directions of fluid flow and important components.

not suitable for use with air-based systems. Frequently, more dampers are required than shown in Fig. 5.14 to prevent leakage and backflows through a cold collector via a leaky damper. Careful design of dampers and actuators is essential.

The collector-to-storage loop consists of insulated ducts, collector manifolds, and a hot-water preheat exchanger. These tube-and-fin exchangers provide some preheat to hot water if placed at the collector outlet to ensure their exposure to the hottest air available. Collector flow balancing is achieved by controlling the pressure drop through each collector and using equal duct lengths for each.

Figure 5.15 shows a hot-air system. During the storage-charging mode (*b*), heated air flows through the rock bed at a low flow rate determined by the desired temperature rise in the collector, usually 10–15 liters/sec − m$_c^2$. As progressively more heat is stored, the interface between the hot and cold regions of storage moves downward in the storage bin. The air returning to the collector is at the temperature of the cool region of storage.

During daytime heating on a sunny day (*a*), air from the collector is diverted to the building instead of to storage. During sunny periods when no heating demand exists, storage is charged by warm air from the collector. During the nighttime or cloudy daytime heating modes, heat is removed from storage by a counterflow of air through the

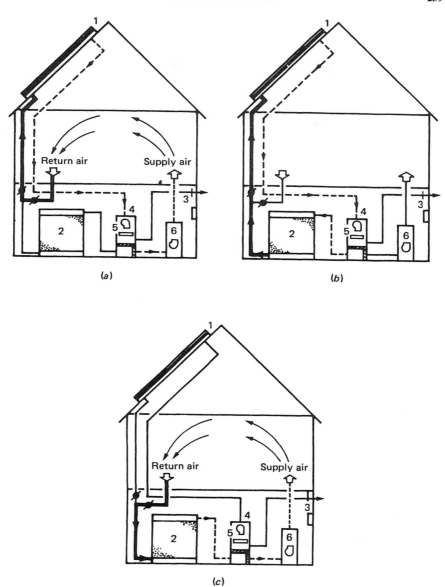

Figure 5.15. Schematic diagram of operating modes of solar air system: (*a*) space heating from collector; (*b*) storing solar heat; and (*c*) space heating from storage. Components are; (1) collector, (2) storage, (3) control, (4) fan, (5) hot-water preheater (optional), and (6) backup heater. Courtesy of the Solaron Corp.

rock bed, as shown in (c). The outlet temperature from storage is close to the daytime collection (inlet) temperature, since the air being heated passes through the hottest zone of storage last. As progressively more heat is removed from storage, the interface between the hot and cold regions of storage moves upward.

The storage medium for air systems has typically been 25–50 mm diameter granite or river-bed rocks. An air filter is required between the heated space and storage to eliminate dust buildup in the gravel bed. Dust would reduce the heat-transfer coefficient to the rock pieces and increase the bed pressure drop. The recommended amount of storage to be used is roughly the same as for a liquid system—about 300 kg of rock per square meter of collector (0.75 ft^3/ft^2_c, 0.25 m^3/m^2_c). Design information on the expected pressure drop through pebble-bed storage is contained in Chapter 4.

5.4.2 Collector Designs

Several air-heating collector designs are shown in Fig. 3.7. However, relatively few performance data for such collectors have been published to provide designers with the information necessary to predict system performance and determine optimal system configurations reliably.

One of the early computer-aided design studies that does provide guidelines for the designer of air-collector systems was conducted by Balcomb and his co-workers [5, 6]. Figure 5.16 shows some of their practical results. Fig. 5.16a depicts the variation of solar delivery with storage volume. It is clear that storage volumes in excess of 60 lb rock/ft^2_c (300 kg/m^2_c) improve the annual solar delivery by a negligible amount. Likewise, Fig. 5.16b indicates that collector air flow rates greater than 15 liters/sec − m^2_c increase delivery very little. Figures 5.16c and 5.16d show temperature profiles for a particular rock bed modeled for various times of day for both the storage-charging and storage-discharging modes. Any properly sized storage will have similar temperature distributions; the key property is that the outlet temperature from storage should be low to permit the inherently inefficient air collector to operate at as low a temperature as possible. The optimal geometric configuration of a storage bed is roughly cubical in order to minimize the air pressure drop.

5.4.3 Fluid Flow Rates

The flow rate through an air collector and storage loop is a compromise between pressure drop and energy transfer to the air stream. As shown in Figs. 5.16 and 5.17, the effect of air capacitance rate $(\dot{m}c_p)_c$ can be significant. Apart from the beneficial effect of increased flow rate on the heat-removal factor F_R, which is usually quite low for air collectors ($\sim$ 0.6–0.7), increased flows tend to cause an offsetting decrease in storage bed stratification. Decreased stratification penalizes collector performance, since the collector inlet temperature is higher under these circumstances.

Nominal flow rates used in air collectors are about 10 liters/m^2_c · sec (2 ft^3/min · ft^2_c). This capacitance rate is well below that for liquid systems but is a compromise between the offsetting effects of stratification reduction and improved heat transfer from

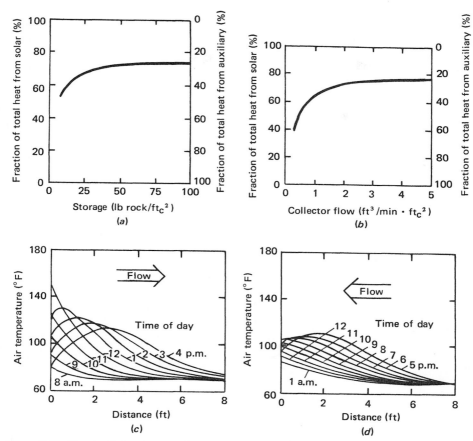

Figure 5.16. Hot-air solar system operating parameters: (*a*) effect of storage size on annual energy delivery; (*b*) effect of collector air flow rate on annual energy delivery; (*c*) typical temperature distribution in rock bed during storage heating by solar energy; and (*d*) typical temperature distribution in rock bed during heat removal from storage. Adapted from [6].

the absorber plate which would result from higher flow rates. This effect is not present in liquid systems, since storage stratification is generally of little consequence in liquid systems.

5.4.4 Other Mechanical Components

Air systems require fewer auxiliary components than liquid systems. Controls in air systems are similar in function to those in liquid systems and are not described in detail here. The controller operates one blower, one pump (preheat), and two dampers, which are the analogs of valves in liquid systems. The dampers must seal tightly. For example, if the middle damper in Fig. 5.15, controlling the air flow to the heated space, should leak, the fan could draw cold air from the nightime collector through the

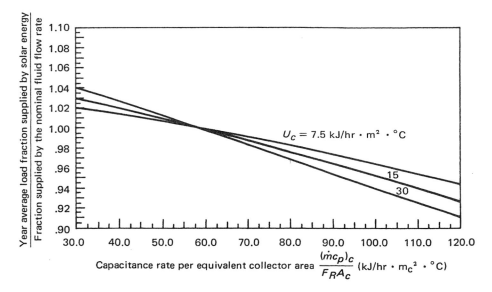

Figure 5.17. Effect of collector fluid flow rate on annual energy delivery in an air-based solar-heating system. The effect of the collector loss coefficient U_c is also shown. The nominal fluid flow rate is 58 kJ/m$_c^2$ · °C · hr, based on the equivalent areas $F_R A_c$. Adapted from [20].

damper to mix with warm air from storage. Such a leak could reduce the heating effect substantially.

Sizing of the blower and ducts in air systems can be done by conventional methods. Pressure-drop data for air ducts are presented in Appendix 4, in a convenient form for such calculations in Fig. A4.4. In some cases a two-speed blower is required in air systems if the system pressure drop in the collector-to-space heating mode is substantially different from the pressure drop through storage. Careful design can frequently avoid this problem, however.

5.4.5 Unglazed Transpired Wall System for Air Preheating

Ventilation air preheating systems using wall-mounted unglazed transpired solar air collectors are the only active solar air heating systems that have found market acceptance in commercial and industrial buildings [26]. Such systems preheat the ventilation air in a once-through mode without any storage. Figure 5.18 shows a transpired wall system in which the air is drawn through a perforated absorber plate by the building ventilation fan. Kutcher and Christensen [27] presented a thermal analysis of this system. From a heat balance on the transpired unglazed collector, the useful heat collected is:

$$q_u = I_c A_c a_s - U_c A_c (T_{out} - T_a) \tag{5.13}$$

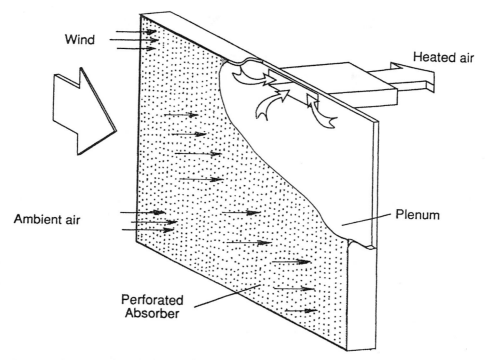

Figure 5.18. Unglazed transpired solar collector system for building ventilation preheat. Intake air is drawn by the building ventilation fan through the perforated absorber plate and up the plenum between the absorber and the south wall of the building. (Adapted from [27])

The overall heat loss coefficient U_c, which is due to radiative and convective losses, is given as:

$$U_c = h_r/\epsilon_{hx} + h_c \qquad (5.14)$$

where ϵ_{hx} is absorber heat exchanger effectiveness, h_r is a linearized radiative heat transfer coefficient and h_c is the convective heat loss coefficient. The heat exchanger effectiveness for air flowing through the absorber plate is defined as:

$$\epsilon_{hx} = \frac{T_{out} - T_a}{T_c - T_a} \qquad (5.15)$$

The forced convective heat loss coefficient due to a wind velocity of U_∞ is given as:

$$h_c = 0.82 \frac{U_\infty \nu c_p}{L \upsilon_0} \qquad (5.16)$$

where ν is the kinematic viscosity of air in m²/s, C_p the specific heat in J/kg-K, v_0 the section velocity and L is the height of the collector.

Radiation heat loss occurs to both the sky and to the ground. Assuming the absorber is gray and diffuse with an emissivity ϵ_c, the radiative loss coefficient h_r is:

$$h_r = \epsilon_c \sigma \; \frac{(T_c^4 - F_{cs}T_{sky}^4 - F_{cg}T_{gnd}^4)}{T_c - T_a} \tag{5.17}$$

where F_{cs} and F_{cg} are the view factors between the collector and the sky, and collector and the ground respectively. For a vertical wall with infinite ground in front of it both F_{cs} and F_{cg} will be 0.5 each. Using the above equations Kutcher and Christensen [27] showed that the predicted performance matches the measured performances well. Figure 5.19 shows their predicted thermal performances.

5.5 METHODS OF MODELING AND DESIGN OF SOLAR HEATING SYSTEMS

Several methods of modeling and design of solar space and water heating have been developed including f-chart, SLR, Utilizability and TRNSYS. The first two of these methods are described briefly in the following sections. Before design of a solar heating system can start, one must know the loads and the availability of solar radiation at the location. Estimation of solar radiation is described in Chapter 2 and estimation of the loads has been described briefly earlier in Chapter 5.

5.5.1 Design of a Liquid-Based Solar Heating System by f-chart

Klein and his co-workers [8,20–23] developed a method of simplified prediction of the performance of a solar heating system based on a large number of detailed simulations for various system configurations in various locations in the United States. The results from these simulations were then correlated with dimensionless parameters on charts that are general in form and usable anywhere. The charts are called f-charts, denoting a parameter f_s, the fraction of monthly load supplied by solar energy. The dimensionless groups used in the f-charts are derived from a nondimensionalization of the equations of governing energy flows; the groups therefore have a physical significance as described below. The f-chart method has been developed for standard solar heating and hot water system configurations.

Schematic diagrams for standard solar heating systems based on liquid and air heat transfer fluids are shown in Figs. 5.8 and 5.14, respectively. Certain deviations from these configurations can be handled in the f-chart method. For example, the collector-to-storage heat exchanger in the liquid based system may be eliminated. The domestic water heater in Fig. 5.14 is shown as a two-tank system, which may be reconfigured as a one-tank system.

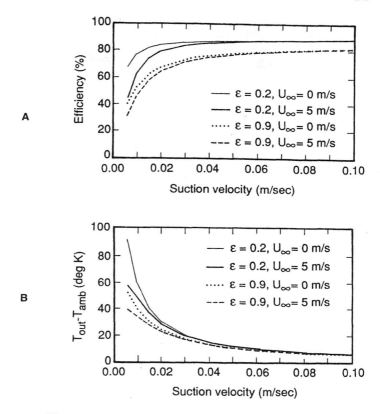

Figure 5.19. Predicted thermal performance of a vertical unglazed transpired solar collector as a function of suction velocity, absorber emissivity, and wind speed. Collector size = 3 m × 3 m, T_{amb} = 10°C, $T_{sky} = T_{amb} - 15°C$, $T_{gnd} = T_{amb}$, and I_c = 700 W/m². (A) Efficiency vs. suction velocity. (B) Suction air temperature rise vs. suction velocity. (From Kutcher and Christensen [27])

In the f-chart method, the fraction of load supplied by solar energy, f_s, is correlated with two dimensionless parameters, called the loss parameter P_L and the solar parameter P_s. Parameters P_L and P_s are defined as:

$$P_L = \frac{A_c F_{hx} F_R U_c \Delta t (T_R - \bar{T}_a)}{L},\tag{5.18}$$

and

$$P_s = \frac{A_c F_{hx} F_R \bar{I}_c (\overline{\tau\alpha})}{L}.\tag{5.19}$$

P_s and P_L are measures, respectively, of the long term insolation gain by the collector and long term thermal loss per unit load. The parameters in Eqs. (5.18) and (5.19) and in the f-chart are described in Table 5.7.

P_s may be rewritten as

$$P_s = \frac{A_c \bar{I}_c F_{hx} F_R(\tau\alpha)_n}{L} \frac{(\overline{\tau\alpha})}{(\tau\alpha)_n} \quad (5.20)$$

As shown in Chapter 3, $F_R U_c$ and $F_R(\tau\alpha)_n$ in the equations above can be determined directly from the slope and intercept of a collector efficiency curve plotted as shown in Fig. 5.20. In addition, Klein suggests that the time-averaged value $F_R(\overline{\tau\alpha})$ is related to the normal incidence value $F_R(\tau\alpha)_n$ as:

$$\frac{F_R(\overline{\tau\alpha})}{F_R(\tau\alpha)_n} = 0.95, \quad (5.21)$$

for a surface tilted within $\pm 20°$ of the local latitude.

Note that Fig. 5.20 is a plot of efficiency versus collector *inlet* fluid temperature, as distinguished from *average* fluid temperature. Such curves are usually available from the collector manufacturers. Figure 5.21 shows the f-chart for a liquid based solar heating and hot water system. The results of the computer model used to generate the f_s curves in Fig. 5.21 can also be expressed in the form of an empirical equation as [21]

$$f_s = 1.029 \, P_s - 0.065 \, P_L - 0.245 \, P_s^2 + 0.0018 \, P_L^2 + 0.0215 \, P_s^3 \quad (5.22)$$

valid for the range

$$0 \leq P_s \leq 3.0; \, 0 \leq P_L \leq 18.0 \text{ and } 0 \leq f_s \leq 1.0.$$

Table 5.7. Definition of parameters in f-chart.

Parameter	Definition
A_c	Net collector aperture area (m²).
f_s	The solar load fraction: percentage of monthly load carried by solar system.
F_R	Collector heat-removal factor. See Chapter 3, Eq. (3.43).
F_{hx}	Collector loop heat-exchanger factor. See Chapter 4, Eq. (4.36).
	$\quad F_{hx} = 1$ if no collector heat exchanger is used.
$\bar{I}_c$	Total monthly collector-plane insolation (J/m²-month). See Chapter 2.
L	Total monthly water-heating load (J/month). See Example 5.3.
T_a	Monthly average ambient temperature (°C).
T_R	Reference temperature with a value of 100°C.
U_c	Collector heat-loss coefficient (W/m² · °C). See Chapter 3.
Δt	Number of seconds per month (sec/month).
$(\overline{\tau\alpha})$	Monthly averaged collector transmittance-absorptance product.

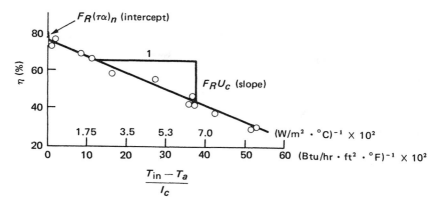

Figure 5.20. Typical flat-plate collector efficiency curve showing method of evaluating $F_R(\tau\alpha)_n$ and $F_R U_c$. From [24].

It is further required that $P_s > P_L/12$ to ensure that the monthly insolation is above the useful threshold and that thermal losses are below an upper bound at which energy absorbed is equal to energy lost from the absorber plate.

The f-chart in Fig. 5.21 is based on nominal values of collector flow rate, thermal storage mass, and load heat-exchanger effectiveness. It is possible to use the f-chart for other values of these important parameters if the loss parameter and a solar parameter

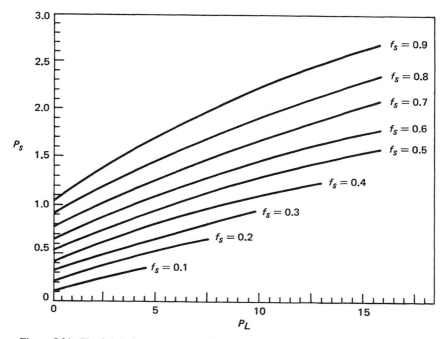

Figure 5.21. The f-chart for solar water-heating systems. From [20].

Table 5.8. Nominal values of physical parameters and modifying groups for *f*-chart use[a] for liquid based systems

Parameter	Nominal value	Modified parameter[b]	
Flow rate[c] $\dfrac{(\dot{m}c_p)_c}{A_c}$	0.0128 liters H_2O equivalent/ sec $\cdot$ m_c^2	$P_L = P_{L,\text{nom}} \dfrac{F_{hx}F_R}{(F_{hx}F_R)_{\text{nom}}}$ $P_s = P_{s,\text{nom}} \dfrac{F_{hx}F_R}{(F_{hx}F_R)_{\text{nom}}}$	(5.23) (5.24)
Storage volume (water) $V_s = \left(\dfrac{M}{\rho A_c}\right)_s$	75 liters H_2O/m_c^2	$P_L = P_{L,\text{nom}} \left(\dfrac{V_s}{75}\right)^{-0.25}$	(5.25)
Load heat exchanger[d] $\dfrac{\epsilon_L(\dot{m}c_p)_{\text{air}}}{Q_L}$	2.0	$P_s = P_{s,\text{nom}}\left\{0.393 + 0.651\exp\left[-0.139\,\dfrac{Q_L}{\epsilon_L(\dot{m}c_p)_{\text{air}}}\right]\right\}$	(5.26)

[a]Table prepared from data and equations presented in [8,20].
[b]Multiply basic definition of P_s and P_L in Eqs. (5.18) and (5.19) by factor for nonnominal group values; $(F_{hx}F_R)_{\text{nom}}$ refers to values of $F_{hx}F_R$ at collector rating or test conditions.
[c]In liquid systems the correction for flow rate is small and can usually be ignored if variation is no more than 50 percent below the nominal value.
[d]$(\dot{m}c_p)_{\text{min}}$ is the minimum fluid capacitance rate, usually that of air for the load heat exchanger; Q_L is the heat load per unit temperature difference between inside and outside of the building.

are appropriately modified. Table 5.8 contains the nominal values used in generating the *f*-chart and the dimensionless groups that are used to modify the loss or the solar parameter for other system values. The *f*-chart in Fig. 5.21 may be used for a water heating only system with the following modification for P_L:

$$P_L = \frac{F_{hx}F_R A_c U_c \Delta t (11.6 + 1.18T_{w,o} + 3.86T_{w,i} - 2.32\overline{T}_a)}{L}, \quad (5.27)$$

where $T_{w,i}$ and $T_{w,o}$ are water supply and delivery temperatures, respectively. Klein [20] has prepared *f*-charts for air based solar heating systems also as shown in Fig. 5.20. Loss and solar parameters, P_L and P_s, are identical to those for liquid systems and are defined in Eqs. (5.18) and (5.19). Since air systems do not use a collector-to-storage heat exchanger, the heat-exchanger factor is given by $F_{hx} \equiv 1.0$.

The data from which the *f*-chart was created can be expressed by an empirical equation as

$$f_s = 1.040\,P_s - 0.065\,P_L - 0.159\,P_s^2 + 0.00187\,P_L^2 - 0.0095\,P_s^3 \quad (5.28)$$

valid for the ranges:

$$0 \le P_s \le 3.0 \quad 0 \le P_L \le 18.0 \quad 0 \le f_s \le 1.0.$$

Table 5.9. Factors for storage capacity and air flow rate for f-chart[a] for air based systems

Parameter	Nominal value[b]	Loss parameter multiplier
Storage capacity V_s	0.25 m³/m²$_c$	$\left(\dfrac{0.25}{V_s}\right)^{0.3}$
Fluid volumetric flow rate Q_c	10.1 liters/sec · m²$_c$	$\left(\dfrac{Q_c}{10.1}\right)^{0.28}$

[a]Adapted from [8,20].
[b]Based on net collector area; fluid volume at standard atmosphere conditions.

It is further required that $P_s > 0.07\,P_L$ to ensure that the monthly insolation is above the useful threshold level and that thermal losses are below an upper bound at which energy absorbed is equal to energy lost from the absorber plate.

The f-chart for air systems is based on nominal values of air flow rate and storage capacity. Table 5.9 shows the nominal value of these two parameters and the dimensionless groups to be used to multiply the loss parameter to correct for flow rate or storage changes from the nominal value used to construct Fig. 5.22.

Example 5.4. Calculate the annual heating energy delivery of a solar space-heating system using a double-glazed, flat-plate collector in Bismarck, ND. The building and solar system specifications are given on page 250.

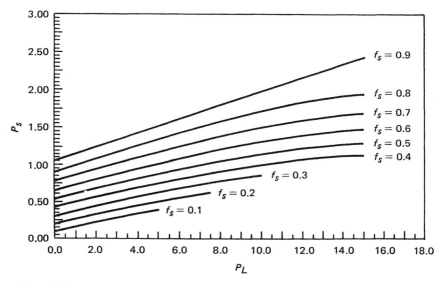

Figure 5.22. The f-chart for air-based solar space-heating systems. From [20].

Table 5.10. Climatic and solar data for Bismarck, ND

Month	Avg. ambient temperature (°C)	Heating degree C-days	Horizontal solar radiation (langleys/day)
Jan.	−13.2	978	157
Feb.	−10.3	801	250
Mar.	−3.8	687	356
Apr.	6.1	368	447
May	12.4	188	550
June	17.7	68	590
July	21.6	10	617
Aug.	20.7	19	516
Sep.	14.2	140	390
Oct.	8.2	313	272
Nov.	−1.7	601	161
Dec.	−9.1	851	124

Building Specifications
Location: 47°N latitude (see Appendix 2, Tables A2.2 and A2.3)
Space-heating load: 28000 kJ/°C · day
Solar System Specifications
Collector loss coefficient: $F_R U_c = 4.44$ W/m² − k
Collector optical efficiency (average): $F_R (\overline{\tau\alpha}) = 0.70$
Collector tilt: $\beta = L + 15° = 62°$
Collector area: $A_c = 60$ m²
Collector fluid flow rate: $\dot{m}_c/A_c = 55$ kg/hr m²
Collector fluid heat capacity: $c_{p_c} = 1.9$ kJ/°C · kg (antifreeze)
Storage capacity: 83 kg of H_2O/m_c^2
Storage fluid flow rate: $\dot{m}_s/A_c = 100$ kg of H_2O/hr m²
Storage fluid heat capacity: $c_{p_s} = 4.187$ kJ/kg °C (water)
Heat-exchanger effectiveness: 0.75
Load heat exchanger: $\epsilon_L (\dot{m}c_p)_{air}/Q_L = 2.0$
Climatic Data
Climatic data from the NWS are tabulated in Table 5.10.

Solution. The *f*-chart method is amenable to a step-by-step application. The following order is suggested.

1. Calculate monthly collector-plane insolation for each month.
2. Calculate solar and loss parameters P_s and P_L [Eqs. (5.18) and (5.19)] for each month and heat-exchanger penalty factor F_{hx} (Eq. 4.36).
3. Evaluate f_s from the *f*-chart for each month.
4. Calculate total annual energy delivery from monthly totals.

Each of these steps is shown in Table 5.11.

Table 5.11. The f-chart summary for Example 5.4

Month	Collector-plane radiation (kJ/m²-day)	Monthly energy demand (GJ)	P_L^*	P_s^*	f_s
Jan.	18378	27.38	2.81	0.83	0.53
Feb.	20654	22.43	3.34	1.14	0.69
Mar.	19327	19.24	3.66	1.24	0.73
Apr.	16491	10.25	6.22	1.99	0.91
May	15913	5.26	11.30	3.75	1.00
June	15356	1.90	29.39	10.01	1.00
July	16604	.28	190.00	73.43	1.00
Aug.	16988	.53	101.53	39.69	1.00
Sep.	17925	3.92	14.85	5.66	1.00
Oct.	18874	8.76	7.11	2.67	1.00
Nov.	14926	16.86	4.09	1.10	0.63
Dec.	15399	23.83	3.11	0.80	0.49
Annual	—	140.64	—	—	0.68

$^*P_s > 3.0$ or $P_L > 18.0$ implies $f_s = 1.0$; no correction for storage size and flow rates is required.

The following are the calculations for the month of January. Results for the other months can be found similarly.
From Table A2.2 (by interpolation):

$$\overline{H}_{o,h} = 3011 \text{ W·hr/m}^2 - \text{day} = 10874 \text{ kJ/m}^2 - \text{day}.$$

From Table 5.10:

$$\overline{H}_h = 157 \text{ langleys/day} = 6572 \text{ kJ/m}^2 - \text{day}.$$

Using Eq. (2.54):

$$\overline{K}_T = \frac{\overline{H}_h}{\overline{H}_{o,h}} = 0.604.$$

By Eq. (2.56):

$$\frac{\overline{D}_h}{\overline{H}_h} = 0.775 + 0.347\left(h_{ss} - \frac{\pi}{2}\right) - \left[0.505 + 0.0261\left(h_{ss} - \frac{\pi}{2}\right)\right]\cos(2\overline{K}_T - 1.8).$$

$$= 0.2156$$

Therefore,

$$\overline{D}_h = \frac{\overline{D}_h}{\overline{H}_h} \times \overline{H}_h = 0.2156 \times 6572 \text{ kJ/m}^2 - \text{day}$$

$$= 1417 \text{ kJ/m}^2 - \text{day}.$$

Using Eq. (2.57):

$$\overline{B}_h = \overline{H}_h - \overline{D}_h = 4662.3 \text{ kJ/m}^2 - \text{day}.$$

From Eq. (2.23) we can find:

$$\delta_s = 23.45° \sin[360(284 + 15)/365]° = -21.27°.$$

Using Eq. (2.30):

$$\begin{aligned} h_{sr(\alpha=0)} &= -\cos^{-1}(-\tan L \cdot \tan \delta_s) \\ &= -\cos^{-1}[-\tan(47°) \cdot \tan(-21\cdot27°)] \\ &= -65.3° \text{ or } -1.14 \text{ rad.} \end{aligned}$$

Therefore, using Eq. (2.59):

$$\begin{aligned} \overline{R}_b &= \frac{\cos(L - \beta)\cos\delta_s \sin h_{sr(\alpha=0)}\ \sin(L - \beta)\sin \delta_s}{\cos L \cos\delta_s \sin h_{sr(\alpha=0)} + h_{sr(\alpha=0)} \sin L \sin \delta_s} \\ &= \frac{\cos(47° - 62°)\cos(-21\cdot27°)\sin(-65.3°) - 1.14 \sin(47° - 62°) \sin(-21\cdot27°)}{\cos 47° \cos(-21\cdot27°) \sin(-65.3°) - 1.14 \sin 47° \sin(-21\cdot27°)} \\ &= 3.36. \end{aligned}$$

Using Eq. (2.61):

$$\overline{R}_d = \cos^2(\beta/2) = \cos^2(62°/2) = 0.7347.$$

Then, using Eq. (2.63), neglecting reflection,

$$\begin{aligned} \overline{H}_c &= \overline{R}_b \cdot \overline{B}_h + \overline{R}_d \cdot \overline{D}_h \\ &= 3.363 \times 5155 \text{ kJ/m}^2 \text{ day} + 0.7347 \times 1417 \text{ kJ/m}^2 \text{ day} \\ &= 18378 \text{ kJ/m}^2 \text{ day.} \end{aligned}$$

The monthly energy demand is:

$$\begin{aligned} \text{Load} &= \overline{U}_a \times \text{degree} - C - \text{days} \\ &= 28{,}000 \text{ kJ/°C day} \times 978°C \text{ day} \\ &= 27.384 \text{ GJ,} \\ (\dot{m}c_p)_c &= (\dot{m}_c/A_c) \times A_c \times c_{pc} \\ &= 55 \text{ kg/hr m}^2 \times 60 \text{ m}^2 \times 1.9 \text{ kJ/kg °K} \\ &= 6270 \text{ kJ/hr °K or 1742 W/K} \end{aligned}$$

$$(\dot{m}c_p)_s = (\dot{m}_s/A_c) \times A_c \times c_{ps}$$
$$= 100 \text{kg/m}^2 \text{ hr} \times 60 \text{m}^2 \times 4.187 \text{ kJ/kg} \ ^\circ\text{K}$$
$$= 25122 \text{ kJ/hr K or } 6978 \text{ W/K}$$
$$(\dot{m}c_p)_{\min} = (\dot{m}c_p)_c$$

Therefore, Eq. (4.36) becomes,

$$F_{hx} = \{1 + [F_R U_c A_c/(\dot{m}c_p)_c][1/\epsilon_{hx} - 1]\}^{-1}$$
$$= \{1 + 4.44 \text{ W/m}^2 - \text{K} \times 60 \text{m}^2/1742 \text{ W/K } [1/0.75 - 1]\}^{-1}$$
$$= 0.951.$$

Using Eq. (5.18),

$$P_L = \frac{A_c F_{hx} F_R U_c \Delta t (T_R - \overline{T}_a)}{\text{Load}}$$

$$= \frac{60 \text{ m}^2 \times 0.951 \times 4.44 \text{ W/m}^2 - \text{K} \times (31 \times 24 \times 3600) \text{hr}}{27.384 \text{ GJ} \times (10^9 \text{kJ/1 GJ})}(100^\circ\text{C} - (-13.2^\circ\text{C})$$

$$= 2.81.$$

Using Eq. (5.12),

$$P_s = \frac{A_c F_{hx} F_R I_c (\overline{\tau\alpha})}{\text{Load}}$$

$$= \frac{60 \text{ m}^2 \times 0.951 \times 0.7 \times (18378 \text{ kJ/m}^2 \text{ day} \times 31 \text{ days})}{27.384 \text{ GJ} \times (10^6 \text{kJ/1 GJ})}$$

$$= 0.83.$$

From Fig. 5.21 or Eq. (5.22),

$$f_s = 0.53.$$

5.5.2 The Utilizability Method

The utilizability method is a method to predict the long-term performance of a solar thermal system used for space heating, hot water, industrial process heat or thermal power systems. It is based on finding the long-term utilizability, ϕ, of a thermal collector, and is defined as the fraction of solar flux absorbed by a collector and delivered to the working fluid. The utilizability method is described in detail in Chapter 8 with applications in solar industrial process heat systems. The ϕ and the ϕ-f methods as described in Chapter 8 (section 8-4) are equally applicable to other solar thermal systems.

PROBLEMS

5.1. The no-load temperature of a building with internal heat sources is given by Eq. (5.6). How would this equation be modified to account for heat losses through the surface of an unheated slab, the heat losses being independent of ambient temperature?

5.2. An unheated garage is placed on the north wall of a building to act as a thermal buffer zone. If the garage has roof area A_r, window area A_{wi}, door area A_d and wall area A_{wa}, what is effective U value for the north wall of the building if its area is A_n? The garage floor is well insulated and has negligible heat loss. Express the effective U value in terms of the U values and areas of the several garage surfaces.

5.3. What is the annual energy demand for a building in Denver, CO, if the peak heat load is 150,000 Btu/hr based on a design temperature difference of 75°F? Internal heat sources are estimated to be 20,000 Btu/hr, and the design building interior temperature is 70°F.

5.4. What is the January solar load fraction for a water-heating system in Washington, D.C., using 100 m² of solar collector if the water demand is 4 m³/day at 65°C with a source temperature of 12°C? No heat exchanger is used and the solar collector efficiency curve is given in Fig. 5.18; the solar collector is tilted at an angle equal to the latitude.

5.5. Repeat Problem 5.4 for Albuquerque, NM, in July if the water source temperature is 17°C.

5.6. Explain how the f-chart (Fig. 5.19) can be used *graphically* to determine the solar load fraction for a range of collector sizes once the solar and loss parameters have been evaluated for only one system size. *Hint*: consider a straight line passing through the origin and the point (P_s, P_L).

5.7. The f-chart was generated using data for flat-plate collectors. What modifications would be necessary to use it for a compound parabolic concentrator collector? Describe the effect on each f-chart parameter in Table 5.7.

5.8. In an attempt to reduce cost, a solar designer has proposed replacing the shell-and-tube heat exchanger in Fig. 5.8 with a tube coil immersed in the storage tank. The shell-and-tube heat exchanger originally specified had a surface area of 10 m² and a U value of 2000 W/m² · K to be used with a 100-m² solar collector. Using Eq. (3.71) to estimate the U value of the submerged coil, how much length of 0.5-inch (1.27-cm) diameter copper pipe would be needed to achieve the same value of UA product as the shell-and-tube heat exchanger? What percentage of the storage tank volume would be consumed by this coil if 50 kg of water is used per square meter of collector? Use a storage water temperature of 60°C and a collector water outlet temperature of 70°C for the calculations.

5.9. If a solar system delivers 2500 MJ/m$^2 \cdot$ yr with a water flow rate of 30 kg/m$_c^2 \cdot$ hr and a plate efficiency factor $F' = 0.93$, how much energy will it deliver if the flow rate is doubled? Neglect the effect of flow rate on F'. The collector has a heat-loss conductance of 4 W/m$^2 \cdot$ °C.

5.10. How large (MJ/hr) should a heat-rejection system be if it must dump the entire heat production of a 1000-m^2 solar collector array in Denver, CO on August 21 if the collector is at 100°C and the ambient temperature is 35°C? Use solar collector data in Fig. 5.18 and hourly solar radiation data in Appendix 2.

5.11. Use the f-chart to determine the amount of solar energy that can be delivered in Little Rock, AR, in January for the following solar and building conditions:
Building:
    Load: 40 million Btu/month
    Latitude: 35°N
Solar System:
Collector tilt: 55°, facing south
Area: 1000 ft^2
Ambient temperature: 40.6°F
Collector efficiency curve: see Fig. 5.18
No heat exchanger used
Nominal storage, flow rate, and load heat-exchanger values used.

5.12. Repeat Problem 5.11 if storage size is doubled and halved.

5.13. Repeat Problem 5.11 for an air-heating system for which $F_R U_c = 0.64$ Btu/hr $\cdot$ ft$^2 \cdot$ °F and $F_R(\tau\alpha)_n = 0.50$ (typical commercial values). Do air or liquid collectors deliver more energy per square meter in this case?

5.14. Repeat Problem 5.13 if collector fluid rate is doubled.

5.15. Using the data in Problem 5.11, calculate system performance for a horizontal and for a vertical collector. Assume Eq. (5.12) applies to both cases.

5.16. The f-chart is based $F_R(\tau\alpha)$ and $F_R U_c$ values, which can be deduced from a plot of collector efficiency versus $(T_{f,\text{in}} - T_a)/I$. If such a plot is not available but (a) a plot of efficiency versus $(\overline{T}_f - T_a)/I$ is available or (b) a plot of efficiency versus $(T_{f,\text{out}} - T_a)/I$ is available, how can $F_R(\tau\alpha)$ and $F_R U_c$ be calculated from the slope and intercept of these two curves? Express your results in terms of the slopes, intercepts, and fluid capacitance rate $\dot{m}c_p/A_c$.

5.17. The schematic diagrams below illustrate the operation of a solar-assisted heat-pump system and a solar system augmented with a heat pump. Discuss the advantages and disadvantages of each system with respect to different climatic conditions.

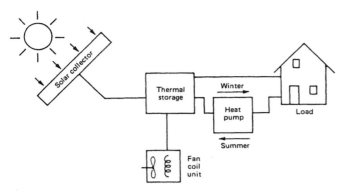

Solar-assisted heat-pump system

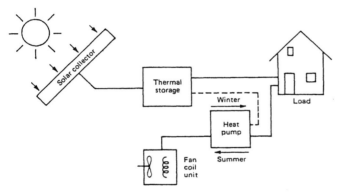

Solar system augmented with heat pump

REFERENCES AND SUGGESTED READINGS

1. Andreassy, S. 1964. *Proc. of the UN Conference on New Sources of Energy,* vol. 5: 20. New York. See also 1959. Root, D.E., A simplified engineering approach to swimming pool heating. *Sol. Energy* 3: 60.

2. ASHRAE. 1997. *Handbook of Fundamentals.* New York: American Society of Heating, Refrigerating and Air-Conditioning Engineers.

3. Balcomb, J.D., R.W. Jones, R.D. McFarland, and W.O. Wray. 1982. Expanding the SLR method. *Passive Solar Journal,* 1.

4. Balcomb, J.D., and J.C. Hedstrom. 1976. A simplified method for calculating required solar array size for space heating. In *Sharing the Sun,* vol. 4: 281–294, Winnipeg: ISES.

5. Balcomb, J.D., and J.D. Hedstrom. 1977. Simulation analysis of passive solar heated buildings—Preliminary Results. *Los Alamos Rept.* LA-UR-76–89, 1976; see also, under the same title, paper in *Sol. Energy* 19, 277–282.

6. Balcomb, J.D., et al. 1975. *Solar Handbook for Los Alamos.* New Mexico: LASL. See also Balcomb, J.D., et al. 1975. Design considerations of air-cooled collector/rock-bin storage solar heating systems, paper presented at the ISES Annual Meeting, Los Angeles, California.

7. Beckman, W.A., J.A. Duffie, and S.A. Klein. 1977. Simulation of Solar Heating Systems. In *Applications of solar energy for heating and cooling buildings.* New York: American Society of Heating, Refrigeration, and Air-Conditioning Engineers.

8. Beckman, W.A., S.A. Klein, and J.A. Duffie. 1977. *Solar Heating Design by the F-Chart Method*. New York: John Wiley & Sons.
9. Claridge, D.E., M. Krarti, and M. Bida. 1987. A validation of variable base degree-day cooling calculations. *ASHRAE Trans.* 93(2): 90–104.
10. Close, D.J. 1962. The performance of solar water heaters with natural circulation. *Sol. Energy* 6: 33.
11. deWinter, F. 1975. *How to Design & Build a Solar Swimming Pool Heater.* New York: Copper Development Association, Inc.
12. Erbs, D.G., S.A. Klein, and W.A. Beckman. 1983. Estimation of degree days and ambient temperature bin data from monthly average temperatures. *ASHRAE Transactions* 25(6): 60.
13. Gilman, S.F. 1975. Solar energy heat pump systems for heating and cooling buildings. *ERDA Rept.*, C00–2560–1.
14. Gutierrez, G., et al. 1974. Simulation of forced circulation water heaters: Effects of auxiliary energy supply load type and storage capacity. *Sol. Energy* 15: 287.
15. Hay, H.R., and J.I. Yellott. 1969. Natural air conditioning with roof-ponds and moving insulation. *Trans. ASHRAE*, vol. 75: 165.
16. Hittman Associates. 1973. Verification of the time-response method for heat load calculation. Rept. 2300–00259. Washington, D.C.: U.S. Government Printing Office.
17. Jacobs, M., and S.R. Peterson. 1974. *Making the Most of Your Energy Dollars*. Washington, D.C.: National Bureau of Standards. (U.S. Government Printing Office Rept. 003–003–01446–0).
18. Karman, V.D., et al. 1976. Simulation study of solar heat pump systems. In *Sharing the Sun,* vol. 3: 324–340. Winnipeg: ISES.
19. Kays, M., and A.L. London. 1964. *Compact Heat Exchangers*. New York: McGraw-Hill Book Co.
20. Klein, S.A. 1976. *A Design Procedure for Solar Heating Systems,* Ph.D. dissertation, University of Wisconsin, Madison. For an approach similar to the *f*-chart for other solar-thermal systems operating above a minimum temperature above that for space-heating ($\sim$ 20°C), see Klein, S.A., and W.A. Beckman. 1977. A general design method for closed loop solar energy systems. *Proc. 1977 ISES Meeting*.
21. Klein, S.A., W.A. Beckman, and J.A. Duffie. 1976. A Design Procedure for Solar Heating Systems. *Sol. Energy* 18: 113.
22. Klein, S.A., W.A. Beckman, and J.A. Duffie. 1976. A design procedure for solar air heating systems. Paper presented at the 1976 ISES Conference, American Section, Winnipeg, Manitoba, August 15–20.
23. Klein, S.A., et al. 1975. A method for simulation of solar processes and its application. *Sol. Energy*, 17: 29–37.
24. Kreider, J.F., and F. Kreith. 1977. *Solar Heating and Cooling,* revised 1st ed. Washington, D.C.: Hemisphere Publ. Corp.
25. Kreider, J.F., and A. Rable. 1994. *Heating and cooling of buildings*. New York: McGraw-Hill Book Co.
26. Kutcher, C.F. 1996. "Transpired solar collector system: A major advance in solar heating." Paper presented at the World Energy Engineering Congress, Nov. 6–8, 1996, Atlanta, GA.
27. Kutcher, C.F., and C.B. Christensen. 1992. "Unglazed transpired solar collectors." Advances in Solar Energy, K. Böer, ed., vol. 7, pp. 283–307.
28. Marvin, W.C., and S.A. Mummer. 1976. Optimum consideration of solar energy and the heat pump for residential heating. In *Sharing the Sun,* vol. 3: 321–323. Winnipeg: ISES.
29. Nyles, P.W.B. 1976. Thermal evaluation of a house using movable-insulation heating and cooling system. *Sol. Energy* 18: 412–419.
30. Phillips, W.F., and R.D. Cook. 1975. Natural circulation from a flat plate collector to a hot liquid storage tank. ASME Paper 75-HT-53.
31. Sfeir, A., et al. 1976. A numerical model for a solar water heater. In *Heliotechnique and Development,* (COMPLES 1975) vol. 2: 38–52, Cambridge, Massachusetts: Development Analysis Associates.
32. Thom, H.C.S. 1966. Normal degree-days above any base by the universal truncation coefficient. *Mon. Weather Rev.* 94: 461–465.
33. Thom, H.C.S. 1954. Normal degree-days below any base. *Mon. Weather Rev.* 82: 111–115.
34. Thom, H.C.S. 1954. The rational relationship between heating degree-days and temperature. *Mon. Weather Rev.* 82: 1–6.
35. Thomas H.C.S. 1952. Seasonal degree-day statistics for the U.S. *Mon. Weather Rev.* 80: 143–147.

36. Trombe, F., et al. 1976. *Some Characteristics of the CNRS Solar House Collectors.* Font Romeu/Odeillo, France: CNRS Solar Laboratory.
37. TRW, Inc. 1974. Solar heating and cooling of buildings (Phase O), Executive Summary. *NSF Rept.* NSF-R4-N-74–022A.
38. U.S. Department of Commerce. 1968. *Climatic atlas of the United States.* Washington, D.C.: U.S. Government Printing Office.
39. Ward, JC., and G.O.G. Löf. 1976. Long-term (18 years) performance of a residential solar heating system. *Sol. Energy* 18: 301–308. See also by the same authors: 1977. Maintenance costs of solar air heating systems. *Proc. 1977 ISES Meeting.*
40. Yellot, J.I., and H.R. Hay. 1969. Thermal analysis of a building with a natural air conditioning. *Trans. ASHRAE,* 75: 78.

SOLAR COOLING AND DEHUMIDIFICATION

The real cycle you're working on is a cycle called yourself.

Robert Pirsig

6.1 SOLAR SPACE COOLING AND REFRIGERATION*

In some ways solar energy is better suited to space cooling and refrigeration than to space heating, but this application of solar energy has not found much commercial success. The seasonal variation of solar energy is extremely well suited to the space-cooling requirements of buildings. The principal factors affecting the temperature in a building are the average quantity of radiation received and the environmental air temperature. Since the warmest seasons of the year correspond to periods of high insolation, solar energy is most available when comfort cooling is most needed. Moreover, as we have seen in Chapter 3, the efficiency of solar collectors increases with increasing insolation and increasing environmental temperature. Consequently, in the summer the amount of energy delivered per unit surface area of collector can be larger than that in winter.

There are several approaches that can be taken to solar space cooling and refrigeration. Because of the limited operating experience with solar-cooling systems, their design must be based on basic principles and experience with conventional cooling systems. The material presented in this chapter will therefore stress the fundamental principles of operation of refrigeration cycles and combine them with special features of the components in a solar system.

---

*In view of the fact that the space comfort industry in the United States uses English units exclusively, some of the examples in this chapter will be worked in these units to facilitate communication with practicing engineers.

The two principal methods of lowering air temperature for comfort cooling are refrigeration with actual removal of energy from the air or evaporation cooling of the air with adiabatic vaporization of moisture into it. Refrigeration systems can be used under any humidity condition of entering air, whereas evaporative cooling can be used only when the entering air has a comparatively low relative humidity.

The most widely used air conditioning method employs a vapor-compression refrigeration cycle. Another method uses an absorption refrigeration cycle similar to that of the gas refrigerator. The vapor compression refrigeration cycle requires energy input into the compressor which may be provided as electricity from a photovoltaic system or as mechanical energy from a solar driven engine. Referring to Figure 6.1, the compressor raises the pressure of the refrigerant which also increases its temperature. The compressed high temperature refrigerant vapor then transfers its heat to the ambient environment in the condenser, where it condenses to a high pressure liquid at a temperature close to (but higher than) the environmental temperature. The liquid refrigerant is then passed through the expansion valve where its pressure is suddenly reduced, resulting in a vapor-liquid mixture at a much lower temperature. The low temperature refrigerant is then used to cool air or water in the evaporator where the liquid refrigerant evaporates by absorbing heat from the medium being cooled. The cycle is completed by the vapor returning to the compressor. If water is cooled by the evaporator, the device is usually called a chiller. The chilled water is then used to cool the air in the building.

In an absorption system, the refrigerant is evaporated or distilled from a less volatile liquid absorbent, the vapor is condensed in a water- or air-cooled condenser, and the resulting liquid is passed through a pressure reducing valve to the cooling section of the unit. There it chills the water as it evaporates, and the resulting vapor flows into a vessel, where it is reabsorbed in the stripped absorbing liquid and pumped back

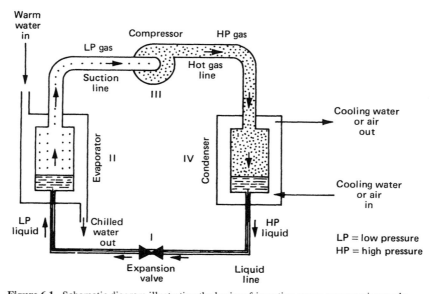

Figure 6.1. Schematic diagram illustrating the basic refrigeration vapor-compression cycle.

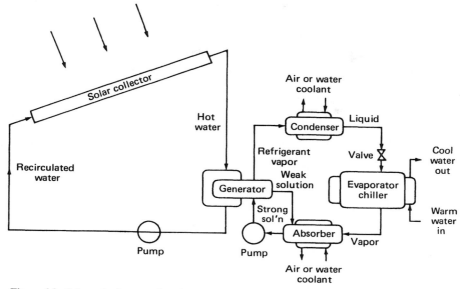

Figure 6.2. Schematic diagram of a solar-powered absorption refrigeration system.

to the heated generator. The heat required to evaporate the refrigerant in the generator can be supplied directly from solar energy as shown in Fig. 6.2.

In humid climates, removal of moisture from the air represents a major portion of the air conditioning load. In such climates, desiccant systems can be used for dehumification, in which solar energy can provide most of the energy requirements. There are several passive space cooling techniques, such as nocturnal cooling (night sky radiation), ground cooling and underground air tunnels. These techniques are described in Chapter 7. The present chapter covers the active solar cooling techniques based on vapor compression and vapor absorption refrigeration cycles and desiccant humidification.

6.1.1 Cooling Requirements for Buildings

The cooling load of a building is the rate at which heat must be removed to maintain the air in a building at a given temperature and humidity. It is usually calculated on the basis of the peak load expected during the cooling season. For a given building the cooling load depends primarily on

1. Inside and outside dry-bulb temperatures and relative humidities,
2. Solar-radiation heat load and wind speed,
3. Infiltration and ventilation, and
4. Internal heat sources.

A method of calculating the cooling load is presented in detail in Ref. [1].

The steps in calculating the cooling load of a building are as follows.

1. Specify the building characteristics: wall area, type of construction, and surface characteristics; roof area, type of construction, and surface characteristics; window area, setback, and glass type; and building location and orientation.
2. Specify the outside and inside wet- and dry-bulb temperatures.
3. Specify the solar heat load and wind speed.
4. Calculate building cooling load resulting from the following: heat transfer through windows; heat transfer through walls; heat transfer through roof; sensible and latent heat gains resulting from infiltration and ventilation, sensible and latent heat gains (water vapor) from internal sources, such as people, lights, cooking, etc.

Equations (6.1)–(6.7) may be used to calculate the various cooling loads for a building. Cooling loads resulting from lights, building occupants, etc. may be estimated from Ref. [1]. For unshaded or partially shaded windows, the load is

$$\dot{Q}_{wi} = A_{wi}\left[F_{sh}\bar{T}_{b,wi}I_{h,b}\frac{\cos i}{\sin \alpha} + \bar{\tau}_{d,wi}I_{h,d} + \bar{\tau}_{r,wi}I_r + U_{wi}(T_{out} - T_{in})\right]. \quad (6.1)$$

For shaded windows, the load (neglecting sky diffuse and reflected radiation) is

$$\dot{Q}_{wi,sh} = A_{wi,sh}U_{wi}(T_{out} - T_{in}). \quad (6.2)$$

For unshaded walls, the load is

$$\dot{Q}_{wa} = A_{wa}\left[\bar{\alpha}_{s,wa}\left(I_r + I_{h,d} + I_{h,b}\frac{\cos i}{\sin \alpha}\right) + U_{wa}(T_{out} - T_{in})\right]. \quad (6.3)$$

For shaded walls, the load (neglecting sky diffuse and reflected radiation) is

$$\dot{Q}_{wa,sh} = A_{wa,sh}[U_{wa}(T_{out} - T_{in})]. \quad (6.4)$$

For the roof, the load is

$$\dot{Q}_{rf} = A_{rf}\left[\bar{\alpha}_{s,rf}\left(I_{h,d} + I_{h,b}\frac{\cos i}{\sin \alpha}\right) + U_{rf}(T_{out} - T_{in})\right]. \quad (6.5)$$

Sensible-cooling load due to infiltration and ventilation is

$$\dot{Q}_i = \dot{m}_a(h_{out} - h_{in}) = \dot{m}_aCp_a(T_{out} - T_{in}). \quad (6.6)$$

Latent load due to infiltration and ventilation is

$$\dot{Q}_w = \dot{m}_a(W_{out} - W_{in})\lambda_w, \quad (6.7)$$

where

$\dot{Q}_{wi}$ = heat flow through unshaded windows of area A_{wi},

$\dot{Q}_{wi,sh}$ = heat flow through shaded windows of area $A_{wi,sh}$,

$\dot{Q}_{wu}$ = heat flow through unshaded walls of area A_{wa},

$\dot{Q}_{wu,sh}$ = heat flow through shaded walls of area $A_{wa,sh}$,

$\dot{Q}_{rf}$ = heat flow through roof of area A_{rf},

$\dot{Q}_i$ = heat load resulting from infiltration and ventilation,

$\dot{Q}_w$ = latent heat load,

$I_{h,b}$ = beam component of insolation on horizontal surface,

$I_{h,d}$ = diffuse component of insolation on horizontal surface,

I_r = ground-reflected component of insolation,

W_{out}, W_{in} = outside and inside humidity ratios,

U_{wi}, U_{wa}, U_{rf} = overall heat-transfer coefficients for windows, walls and roof, including radiation,

$\dot{m}_a$ = net infiltration and ventilation mass flow rate of dry air,

Cp_a = specific heat of air (approximately 1.025 kJ/kg-K for moist air),

T_{out} = outside dry-bulb temperature,

T_{in} = indoor dry-bulb temperature,

F_{sh} = shading factor (1.0 = unshaded, 0.0 = fully shaded),

$\overline{\alpha}_{s,wa}$ = wall solar absorptance,

$\overline{\alpha}_{s,rf}$ = roof solar absorptance,

i = solar-incidence angle on walls, windows and roof,

h_{out}, h_{in} = outside and inside air enthalpy,

α = solar-altitude angle,

λ_w = latent heat of water vapor,

$\overline{\tau}_{b,wi}$ = window transmittance for beam (direct) insolation,

$\overline{\tau}_{d,wi}$ = window transmittance for diffuse insolation, and

$\overline{\tau}_{r,wi}$ = window transmittance for ground-reflected insolation.

Recent ASHRAE handbooks recommend the use of the CLTD (Cooling Load Temperature Difference) method. For more details of the method one should refer to Ref. [1].

Example 6.1. Determine the cooling load for a building in Phoenix, AZ, with the specifications tabulated in Table 6.1.

Solution. To determine the cooling load for the building just described, calculate the following factors in the order listed.

1. Incidence angle for the south wall i at solar noon can be written from Eqs. (2.44) and (2.28) as

$$\cos i = \cos \beta \cos (L - \delta_s) + \sin \beta \sin (L - \delta_s) \tag{6.8}$$

$$= 0.26.$$

Table 6.1.

Factor	Description or specification
Building characteristics:	
Roof:	
Type of roof	Flat, shaded
Area $A_{rf,sh}$, ft²	1700
Walls (painted white):	
Size, north and south, ft	8 × 60 (two)
Size, east and west, ft	8 × 40 (two)
Area A_{wa}, north and south walls, ft²	$480 - A_{wi} = 480 - 40 = 440$ (two)
Area A_{wa}, east and west walls, ft²	$320 - A_{wi} = 320 - 40 = 280$ (two)
Absorptance $\overline{\alpha}_{s,wa}$ of white paint	0.12
Windows:	
Size, north and south, ft	4 × 5 (two)
Size, east and west, ft	4 × 5 (two)
Shading factor F_{sh}	0.20
Insolation transmittance	$\overline{\tau}_{b,wi} = 0.60; \overline{\tau}_{d,wi} = 0.81; \overline{\tau}_{r,wi} = 0.60$
Location and latitude	Phoenix, AZ.; 33°N
Date	August 1
Time and local-solar-hour angle H_s	Noon; $H_s = 0$
Solar declination δ_s, deg	17° − 55′
Wall surface tilt from horizontal β	90°
Temperature, outside and inside, °F	$T_{out} = 100; T_{in} = 75$
Insolation I, Btu/hr · ft²	$I_{h,b} = 185; I_{h,d} = 80; I_r = 70$
U factor for walls, windows and roof	$U_{wa} = 0.19; U_{wi} = 1.09; U_{rf} = 0.061$
Infiltration, lbm dry air/hr	Neglect
Ventilation, lbm dry air/hr	Neglect
Internal loads	Neglect
Latent heat load Q_w, percent	30 percent of wall sensible heat load[a]

[a] Approximate rule of thumb for Phoenix.

2. Solar altitude α at solar noon (from Eq. (2.28))

$$\sin \alpha = \sin \delta_s \sin L + \cos \delta_s \cos L \cos h_s = \cos (L - \delta_s) = \cos 15° = 0.966.$$

3. South-facing window load [from Eq. (6.1)]

$$\dot{Q}_{wi} = 40\left\{\left(0.20 \times 0.6 \times 185 \frac{0.26}{0.966}\right) + (0.81 \times 80) + (0.60 \times 70).\right.$$
$$\left. + [1.09(100 - 75)]\right\} = 5600\text{Btu/hr.}$$

4. Shaded-window load [from Eq. (6.2)]

$$\dot{Q}_{wi,sh} = (3 \times 40)[1.09(100 - 75)] = 3270 \text{ Btu/hr.}$$

5. South-facing wall load [from Eq. (6.3)]

$$\dot{Q}_{wa} = (480 - 40)\left\{0.12\left[70 + 80 + \left(185\frac{0.26}{0.966}\right)\right] + 0.19(100 - 75)\right\}$$

$$= 12{,}610\,\text{Btu/hr}.$$

6. Shaded-wall load [from Eq. (6.4)]

$$\dot{Q}_{wa,sh} = [(480 + 320 + 320) - (3 \times 40)][0.19(100 - 75)] = 4750\,\text{Btu/hr}.$$

7. Roof load [from Eq. (6.5)]

$$\dot{Q}_{rf} = 1700[\overline{\alpha}_{s,rf} \times 0 + 0.061(100 - 75)] = 2600\,\text{Btu/hr}.$$

8. Latent-heat load (30 percent of sensible wall load)

$$\dot{Q}_w = 0.3[(480 + 480 + 320 + 320) - (4 \times 40)][0.19(100 - 75)] = 2050\,\text{Btu/hr}.$$

9. Infiltration load

$$\dot{Q}_i = 0.$$

10. Total cooling load for the building described in the example

$$\dot{Q}_{tot} = \dot{Q}_{wi} + \dot{Q}_{wi,sh} + \dot{Q}_{wa} + \dot{Q}_{wa,sh} + \dot{Q}_{rf} + \dot{Q}_w + \dot{Q}_i,$$

$$\dot{Q}_{tot} = 30{,}880\,\text{Btu/hr}.$$

6.1.2 Vapor-Compression Cycle

The principle of operation of a vapor-compression refrigeration cycle can be illustrated conveniently with the aid of a pressure-enthalpy diagram as shown in Fig. 6.3. The ordinate is the pressure of the refrigerant in N/m^2 absolute, and the abscissa its enthalpy in kJ/kg. The roman numerals in Fig. 6.3 correspond to the physical locations in the schematic diagram of Fig. 6.1.

Process I is a throttling process in which hot liquid refrigerant at the condensing pressure p_c passes through the expansion valve, where its pressure is reduced to the evaporator pressure p_e. This is an isenthalpic (constant enthalpy) process, in which the temperature of the refrigerant decreases. In this process, some vapor is produced and the state of the mixture of liquid refrigerant and vapor entering the evaporator is shown by point A. Since the expansion process is isenthalpic, the following relation holds:

$$h_{ve}f + h_{le}(1 - f) = h_{lc}, \tag{6.9}$$

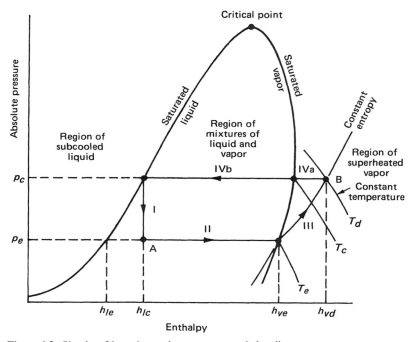

Figure 6.3. Simple refrigeration cycle on pressure-enthalpy diagram.

where f is the fraction of mass in vapor state, subscripts v and l refer to vapor and liquid states, respectively, and c and e refer to states corresponding to condenser and evaporator pressures, respectively. And,

$$f = \frac{h_{lc} - h_{le}}{h_{ve} - h_{le}}. \tag{6.10}$$

Process II represents the vaporization of the remaining liquid. This is the process during which heat is removed from the chiller. Thus, the specific refrigeration effect per kilogram of refrigerant q_r is

$$q_r = h_{ve} - h_{lc} \text{ in kJ/kg (Btu/lb)} \tag{6.11}$$

In the United States, it is still common practice to measure refrigeration in terms of *tons*. One ton is the amount of cooling produced if 1 ton of ice is melted over a period of 24 hr. Since 1 ton = 907.2 kg and the latent heat of fusion of water is 334.9 kJ/kg,

$$1 \text{ ton} = \frac{(907.2 \text{ kg}) \times (334.9 \text{ kJ/kg})}{(24 \text{ hr}) \times (3600 \text{ sec/hr})} = 3.516 \text{ kW} = 12{,}000 \text{ Btu/hr}. \tag{6.12}$$

If the desired rate of refrigeration requires a heat-transfer rate of $\dot{Q}_r$, the rate of mass flow of refrigerant necessary $\dot{m}_r$ is

$$\dot{m}_r = \frac{\dot{Q}_r}{(h_{ve} - h_{lc})}. \tag{6.13}$$

Process III in Fig. 6.3 represents the compression of refrigerant from pressure p_e to p_c. The process requires work input from an external source, which may be obtained from a solar-driven expander-turbine or a solar electrical system. In general, if the heated vapor leaving the compressor is at the condition represented by point B in Fig. 6.3, the work of compression W_c is

$$W_c = \dot{m}_r(h_{vd} - h_{ve}). \tag{6.14}$$

In an idealized cycle analysis, the compression process is usually assumed to be isentropic.

Process IV represents the condensation of the refrigerant. Actually, sensible heat is first removed in the subprocess IVa as the vapor is cooled at constant pressure from T_d to T_c and latent heat is removed at the condensation temperature T_c, corresponding to the saturation pressure p_c in the condenser. The heat-transfer rate in the condenser $\dot{Q}_c$ is

$$\dot{Q}_c = \dot{m}_r(h_{vd} - h_{lc}) \tag{6.15}$$

This heat must be rejected into the environment, either to cooling water or to the atmosphere if no water is available.

The overall performance of a refrigeration machine is usually expressed as the ratio of the heat transferred in the evaporator $\dot{Q}_r$ to the shaft work supplied to the compressor. This ratio is called the **coefficient of performance** (*COP*), defined by

$$COP = \frac{\dot{Q}_r}{W_c} = \frac{h_{ve} - h_{lc}}{h_{vd} - h_{ve}}. \tag{6.16}$$

The highest coefficient of performance for any given evaporator and condenser temperatures would be obtained if the system were operating on a reversible Carnot cycle. Under these conditions [1]

$$COP(\text{Carnot}) = \frac{T_e}{T_d - T_e}. \tag{6.17}$$

However, frictional effects and irreversible heat losses reduce the *COP* of real cycles much below this maximum.

Example 6.2. Calculate the amount of shaft work to be supplied to a 1-ton (3.52-kW) refrigeration plant operation at evaporater and condenser temperatures of

Table 6.2. Properties of refrigerant 134a for Example 6.2

Temperature (K)	Absolute pressure (kP$_a$)	Vapor specific volume (m³/kg)	Liquid enthalpy (kJ/kg)	Vapor enthalpy (kJ/kg)	Vapor entropy (KJ/kg · K)
		Saturated			
273	292.8	0.0689	50.02	247.2	0.919
309	911.7	0.0223	100.25	266.4	0.9053
		Superheated			
308.5	900	0.0226	—	266.18	0.9054
313	900	0.0233	—	271.3	0.9217
312.4	1000	0.0202	—	268.0	0.9043
313	1000	0.0203	—	268.7	0.9066

273 K and 309 K, respectively, using Refrigerant 134a (R-134a) as the working fluid. The properties of Refrigerant 134a are tabulated in Table 6.2. (More complete data are given in Appendix 6.) Also calculate the *COP* and the mass flow rate of the refrigerant.

Solution. From the property table the enthalpies for process I are

saturated vapor at 273 K $h_{ve} = 247.2$ kJ/kg,
saturated liquid at 309 K $h_{lc} = 100.3$ kJ/kg, and
saturated liquid at 273 K $h_{le} = 50.0$ kJ/kg.

Therefore, from Eq. (6.10)

$$f = \frac{100.3 - 50.0}{247.2 - 50.0} = 0.255.$$

The mass flow rate of refrigerant $\dot{m}_r$ is obtained from Eq. (6.13) and the enthalpies above, or

$$\dot{m}_r = \frac{3.52 \text{ kW}}{(247.2 - 100.3) \text{ kJ/kg}} = 0.024 \text{ kg/sec}.$$

The specific shaft-work input required is

$$\frac{W_c}{\dot{m}_r} = h_{vd} - h_{ve}.$$

The entropy s_e of the saturated vapor entering the compressor at 273 K and 292.8 kPa is 0.919 kJ/kg · K. From the property table, superheated vapor at a pressure of 911.7 kPa has an entropy of 0.919 kJ/kg · K at a temperature of 313 K with an enthalpy of 270.8 kJ/kg. Thus, the energy input to the working fluid by the compressor is

$$W_c = 0.024(270.8 - 247.2) = 0.566 \text{ kW}.$$

Finally, the heat-transfer rate from the refrigerant to the sink, or cooling water in the condensor, is from Eq. (6.15)

$$\dot{Q}_c = \dot{m}_r(h_{vd} - h_{lc}) = 0.024(270.8 - 100.3) = 4.09 \text{ kW}.$$

The *COP* of the thermodynamic cycle is

$$COP = \frac{247.2 - 100.3}{270.8 - 247.2} = 6.2,$$

whereas the Carnot *COP* is 273/36 or 7.6.

The above cycle has been idealized. In practice, the liquid entering the expansion valve is several degrees below the condensing temperature, while the vapor entering the compressor is several degrees above the evaporation temperature. In addition, pressure drops occur in the suction, discharge and liquid pipelines, and the compression is not truly isentropic. Finally, the work required to drive the compressor is somewhat larger than W_c above, because of frictional losses. All of these factors must be taken into account in a realistic engineering design.

6.1.3 Absorption Air-Conditioning

Absorption air-conditioning is compatible with solar energy since a large fraction of the energy required is thermal energy at temperatures that currently available flat-plate collectors can provide.

Solar absorption air conditioning has been the subject of investigation by a number of researchers [5, 15, 19, 24–26, 35, 39]. Fig. 6.4 shows a schematic of an absorption refrigeration system. Absorption refrigeration differs from vapor-compression air-conditioning only in the method of compressing the refrigerant (left of the dashed line in Fig. 6.4). In absorption air-conditioning systems, the pressurization is accomplished by first dissolving the refrigerant in a liquid (the absorbent) in the absorber section, then pumping the solution to a high pressure with an ordinary liquid pump. The low-boiling refrigerant is then driven from solution by the addition of heat in the generator. By this means, the refrigerant vapor is compressed without the large input of high-grade shaft work that the vapor-compression air-conditioning demands.

The effective performance of an absorption cycle depends on the two materials that comprise the refrigerant-absorbent pair. Desirable characteristics for the refrigerant-absorbent pair follow.

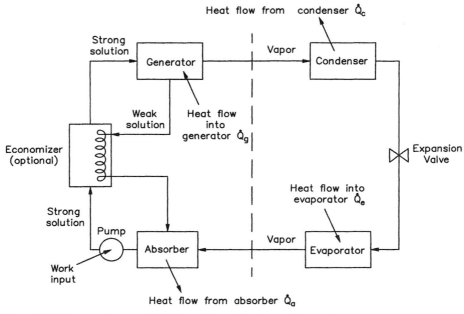

Figure 6.4. Diagram of heat and fluid flow of absorption air conditioner, with economizer. From [20].

1. The absence of a solid-phase absorbent.
2. A refrigerant more volatile than the absorbent so that separation from the absorbent occurs easily in the generator.
3. An absorbent that has a strong affinity for the refrigerant under conditions in which absorption takes place.
4. A high degree of stability for long-term operations.
5. Nontoxic and nonflammable fluids for residential applications. This requirement is less critical in industrial refrigeration.
6. A refrigerant that has a large latent heat so that the circulation rate can be kept low.
7. A low fluid viscosity that improves heat and mass transfer and reduces pumping power.
8. Fluids that must not cause long term environmental effects.

Lithium Bromide-Water (LiBr-H_2O) and Ammonia-Water (NH$_3$-H_2O) are the two pairs that meet most of the requirements. In the LiBr-H_2O system, water is the refrigerant and LiBr is the absorber, while in the Ammonia-Water system, ammonia is the refrigerant and water is the absorber. Because the LiBr-H_2O system has high volatility ratio, it can operate at lower pressures and, therefore, at the lower generator temperatures achievable by flat-plate collectors. A disadvantage of this system is that the pair tends to form solids. LiBr has a tendency to crystallize when air cooled, and the system cannot be operated at or below the freezing point of water. Therefore, the LiBr-H_2O

system is operated at evaporator temperatures of 5°C or higher. Using a mixture of LiBr with some other salt as the absorbent can overcome the crystallization problem. The ammonia-water system has the advantage that it can be operated down to very low temperatures. However, for temperatures much below 0°C, water vapor must be removed from ammonia as much as possible to prevent ice crystals from forming. This requires a rectifying column after the boiler. Also, ammonia is a safety Code Group B2 fluid (ASHRAE Standard 34–1992) which restricts its use indoors [1].

Other refrigerant-absorbent pairs include [24]

- Ammonia-salt,
- Methylamine-salt,
- Alcohol-salt,
- Ammonia-organic solvent,
- Sulfur dioxide-organic solvent,
- Halogenated hydrocarbons-organic solvent,
- Water-alkali nitrate, and
- Ammonia-water-salt.

If the pump work is neglected, the *COP* of an absorption air-conditioner can be calculated from Fig. 6.4:

$$COP = \frac{\text{cooling effect}}{\text{heat input}} = \frac{\dot{Q}_e}{\dot{Q}_g} \tag{6.18}$$

The *COP* values for absorption air-conditioning range from 0.5 for a small, single-stage unit to 0.85 for a double-stage, steam-fired unit. These values are about 15 percent of the *COP* values that can be achieved by a vapor-compression air-conditioner. It is difficult to compare the *COP* of an absorption air-conditioner with that of a vapor-compression air conditioner directly because the efficiency of electric power generation or transmission is not included in the *COP* of the vapor-compression air conditioning. The following example illustrates the thermodynamics of an LiBr-H_2O absorption refrigeration system.

Example 6.3. A water-lithium bromide, absorption refrigeration system such as that shown in Fig. 6.5 is to be analyzed for the following requirements:

1. The machine is to provide 352 kW of refrigeration with an evaporator temperature of 5°C, an absorber outlet temperature of 32°C, and a condenser temperature of 43°C.
2. The approach at the low-temperature end of the liquid heat exchanger is to be 6°C.
3. The generator is heated by a flat-plate solar collector capable of providing a temperature level of 90°C.

Determine the *COP*, absorbent and refrigerant flow rates and heat input.

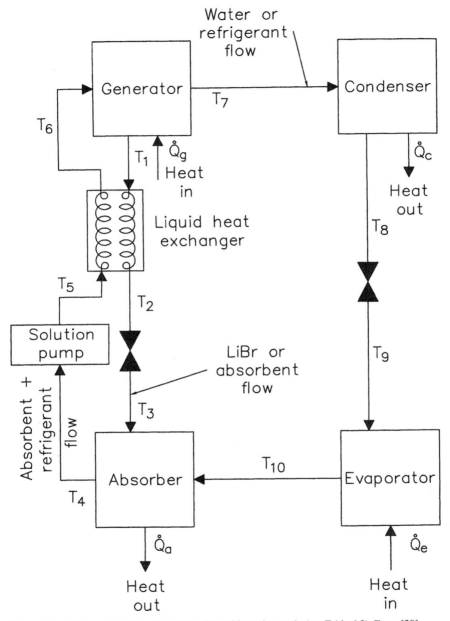

Figure 6.5. Lithium bromide-water, absorption refrigeration cycle (see Table 6.3). From [20].

Solution. For analytical evaluation of the LiBr-H_2O cycle, the following simplifying assumptions are made.

1. At those points in the cycle for which temperatures are specified, the refrigerant and absorbent phases are in equilibrium.
2. With the exception of pressure reductions across the expansion device between points 2 and 3, and 8 and 9 in Fig. 6.5, pressure reductions in the lines and heat exchangers are neglected.
3. Pressures at the evaporator and condenser are equal to the vapor pressure of the refrigerant, i.e., water, as found in steam tables.
4. Enthalpies for LiBr-H_2O mixtures are given in Fig. 6.6.

As a first step in solving the problem, set up a table (Table 6.3) of properties; for example, given

Generator Temp. $= 90°C = T_1 = T_7$,
Evaporator Temp. $= 5°C = T_9 = T_{10}$,
Condenser Temp. $= 43°C = T_8$,
Absorber Temp. $= 32°C = T_4$.

Neglecting the pump work $T_5 \approx T_4 = 32°C$.
Since the approach at the low temperature end of the heat exchanger is $6°C$,

$$T_2 = T_5 + 6°C = 38°C, \text{ and}$$
$$T_3 \approx T_2 = 38°C.$$

Table 6.3. Thermodynamic properties of refrigerant and absorbent for Fig. 6.5[a]

Condition No. in Fig. 6.5	Temperature (°C)	Pressure (kPa)	LiBr weight fraction	Flow (kg/kg H_2O)	Enthalpy (kJ/kg)
1	90	8.65	0.605	11.2	215
2	38	8.65	0.605	11.2	110
3	38	0.872	0.605	11.2	110
4	32	0.872	0.53	12.2	70
5	32	0.872	0.53	12.2	70
6	74	8.65	0.53	12.2	162
7	90	8.65	0	1.0	2670
8	43	8.65	0	1.0	180
9	5	0.872	0	1.0	180
10	5	0.872	0	1.0	2510

[a]From [20].

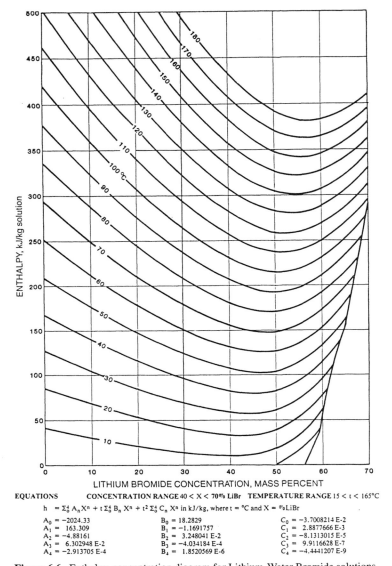

EQUATIONS   CONCENTRATION RANGE 40 < X < 70% LiBr   TEMPERATURE RANGE 15 < t < 165°C

$$h = \sum_0^4 A_n X^n + t \sum_0^4 B_n X^n + t^2 \sum_0^4 C_n X^n \text{ in kJ/kg, where } t = °C \text{ and } X = \%LiBr$$

$A_0 = -2024.33$	$B_0 = 18.2829$	$C_0 = -3.7008214 \text{ E-2}$
$A_1 = 163.309$	$B_1 = -1.1691757$	$C_1 = 2.8877666 \text{ E-3}$
$A_2 = -4.88161$	$B_2 = 3.248041 \text{ E-2}$	$C_2 = -8.1313015 \text{ E-5}$
$A_3 = 6.302948 \text{ E-2}$	$B_3 = -4.034184 \text{ E-4}$	$C_3 = 9.9116628 \text{ E-7}$
$A_4 = -2.913705 \text{ E-4}$	$B_4 = 1.8520569 \text{ E-6}$	$C_4 = -4.4441207 \text{ E-9}$

Figure 6.6. Enthalpy-concentration diagram for Lithium-Water Bromide solutions.

Since the fluid at conditions 7, 8, 9 and 10 is pure water, the properties can be found from the steam tables. Therefore,

$$P_7 = P_8 = \text{Saturation pressure of } H_2O \text{ at } 43°C = 8.65 \text{ kPa,}$$

and

$$P_9 = P_{10} = \text{Saturation pressure of } H_2O \text{ at } 5°C = 0.872 \text{ kPa.}$$

Therefore,

$$P_1 = P_2 = P_5 = P_6 = P_7 = 8.65 \text{ kPa},$$

and

$$P_3 = P_4 = P_{10} = 0.872 \text{ kPa}.$$

Enthalpy,

$$h_9 = h_8 = 180 \text{ kJ/kg (Saturated liquid at 43°C),}$$

$$h_{10} = 2510 \text{ kJ/kg (Saturated vapor enthalpy at 6°C), and}$$

$$h_7 = 2760 \text{ kJ/kg (Superheated vapor @ 8.65 kPa, 90°C).}$$

For the LiBr–H_2O mixture, conditions 1 and 4 may be considered equilibrium saturation conditions which may be found from Fig. 6.6 and 6.7 as:
   For

$$T_4 = 32°C \text{ and } P_4 = 0.872 \text{ kPa}, \quad Xr = 0.53, h_4 = 70 \text{ kJ/kg} - \text{Sol.}$$

Therefore,

$$h_5 \approx 70 \text{ kJ/kg} - \text{Sol.}$$

And for

$$T_1 = 90°C \text{ and } P_1 = 8.65 \text{ kPa}, \quad X_{ab} = 0.605, h_1 = 215 \text{ kJ/kg} - \text{Sol.},$$

for

$$T_3 = 38°C, \qquad\qquad X_3 = 0.605, h_3 = 110 \text{ kJ/kg} - \text{Sol.},$$
$$h_2 = h_3 = 110 \text{ kJ/kg}$$

6.1.4 Mass Balance Equations

Relative flow rates for the absorbent (LiBr) and the refrigerant (H_2O) are obtained from material balances. A total material balance on the generator gives

$$\dot{m}_6 = \dot{m}_1 + \dot{m}_7,$$

while a LiBr balance gives

$$\dot{m}_6 X_s = \dot{m}_1 X_{ab},$$

where X_{ab} = concentration of LiBr in absorbent of solution,
       X_r = concentration of LiBr in refrigerant-absorbent of solution.

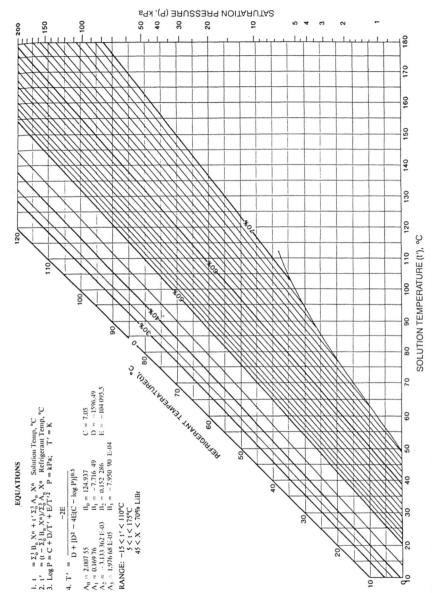

Figure 6.7. Equilibrium chart for Lithium Bromide–Water solutions.

Substituting $(\dot{m}_1 + \dot{m}_7)$ for $\dot{m}_6$ gives

$$\dot{m}_1 X_s + \dot{m}_7 X_s = \dot{m}_1 X_{ab}.$$

Since the fluid entering the condenser is pure refrigerant, that is, water, $\dot{m}_7$ is the same as the flow rate of the refrigerant $\dot{m}_r$:

$$\frac{\dot{m}_1}{\dot{m}_7} = \frac{X_s}{X_{ab} - X_s} = \frac{\dot{m}_{ab}}{\dot{m}_r},$$

where $\dot{m}_{ab}$ = flow rate of absorbent,

$$\dot{m}_r = \text{flow rate of refrigerant.}$$

Substituting for X_s and X_{ab} from the table gives the ratio of absorbent-to-refrigerant flow rate:

$$\frac{\dot{m}_{ab}}{\dot{m}_r} = \frac{0.53}{0.605 - 0.53} = 7.07.$$

The ratio of the refrigerant-absorbent solution flow rate $\dot{m}_s$ to the refrigerant-solution flow rate $\dot{m}_r$ is

$$\frac{\dot{m}_s}{\dot{m}_r} = \frac{\dot{m}_{ab} + \dot{m}_r}{\dot{m}_r} = 7.07 + 1 = 8.07.$$

Now Table 6.3 is complete except for T_6 and h_6 which may be found from an energy balance at the heat exchanger.

$$\dot{m}_s h_5 + \dot{m}_{ab} h_1 = \dot{m}_{ab} h_2 + \dot{m}_s h_6.$$

Hence,

$$h_6 = h_5 + \left[\frac{\dot{m}_{ab}}{\dot{m}_s}(h_1 - h_2)\right] = 70 + \frac{7.07}{8.07}[215 - 110] = 162 \text{ kJ/kg of solution.}$$

The temperature corresponding to this value of enthalpy and a LiBr mass fraction 0.53 is found from Fig. 6.7 to be 74°C.

The flow rate of refrigerant required to produce the desired 352 kW of refrigeration is

$$\dot{Q}_e = \dot{m}_r(h_{10} - h_9),$$

where $\dot{Q}_e$ is the cooling effect produced by the refrigeration unit, and

$$\dot{m}_r = \frac{352}{2510 - 180} = 0.15 \text{ kg/sec.}$$

The flow rate of the absorbent is

$$\dot{m}_{ab} = \frac{\dot{m}_{ab}}{\dot{m}_r} \dot{m}_r = 7.07 \times 0.15 = 1.06 \text{ kg/sec,}$$

while the flow rate of the solution is

$$\dot{m}_s = \dot{m}_{ab} + \dot{m}_r = 1.06 + 0.15 = 1.21 \text{ kg/sec.}$$

The rate at which heat must be supplied to the generator $\dot{Q}_g$ is obtained from the heat balance

$$\dot{Q}_g = \dot{m}_r h_7 + \dot{m}_{ab} h_1 - \dot{m}_s h_6$$
$$= [(0.15 \times 2670) + (1.06 \times 215)] - (1.21 \times 1.62)$$
$$= 432 \text{ kW.}$$

This requirement, which determines the size of the solar collector, probably represents the maximum heat load that the collector unit must supply during the hottest part of the day.

The coefficient of performance COP is

$$COP = \frac{\dot{Q}_e}{\dot{Q}_g} = \frac{352}{432} = 0.81$$

The rate of heat transfer in the other three heat-exchanger units—the liquid heat exchanger, the water condenser and the absorber—is obtained from heat balances. For the liquid heat exchanger this gives

$$\dot{Q}_{1-2} = \dot{m}_{ab}(h_1 - h_2) = 1.06[215 - 110] = 111 \text{ kW,}$$

where $\dot{Q}_{1-2}$ is the rate of heat transferred from the absorbent stream to the refrigerant-absorbent stream. For the water condenser the rate of heat transfer $\dot{Q}_{7-8}$ rejected to the environment is

$$\dot{Q}_{7-8} = \dot{m}_r(h_7 - h_8) = 0.15(2670 - 180) = 374 \text{ kW.}$$

The rate of heat removal from the absorber can be calculated from an overall heat balance on this system:

$$\dot{Q}_a = \dot{Q}_{7-8} - \dot{Q}_g - \dot{Q}_e = 374 - 432 - 352 = -410 \, \text{kW}.$$

Explicit procedures for the mechanical and thermal design as well as the sizing of the heat exchangers are presented in standard heat-transfer texts. In large commercial units, it may be possible to use higher concentrations of LiBr, operate at a higher absorber temperature, and thus save on heat-exchanger cost. In a solar-driven unit, this approach would require concentrator-type or high efficiency flat-plate solar collectors.

6.1.5 Ammonia-Water Refrigeration System

The main difference between an ammonia-water system and a water-lithium bromide system is that the ammonia-water system has a rectifier (also called Dephlagmator) after the boiler to condense as much water vapor out of the mixture vapor as possible. Figure 6.8 shows a schematic of an NH_3-H_2O absorption refrigeration system. Since ammonia has a much lower boiling point than water, a very high fraction of ammonia and a very small fraction of water are boiled off in the boiler. The vapor is cooled, as it rises in the rectifier, by the countercurrent flow of the strong NH_3-H_2O) solution from the absorber, therefore, some moisture is condensed. The weak ammonia-water solution from the boiler goes through a pressure reducing valve to the absorber, where it absorbs the ammonia vapor from the evaporator. The high pressure, high temperature ammonia from the rectifier is condensed by rejecting heat to the atmosphere. It may be further sub-cooled before expanding in a throttle valve. The low pressure, low temper-

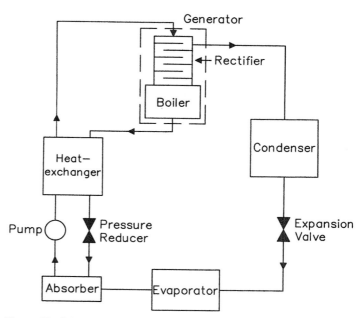

Figure 6.8. Schematic of an ammonia-water absorption refrigeration system.

ature ammonia from the throttle valve provides refrigeration in the evaporator. The vapor from the evaporator is reunited with the weak ammonia solution in the absorber.
Operating pressures are primarily controlled by the ambient air temperature for an air
cooled condenser, the evaporator temperature, and the concentration of the ammonia
solution in the absorber. Thermodynamic analysis of an ammonia-water absorption
cooling system will become clear from the following example.

Example 6.4. A 10.5 kW, gas fired, ammonia-water absorption chiller is shown
schematically in Fig. 6.9. The chiller is operating with the following conditions:

Evaporator temperature	2°C,
Evaporator pressure	4.7 Bar,
Condenser and Generator Pressure	21.7 Bar,
Concentration of ammonia in refrigerant	$(X_r) = 0.985$,

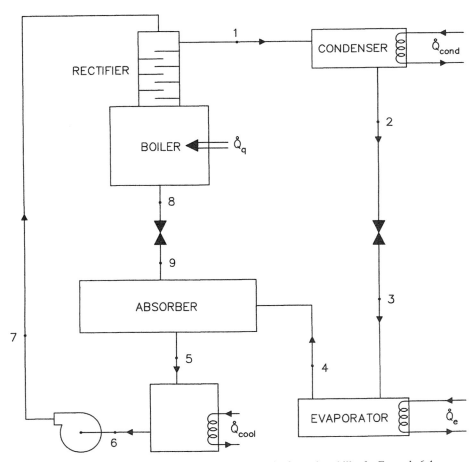

Figure 6.9. Schematic of a 10.5 kW, gas-fired, aqua-ammonia absorption chiller for Example 6.4.

Concentration of ammonia in strong solution $(X_s) = 0.415$,
Concentration of ammonia in weak solution $(X_w) = 0.385$,

Find the mass flow rates, temperatures, and enthalpies at various points in the cycle, and find the COP of the cycle.

Solution. States are defined using Fig. 6.10. Properties at all the points are summarized in Table 6.4 and they are calculated as follows.

State Point 2. Assume equilibrium saturated liquid condition

$$P_2 = 21.7 \text{ Bar}, \quad h_2 = 250 \text{ kJ/kg},$$
$$X_2 = 0.985, \quad T_2 = 54°C.$$

State Point 3. From an isenthalpic process at the valve,

$$h_3 = h_2 = 250 \text{ kJ/kg}.$$

State Point 4. Assume equilibrium saturated vapor condition,

$$P_4 = 4.7 \text{ Bar}, \quad h_4 = 1425 \text{ kJ/kg},$$
$$X_4 = 0.985, \quad T_4 = 2°C.$$

Energy balance at the evaporator,

$$\dot{Q}_e = \dot{m}_r(h_4 - h_3)$$

$$\dot{m}_r = \frac{\dot{Q}_e}{(h_4 - h_3)} = \frac{10.5 \text{ kW}}{(1425 - 250)\text{kJ/kg}} = 0.01 \text{ kg/sec}.$$

Ammonia mass balance at the absorber,

$$\dot{m}_s X_s = \dot{m}_r X_r + \dot{m}_w X_w, \text{ and } \dot{m}_w = \dot{m}_s - \dot{m}_r,$$

$$\dot{m}_s = \dot{m}_r \frac{(X_r - X_w)}{(X_s - X_w)} = 0.01 \frac{(0.985 - 0.385)}{(0.415 - 0.385)} = 0.2 \text{ kg/sec},$$

$$\dot{m}_w = \dot{m}_s - \dot{m}_r = 0.2 - 0.01 = 0.19 \text{ kg/sec}.$$

State Point 1: Assume equilibrium saturated vapor condition,

$$P_1 = 21.7 \text{ Bar}, \quad h_1 = 1430 \text{ kJ/kg},$$
$$X_1 = 0.985, \quad T_1 = 90°C.$$

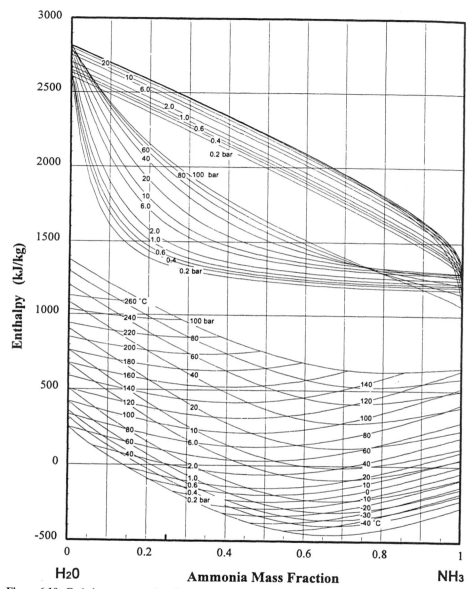

Figure 6.10. Enthalpy-concentration diagram for ammonia/water.

Table 6.4. Thermodynamic properties of ammonia-water in Example 6.4.

State Point	Press. P (Bar)	Temp. T (°C)	Enthalpy h (kJ/kg)	Conc. X (kg NH₃/ kg mix)	Flow rate $\dot{m}$ (kg/sec)	Equilib. Saturated Condition	Quality $(\dot{m}_g/\dot{m}_T)$
1	21.7	90	1430	0.985	0.01	✓	1
2	21.7	≈54	≈250	0.985	0.01	✓	0
3	4.7	2	250	0.985	0.01		
4	4.7	2	1265	0.985	0.01	✓	1
5	4.7	52	361	0.415	0.2		
6	4.7	52	0	0.415	0.2	✓	0
7	21.7	≈52	≈0	0.415	0.2		
8	21.7	120	305	0.385	0.19	✓	0
9	4.7	59	305	0.385	0.19		

State Point 6: Assume equilibrium saturated liquid condition,

$$P_6 = 4.7 \text{ Bar}, \quad h_6 = 0 \text{ kJ/kg},$$
$$X_6 = 0.415, \quad T_6 = 52°C.$$

State point 7: Neglecting the pump work,

$$P_7 = 21.7 \text{ Bar}, \quad h_7 \approx h_6 = 0 \text{ kJ/kg},$$
$$X_7 = 0.415, \quad T_7 \approx T_6 = 52°C.$$

State Point 8: Assume equilibrium saturated liquid condition,

$$P_8 = 21.7 \text{ Bar}, \quad h_8 = 305 \text{ kJ/kg},$$
$$X_8 = 0.385, \quad T_8 = 120°C.$$

Energy balance at the generator,

$$\dot{Q}_g = \dot{m}_r h_1 + \dot{m}_w h_8 - \dot{m}_s h_7$$
$$\dot{Q}_g = 0.01 \times 1430 + 0.19 \times 305 - 0.2 \times 0 = 72.3 \text{ kW}.$$

Coefficient of performance,

$$COP = \frac{\dot{Q}_e}{\dot{Q}_g} = \frac{10.5 \text{ kW}}{72.3 \text{ kW}} = 0.15.$$

State point 9: From an isoenthalpic process at the valve,

$$h_9 = h_8 = 305 \text{ kJ/kg}.$$

Energy balance at the absorber

$$h_5 = \frac{\dot{m}_r h_4 + \dot{m}_w h_9}{\dot{m}_s} = \frac{0.01 \times 1425 + 0.19 \times 305}{0.2} = 361 \text{ kJ/kg}.$$

Example 6.5. It is proposed to convert the gas fired chiller of Example 6.4 to a solar based chiller. Propose the modifications and analyze the performance.

Solution. Assuming that we can use a flat-plate solar collector system to provide heat for the generator, which will operate at 75°C, the high pressure in the cycle would have to be reduced. Taking the high pressure to be 17 Bar and following the procedure of Example 6.4, the properties are as in Table 6.5.

Note that the values of concentration in the cycle must change in order to obtain reasonable values of temperatures. For example, in order to achieve condensation (State 2) at a temperature above the ambient (we choose 40°C) the concentration at State 8 must be increased to 0.55. This will allow us to get ammonia vapor of concentration 0.99 at 70°C at State 1 (pressure 17 Bars) which, in turn, will condense at 40°C. The concentration of 0.55 at State 8 is found by an iterative procedure.

6.2 SOLAR DESICCANT DEHUMIDIFICATION

In hot and humid regions of the world experiencing significant latent cooling demand, solar energy may be used for dehumidification using liquid or solid desiccants. Rangarajan et al [34] compared a number of strategies for ventilation air-conditioning for Miami, FL and found that a conventional vapor compression system could not even

Table 6.5. Thermodynamic properties of ammonia-water in Example 6.5.

State Point	Pressure P (Bar)	Temp. T (°C)	Enthalpy H (kJ/kg)	Conc. X (kg NH₃/ kg mix)	Flow rate $\dot{m}$ (kg/sec)	Equilib. Saturated Condition	Quality $(\dot{m}_g/\dot{m}_T)$
1	17	70	1400	0.99	0.0086	✓	1
2	17	≈40	≈180	0.99	0.0086	✓	0
3	4.7	2	180	0.99	0.0086		
4	4.7	2	1400	0.99	0.0086	✓	1
5	4.7	28	156	0.57	0.2		
6	4.7	28	−100	0.57	0.2	✓	0
7	17	≈28	≈−100	0.57	0.2		
8	17	75	100	0.55	0.1914	✓	0
9	4.7	25	100	0.55	0.1914		

$$\dot{Q}_g = 51.2 \text{ kW}, \quad \text{and} \quad COP = 0.205.$$

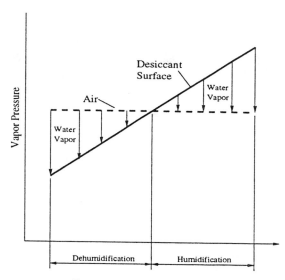

Figure 6.11. Vapor pressure versus temperature and water content for dessicant and air.

meet the increased ventilation requirements of ASHRAE Standard 62-1989. By pretreating the ventilation air with a desiccant system, proper indoor humidity conditions could be maintained and significant electrical energy could be saved. A number of researchers have shown that a combination of a solar desiccant and a vapor compression system can save from 15 to 80 percent of the electrical energy requirements in commercial applications, such as supermarkets [27–32, 36–38].

In a desiccant air-conditioning system, moisture is removed from the air by bringing it in contact with the desiccant and followed with sensible cooling of the air by a vapor compression cooling system, vapor absorption cooling systems, or evaporative cooling system. The driving force for the process is the water vapor pressure. When the vapor pressure in air is higher than on the desiccant surface, moisture is transferred from the air to the desiccant until an equilibrium is reached (see Fig. 6.11). In order to regenerate the desiccant for reuse, the desiccant is heated, which increases the water vapor pressure on its surface. If air with lower vapor pressure is brought in contact with this desiccant, the moisture passes from the desiccant to the air (Fig. 6.11). Two types of desiccants are used: solids, such as silica gel and lithium chloride; or liquids, such as salt solutions and glycols.

6.2.1 Solid Desiccant Cooling System

The two solid desiccant materials that have been used in solar systems are silica gel and the molecular sieve, a selective absorber. Fig. 6.12 shows the equilibrium absorption capacity of several substances. Note that the molecular sieve has the highest capacity up to 30 percent humidity, and silica gel is optimal between 30 and 75 percent— the typical humidity range for buildings.

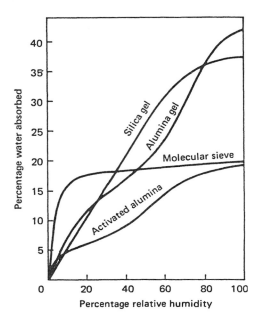

Figure 6.12. Equilibrium capacities of common water absorbents.

Figure 6.13 is a schematic diagram of a desiccant cooling ventilation cycle (also known as Pennington cycle), which achieves both dehumidification and cooling. The desiccant bed is normally a rotary wheel of a honeycomb type substrate impregnated with the desiccant. As the air passes through the rotating wheel, it is dehumidified while its temperature increases (processes 1 and 2) due to the latent heat of condensation. Simultaneously, a hot air stream passes through the opposite side of the rotating wheel which removes moisture from the wheel. The hot and dry air at state 2 is cooled in a heat exchanger wheel to condition 3 and further cooled by evaporative cooling to condition 4. Air at condition 3 may be further cooled by vapor compression or vapor absorption systems instead of evaporative cooling. The return air from the conditioned space is cooled by evaporative cooling (processes 5 and 6), which in turn cools the heat exchanger wheel. This air is then heated to condition 7. Using solar heat, it is further heated to condition 8 before going through the desiccant wheel to regenerate the desiccant. A number of researchers have studied this cycle, or an innovative variation of it, and have found thermal *COP*s in the range of 0.5 to 2.58 [33].

6.2.2 Liquid Desiccant Cooling System

Liquid desiccants offer a number of advantages over solid desiccants. The ability to pump a liquid desiccant makes it possible to use solar energy for regeneration more efficiently. It also allows several small dehumidifiers to be connected to a single regeneration unit. Since a liquid desiccant does not require simultaneous regeneration, the liquid may be stored for later regeneration when solar heat is available. A major disadvantage is that the vapor pressure of the desiccant itself may be enough to cause some

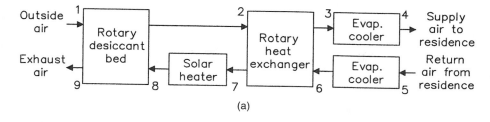

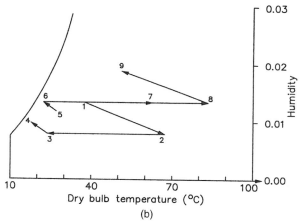

Figure 6.13. Schematic of a dessicant cooling ventilation cycle, a) schematic of airflow, b) process on a psychromatic chart.

desiccant vapors to mix with the air. This disadvantage, however, may be overcome by proper choice of the desiccant material.

A schematic of a liquid desiccant system is shown in Fig. 6.14. Air is brought in contact with concentrated desiccant in a countercurrent flow in a dehumidifier. The dehumidifier may be a spray column or packed bed. The packings provide a very large area for heat and mass transfer between the air and the desiccant. After dehumidification, the air is sensibly cooled before entering the conditioned space. The dilute desiccant exiting the dehumidifier is regenerated by heating and exposing it to a countercurrent flow of a moisture scavenging air stream.

Liquid desiccants commonly used are aqueous solutions of lithium bromide, lithium chloride, calcium chloride, mixtures of these solutions and triethylene glycol (TEG). (See Oberg and Goswami [32]). Vapor pressures of these common desiccants are shown in Fig. 6.15 as a function of concentration and temperature, based on a number of references [8–10, 12, 41]. Other physical properties important in the selection of desiccant materials are listed in Table 6.6. Although salt solutions and TEG have similar vapor pressures, the salt solutions are corrosive and have higher surface tension. The disadvantage of TEG is that it requires higher pumping power because of higher viscosity.

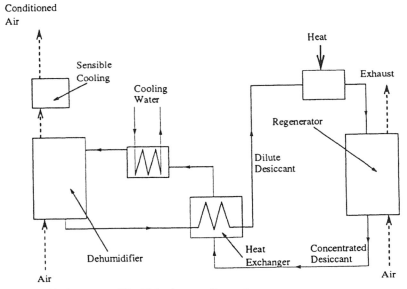

Figure 6.14. A conceptual liquid dessicant cooling system.

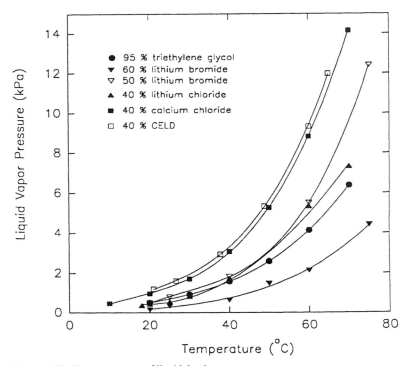

Figure 6.15. Vapor pressures of liquid dessicants.

Table 6.6. Physical properties of liquid dessicants at 25°C [32].

Dessicant	Density $\rho \cdot 10^{-3}$ (kg/m³)	Viscosity $\mu \cdot 10^{3}$ (Ns/m²)	Surface Tension $\gamma \cdot 10^{3}$ (N/m)	Sp. Heat c_p (kJ/kg−°C)	Reference
95% by weight triethylene glycol	1.1	28	46	2.3	[20]
55% by weight lithium bromide	1.6	6	89	2.1	[17, 41]
40% calcium chloride	1.4	7	93	2.5	[7,18, 41]
40% by weight lithium chloride	1.2	9	96	2.5	[41]
40% by weight CELD	1.3	5	—	—	[22]

Oberg and Goswami [32] have presented an in-depth review of liquid desiccant cooling systems. Based on an extensive numerical modeling and on experimental studies, they have presented correlations for the performance of a packed bed liquid desiccant dehumidifier and a regenerator.

The performance of a packed bed dehumidifier or a regenerator may be represented by a humidity effectiveness ϵ_y defined as the ratio of the actual change in humidity of the air to the maximum possible for the operating conditions [6, 18, 40].

$$\epsilon_y = \frac{Y_{in} - Y_{out}}{Y_{in} - Y_{eq}}, \qquad (6.19)$$

where Y_{in} and Y_{out} are the humidity ratios of the air inlet and outlet, respectively, and Y_{eq} is the humidity ratio in equilibrium with the desiccant solution at the local temperature and concentration. (See Fig. 6.16.)

In addition to the humidity effectiveness, an Enthalpy Effectiveness ϵ_H is also used as a performance parameter [17,18]:

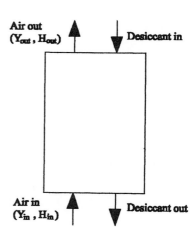

Figure 6.16. Exchange of humidity and moisture between dessicant and air in the tower.

Table 6.7. Constants for performance correlations

	C_1	b	k_1	m_1	k_2	m_2
ϵ_y	48.345	-0.751	0.396	-1.573	0.033	-0.906
ϵ_H	3.766	-0.528	0.289	-1.116	-0.004	-0.365

$$\epsilon_H = \frac{H_{a,\text{in}} - H_{a,\text{out}}}{H_{a,\text{in}} - H_{a,\text{eq}}}, \tag{6.20}$$

where $H_{a,\text{in}}$, $H_{a,\text{out}}$, and $H_{a,\text{eq}}$ are the enthalpies of the air at the inlet and outlet, and in equilibrium with the desiccant, respectively. Oberg and Goswami [31] found the following correlation for ϵ_y and ϵ_H:

$$\epsilon_y, \epsilon_H = 1 - C_1 \left(\frac{L}{G}\right)^a \left(\frac{H_{a,\text{in}}}{H_{L,\text{in}}}\right)^b (aZ)^c, \tag{6.21}$$

where

$$a = k_1 \frac{\gamma_L}{\gamma_c} + m_1, \text{ and}$$

$$c = k_2 \frac{\gamma_L}{\gamma_c} + m_2.$$

Here C_1, b, k_1, m_1, and m_2 are constants listed in Table 6.7. L and G are the liquid and air mass flow rates, respectively; a is the packing surface area per unit volume for heat and mass transfer in m^2/m^3; Z is the tower height in meters; γ_L is the surface tension of the liquid dessicant; and γ_c is the critical surface tension for the packing material.

SUMMARY

The techniques useful for active solar cooling, refrigeration and dehumidification have been described in this chapter. The absorption method provides a suitable thermodynamic match of flat-plate collector to cooling machine because of the usability of low temperatures by the H_2O-LiBr and NH_3-H_2O absorption methods.

Air dehumidification by desiccants and solar regeneration of desiccants is also a method whereby a large part of the cooling load can be met by solar energy. Both solid and liquid desiccants have been described.

PROBLEMS

6.1. The table below gives the characteristics of a building in Houston, Texas. Determine the cooling load for July 30 at solar noon. Any information regarding the load not given may be assumed or neglected.

Factor	Description or specification
Roof:	
Type	Flat, shaded
Area $A_{rf,sh}$, (m²)	250
U factor (W/m²K)	$U_{rf} = 0.35$
Walls:	
Type	Vertical, painted white
Orientation, Size (m × m)	North, South, 3 × 10; East, West, 3 × 25
U factor (W/m²K)	$U_{wa} = 1.08$
Windows:	
Orientation, Area (m²)	North, 8; South, 8; East, 20; West, 25
U factor (W/m²K)	$U_{wi} = 6.2$
Insolation transmittance	$\bar{\tau}_{b,wi} = 0.60;\ \bar{\tau}_{d,wi} = 0.80;\ \bar{\tau}_{r,wi} = 0.55$
Temperature:	
Inside, Outside (°C)	$T_{in} = 24;\ T_{out} = 37$
Insolation:	
Beam, Diffuse, Reflected (W/m²)	$I_{h,b} = 580;\ I_{h,d} = 250;\ I_r = 200$

6.2. An air-conditioning system working in a vapor-compression cycle is used to manage the load for the building in Problem 6.1. If the high and low pressure in the cycle are 915 kPa and 290 kPa, respectively, and the efficiency of the compressor is 90 percent, find the flow rate of Refrigerant 134a (R-134a) used for the equipment and the COP of the cycle.

6.3. Consider the absorption refrigeration cycle, shown in the line diagram below, that uses lithium bromide as carrier and water as refrigerant to provide 1 kW of cooling. By using steam tables and the chart giving the properties of lithium bromide and water, calculate first:

a. Heat removed from the absorber,
b. Heat removed from the condenser,
c. Heat added to the evaporator, and
d. *COP* of the cycle.

Then calculate, for a flat-plate collector with a $F_R(\tau\alpha)$ intercept of 0.81 and a $F_R U_c$ of 3 W/m² · K, the area required for operation in Arizona at noon in August for a 3-ton unit. Assume that the enthalpy of the water vapor leaving the condenser can be approximated by the equation

$$h_{vc} = 2463 + 1.9\,T_c \text{ kJ/kg},$$

and that the enthalpy of the liquid water is

$$h_{\ell c} = 4.2\,T_c \text{ kJ/kg},$$

where T_c is the temperature of the evaporator in °C. In the analysis assume that evaporation occurs at 1°C and condensation at 32°C. Answers for letters a, b, c and d of the list above: 1.85 kW, 1.0 kW, 0.96 kW, 0.51, respectively.

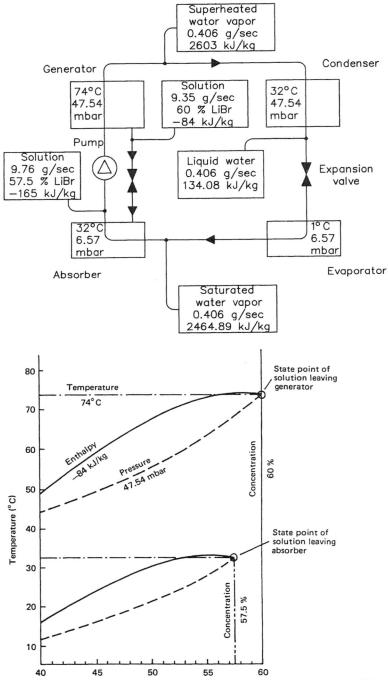

Figure for Problem 6.3.

6.4. Make a preliminary design for a solar-driven Rankine refrigeration machine to provide the temperature environment required below the surface of a 20 × 40 m ice skating rink that is to operate all year in the vicinity of Denver, CO. State all your assumptions.

6.5. An inventor has proposed a desiccant heat pump to augment the heat delivered by a solar collector, and a sketch of the system is shown below. The collector bottom is covered with a desiccant, such as silica gel or zeolite, that contains water, and during the day the solar radiation heats up the desiccant bed and vaporizes a significant fraction of the water. The water vapor thus driven off passes through a heat exchanger in the building, and the vapor condenses with the release of heat into the building. At night, the liquid water stored in a tank passes through a heat exchanger outside the building where it absorbs heat and evaporates. The vapor produced then condenses in the desiccant at a temperature of 150°F, and air from the building is circulated between the building and collector to supply heat to the building at night. Comment on the feasibility of the system proposed and estimate the effective *COP*.

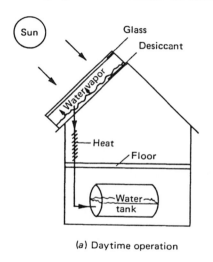

(a) Daytime operation

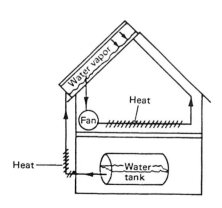

(b) Nighttime operation

6.6. A 5-ton, ammonia-water refrigeration system as shown in Fig. 6.9 was designed to work with solar energy. The low and high pressures are 3 and 15 Bar, respectively. The concentrations are $X_r = 0.98$, $X_w = 0.6$, and $X_s = 0.58$, and the *COP* is expected to be 0.3. Determine the area and the required minimum temperature for a flat-plate solar collector system. Choose a site of interest to you.

6.7. The table below shows data from a dehumidification process using aqueous solution of Lithium Chloride (LiCl) in a packed tower. The dessicant leaving the dehumidifier is passed through another packed tower for regeneration. For the regeneration process it is known that the rate of evaporation of water as a function of the concentration X (kg_{LiCL}/kg_{sol}) and temperature, T ($°C$) of the desiccant is given by the following equation

$$m_{evap} = (a_0 + a_1 T + a_2 T^2) + (b_0 + b_1 T + b_2 T^2)X + (c_0 + c_1 T + c_2 T^2)X^2 \text{ (g/sec)},$$

where

$a_0 = 285077,$ $b_0 = -1658652,$ $c_0 = 2412282,$
$a_1 = -8992,$ $b_1 = 52326,$ $c_1 = -76112,$ and
$a_2 = 70.88,$ $b_2 = -412,$ $c_2 = 600.$

Find the temperature to which a flat-plate collector must raise the temperature of the desiccant for the regeneration process. Assume that the flow rate of liquid desiccant in both processes (dehumidification, regeneration) is the same.

	AIR		DESICCANT		
Variable	Inlet	Outlet	Inlet	Outlet	Variable
Temp (°C)	30.4	32.6	30	33.1	Temp (°C)
RH (%)	66.7	36.7	35	—	Concent. (%)
Mass (kg/h)	260	—	850	—	Mass (kg/h)

REFERENCES AND SUGGESTED READINGS

1. ASHRAE. 1997. *Handbook of Fundamentals.* Atlanta, GA: American Society of Heating, Refrigerating and Air-Conditioning Engineers.

2. Barber, R. 1975. Solar organic Rankine cycle powered three-ton cooling system. In Francis deWinter, ed., *Solar Cooling for Buildings USGPO* 3800-00189: 206–214. Washington, D.C.: U.S. Government Printing Office.

3. Beckman, W.A., and J.A. Duffie. 1973. *Modeling of Solar Heating and Air Conditioning.* Springfield, Va.: National Technical Information Service.

4. Burriss, W. 1975. Solar powered Rankine cycle cooling system. In Francis deWinter, ed., *Solar Cooling for Buildings, USGPO* 3800-0189: pp. 186–189. Washington, D.C.: U.S. Government Printing Office.

5. Chinnappa, J.C.V., and N.E. Wijeysundera. 1992. Simulation of solar-powered ammonia-water integrated hybrid cooling system. *Journal of Solar Energy Engineering, Transactions of ASME* 114: 125–127.

6. Chung, T.-W. 1989. Predictions of the moisture removal efficiencies for packed-bed dehumidification systems. *Solar Engineering—1989, Proc. of the 11ᵗʰ Annual ASME Solar Energy Conf.*, San Diego, California, 371–377.

7. Close, D.J. Rock pile thermal storage for comfort air conditioning, *Mech. Chem. Eng. Trans. Inst. Eng. (Australia)* MC-1: 11.

8. Cyprus Foote Mineral Company. *Technical data on lithium bromide and lithium chloride, Bulletins 145 and 151*, Kings Mountain, NC: Cyprus Foote Mineral Company.

9. Dow Chemical Company. 1996. *Calcium Chloride Handbook*, Midland, MI: Dow Chemical Company.

10. Dow Chemical Company. 1992. *A guide to glycols*. Midland, MI: Dow Chemical Company.

11. Dunkle, R.V. 1965. "A method of solar air conditioning." *Mech. Chem. Eng. Trans. Inst. Eng. (Australia)* MC-1, 74.

12. Ertas, A., E.E. Anderson, and I. Kiris. 1992. Properties of a new liquid desiccant solution—lithium chloride and calcium chloride mixture. *Sol. Energy* 49: 205–212.

13. Hay, H.R., and J.I. Yellott. 1970. A naturally air conditioned building. *Mech. Eng.* 92: 19.

14. Hay, H.R. 1973. Energy technology and solarchitecture. *Mech. Eng.* 94: 18.

15. Hewett, R. 1995. Solar absorption cooling: An innovative use of solar energy. *AIChE Symposium Series*, No. 306, vol. 91, Heat Transfer.

16. Keenan, J.H., and F.G. Keyes. 1936. *Thermodynamic Properties of Steam*. New York: John Wiley & Sons.

17. Kettleborough, C.F., and D.G. Waugaman. 1995. An alternative desiccant cooling cycle. *J. of Sol. Energy Eng.* 117: 251–255.

18. Khan, A.Y. 1994. Sensitivity analysis and component modeling of a packed-type liquid desiccant system at partial load operating conditions. *Int. Jour. of Energy Res.* 18: 643–655.

19. Kochhar, G.S., and S. Satcunanathan. 1981. Optimum operating conditions of absorption refrigeration systems for flat plate collector temperatures. *Sol. Energy* 260–267.

20. Kreider, J.F. and F. Kreith. 1977. *Solar heating and cooling,* revised 1st ed. Washington, DC: Hemisphere Publ. Corp.

21. Löf, G.O.G. 1956. Cooling with solar energy. *Proc. World Symp. Applied Solar Energy (1955).*

22. Löf, G.O.G., and R.A. Tybout. 1974. Design and cost of optimal systems for residential heating and cooling by solar energy. *Sol. Energy* 16: 9.

23. Löf, G.O.G. 1974. *Design and construction of a residential solar heating and cooling system.* Springfield, VA.: National Technical Information Service.

24. Macriss, R.A., and T.S. Zawacki. 1989. Absorption fluid data survey: 1989 update. *Oak Ridge National Laboratory Report,* ORNL/Sub84-47989/4.

25. Manrique, J.A. 1991. Thermal performance of ammonia-water refrigeration system. *Int. Comm., Heat Mass Transfer* 19(6): 779–789.

26. Mathur, G.D. 1989. Solar-operated absorption coolers. *Heating/Piping/Air Conditioning.* November.

27. Meckler, H. 1994. Desiccant-assisted air conditioner improves IAQ and comfort. *Heating, Piping & Air Conditioning* 66(10): 75–84.

28. Meckler, M. 1995. Desiccant outdoor air preconditioners maximize heat recovery ventilation potentials. *ASHRAE Transactions,* 101, pt. 2, 992–1000.

29. Meckler, M. 1988. "Off-Peak desiccant cooling and cogeneration combine to maximize gas utilization." *ASHRAE Transactions,* 94, pt. 1: 575–596.

30. Meckler, M., Y.O. Parent, and A.A. Pesaran. 1993. Evaluation of dehumidifiers with polymeric desiccants." *Gas Institute Report,* Contract No. 5091-246-2247. Chicago, Illinois: Gas Research Institute.

31. Oberg, V., and D.Y. Goswami. 1998. Experimental study of heat and mass transfer in a packed bed liquid desiccant air dehumidifier. In J.H. Morehouse and R.E. Hogan, eds. *Solar Engineering.* 155–166.

32. Oberg, V., and D.Y. Goswami. 1998. A review of liquid desiccant cooling. *Advances in Solar Energy,* ASES 12: 431–470.

33. Pesaran, A.A., T.R. Penney, and A.W. Czanderna. 1992. Desiccant cooling: State-of-the-art assessment. *National Renewable Laboratory,* Golden Colorado, NREL, Report No. NREL/TP-254-4147, October.

34. Rangarajan, K., D.B. Shirley, III, and R.A. Raustad. 1989. Cost-effective HVAC technologies to meet ASHRAE Standard 62-1989 in hot and human climates." *ASHRAE Trans.,* pt. 1: 166–182.

35. Siddiqui, A.M. 1993. Optimum generator temperatures in four absorption cycles using different sources of energy. *Energy Conversion and Management* 34, (4) 251–266.

36. Spears, J.W., and J. Judge. 1997. Gas-fired desiccant system for retail super center. *ASHRAE Journal* 39: 65–69.

37. Thornbloom, M., and B. Nimmo. 1994. Modification of the absorption cycle for low generator firing temperatures. *Joint Solar Engineering Conf.* ASME.

38. Thornbloom, M., and B. Nimmo. 1995. An economic analysis of a solar open cycle desiccant dehumidification system. *Solar Engineering—1995, Proc. of the 13th Annual ASME Conference,* Hawaii 1: 705–709.

39. Thornbloom, M., and B. Nimmo. 1996. Impact of design parameters on solar open cycle liquid desiccant regenerator performance. *SOLAR '96, Proc. of 1996 Annual Conf. of the American Solar Energy Society,* Asheville, NC, 107–111.

40. Ullah, M.R., C.F. Kettleborough, and P. Gandhidasan. 1988. Effectiveness of moisture removal for an adiabatic counterflow packed tower absorber operating with $CaCl_2$-air contact system. *Jour. of Solar Energy Eng.* 110: 98–101.

41. Zaytsev, I.O., and G.G. Aseyev. 1992. *Properties of aqueous solutions of electrolytes.* Boca Raton, Florida: CRC Press.

PASSIVE METHODS FOR HEATING, COOLING AND DAYLIGHTING

7

7.1 INTRODUCTION

Passive systems are defined, quite generally, as systems in which the thermal energy flow is by natural means: by conduction, radiation, and natural convection. Passive features increase the use of solar energy to meet heating and lighting loads and the use of ambient air for cooling. For example, window placement can enhance solar gains to meet winter heating loads, to provide daylighting, or to do both, and this is passive solar use. Using a thermal chimney to draw air through a building in order to provide cooling is another passive effect.

A distinction is made between energy conservation techniques and passive solar measures. Energy conservation features are designed to reduce the heating and cooling energy required to thermally condition a building. Such features would include the use of insulation to reduce heating or cooling loads. Similarly, window shading or appropriate window placement could lower solar gains, thus reducing summer cooling loads.

7.1.1 Definition of Passive Systems

A *passive heating system* is one in which the sun's radiant energy is converted to heat upon absorption by the building. The absorbed heat can be transferred to thermal storage by natural means or it can be used to directly heat the building. *Passive cooling systems* use natural energy flows to transfer heat to the environmental sinks: the ground, air, and sky. If one of the major heat transfer paths employs a pump or fan to force flow of a heat transfer fluid, then the system is referred to as having an *active* component or subsystem. *Hybrid systems*, either for heating or cooling, have both passive and active energy flows. The use of the sun's radiant energy for the natural illumination of a building's interior spaces is called *daylighting*.

7.1.2 Current Applications

Almost one-half million buildings in the U.S. were constructed or retrofitted with passive features in the twenty years after 1980. Passive heating applications are primarily in single-family dwellings and secondarily in small commercial buildings. Daylighting features, which reduce lighting loads and the associated cooling loads, are usually more appropriate for large office buildings.

A typical passive heating design in a favorable climate might supply up to one-third of a home's original load at a cost of $5 to $10 per million Btu (1.7¢ to 3.5¢ per kWh) net energy saved. An appropriately designed daylighting system can supply lighting at a cost of 2.5¢ to 5¢ per kWh [2].

7.1.3 Economic Basis for Passive Systems

The distinction between passive systems, active systems, and energy conservation is not critical for economic calculations. They are the same in all cases, a trade-off between the life cycle cost of the energy saved (performance) and the life cycle cost of the initial investment plus operating and maintenance costs (cost).

The key performance parameter to be determined is the net annual energy saved by the installation of the passive system. The basis for calculating the economics of any solar energy system is to compare it against a normal building. Thus, the actual difference in the annual cost of fuel is the difference in auxiliary energy that would be used with and without solar. The energy saved must be determined, rather than energy delivered, energy collected, useful energy, or some other energy measure. The other significant part of the economic trade-off involves determining the difference between the cost of construction of the passive building and that of the normal building against which it is to be compared, this is the "solar add-on cost." Again, this may be a difficult definition in the case of passive designs, because the building can be significantly altered compared to typical construction.

Passive solar water heaters are described in Chapter 5. This chapter describes passive space heating and cooling systems and daylighting.

7.2 PASSIVE SPACE-HEATING SYSTEMS

Passive heating systems contain the five basic components of all solar systems described in Chapters 4 and 5. Typically these are:

1. Collector—windows, walls and floors.
2. Storage—walls and floors, large interior masses (often integrated with the collector absorption function).
3. Distribution system—radiation, free convection, simple circulation fans.
4. Controls—moveable window insulation, vents both to other inside spaces or to ambient.
5. Backup system—any nonsolar heating system.

The design of passive heating systems requires the strategic placement of windows, storage masses, and the occupied spaces themselves. The fundamental principles of solar radiation geometry and availability are instrumental in the proper location and sizing of the system's collectors (windows). Storage devices are usually more massive than those used in active systems and are frequently an integral part of the collection and distribution system.

7.2.1 Types of Passive Heating Systems

A common method of cataloging the various passive system concepts distinguishes three general categories: direct, indirect, and isolated gain. Most of the physical configurations of passive heating systems are seen to fit within one or another of these three categories.

For direct gain systems (Figure 7.1), sunlight enters the heated space and is converted to heat at absorbing surfaces. This heat is then distributed throughout the space and to the various enclosing surfaces and room contents.

For indirect gain systems, sunlight is absorbed and stored by a mass interposed between the glazing and the conditioned space. The conditioned space is partially enclosed and bounded by this thermal storage mass, so a natural thermal coupling is achieved. Examples of the indirect approach are the thermal storage wall, the thermal storage roof, and the northerly room of an attached sunspace.

In the thermal storage wall (Figure 7.2a), sunlight penetrates the glazing and is absorbed and converted to heat at a wall surface interposed between the glazing and the heated space. The wall is usually masonry (Trombe wall) or containers filled with water (water wall), although it might contain phase-change material. The attached sunspace (Figure 7.2b) is actually a two-zone combination of direct gain and thermal storage wall. Sunlight enters and heats a direct gain southerly sunspace and a mass wall separating the northerly buffered space, which is heated indirectly. The sunspace is frequently used as a greenhouse, in which case the system is called an attached

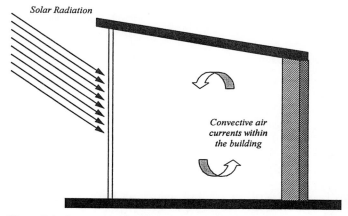

Figure 7.1. Concept of a direct gain passive heating system.

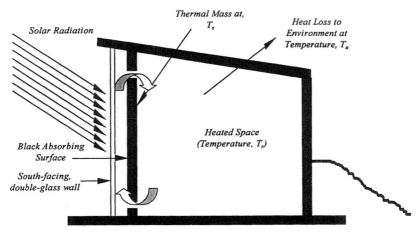

Figure 7.2a. Thermal Storage Wall.

greenhouse. The thermal storage roof (Figure 7.2c) is similar to the thermal storage wall except that the interposed thermal storage mass is located on the building roof. A thermal storage roof using water for storage and movable insulation on the top was developed by Hay [12] and is also known as the Roof-Pond system.

The isolated gain category concept is an indirect system, except that there is a distinct thermal separation (by means of either insulation or physical separation) between the thermal storage and the heated space. The convective (thermosyphon) loop, as depicted in Figure 7.3, is in this category and is often used to heat domestic water. It is most akin to conventional active systems in that there is a separate collector and separate thermal storage. It is a passive approach, however, because the thermal energy flow is by natural convection. The thermal storage wall, thermal storage roof, and attached sunspace approaches can also be made into isolated systems by insulating between the thermal storage and the heated space.

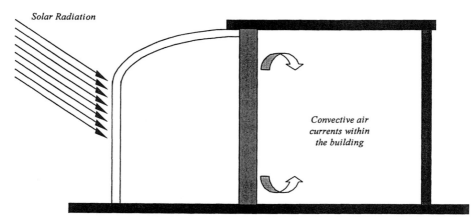

Figure 7.2b. Attached SunSpace.

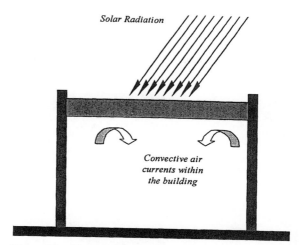

Figure 7.2c. Thermal Storage Roof.

7.2.2 Fundamental Concepts for Passive Heating Design

Figure 7.4 is an equivalent thermal circuit for the building illustrated in Figure 7.2a, the Trombe wall type system. For the heat transfer analysis of the building, three temperature nodes can be identified—room temperature, storage wall temperature, and ambient temperature. The circuit responds to climatic variables represented by a current injection I_s (solar radiation) and by the ambient temperature T_a. The storage temperature T_s and room temperature T_r are determined by current flows in the equivalent circuit. By using seasonal and annual climatic data, the performance of a passive structure can be simulated, and the results of many such simulations correlated to give the design approaches described below.

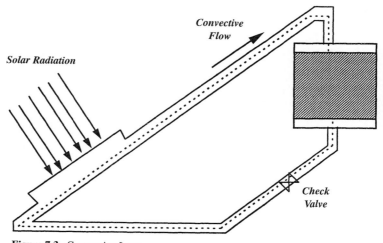

Figure 7.3. Convective Loop

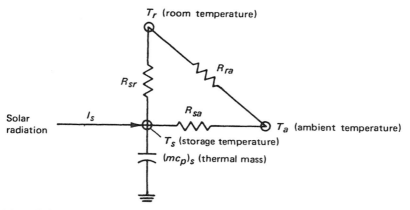

Figure 7.4. Equivalent thermal circuit for passively heated solar building in Fig. 7.2a.

7.2.3 Generalized Methods

Design of a passive heating system involves selection and sizing of the passive feature type(s), determination of thermal performance, and cost estimation. Ideally, a cost/performance optimization would be performed by the designer. Owner and architect ideas usually establish the passive feature type, with general size and cost estimation available. However, the thermal performance of a passive heating system has to be calculated.

There are several levels of methods that can be used to estimate the thermal performance of passive designs. First-level methods involve a rule of thumb and/or generalized calculation to get a starting estimate for size and/or annual performance. A second-level method involves climate, building, and passive system details. These details allow annual performance determination, plus some sensitivity to passive system design changes. Third-level methods involve periodic calculations (hourly, monthly) of performance and permit more detailed variations of climatic, building, and passive solar system design parameters.

These three levels of design methods have a common basis in that they all are derived from correlations of a multitude of computer simulations of passive systems [25,26]. As a result, a similar set of defined terms is used in many passive design approaches:

- A_p, solar projected area, m² (ft²). The net south-facing passive solar glazing area projected onto a vertical plane.
- *NLC*, net building load coefficient, kJ/C-day (Btu/F-day). Net load of the nonsolar portion of the building per day per degree of indoor-outdoor temperature difference. The C-day and F-day terms refer to Centigrade and Fahrenheit degree days, respectively.
- Q_{net}, net reference load, Wh (Btu). Heat loss from nonsolar portion of building as calculated by

$$Q_{net} = NLC \times (\text{No. of degree days}). \tag{7.1}$$

- LCR, load collector ratio, kJ/m² C-day (Btu/ft² F-day) is the ratio of NLC to A_p,

$$LCR = NLC/A_p. \qquad (7.2)$$

- SSF, solar savings fraction, is the fractional reduction in required auxiliary heating (Q_{aux}) relative to net reference load (Q_{net}),

$$SSF = 1 - \frac{Q_{aux}}{Q_{net}}. \qquad (7.3)$$

So, using Eq. (7.1),

Auxiliary heat required, $Q_{aux} = (1 - SSF) \times NLC \times$ (No. of degree days). (7.4)

The amount of auxiliary heat required is often a basis of comparison between possible solar designs. It is also the basis for determining building energy operating costs. Thus, many passive design methods are based on determining SSF, NLC, and the number of degree days in order to calculate the auxiliary heat required for a particular passive system by using Eq. (7.4).

7.2.4 The First Level: Rules of Thumb

A first estimate or starting value is needed to begin the passive system design process. Rules of thumb have been developed to generate initial values for solar aperture size, storage size, solar savings fraction, auxiliary heat required, and other size and performance characteristics. The following rules of thumb are meant to be used with the defined terms presented above.

Load. A rule of thumb used in conventional building design is that a design heating load of 120 to 160 kJ/C-day per m² of floor area (6 to 8 Btu/F-day ft²) is considered an energy conservative design. Reducing these nonsolar values by 20 percent to solarize the proposed south-facing solar wall gives rule-of-thumb NLC values per unit of floor area:

$$NLC/\text{Floor area} = 100 \text{ to } 130 \text{ kJ/C-day m}^2 \ (4.8 \text{ to } 6.4 \text{ Btu/F-day ft}^2). \quad (7.5)$$

Solar savings fraction. A method of getting starting-point values for the solar savings fraction in the United States is presented in Figure 7.5 [26]. The map values represent optimum SSF in percent for a particular set of conservation and passive-solar costs for different climates across the United States. With the Q_{net} generated from the NLC rule of thumb above and the SSF read from the map, the Q_{aux} can be determined.

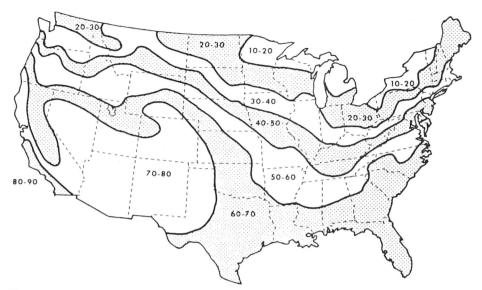

Figure 7.5. Starting-point values of solar savings fraction (SSF) in percent. *Source:* PSDH [26].

LCR. The A_p can be determined using the *NLC* from above if the *LCR* is known. The rule of thumb associated with good values of *LCR* [27] differs depending on whether the design is for a cold or warm climate:

$$Good\ LCR = \begin{cases} \text{For cold climate: } 410 \text{ kJ/m}^2 \text{ C-day (20 Btu/ft}^2 \text{ F-day)} \\ \text{For warm climate: } 610 \text{ kJ/m}^2 \text{ C-day (30 Btu/ft}^2 \text{ F-day)} \end{cases} \quad (7.6)$$

Storage. Rules of thumb for thermal mass storage relate storage material total heat capacity to the solar projected area [26]. The use of the storage mass is to provide for heating on cloudy days and to regulate sunny day room air temperature swing. When the thermal mass direct absorbs the solar radiation, each square meter of the projected glazing area requires enough mass to store 613 kJ/C. If the storage material is not in direct sunlight, but heated from room air only, then four times as much mass is needed. In a room with a direct sunlight heated storage mass, the room air temperature swing will be approximately one-half the storage mass temperature swing. For room air heated storage, the air temperature swing is twice that of the storage mass.

A more location-dependent set of rules of thumb is presented in PSDH [25]. Comparing the results of that method to those of the method presented above, the two rules of thumb are seen to produce roughly similar answers. General system cost and performance information can be generated with results from rule-of-thumb calculations, but a more detailed level of information is needed to determine design-ready passive system type (direct gain, thermal wall, sunspace), size, performance, and costs.

7.2.5 The Second Level: *LCR* Method

The *LCR* method is useful for making estimates of the annual performance of specific types of passive system(s) combinations. The *LCR* method was developed by calculating the annual *SSF* for 94 reference passive solar systems for 219 U.S. and Canadian locations over a range of *LCR* values. Appendix Table A7.1 includes the description of these 94 reference systems for use both with the *LCR* method and with the *SLR* method described below. Tables were constructed for each city with *LCR* versus *SSF* listed for each of the 94 reference passive systems. (*Note:* The solar load ratio (*SLR*) method was used to make the *LCR* calculations. This *SLR* method is described in the next section as the third-level method.) While the complete *LCR* tables (PSDH, 1984) includes 219 locations, Appendix Table A7.2 includes only six representative cities (Albuquerque, Boston, Madison, Medford, Nashville, Santa Maria), due to space restrictions. The *LCR* method consists of the following steps [26]:

1. Determine the building parameters:
   a. Building load coefficient, *NLC*,
   b. Solar projected area, A_p, and
   c. Load collector ratio, $LCR = NLC/A_p$.
2. Find the short designation of the reference system closest to the passive system design (Table A7.1).
3. Enter the *LCR* Tables (Table A7.2):
   a. Find the city,
   b. Find the reference system designation,
   c. Determine annual *SSF* by interpolation using the *LCR* value from above, and
   d. Note the annual heating degree days (No. of degree days).
4. Calculate the annual auxiliary heat required:
   Auxiliary heat required $= (1 - SSF) \times NLC \times$ (No. of degree days)

If more than one reference solar system is being used, then find the aperture area weighted *SSF* for the combination. Determine each individual reference system *SSF* using the total aperture area *LCR*, then take the area weighted average of the individual *SSF*s.

The *LCR* method allows no variation from the 94 reference passive designs. To treat off-reference designs, sensitivity curves have been produced that illustrate the effect on *SSF* of varying one or two design variables. These curves were produced for the six representative cities, which were chosen for their wide geographical and climatological ranges. Several of these sensitivity curves are presented in Appendix Figure A7.1.

Using the *LCR* method allows a basic design of passive system types for the 94 reference systems and the resulting annual performance. A bit more design variation can be obtained by using the sensitivity curves of Figure A7.1 to modify the *SSF* of a particular reference system. For instance, a storage wall system (Albuquerque, LCR = 72, thickness of 4 inches) *SSF* of 0.41 would increase by approximately 0.05, if the mass-area-to-glazing-area ratio (Am/Ag), assumed 6, were increased to 10, and would decrease by about 0.06 if the mass-glazing-area ratio were decreased to 3. This information provides a designer with quantitative information for making trade-offs.

7.2.6 The Third Level: SLR Method

The solar-load ratio (SLR) method calculates monthly performance; the terms and values used are monthly based. The method allows the use of specific location weather data and the 94 reference design passive systems (Appendix Table A7.1). In addition, the sensitivity curves (Appendix Figure A7.1) can again be used to define performance outside the reference design systems. The result of the *SLR* method is the determination of the monthly heating auxiliary energy required, which is then summed to give the annual requirement for auxiliary heating energy. Generally, the *SLR* method gives annual values within ±3% of detailed simulation results, but the monthly values may vary more [7,26]. Thus, the monthly *SLR* method is more accurate than the rule-of-thumb methods, and it provides the designer with system performance on a month-by-month basis.

The *SLR* method uses equations and correlation parameters for each of the 94 reference systems in combination with the insolation absorbed by the system, the monthly degree days, and the system's *LCR* to determine the monthly *SSF*. These correlation parameters are listed in Appendix Table A7.3 as *A, B, C, D, R, G, H*, and *LCR*s for each reference system [26]. The correlation equations are

$$SSF = 1 - K(1 - F), \tag{7.7}$$

where

$$K = 1 + G/LCR, \tag{7.8}$$

$$F = \begin{cases} AX, & \text{when } X < R \\ B - C\exp(-DX), & \text{when } X > R \end{cases} \tag{7.9}$$

$$X = \frac{S/DD - (LCRs)H}{(LCR)K}, \tag{7.10}$$

and X is called the generalized solar load ratio. The DD term is the monthly number of degree-days. The term S is the monthly insolation absorbed by the system per unit of solar projected area. Monthly average daily insolation data on a vertical south facing surface can be found or calculated using various sources [21,26]. The S term can be determined by multiplying by a transmission and an absorption factor and the number of days in the month. Absorption factors for all systems are close to 0.96 [21], whereas the transmission is approximately 0.9 for single glazing, 0.8 for double glazing, and 0.7 for triple glazing.

Example 7.1. For a vented, 180 ft², double-glazed with night insulation, 12″ thick Trombe wall system (TWD4) in a *NLC* = 11,800 Btu/F-day house in Medford, Oregon, determine the auxiliary energy required in January.

Solution. Weather data for Medford, Oregon [26], yields for January (N = 31 days): daily vertical surface insolation = 565 Btu/ft² and DD = 880 F-days; so S = (31)(565)(0.8)(0.96) = 13,452 Btu/ft² -month.

$$LCR = NLC/A_p = 11,800/180 = 65.6 \text{ Btu/F-day ft}^2.$$

From Table A7.3 at TWD4: A = 0, B = 1, C = 1.0606, D = 0.977, R = −9, G = 0, H = 0.85, LCRs = 5.8 Btu/F-day ft².
Substituting into Eq. (7.8) gives

$$K = 1 + 0/65.6 = 1.$$

Eq. (7.10) gives

$$X = \frac{13{,}452/880 - (5.8 \times 0.85)}{65.6 \times 1} = 0.16.$$

Eq. (7.9) gives

$$F = 1 - 1.0606\, e^{-0.977 \times 0.16} = 0.09,$$

and Eq. (7.7) gives

$$SSF = 1 - 1(1 - 0.09) = 0.09.$$

The January auxiliary energy required can be calculated using Eq. (7.4):

$$Q_{\text{aux}}(\text{Jan}) = (1 - SSF) \times NLC \times (\text{No. of degree days})$$
$$= (1 - 0.09) \times 11{,}800 \times 880$$
$$= 9{,}450{,}000 \text{ Btu.}$$

As mentioned, the use of sensitivity curves [26] as in Figure A7.1 will allow SSF to be determined for many off-reference system design conditions involving storage mass, number of glazings, and other more esoteric parameters. Also, the use of multiple passive system types within one building would be approached by calculating the SSF for each type system individually using a combined area LCR, and then a weighted-area (aperture) average SSF would be determined for the building.

7.3 PASSIVE SPACE-COOLING SYSTEMS

Passive cooling systems are designed to use natural means to transfer heat from buildings by means of convection/ventilation, evaporation, radiation, and conduction. However, the most important element in both passive and conventional cooling design is to prevent heat from entering the building in the first place. Cooling conservation techniques involve building surface colors, insulation, special window glazings, overhangs and orientation, and numerous other architectural and engineering features.

7.3.1 Controlling the Solar Input

Controlling the solar energy input to reduce the cooling load is usually considered a passive (versus a conservation) design concern because solar input may be needed for other purposes, such as daylighting throughout the year or heating during the winter or both. Basic architectural solar control is normally designed in via the shading of the solar windows, where direct radiation is desired for winter heating and needs to be excluded during the cooling season.

The shading control of the windows can be of various types and controlability, ranging from drapes and blinds, to use of deciduous trees, to commonly used overhangs and vertical louvers. A rule-of-thumb design for determining proper south-facing window overhang for both winter heating and summer shading is presented in Table 7.1. Technical details on calculating shading from various devices and orientations are found in ASHRAE [3] and Olgyay and Olgyay [23].

Table 7.1. South-facing window overhang rule of thumb

$$\text{Length of the Overhang} = \frac{\text{Window Height}}{F}$$

(a) Overhang Factors		(b) Roof Overhang Geometry
North Latitude	F*	
28	5.6–11.1	
32	4.0–6.3	
36	3.0–4.5	
40	2.5–3.4	
44	2.0–2.7	
48	1.7–2.2	
52	1.5–1.8	
56	1.3–1.5	

Properly sized overhangs shade out hot summer sun but allow winter sun (which is lower in the sky) to penetrate windows.

*Select a factor according to your latitude. Higher values provide complete shading at noon on June 21; lower values, until August 1.

Source: Halacy [11].

7.3.2 Movement of Air

Air movement provides cooling comfort through convection and evaporation from human skin. ASHRAE [3] places the comfort limit at 79°F (26°C) for an air velocity of 50 ft/min (0.25 m/s) (fpm), 82°F for 160 fpm, and 85°F for 200 fpm. To determine whether or not comfort conditions can be obtained, a designer must calculate the volumetric flow rate, Q, which is passing through the occupied space. Using the cross-sectional area, A_x, of the space and the room air velocity, V_a, required, the flow is determined by

$$Q = A_x V_a. \tag{7.11}$$

The proper placement of windows, narrow building shape, and open landscaping can enhance natural wind flow to provide ventilation. The air flow rate through open windows for wind-driven ventilation is given by ASHRAE [3]:

$$Q = C_v V_w A_w, \tag{7.12}$$

where Q = air flow rate, m³/s
   A_w = free area of inlet opening, m²
   V_w = wind velocity, m/s
   C_v = effectiveness of opening = 0.5 to 0.6 for wind perpendicular to opening, and 0.25 to 0.35 for wind diagonal to opening

The stack effect can induce ventilation when warm air rises to the top of a structure and exhausts outside, while cooler outside air enters the structure to replace it. Figure 7.6 illustrates the solar chimney concept, which can easily be adapted to a thermal storage wall system. The greatest stack-effect flow rate is produced by maximizing the stack height and the air temperature in the stack, as given by

$$Q = 0.116 A_j \sqrt{h(T_s - T_o)}, \tag{7.13}$$

where Q = stack flow rate, m³/s,
   A_j = area of inlets or outlets (whichever is smaller), m²,
   h = inlet to outlet height, m,
   T_s = average temperature in stack, °C, and
   T_o = outdoor air temperature, °C.

If either the inlet or outlet area is twice the other, the flow rate will increase by 25 percent, and if the area ratio is 3:1 or larger, it will increase by 35 percent.

Example 7.2. A two-story (5 m) solar chimney is being designed to produce a flow of 0.25 m³/s through a space. The preliminary design features include a 25 cm × 1.5 m inlet, a 50 cm × 1.5 m outlet, and an estimated 35°C average stack temperature on a sunny 30°C day. Can this design produce the desired flow?

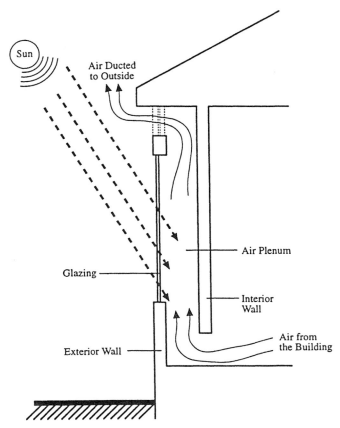

Figure 7.6. The stack-effect/solar chimney concept to induce convection/ventilation. *Source:* PSDH [25].

Solution. Substituting the design data into Eq. (7.13):

$$Q = 0.116(.25 \times 1.5)[5(5)]^{1/2}$$
$$= 0.2 \text{ m}^3/\text{s}.$$

Since the outlet area is twice the inlet area, the 25 percent flow increase can be used:

$$Q = 0.2 (1.25) = 0.25 \text{ m}^3/\text{s} \quad \text{(Answer: Yes, the proper flow rate is obtained.)}$$

7.3.3 Evaporative Cooling

When air with less than 100 percent relative humidity moves over a water surface, the evaporation of water causes both the air and the water itself to cool. The lowest temperature that can be reached by this direct evaporative cooling effect is the wet-bulb temperature of the air, which is directly related to the relative humidity, lower wet-bulb temperature being associated with lower relative humidity. Thus, dry air (low relative humidity) has a low wet-bulb temperature and will undergo a large temperature drop with evaporative cooling, while humid air (high relative humidity) can only be slightly cooled evaporatively. The wet-bulb temperature for various relative humidity and air temperature conditions can be found via the psychrometric chart available in most thermodynamic texts. Normally, an evaporative cooling process cools the air only part of the way down to the wet-bulb temperature. To get the maximum temperature decrease, it is necessary to have a large water surface area in contact with the air for a long time. Interior ponds and fountain sprays are often used to provide this air-water contact area.

The use of water sprays and open ponds on roofs provides cooling primarily via evaporation. The hybrid system involving a fan and wetted mat, the swamp cooler, is by far the most widely used evaporative cooling combined design. Features are described in ASHRAE [3,4].

7.3.4 Nocturnal Cooling Systems

Another approach to passive convective/ventilative cooling involves using cooler night air to reduce the temperature of the building or of a storage mass. Thus, the building/storage mass is prepared to accept part of the heat load during the hotter daytime. This type of convective system can also be combined with evaporative and radiative modes of heat transfer, utilizing air or water, or both, as the convective fluid. Work in Australia [5] investigated rock storage beds that were chilled using evaporatively cooled night air. Room air was then circulated through the bed during the day to provide space cooling. The use of encapsulated roof ponds as a thermal cooling mass has been tried by several investigators [12,19] and is often linked with nighttime radiative cooling.

All warm objects emit thermal infrared radiation; the hotter the body, the more energy it emits. A passive cooling scheme uses the cooler night sky as a sink for thermal radiation emitted by a warm storage mass, thus chilling the mass for cooling use the next day. The net radiative cooling rate, Q_r, for a horizontal unit surface [3] is

$$Q_r = \epsilon\sigma\,(T_{body}^4 - T_{sky}^4), \tag{7.14}$$

where Q_r = net radiative cooling rate, W/m² (Btu/h ft²),
$\quad\epsilon$ = surface emissivity (usually 0.9 for water),
$\quad\sigma$ = 5.67×10^{-8} W/m² K⁴ (1.714×10^{-9} Btu/h ft² R⁴),
$\quad T_{body}$ = warm body temperature, Kelvin (Rankine), and
$\quad T_{sky}$ = effective sky temperature, Kelvin (Rankine).

The monthly average air-sky temperature difference has been determined [20] and Figure 7.7 presents these values for July (in °F) for the United States.

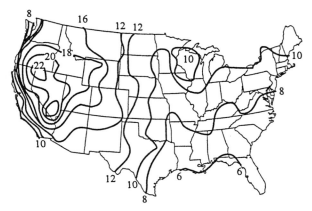

Figure 7.7. Average monthly sky temperature depression $(T_{air} - T_{sky})$ for July in °F. Adapted from Martin and Berdahl [20].

Example 7.3. Estimate the overnight cooling possible for a 10 m², 85°F water thermal storage roof during July in Los Angeles.

Solution. Assume the roof storage unit is black with $\epsilon = 0.9$. From Figure 7.7, $T_{air} - T_{sky}$ is approximately 10°F for Los Angeles. From weather data for LA airport, [3,27], the July average temperature is 69°F with a range of 15°F. Assuming night temperatures vary from the average (69°F) down to half the daily range (15 × 1/2), then the average nighttime temperature is chosen as 69 − (1/2)(15/2) = 65°F. So, T_{sky} = 65 − 10 = 55°F. From Eq. (7.14),

$$Q_r = 0.9 \, (1.714 \times 10^{-9})[(460 + 85)^4 - (460 + 55)^4]$$
$$= 27.6 \text{ Btu/h ft}^2$$

For a 10-hour night and 10 m² (107.6 ft²) roof area,

$$\text{Total radiative cooling} = 27.6 \, (10)(107.6)$$
$$= 29,700 \text{ Btu}$$

Note: This does not include the convective cooling possible which can be approximated (at its maximum rate) for still air [3] by

$$\text{Maximum total } Q_{conv} = hA(T_{roof} - T_{air})(\text{Time})$$
$$= 5(129)(85 - 55)(10)$$
$$= 161,000 \text{ Btu}$$

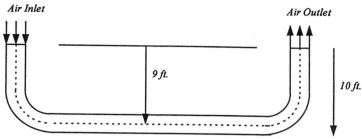

Figure 7.8a. Open loop underground air tunnel system.

This is a maximum since the 85°F storage temperature will drop as it cools, which is also the case for the radiative cooling calculation. However, convection is seen to usually be the more dominant mode of nighttime cooling.

7.3.5 Earth Contact Cooling (or Heating)

Earth contact cooling or heating is a passive summer cooling and winter heating technique that utilizes underground soil as the heat sink or source. By installing a pipe underground and passing air through the pipe, the air will be cooled or warmed depending on the season. Schematics of an open loop system and a closed loop air-conditioning system are presented in Figures 7.8a and 7.8b, respectively [8].

The use of this technique can be traced back to 3000 B.C. when Iranian architects designed some buildings to be cooled by natural resources only. In the 19th century, Wilkinson [29] designed a barn for 148 cows where a 500-ft long underground passage was used for cooling during the summertime. Since that time, a number of experimental and analytical studies of this technique have appeared in the literature. Goswami and Dhaliwal [9] have given a brief review of the literature and presented an analytical solution to the problem of transient heat transfer between the air and the surrounding soil as the air is made to pass through a pipe buried underground. More recently, Krarti and Kreider [14] have also presented an analytical model for heat transfer in an underground air tunnel.

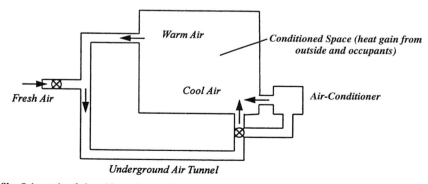

Figure 7.8b. Schematic of closed loop air-conditioning system using air-tunnel.

Heat transfer analysis. The transient thermal analysis of the air and soil temperature fields [9] is conducted using finite elements with the convective heat transfer between the air and the pipe and using semi-infinite cylindrical conductive heat transfer to the soil from the pipe. It should be noted that the thermal resistance of the pipe (whether of metal, plastic or ceramic) is negligible relative to the surrounding soil.

Air and pipe heat transfer. The pipe is divided into a large number of elements and a psychrometric energy balance written for each, depending on whether the air leaves the element (a) unsaturated, or (b) saturated.

(a) If the air leaves an element as unsaturated, the energy balance on the element is:

$$mc_p(T_1 - T_2) = hA_p(T_{air} - T_{pipe}),   \tag{7.15}$$

$$T_{air} \text{ can be taken as } \frac{T_1 + T_2}{2}.$$

Substituting and simplifying we get

$$T_2 = \left[\left(1 - \frac{U}{2}\right)T_1 + UT_{pipe}\right]\bigg/\left(1 + \frac{U}{2}\right),   \tag{7.16}$$

where U is defined as

$$U = \frac{A_p h}{mc_p}.$$

(b) If the air leaving the element is saturated, the energy balance is

$$mC_pT_1 + m(W_1 - W_2)H_{fg} = mC_pT_2 + hA_p(T_{air} - T_{pipe}).   \tag{7.17}$$

Simplifying we get:

$$T_2 = \left(1 - \frac{U}{2}T_1\right) + \frac{W_1 - W_2}{C_p}H_{fg} + UT_{Pipe}\bigg/\left(1 + \frac{U}{2}\right).   \tag{7.18}$$

The convective heat transfer coefficient h in the preceding equations depends on Reynolds Number, the shape, and roughness of the pipe.

Using the exit temperature from the first element as the inlet temperature for the next element, the exit temperature for the second element can be calculated in a similar way. Continuing this way from one element to the next, the temperature of air at the exit from the pipe can be calculated.

Soil heat transfer. The heat transfer from the pipe to the soil is analyzed by considering the heat flux at the internal radius of a semi-infinite cylinder formed by the soil around the pipe. For a small element the problem can be formulated as

$$\frac{\partial T(r,t)}{\partial r^2} + \frac{1}{r}\frac{\partial T(r,t)}{\partial r} = \frac{1}{\alpha}\frac{\partial T(r,t)}{\partial t}, \tag{7.19}$$

with initial and boundary conditions as

$$T(r,0) = T_e, \tag{7.20}$$

$$T(\infty,t) = T_e, \tag{7.21}$$

$$-K\frac{\partial T}{\partial r}(r,t) = q'', \tag{7.22}$$

where T_e is the bulk earth temperature and q'' is also given by the amount of heat transferred to the pipe from the air by convection, i.e., $q'' = h(T_{air} - T_{pipe})$.

Soil temperatures and properties. Labs [17] studied the earth temperatures in the United States. According to this study, temperature swings in the soil during the year are dampened with depth below the ground. There is also a phase lag between the soil temperature and the ambient air temperature. This phase lag increases with depth below the surface. For example, the soil temperature for light dry soil at a depth of about 10 ft (3.05m) varies by approximately ± 5°F (2.8°C) from the mean temperature (approximately equal to mean annual air temperature) and has a phase lag of approximately 75 days behind ambient air temperature [17].

The thermal properties of the soil are difficult to determine. The thermal conductivity and diffusivity both change with the moisture content of the soil itself, which is directly affected by the temperature of and heat flux from and to the buried pipe. Most researchers have found that using constant property values for soil taken from standard references gives reasonable predictive results [10].

Generalized results from experiments. Figure 7.9 presents data from Goswami and Biseli [8] for an open system, 100-foot long, 12-inch diameter pipe, buried 9 feet deep. The figure shows the relationship between pipe inlet-to-outlet temperature reduction $(T_{in} - T_{out})$ and the initial soil temperature with ambient air inlet conditions of 90°F, 55 percent relative humidity for various pipe flow rates.

Other relations from this same report which can be used with the Figure 7.9 data include: (1) the effect of increasing pipe/tunnel length on increasing the inlet-to-outlet air temperature difference is fairly linear up to 250 feet; and (2) the effect of decreasing pipe diameter on lowering the outlet air temperature is slight, and only marginally effective for pipes less than 12 inches in diameter.

Example 7.4. Provide the necessary 12-inch diameter pipe length(s) which will deliver 1500 cfm of 75°F air if the ambient temperature is 85°F and the soil at 9 feet is 65°F.

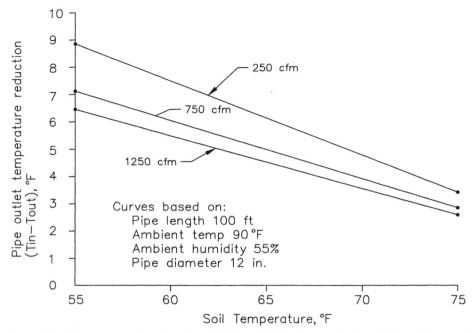

Figure 7.9. Air temperature drop through a 100-ft long, 12-inch diameter pipe buried 9 feet underground.

Solution. From Fig. 7.9, for 100 ft. of pipe at 70°F soil temperature (use 70°F to keep the same ambient-to-soil temperature as is used in Fig. 7.9), the pipe temperature reduction is

$$T_{in} - T_{out} = 5°F \text{ (at 250 cfm)},$$

$$= 4°F \text{ (at 750 cfm), and}$$

$$= 3.5°F \text{ (at 1250 cfm)}.$$

Since the length versus temperature reduction is linear (see text above), the 10°F reduction required (85°F down to 75°F) would be met by the 750 cfm case (4°F for 100 ft) if 250 ft of pipe is used. Then, two 12-inch diameter pipes would be required to meet the 1500 cfm requirement.

Answer: Two 12-inch diameter pipes, each 250 ft long. (Note: See what would be needed if the 250 cfm or the 1250 cfm cases had been chosen. Which of the three flow rate cases leads to the least expensive installation?)

7.4 DAYLIGHTING FUNDAMENTALS

Daylighting is the use of the sun's radiant energy to illuminate the interior spaces in a building. In the last century, electric lighting was considered an alternative technology to daylighting. Today the situation is reversed, primarily due to the economics of energy use and conservation. However, there are good physiological reasons for using daylight as an illuminant. The quality of daylight matches the human eye's response, thus permitting lower light levels for task comfort, better color rendering, and clearer object discrimination [27].

Lighting terms and units. Measurement of lighting level is based on the standard candle, where the lumen (lm), the unit of luminous flux (Φ), is defined as the rate of luminous energy passing through a 1 square meter area located 1 meter from the candle. Thus, a standard candle generates 4π lumens which radiate away in all directions. The illuminance (E) on a surface is defined as the luminous flux on the surface divided by the surface area, $E = \Phi/A$. Illuminance is measured in either lux (lx), as lumens per square meter, or footcandles (fc), as lumens per square foot.

Determination of the daylighting available in a building space at a given location and time is important in order to evaluate the reduction possible in electric lighting and to determine the associated impact on heating and cooling loads. Daylight provides about 110 lm/W of solar radiation, fluorescent lamps about 75 lm/W of electrical input, and incandescent lamps about 20 lm/W; thus daylighting generates only 1/2 to 1/5 the heating that equivalent electric lighting does, significantly reducing the building cooling load.

7.4.1 Economics of Daylighting

The economic benefit of daylighting is directly tied to the reduction in lighting electrical energy operating costs. Also, lower cooling system operating costs are possible due to the reduction in heating caused by the reduced electrical lighting load. The reduction in lighting and cooling system electrical power during peak demand periods could also beneficially affect demand charges.

The reduction of the design cooling load through the use of daylighting can also lead to the reduction of installed or first-cost cooling system dollars. Normally, economics dictate that an automatic lighting control system must take advantage of the reduced lighting/cooling effect, and the control system cost, minus any cooling system cost, savings should be expressed as a net first cost. A payback time for the lighting control system (net or not) can be calculated from the ratio of first costs to yearly operating savings. In some cases, these paybacks for daylighting controls have been found to be in the range of 1 to 5 years for office building spaces [28].

Controls, both aperture and lighting, directly affect the efficacy of the daylighting system. As shown in Figure 7.10, aperture controls can be architectural (overhangs, light shelves, etc.) or window shading devices (blinds, automated louvers, etc.). The aperture controls generally moderate the sunlight entering the space to maximize or minimize solar thermal gain, permit the proper amount of light for visibility, and prevent glare and beam radiation onto the workplace. Photosensor control of electric lighting allows the dimming (or shutting off) of the lights in proportion to the amount of available daylighting illuminance.

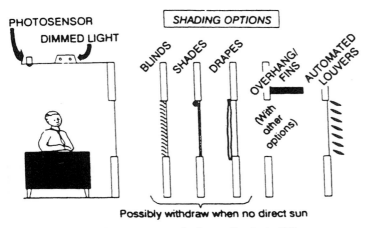

Figure 7.10. Daylighting system controls. *Source:* Rundquist [29].

   In most cases, using daylighting controls to increase the solar gain for daylighting purposes saves more in electrical lighting energy and in cooling energy associated with the lighting than is incurred with the added solar gain [28]. In determining the annual energy savings from daylighting (ES_T), the annual lighting energy saved from day-lighting (ES_L) is added with the reduction in cooling system energy (∇ES_C) and with the negative of the heating system energy increase (∇ES_H):

$$ES_T = ES_L + \nabla ES_C - \nabla ES_H. \tag{7.23}$$

A simple approach to estimating the heating and cooling energy changes associated with lighting energy reduction is to use the fraction of the year associated with the cooling or heating season (f_c, f_H) and the seasonal *COP* of the cooling or heating equipment. Thus, Eq. (7.23) can be expressed as

$$ES_T = ES_L + \frac{f_c ES_L}{COP_c} - \frac{f_H ES_L}{COP_H},$$

$$= ES_L\left(1 + \frac{f_c}{COP_c} - \frac{f_H}{COP_H}\right). \tag{7.24}$$

It should be noted that the increased solar gain due to daylighting has not been in-cluded here but would reduce summer savings and increase winter savings. If it is as-sumed that the increased wintertime daylighting solar gain approximately offsets the reduced lighting heat gain, then the last term in Eq. (7.24) becomes negligible.

7.4.2 Daylighting Design Fundamentals

As mentioned, aperture controls such as blinds and drapes are used to moderate the amount of daylight entering the space, as are the architectural features of the building itself (glazing type, area, orientation; overhangs and wingwalls, lightshelves, etc.). Dimming controls are used to adjust the electric light level based on the quantity of the daylighting. With these two types of controls (aperture and lighting), the electric lighting and cooling energy use and demand, as well as cooling system size, can be reduced. However, the determination of the daylighting position and time **illuminance** value within the space is required before energy usage and demand reduction calculations can be made.

Architectural features. Daylighting design approaches use both solar beam radiation (sunlight) and the diffuse radiation scattered by the atmosphere (skylight) as sources for interior lighting, with historical design emphasis being on utilizing skylight. Daylighting is provided through a variety of glazing features which can be grouped as sidelighting (light entering via the side of the space) and toplighting (light entering from the ceiling area). Figure 7.11 illustrates several architectural forms producing sidelighting and toplighting. The dashed lines represent the illuminance distribution within the space. Calculation of workplane illuminance depends on whether sidelighting or toplighting features are used. The combined illuminance values are additive.

Daylighting geometry. The solar illuminance on a vertical or horizontal window depends on the position of the sun relative to that window. In the method described here, the sun and sky illuminance values are determined using the sun's altitude angle (α) and the sun-window azimuth angle difference (a_{sw}). These angles need to be determined for the particular time of day, day of year, and window placement under investigation.

Solar altitude angle (α). The solar altitude angle is the angle swept out by a person's arm when pointing to the horizon directly below the sun and then raising the arm to point at the sun. Equation (2.28) can be used to calculate solar altitude, α, as

$$\sin \alpha = \cos L \cos \delta_s \cos h_s + \sin L \sin \delta_s. \tag{2.28}$$

Sun-window azimuth angle difference (a_{sw}). The difference between the sun's azimuth and the window's azimuth needs to be calculated for vertical window illuminance. The window's azimuth angle, a_w, is determined by which way it faces, as measured from south (east of south is negative, westward is positive). The solar azimuth angle, a_s, is calculated using Eq. (2.29):

$$\sin a_s = \cos \delta_s \sin h_s / \cos \alpha. \tag{2.29}$$

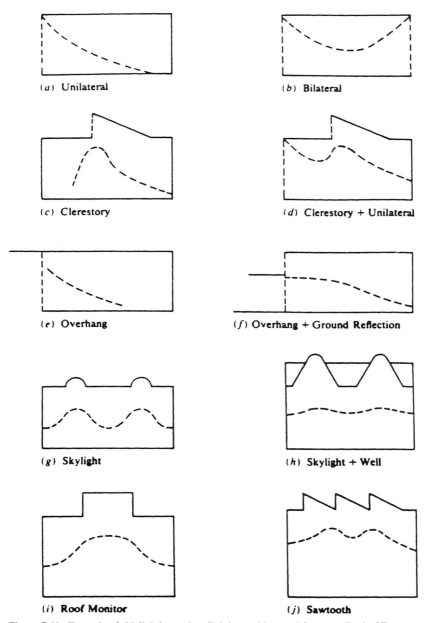

Figure 7.11. Example of sidelighting and toplighting architectural features. (Dashed lines represent illuminance distributions.) *Source:* Murdoch [22].

The sun-window azimuth angle difference, a_{sw}, is given by the absolute value of the difference between a_s and a_w:

$$a_{sw} = |a_s - a_w|. \tag{7.25}$$

7.4.3 Design Methods

To determine the annual lighting energy saved (ES_L) for the space under investigation, calculations using the lumen method described below should be performed on a monthly basis for both clear and overcast days. Monthly weather data for the site would then be used to prorate clear and overcast lighting energy demands each month. Subtracting the calculated daylighting illuminance from the design illuminance leaves the supplementary lighting needed, which determines the lighting energy required.

The approach in the method below is to calculate the sidelighting and the skylighting of the space separately and then combine the results. This procedure has been computerized and includes many details of controls, daylighting methods, weather, and heating and cooling load calculations. ASHRAE [3] lists many of the methods and simulation techniques currently used with daylighting and its associated energy effects.

7.4.4 Lumen Method of Sidelighting (Vertical Windows)

The lumen method of sidelighting calculates interior horizontal illuminance at three points, as shown in Figure 7.12, along the 30-inch (0.76 m) work plane on the room-and-window centerline. A vertical window is assumed to extend from 36 inches (0.91 m) above the floor to the ceiling. The method accounts for both direct and ground-

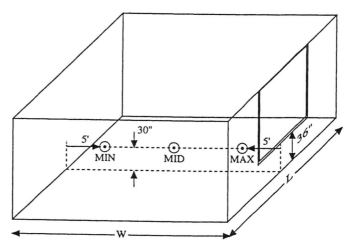

Figure 7.12. Location of illumination points within the room (along centerline of window) determined by lumen method of sidelighting.

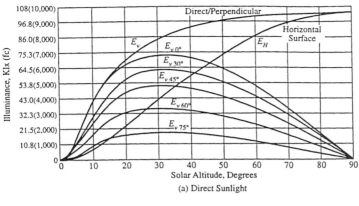

(a) Direct Sunlight

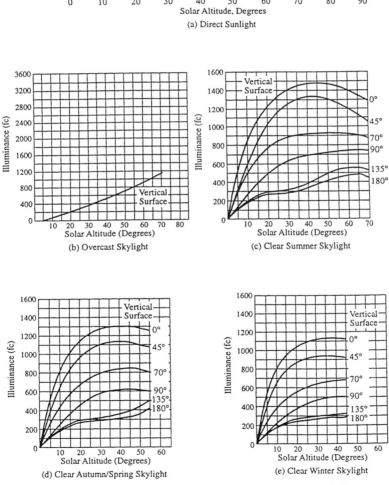

(b) Overcast Skylight

(c) Clear Summer Skylight

(d) Clear Autumn/Spring Skylight

(e) Clear Winter Skylight

Figure 7.13. Vertical illuminance from (a) direct sunlight and (b–e) skylight, for various sun-window azimuth angle differences. *Source:* IES [13].

reflected sunlight and skylight, so both horizontal and vertical illuminances from sun and sky are needed. The steps in the lumen method of sidelighting are presented next.

The incident direct and ground-reflected window illuminance are normally calculated for both a cloudy and a clear day for representative days during the year (various months), as well as for clear or cloudy times during a given day. Thus, the interior illumination due to sidelighting and skylighting can then be examined for effectiveness throughout the year.

Step 1: Incident direct sky and sun illuminances. The solar altitude and sun-window azimuth angle difference are calculated for the desired latitude, date, and time using Eqs. (2.28) and (7.25), respectively. Using these two angles, the total illuminance on the window (E_{sw}) can be determined by summing the direct sun illuminance (E_{uw}) and the direct sky illuminance (E_{kw}), each determined from the appropriate graph in Figure 7.13.

Step 2: Incident ground-reflected illuminance. The sun illuminance on the ground (E_{ug}), plus the overcast or clear sky illuminance (E_{kg}) on the ground, make up the total horizontal illuminance on the ground surface (E_{sg}). A fraction of the ground surface illuminance is then considered diffusely reflected onto the vertical window surface (E_{gw}), where gw indicates from the ground to the window.

The horizontal ground illuminances can be determined using Figure 7.14, where the clear sky plus sun case and the overcast sky case are functions of solar altitude. The fractions of the ground illuminance diffusely reflected onto the window depend on the reflectivity (ρ) of the ground surface (see Table 7.2) and the window-to-ground surface geometry.

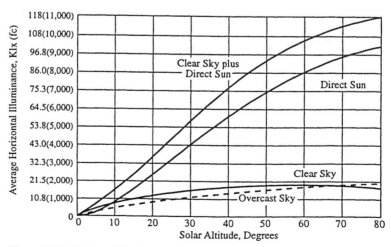

Figure 7.14. Horizontal illuminance for overcast sky, clear sky, direct sun, and clear sky plus direct sun. *Source:* Murdoch [23].

Table 7.2. Ground reflectivities

Material	ρ
Cement	.27
Concrete	.20–.40
Asphalt	.07–.14
Earth	.10
Grass	.06–.20
Vegetation	.25
Snow	.70
Red brick	.30
Gravel	.15
White paint	.55–.75

Murdoch [23].

If the ground surface is considered uniformly reflective from the window outward to the horizon, then the illuminance on the window from ground reflection is

$$E_{gw} = \frac{\rho E_{sg}}{2}. \tag{7.26}$$

A more complicated ground-reflection case is illustrated in Figure 7.15, with multiple strips of differently reflecting ground being handled using the angles to the window, where a strip's illuminance on a window is calculated,

$$E_{gw(strip)} = \frac{\rho_{strip} E_{sg}}{2}(\cos\theta_1 - \cos\theta_2). \tag{7.27}$$

And the total reflected onto the window is the sum of the strip illuminances:

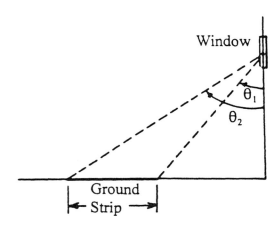

Figure 7.15. Geometry for ground strips. *Source:* Murdoch [22].

$$E_{gw} = \frac{E_{sg}}{2}[\rho_1 (\cos\theta_0 - \cos\theta_1) + \rho_2 (\cos\theta_1 - \cos\theta_2) + \cdots + \rho_n (\cos\theta_{n-1} - \cos90)].$$

$$(7.28)$$

Step 3: Luminous flux entering space. The direct sky-sun and ground reflected luminous fluxes entering the building are attenuated by the transmissivity of the window. Table 7.3 presents the transmittance fraction (τ) of several window glasses. The fluxes entering the space are calculated from the total sun-sky and the ground reflected illuminances by using the area of the glass, A_w

$$\phi_{sw} = E_{sw}\tau A_w,$$
$$\phi_{gw} = E_{gw}\tau A_w.$$

$$(7.29)$$

Step 4: Light loss factor. The light loss factor (K_m) accounts for the attenuation of luminous flux due to dirt on the window (WDD, window dirt depreciation) and on the room surfaces (RSDD, room surface dirt depreciation). WDD depends on how often the window is cleaned, but a 6-month average for offices is 0.83 and for factories is 0.71 [22].

The RSDD is a more complex calculation involving time between cleanings, the direct-indirect flux distribution, and room proportions. However, for rooms cleaned regularly, RSDD is around 0.94 and for once-a-year-cleaned dirty rooms, the RSDD would be around 0.84.

Table 7.3. Glass transmittances

Glass	Thickness (in.)	τ
Clear	$\frac{1}{8}$	.89
Clear	$\frac{3}{16}$	.88
Clear	$\frac{1}{4}$	.87
Clear	$\frac{5}{16}$	.86
Grey	$\frac{1}{8}$	.61
Grey	$\frac{3}{16}$	.51
Grey	$\frac{1}{4}$	.44
Grey	$\frac{5}{16}$	.35
Bronze	$\frac{1}{8}$	.68
Bronze	$\frac{3}{16}$	.59
Bronze	$\frac{1}{4}$	.52
Bronze	$\frac{5}{16}$	.44
Thermopane	$\frac{1}{8}$	.80
Thermopane	$\frac{3}{16}$	.79
Thermopane	$\frac{1}{4}$	.77

Murdoch [23].

The light loss factor is the product of the preceding two fractions:

$$K_m = (WDD)\,(RSDD). \tag{7.30}$$

Step 5: Work-plane illuminances. As discussed earlier, Figure 7.12 illustrates the location of the work-plane illuminances determined with this lumen method of sidelighting. The three illuminances (max, mid, min) are determined using two coefficients of utilization, the C factor and K factor. The C factor depends on room length and width and wall reflectance. The K factor depends on ceiling-floor height, room width, and wall reflectance. Table 7.4 presents C and K values for the three cases of incoming fluxes: (a) sun plus clear sky, (b) overcast sky, and (c) ground reflected. Assumed ceiling and floor reflectances are given for this last case with no window controls (shades, blinds, overhangs, etc.). These further window control complexities can be found in IES [14], LOF [18], and others. A reflectance of 70 percent represents light-colored walls, and 30 percent represents darker walls.

The work-plane max, mid, and min illuminance are each calculated by adding the sun-sky and ground reflected illuminances, which are given by

$$E_{sp} = \phi_{sw} C_s K_s K_m,$$
$$E_{gp} = \phi_{gw} C_g K_g K_m, \tag{7.31}$$

where the *sp* and *gp* refer to the sky-to-work-plane and ground-to-work-plane illuminances.

7.4.5 Lumen Method of Skylighting

The lumen method of skylighting calculates the average illuminance at the interior work plane provided by horizontal skylights mounted on the roof. The procedure for skylighting is generally the same as that described above for sidelighting. As with windows, the illuminance from both overcast sky and clear sky plus sun cases are determined for specific days in different seasons and for different times of the day, and a judgment is then made as to the number and size of skylights and any controls needed.

The procedure is presented in four steps: (1) finding the horizontal illuminance on the outside of the skylight; (2) calculating the effective transmittance through the skylight and its well; (3) figuring the interior space light loss factor and the utilization coefficient; and finally, (4) calculating illuminance on the work plane.

Step 1: Horizontal sky and sun illuminances. The horizontal illuminance value for an overcast sky or a clear sky plus sun situation can be determined from Figure 7.14 knowing only the solar altitude (see Eq. (2.28)).

Step 2: Net skylight transmittance. The transmittance of the skylight is determined by the transmittance of the skylight cover(s), the reflective efficiency of the skylight

Table 7.4a. C and K factors for no window controls for Overcast Sky

Illumination by Overcast Sky
C: Coefficient of Utilization

Room Length		20'		30'		40'	
Wall Reflectance		70%	30%	70%	30%	70%	30%
Room Width							
Max	20'	.0276	.0251	.0191	.0173	.0143	.0137
	30'	.0272	.0248	.0188	.0172	.0137	.0131
	40'	.0269	.0246	.0182	.0171	.0133	.0130
Mid	20'	.0159	.0177	.0101	.0087	.0081	.0071
	30'	.0058	.0050	.0054	.0040	.0034	.0033
	40'	.0039	.0027	.0030	.0023	.0022	.0019
Min	20'	.0087	.0053	.0063	.0043	.0050	.0037
	30'	.0032	.0019	.0029	.0017	.0020	.0014
	40'	.0019	.0009	.0016	.0009	.0012	.0008

K: Coefficient of Utilization

Ceiling Height		8'		10'		12'		14'	
Wall Reflectance		70%	30%	70%	30%	70%	30%	70%	30%
Room Width									
Max	20'	.125	.129	.121	.123	.111	.111	.0091	.0973
	30'	.122	.131	.122	.121	.111	.111	.0945	.0973
	40'	.145	.133	.131	.126	.111	.111	.0973	.0982
Mid	20'	.0908	.0982	.107	.115	.111	.111	.105	.122
	30'	.156	.102	.0939	.113	.111	.111	.121	.134
	40'	.106	.0948	.123	.107	.111	.111	.135	.127
Min	20'	.0908	.102	.0951	.114	.111	.111	.118	.134
	30'	.0924	.119	.101	.114	.111	.111	.125	.126
	40'	.111	.0926	.125	.109	.111	.111	.133	.130

Source: IES, 1979.

Table 7.4b. C and K factors for no window controls for Clear Sky

Illumination by Clear Sky
C: Coefficient of Utilization

Room Length		20'		30'		40'	
Wall Reflectance		70%	30%	70%	30%	70%	30%
	Room Width						
Max	20'	.0206	.0173	.0143	.0123	.0110	.0098
	30'	.0203	.0173	.0137	.0120	.0098	.0092
	40'	.0200	.0168	.0131	.0119	.0096	.0091
Mid	20'	.0153	.0104	.0100	.0079	.0083	.0067
	30'	.0082	.0054	.0062	.0043	.0046	.0037
	40'	.0052	.0032	.0040	.0028	.0029	.0023
Min	20'	.0106	.0060	.0079	.0049	.0067	.0043
	30'	.0054	.0028	.0047	.0023	.0032	.0021
	40'	.0031	.0014	.0027	.0013	.0021	.0012

K: Coefficient of Utilization

Ceiling Height		8'		10'		12'		14'	
Wall Reflectance		70%	30%	70%	30%	70%	30%	70%	30%
	Room Width								
Max	20'	.145	.155	.129	.132	.111	.111	.101	.0982
	30'	.141	.149	.125	.130	.111	.111	.0954	.101
	40'	.157	.157	.135	.134	.111	.111	.0964	.0991
Mid	20'	.110	.128	.116	.126	.111	.111	.103	.108
	30'	.106	.125	.110	.129	.111	.111	.112	.120
	40'	.117	.118	.122	.118	.111	.111	.123	.122
Min	20'	.105	.129	.112	.130	.111	.111	.111	.116
	30'	.0994	.144	.107	.126	.111	.111	.107	.124
	40'	.119	.116	.130	.118	.111	.111	.120	.118

Source: IES [13].

Table 7.4c. C and K factors for no window controls for Ground Illumination (Ceiling Reflectance, 80%; Floor Reflectance, 30%)

Ground Illumination
C: Coefficient of Utilization

Room Length		20'		30'		40'	
Wall Reflectance		70%	30%	70%	30%	70%	30%
Room Width							
Max	20'	.0147	.0112	.0102	.0088	.0081	.0071
	30'	.0141	.0012	.0098	.0088	.0077	.0070
	40'	.0137	.0112	.0093	.0086	.0072	.0069
Mid	20'	.0128	.0090	.0094	.0071	.0073	.0060
	30'	.0083	.0057	.0062	.0048	.0050	.0041
	40'	.0055	.0037	.0044	.0033	.0042	.0026
Min	20'	.0106	.0071	.0082	.0054	.0067	.0044
	30'	.0051	.0026	.0041	.0023	.0033	.0021
	40'	.0029	.0018	.0026	.0012	.0022	.0011

K: Coefficient of Utilization

Ceiling Height		8'		10'		12'		14'	
Wall Reflectance		70%	30%	70%	30%	70%	30%	70%	30%
Room Width									
Max	20'	.124	.206	.140	.135	.111	.111	.0909	.0859
	30'	.182	.188	.140	.143	.111	.111	.0918	.0878
	40'	.124	.182	.140	.142	.111	.111	.0936	.0879
Mid	20'	.123	.145	.122	.129	.111	.111	.100	.0945
	30'	.0966	.104	.107	.112	.111	.111	.110	.105
	40'	.0790	.0786	.0999	.106	.111	.111	.118	.118
Min	20'	.0994	.108	.110	.114	.111	.111	.107	.104
	30'	.0816	.0822	.0984	.105	.111	.111	.121	.116
	40'	.0700	.0656	.0946	.0986	.111	.111	.125	.132

Source: IES [13].

well, the net-to-gross skylight area, and the transmittance of any light-control devices (lenses, louvers, etc.).

The transmittance for several flat-sheet plastic materials used in skylight domes is presented in Table 7.5. To get the effective dome transmittance (T_D) from the flat-plate transmittance (T_F) value [1], use

$$T_D = 1.25 \, T_F(1.18 - 0.416 \, T_F). \tag{7.32}$$

If a double-domed skylight is used, then the single-dome transmittances are combined as follows [24]:

$$T_D = \frac{T_{D_1} T_{D_2}}{T_{D_1} T_{D_2} - T_{D_1} T_{D_2}}. \tag{7.33}$$

If the diffuse and direct transmittances for solar radiation are available for the skylight glazing material, it is possible to follow this procedure and determine diffuse and direct dome transmittances separately. However, this difference is usually not a significant factor in the overall calculations.

The efficiency of the skylight well (N_w) is the fraction of the luminous flux from the dome that enters the room from the well. The well index (WI) is a geometric index (height, h; length, l; width, w) given by

$$WI = \frac{h(w + l)}{2wl}, \tag{7.34}$$

and WI is used with the well-wall reflectance value in Figure 7.16 to determine well efficiency, N_w.

With T_D and N_w determined, the net skylight transmittance for the skylight and well is given by:

$$T_n = T_D N_W R_A T_C, \tag{7.35}$$

Table 7.5. Flat-plate plastic material transmittance for skylights

Type	Thickness (in.)	Transmittance
Transparent	$\frac{1}{8} - \frac{3}{16}$	.92
Dense translucent	$\frac{1}{8}$	.32
Dense translucent	$\frac{3}{16}$	.24
Medium translucent	$\frac{1}{8}$	.56
Medium translucent	$\frac{3}{16}$	.52
Light translucent	$\frac{1}{8}$	.72
Light translucent	$\frac{3}{16}$	.68

Source: Murdoch [22].

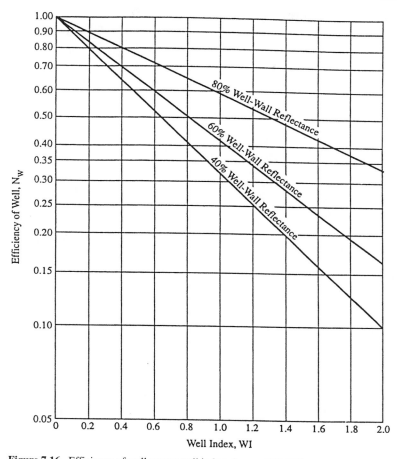

Figure 7.16. Efficiency of well versus well index. *Source:* IES [15].

where R_A = ratio of net to gross skylight areas, and T_C = transmittance of any light-controlling devices.

Step 3: Light loss factor and utilization coefficient. The light loss factor (K_m) is again defined as the product of the room surface direct depreciation (RSDD) and the skylight direct depreciation (SDD) fractions, similarly to Eq. (7.30). Following the reasoning for the sidelighting case, the RSDD value for clean rooms is around 0.94, and 0.84 for dirty rooms. Without specific data indicating otherwise, the SDD fraction is often taken as 0.75 for office buildings and 0.65 for industrial areas.

The fraction of the luminous flux on the skylight that reaches the work plane (K_u) is the product of the net transmittance (T_n) and the room coefficient of utilization (*RCU*). Dietz et al. [6] developed *RCU* equations for office and warehouse interiors with ceiling, wall, and floor reflectances of 75%, 50%, and 30%, and 50%, 30%, and 20%, respectively.

$$RCU = \frac{1}{1 + A(RCR)^B}, \text{ if } RCR < 8, \tag{7.36}$$

where $A = 0.0288$ and $B = 1.560$ (offices), and
$\quad A = 0.0995$ and $B = 1.087$ (warehouses),

and room cavity ratio (RCR) is given by

$$RCR = \frac{5h_c(l + w)}{lw}, \tag{7.37}$$

with h_c the ceiling height above the work plane and l and w being room length and width, respectively.

The RCU is then multiplied by the previously determined T_n to give the fraction of the external luminous flux passing through the skylight and incident on the workplace:

$$K_u = T_n(RCU). \tag{7.38}$$

Step 4: Work-plane illuminance. The illuminance at the work plane (E_{TWP}) is given by

$$E_{TWP} = E_H\left(\frac{A_T}{A_{WP}}\right)K_u K_m, \tag{7.39}$$

where E_H is the horizontal overcast or clear sky plus sun illuminance from Step 1, A_T is total gross area of the skylights (number of skylights times skylight gross area), and A_{WP} is the work-plane area (generally room length times width). Note that in Eq. (7.39), it is also possible to fix the E_{TWP} at some desired value and determine the skylight area required.

Rules of thumb for skylight placement for uniform illumination include 4 to 8% of roof area and spacing less than 1 1/2 times ceiling-to-work-plane distance between skylights.

Example 7.5. Determine the work-plane clear sky plus sun illuminance for a 30′ × 30′ × 10′ office with 75% ceiling, 50% wall, and 30% floor reflectance and with four 4′ × 4′ double-domed skylights at 2 PM on January 15th at 32° latitude. The skylight well is 1′ deep with 60% reflectance walls, and the outer and inner dome flat-plastic transmittances are 0.85 and 0.45, respectively. The net skylight area is 90%.

Solution. Follow the four steps in the lumen method for skylighting.
Step 1: Use Figure 7.14 with the solar altitude of 41.7° (calculated from Eq. (2.28)) for the clear sky plus sun curve to get horizontal illuminance:

$$E_H = 7,400 \text{ fc}$$

Step 2: Use Eq. (7.32) to get domed transmittances from the flat plate plastic t transmittances given,

$$T_{D_1} = 1.25(0.85)[1.18 - 0.416(0.85)] = 0.89,$$

$$T_{D_2} = (T_F = 0.45) = 0.56,$$

and Eq. (7.33) to get total dome transmittance from the individual dome transmittances:

$$T_D = \frac{(0.89)(0.56)}{(0.89) + (0.56) - (0.89)(0.56)} = 0.52.$$

To get well efficiency, use $WI = 0.25$ from Eq. (7.34) with 60% wall reflectance in Figure 7.16 to give $N_w = 0.80$. With $R_A = 0.90$, use Eq. (7.35) to calculate net transmittance:

$$T_n = (0.52)(0.80)(0.90)(1.0) = 0.37.$$

Step 3: The light loss factor is assumed to be from typical values: $K_m = (0.94)(0.75) = 0.70$. The room utilization coefficient is determined using Eqs. (7.36) and (7.37):

$$RCR = \frac{5(7.5)(30 + 30)}{(30)(30)} = 2.5,$$

$$RCU = [1 + 0.0288(2.5)^{1.560}]^{-1} = 0.89,$$

and Eq. (7.38) yields $K_u = (0.37)(0.89) = 0.33$.

Step 4: The work-plane illuminance is calculated by substituting the above values into Eq. (7.39):

$$E_{TWP} = 7,400\left[\frac{4(16)2}{30(30)}\right]0.33(0.70).$$

$$= 122\,fc$$

PROBLEMS

7.1. Explain how window placement in a building could be defined as: (a) a passive solar feature, (b) an energy conservation technique, and (c) both of the above.

7.2. Write an equation for calculating the cost and savings life-cycle economics of a proposed passive solar system. Explain why it is important to be able to determine the auxiliary energy required for any given passive (or active) system design.

7.3. Referring to the thermal circuit diagram of Figure 7.4 for the thermal storage (Trombe) wall building, construct appropriate thermal circuits for: (a) attached sunspace; (b) thermal storage roof; and (c) direct gain buildings.

7.4. Using rules of thumb for a 200 m² floor area Denver residence, determine: (a) the auxiliary heating energy required; (b) the solar projected area; and (c) the concrete storage mass needed for a maximum 10°C daily temperature swing.

7.5. A 2000 ft² house in Boston is being designed with NLC = 12,000 Btu/F-day and 150 ft² of direct gain. The direct gain system includes double glazing, night-time insulation, and 30 Btu/ft²F thermal storage capacity. Using the LCR method, determine: (a) the annual auxiliary heating energy needed by this design; and (b) the storage mass and dimensions required.

7.6. Compare the annual SSF for 150 ft² of the following passive systems for the house in Problem 7.5: (a) direct gain (DGA3); (b) vented Trombe wall (TWD4); (c) unvented Trombe wall (TWI4); (d) waterwall (WWB4); and (e) sunspace (SSB4).

7.7. A design modification to the house in Problem 7.5 is desired. A 200 ft², vented, 12″ thick Trombe wall is to be added to the direct gain system. Assuming the same types of glazing and storage as described above, determine: (a) the annual heating auxiliary energy needed; and (b) the Trombe wall mass.

7.8. Using the SLR method, calculate the auxiliary energy required in March for a 2000 ft², NLC = 12,000 Btu/F-day, house in Boston with a 150 ft², night-insulated, double glazed direct gain system with 6″ thick storage floors of 45 Btu/ft²F capacity.

7.9. Calculate the heating season auxiliary energy required for the Boston house in Problem 7.8.

7.10. Determine the length of overhang needed to shade a south-facing 2 m high window in Dallas, TX (latitude 32°51′) to allow for both winter heating and summer shading.

7.11. A 10 mph wind is blowing directly into an open 3 ft × 5 ft window which is mounted in a room's 8 ft high by 12 ft wide wall. If the wind's temperature is 80°F, are the room's occupants thermally comfortable?

7.12. Design a stack effect/solar chimney (vented Trombe wall) to produce an average velocity of 0.3 m/s within a 4 m wide by 5 m long by 3 m high room. Justify your assumptions.

7.13. Estimate the overnight radiant cooling possible from an open, 30°C, 8 m diameter water tank during July in Chicago. What would you expect for convective and evaporative cooling values?

7.14. For the Buried Pipe example (7.4) in the text, determine which of the three flow rate cases leads to the least expensive installation.

7.15. Using Figure 7.9 data, design a 9-ft deep ground-pipe system for Dallas in June to deliver 1000 cfm at 75°F when the outside air temperature is 90°F.

7.16. A 30 ft by 20 ft office space has a photosensor dimmer control working with installed lighting of 2 W/ft². The required work-place illuminance is 60 fc and the available daylighting is calculated as 40 fc on the summer peak afternoon. Determine the payback period for the dimmer control system assuming the following: 1½ tons cooling installed for 600 ft² at $2,200/ton; lighting control system cost at $1/ft²; 30% reduction in annual lighting due to daylighting; $0.10/kWh electricity cost; and cooling for 6 months at a $COP_c = 2.5$.

7.17. Determine the illuminances (sun, sky, and ground-reflected) on a vertical, south-facing window at solar noon at 36°N latitude on June 21st and December 21st for: (a) a clear day; and (b) an overcast day.

7.18. Determine the sidelighting work-plane illuminances for a 20 ft long, 15 ft wide (deep), 8 ft high light-colored room with a 15 ft long by 5 ft high window. Assume the direct sun plus clear sky illuminance is 3,000 fc and the ground-reflected illuminance is 200 fc.

7.19. Determine the clear sky day and the cloudy day work-plane illuminances for a 30 ft long, 30 ft wide, 10 ft high light-colored room. A 20 ft long by 7 ft high window with ¼-inch clear glass faces 10°E of South, the building is at 32°N latitude, and it is January 15th at 2 PM solar. The ground outside is covered by dead grass!

7.20. Determine the clear day and cloudy day illuminances on a horizontal skylight at noon on June 21st and December 21st in: (a) Miami; (b) Los Angeles; (c) Denver; (d) Boston; and (e) Seattle.

7.21. A 3 ft by 5 ft double-domed skylight has outer and inner flat-plate plastic transmittances of 0.8 and 0.7, respectively, has a 2 ft deep well with 80% reflectance walls, and has a 90% net skylight area. Calculate the net transmittance of the skylight.

7.22. Determine the number and roof placement of 10 ft by 4 ft skylights needed for a 50 ft by 50 ft by 10 ft high office when the horizontal illuminance is 6,000 fc, the skylight has 45% net transmittance, and the required work-plane illuminance is 100 fc.

7.23. What would be the procedure for producing uniform work-place illuminance when both sidelighting and skylighting are used simultaneously?

REFERENCES AND SUGGESTED READINGS

1. AAMA. 1977. *Voluntary Standard Procedure for Calculating Skylight Annual Energy Balance.* Chicago, IL: Architectural Aluminum Manufacturers Association Publication. 1602.1.1977.

2. ASES. 1992. *Economics of Solar Energy Technologies.* Eds. R. Larson, F. Vignola, and R. West, American Solar Energy Society, December 1992.

3. ASHRAE. 1989. *Fundamentals: I-P Edition.* Atlanta, GA: American Society of Heating, Refrigerating and Air-Conditioning Engineers.

4. ASHRAE. 1991. *HVAC Applications: I-P Edition.* Atlanta, GA: American Society of Heating, Refrigerating and Air-Conditioning Engineers.

5. Close, D.J., Dunkle, R.V., and Robeson, K.A. 1968. *Design and Performance of a Thermal Storage Air Conditioning System.* Mech. and Eng. Trans., Institute Eng. Australia, MC4, 45.

6. Dietz, P., Murdoch, J., Pokoski, J., and Boyle, J. 1981. "A skylight energy balance analysis procedure." *J. of the Illum. Eng. Soc.* October.

7. Duffie, J.A. and Beckman, W.A. 1991. *Solar Engineering of Thermal Processes.* 2nd Ed., New York: John Wiley and Sons, Inc.

8. Goswami, D.Y. and Dhaliwal, A.S. 1985. Heat transfer analysis in environmental control using an underground air tunnel. *J. of Solar Energy Eng.,* 107 (May): 141–45.

9. Goswami, D.Y. and Ileslamlou, S. 1990. Performance analysis of a closed-loop climate control system using underground air tunnel. *J. of Solar Energy Eng.* 112 (May): 76–81.

10. Goswami, D.Y. and Biseli, K.M. 1994. Use of underground air tunnels for heating and cooling agricultural and residential buildings. Report EES-78, Florida Energy Extension service, University of Florida, Gainesville, Fl. August.
11. Halacy, D.S. 1984. *Home Energy,* Emmaus, PA: Rodale Press, Inc.
12. Hay, H. and Yellot, J. 1969. Natural air conditioning with roof ponds and movable insulation. ASHRAE Transactions 75(1): 165–77.
13. IES. 1979. *Lighting Handbook, Application Volume.* New York: Illumination Engineering Society.
14. Krarti, M. and Kreider, J.F. 1996. Analytical model for heat transfer in an underground air tunnel. Energy Conversion and Management 30(10):1561–74.
15. IES. 1987. *Lighting Handbook, Application Volume.* New York: Illumination Engineering Society.
16. Kusuda, T. and Achenbach, P.R. 1965. Earth temperature and thermal diffusivity at selected stations in the United States. ASHRAE Transactions, 71 (1): 965.
17. Labs, K. 1981. Regional analysis of ground and above ground climate. Report ORNL/Sub-81/40451/1, Oak Ridge National Laboratory, Oak Ridge, TN.
18. LOF. 1976. *How to Predict Interior Daylight Illumination.* Toledo, OH: Libbey-Owens-Ford Co.
19. Marlatt, W., Murray, C. and Squire, S. 1984. Roof pond systems energy technology engineering center. Rockwell International Report No. ETEC 6, April.
20. Martin, M. and Berdahl, P. 1984. Characteristics of infrared sky radiation in the United States, Solar Energy 33(314): 321–36.
21. McQuiston, P.C. and Parker, J.D. 1994. *Heating, Ventilating, and Air Conditioning.* 4th Ed., Wiley, New York.
22. Murdoch, J.B. 1985. *Illumination Engineering: From Edison's Lamp to the Laser.* New York: Macmillan Publishing Co.
23. Olgyay, A. and Olgyay, V. 1967. *Solar Control and Shading Devices.* Princeton, NJ: Princeton University Press.
24. Pierson, O. 1962. *Acrylics for the Architectural Control of Solar Energy.* Philadelphia: Rohm and Haas.
25. PSDH. 1980. *Passive Solar Design Handbook. Volume One: Passive Solar Design Concepts,* DOE/CS-0127/1 March: *Volume Two: Passive Solar Design Analysis,* DOE/CS-0127/2 January. Washington, DC: U.S. Department of Energy.
26. PSDH. 1984. *Passive Solar Design Handbook.* New York: Van Nostrand Reinhold Co.
27. Robbins, C.L. 1986. *Daylighting: Design and Analysis.* New York: Van Nostrand Company, Inc.
28. Rundquist, R.A. 1991. *Daylighting controls: Orphan of HVAC design. ASHRAE J.* November: 30–34.
29. U.S.D.A. 1960. Power to produce. *1960 Yearbook of Agriculture.* U.S. Department of Agriculture.

SOLAR THERMAL POWER AND PROCESS HEAT

The illusion of unlimited powers, nourished by astonishing scientific and technological achievements, has produced the concurrent illusion of having solved the problem of production.

E. F. Schumacher

8.1 HISTORICAL PERSPECTIVE

Attempts to harness the sun's energy for power production date back to at least 1774 [56], when the French chemist Lavoisier and the English scientist Joseph Priestley discovered oxygen and developed the theory of combustion by concentrating the rays of the sun on mercuric oxide in a test tube, collecting the gas produced with the aid of solar energy, and burning a candle in the gas. Also, during the same year an impressive picture of Lavoisier was published in which he stands on a platform near the focus of a large glass lens and is carrying out other experiments with focused sunlight (see schematic, Fig. 8.1).

A century later, in 1878, a small solar power plant was exhibited at the World's Fair in Paris (Fig. 8.2). To drive this solar steam engine, sunlight was focused from a parabolic reflector onto a steam boiler located at its focus; this produced the steam that operated a small reciprocating steam engine that ran a printing press. In 1901, a 10-hp solar steam engine was operated by A.G. Eneas in Pasadena, California [6]. It used a 700-ft^2 focusing collector the shape of a truncated cone as shown in Fig. 8.3. Between 1907 and 1913 the American engineer F. Shuman developed solar-driven hydraulic pumps; in 1913 he built, jointly with C.V. Boys, a 50-hp solar engine for pumping irrigation water from the Nile near Cairo in Egypt (Fig. 8.4). This device used long parabolic troughs that focused solar radiation onto a central pipe with a concentration ratio of 4.5:1.

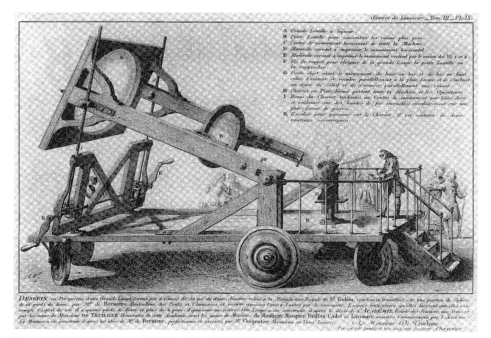

Figure 8.1. Solar furnace used by Lavoisier in 1774. Illustration courtesy of Bibliotheque Nationale de Paris. Lavoisier, *Oeuvres*, vol. 3.

Figure 8.2. Parabolic collector powered a printing press at the 1878 Paris Exposition.

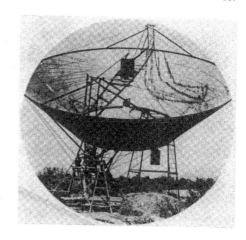

Figure 8.3. Irrigation pumps were run by a solar-powered steam engine in Arizona in the early 1900s. The system consisted of an inverted cone that focused rays of the sun on the boiler.

With the increasing availability of low-cost oil and natural gas, interest in solar energy for power production waned. Except for C.G. Abbott, who exhibited in 1936 a ½-hp solar-powered engine at an International Power Conference in Washington, D.C. and in 1938 in Florida, an improved, somewhat smaller version with a flash boiler, there was very little activity in the field of solar power between 1915 and 1950. Interest in solar power revived in 1949 when, at the centennial meeting of the American Association for the Advancement of Science in Washington, D.C., one session was devoted to future energy sources. At that time, the potentials as well as the economic problems of solar energy utilization were clearly presented by Daniels [7]. Some important conferences that considered solar power generation were held by UNESCO in 1954, the Association for Applied Solar Energy in 1955, the U.S. National Academy of Sciences in 1961, and the United Nations in 1961. In addition, a research and development program supported by the National Aeronautics and Space Administration to build a solar electric power system capable of supplying electricity for the U.S. space

Figure 8.4. Solar irrigation pump (50-hp) operating in 1913 in Egypt.

program was undertaken in the 1960s. However, widespread interest developed only after research funds became available for the development of earth-bound solar electric power and process heat after the oil embargo in 1973.

8.2 SOLAR INDUSTRIAL PROCESS HEAT

Industrial process heat consumed over 15 quadrillion (10^{15}) Btus of energy in 1972 and this amount is expected to double before the end of the century. A study conducted for the Energy Research and Development Administration by the InterTechnology Corporation [12] indicated that solar energy has the potential of providing about 20 percent of this energy. The economic outlook for industrial solar heat appears to be extremely favorable because process heat solar collectors could be used throughout the year and each system can be designed to fit the temperature level required for its specific applications, which is particularly important in the use of process heat. Table 8.1 shows the amount of heat used by selected industries in the U.S.A. A majority of the heat is used in the mining, food, textiles, lumber, paper, chemicals, petroleum products, stone-clay-glass, and primary metals [33]. The breakdown of industrial energy usage is as follows

Table 8.1. Summary of U.S. industrial heat usage by SIC category for 1971 and 1994[a]

		Quantities in 10^{12} kJ	
SIC group		1971	1994
20	Food and kindred products	779	1,254
21	Tobacco products	14	W[b]
22	Textile mills	268	327
23	Apparel	22	W[b]
24	Lumber and wood products	188	518
25	Furniture	37	73
26	Paper and allied products	2,006	2,812
27	Printing and publishing	16	118
28	Chemicals	2,536	5,621
29	Petroleum products	2,576	6,688
30	Rubber	158	303
31	Leather	19	W[b]
32	Stone, clay and glass	1,541	996
33	Primary metals	3,468	2,597
34	Fabricated metal products	295	387
35	Machinery	283	260
36	Electrical equipment	213	256
37	Transportation	310	383
38	Instruments	53	113
39	Miscellaneous	72	W[b]
	Subtotal	14,854	22,854

[a]From [23].
[b]W = Withheld to avoid disclosing data for individual establishments.

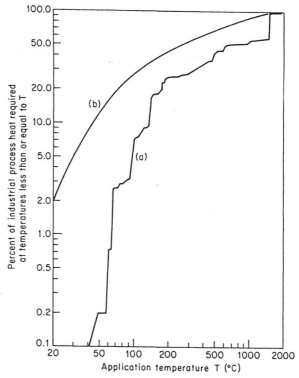

Figure 8.5 Distribution of U.S. process heat use by required temperature level: (a) heat requirements; (b) IPH requirements plus preheat from 15°C. From [23].

[23]: process steam, 41 percent; direct process heat, 28 percent; shaft drive, 19 percent; feedstock, 9 percent; and other 3 percent. In addition to the quantity of heat, quality (i.e., temperature) is also very important to match the proper solar collection system to the application. Figure 8.5 shows the cumulative process heat use by temperature requirement [23]. It is seen that about 25 percent of the heat is used at temperatures below 100°C which may be provided by flat-plate collectors, CPC's, or solar ponds, about 50 percent of the heat is used at temperatures below 260°C, and 60 percent below 370°C. Therefore, 50–60 percent of all the U.S. process heat could be delivered by parabolic trough collectors (PTC).

Since 50 percent of the industrial process heat requirements which are below 260°C are provided by fossil fuels, it represents enormous waste of availability. Table 8.2 shows approximate second law efficiencies for IPH systems below 260°C assuming 80 percent first law efficiency for fossil fuel systems. Although the numbers in Table 8.2 are not precise, they do point out the potential to increase the efficiency of energy use by replacing the high quality fossil fuels with Solar IPH. Solar IPH (SIPH) systems are quite simple and are based on the solar heating systems already discussed

Table 8.2. Second law efficiencies for U.S. industrial processes below 260°C (500°F)[a]

Temperature	Fraction of U.S. process heat (%)	Fossil fuel (%) (η_2)	Solar (%) (η_2)
29.4°C (85°F)	10	<1	12
49°C (120°F)	5	6	52
65.6°C (150°F)	5	10	65
79.4°C (175°F)	5	13	71
98.9°C (210°F)	5	16	72
121.1°C (250°F)	5	20	77
148.9°C (300°F)	5	25	83
187.8°C (370°F)	5	30	85
237.8°C (460°F)	5	35	85
Total/Averages	50	16	61

[a]Adapted from [33].

in Chapter 5. The selection of the type of solar collectors depends on the process temperature requirements. Table 8.3 gives the temperature requirements and the type of solar collectors suitable for the process.

The material and type of storage depends on the temperature requirement, the design storage duration, the required energy density (space constraints), and the charging and discharging characteristics. These topics are discussed in detail in Chapter 4. The storage duration for SIPH systems is rarely more than one day since the solar systems are designed to displace part of the fossil fuel requirements. The size of storage must be evaluated based on a cost-benefit analysis. Land availability can be critical for SIPH for existing industries. However, in many cases roofs of industrial buildings can be utilized for this purpose.

Since SIPH systems use components and systems already described in Chapters 3, 4, and 5, this section will give some examples of SIPH systems and a methodology for long-term performance prediction.

8.3 EXAMPLES OF SIPH SYSTEMS

Most of the industrial process heat systems below 200°C require hot water, steam or hot air. A typical low-temperature SIPH system is shown schematically in Fig. 8.6. If the heat is needed in process air, water-to-air heat exchangers may be used or air heating collectors and rock storage may be used. Hot air is needed typically in agricultural drying, which may be provided by passive solar air heaters. If a large body of water is available, solar ponds may be used for low-temperature applications. Solar ponds are described later in this chapter.

A high-temperature SIPH system may be designed to use low-temperature, medium-temperature and high-temperature collectors in stages in order to minimize the cost and maximize the efficiency. A schematic of such a system designed for textile dyeing process is shown in Fig. 8.7.

Table 8.3. Estimated date of technical readiness for various solar IPH systems[a]

Industry/process	Energy form[b]	Temperature (°C)	Shallow ponds or simple air heaters	Flat plates	Fixed compound surfaces	Single-tracking troughs	Central receivers
Aluminum							
Bayer process digestion	Steam	216				X	
Automobile and truck manufacturing							
Heating solutions	Steam (water)	49–82	X	X			
Heating makeup air in paint booths	Air	21–29	X				
Drying and baking	Air	163–218			X	X	
Concrete block and brick							
Curing product	Steam	74–177			X	X	
Gypsum							
Calcining	Air	160			X	X	
Curing platerboard	Steam (air)	299				X	X
Chemicals							
Borax, dissolving and thickening	Steam	82–99		X	X		
Borax, drying	Air	60–77	X	X			
Bromine, blowing brine/distillation	Steam	107			X		
Chlorine, brine heating	Steam (water)	66–93	X	X			
Chlorine, caustic evaporation	Steam	143–149			X	X	
Phosphoric acid, drying	Air	121			X		
Phosphoric acid, evaporation	Steam	160			X	X	
Potassium chloride, leaching	Steam	93		X	X	X	
Potassium chloride, drying	Air	121			X	X	
Sodium metal, salt purification	Steam	135			X	X	
Sodium metal, drying	Steam (air)	116			X	X	
Food							
Washing	Water	49–71	X	X			
Concentration	Steam (water)	38–43	X	X			
Cooking	Steam	121–188			X	X	
Drying	Steam (air)	121–232			X	X	

Table 8.3. (*Continued*)

Industry/process	Energy form[b]	Temperature (°C)	Shallow ponds or simple air heaters	Flat plates	Fixed compound surfaces	Single-tracking troughs	Central receivers
Glass							
Washing and rinsing	Water	71–93	X	X			
Laminating	Air	100–177			X	X	
Drying glass fiber	Air	135–141			X	X	
Decorating	Air	21–93	X	X	X		
Lumber							
Kiln drying	Air	66–99	X	X			
Glue preparation/plywood	Steam	99–177			X	X	
Hot pressing/fiberboard	Steam	199				X	
Log conditioning	Water	82		X			
Mining (Frasch sulfur)							
Extraction	Pressurized Water	160–166			X	X	
Paper and pulp							
Kraft pulping	Steam	182–188				X	
Kraft liquor evaporation	Steam	138–143			X	X	
Kraft bleaching	Steam	138–143			X	X	
Papermaking (drying)	Steam	177				X	
Plastics							
Initiation	Steam	121–146			X	X	
Steam distillation	Steam	146			X	X	
Flash separation	Steam	216				X	
Extrusion	Steam	146			X	X	
Drying	Steam	188				X	
Blending	Steam	121			X	X	

	Medium	Temperature (°C)					
Synthetic rubber							
Initiation	Steam (water)	121			X		
Monomer recovery	Steam	121			X		
Drying	Steam (air)	121			X	X	X
Steel							
Pickling	Steam	66–104	X		X	X	
Cleaning	Steam	82–93			X	X	
Textiles							
Washing	Water	71–82	X	X	X		
Preparation	Steam	49–113	X	X	X	X	
Mercerizing	Steam	21–99	X	X	X	X	
Drying	Steam	60–135	X	X	X	X	
Finishing	Steam	60–149	X	X	X	X	X

aFrom [33].
bPreferred form (secondary form).
cDemonstrated thermal efficiency and reliability for 10 yr.

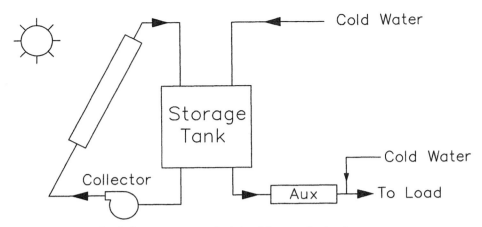

Figure 8.6. Schematic of a low-temperature solar industrial process heat system.

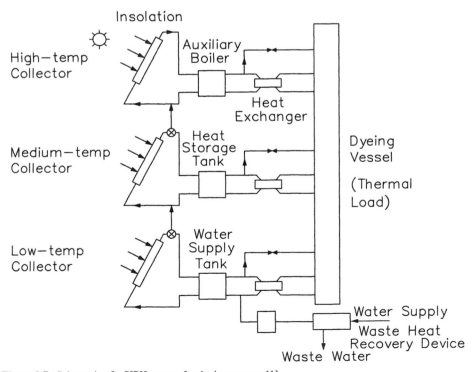

Figure 8.7. Schematic of a SIPH system for dyeing process [1].

8.3.1 SIPH for Textile Industries

The textile industry is one of the ten largest energy consuming industries. Of all the energy used in the textile industry 60–65 percent is used in wet processing, including dyeing, finishing, drying and curing. The energy for wet processing is used as hot water and steam. The textile industry in the U.S. uses approximately 500 billion liters of water per day and approximately 25 percent of this water is used at an average temperature of 60°C. Tables 8.4 and 8.5 show typical calculations for determining energy consumption for jet dyeing [57] and tenter frame drying [21], respectively, for 100 percent textured polyester circular knit fabric [13]. In the analysis in Table 8.4 Wagner assumed 40 percent moisture content in the fabric. In carpet dyeing, moisture may be as much as 300 percent [35]. Drying involves the use of high pressure (6 atm) steam so that the condensed water may be used for other processes. There are no known examples of SIPH for textile drying, but there are a number of examples of SIPH for textile dyeing. One of those is for a dyeing operation at the Riegel Textile Corp. plant in LaFrance, SC [13,15]. Figure 8.8 shows a schematic of the system.

The original system used evacuated tube collectors made by General Electric Corp. That system failed because of repeated tube breakage. Since the dyeing process used water at a temperature of 70°C, the evacuated tube collectors were replaced with flat-plate collectors. The flat-plate collectors at this plant have copper absorbers and tubes painted flat black, low iron textured and tempered glass cover, and bronze enameled steel frames. The system has 621 m² of collector area.

The dyeing process at this plant is an atmospheric Dye Beck batch process at a maximum temperature of 90°C which is also typical of the textile industry in the U.S. [13]. The batch dyeing process involves heating of approximately 4500 liters of inlet water from 10°C–25°C to 90°C. The typical dye cycle is shown in Table 8.6. The refurbished SIPH system at Riegel Textile plant has been operating successfully. The system was simulated using TRNSYS computer model [55]. Figure 8.9 shows the actual performance of the system and that predicted by TRNSYS. According to mea-

Table 8.4. Sample calculations*: heat for pressure jet dyeing [57,13]

Scour	Energy Consumption GJ/100 kg Fabric
Heat Bath, 21°C–60°C	163
Heat Cloth, 21°C–60°C	9
Raise Bath to 129°C	293
Raise Cloth to 129°C	14
Replace Heat Loss from Radiation	
During Cycle 21°C–129°C	23
During Dyeing at 129°C	65
Scour at 60°C	172
TOTAL	739

*Fabric Load = 227 kg, Bath Ratio 10 of 1
Cycle Time = 2.75 hours

Table 8.5. Energy consumption during tenter frame drying [21,13]

A. Steps	Energy Requirements (MJ/100 kg H$_2$O)
Evaporate Water	
100 kg × 4.18 kJ/kg°C (100°C–21°C)	33
Latent Heat of Vaporization	226
Raise Steam of 121°C	
100 kg × 1.9 kJ/kg°C (121°C–100°C)	4
Heat Air (24 lb Air/lb H$_2$O)	
2400 kg × 1.015 kJ/kg°C (121°C–21°C)	244
Heat Fabric to 250	
100 kg × 100/40 × 2.08 × (121°C–21°C)	52
Dryer Run at 121°C	559
Dryer Run at 149°C	647 (+15%)

Effect of fabric moisture content on tenter frame energy demand

B. Steps	Energy Requirements (MJ/100 kg Fabric) 30%	80%
Evaporate Water		
Raise Temperature (21°C–100°C)	10.0	26.5
Latent Heat of Vaporization	67.7	180.5
Raise Steam Temperature to 121°C	1.4	3.5
Raise Air to 121°C	73.3	195.2
TOTAL	152.4	405.7

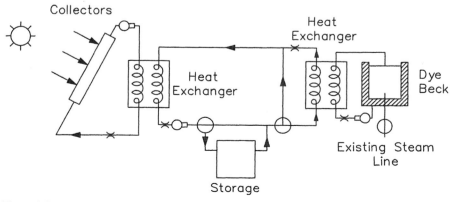

Figure 8.8. Schematic of solar energy system at Riegel Textile Corporation plant at LaFrance, South Carolina [16].

Table 8.6. Typical Dye Beck heat requirements

Dye process operation	Temperature of Dye Beck (°C)	Time interval of cycle (hours)	Percent of total process energy (%)
Heat the Initial Load	16–32	0.5	7
Dyeing Preparation	32–88	0.5	22
Dye Period	88	7–9	68
Cool and Reheat for Dye Fixation	32–43	1.0	3

sured performance of the system from Aug. 14 to Oct. 8, 1983, the system operated with an efficiency of 46 percent [13].

8.3.2 SIPH System for Milk Processing

Food systems uses approximately 17 percent of the U.S. energy [46], almost 50 percent of which is for food processing as hot water (< 100°C) or hot air. Proctor and Morse [40] show that over 40 percent of the energy demand in the beverage industry in Australia was in the form of hot water between 60°C and 80°C. Considering the temperature and heat requirements of food processing one would be tempted to conclude that SIPH would be ideal. However, many food processing requirements are seasonal, which may not be economical considering the present price of the conventional fuels unless the SIPH system can be used for a majority of the year. One application that is year-round is milk processing. Singh et al. [46] simulated a SIPH system for milk pro-

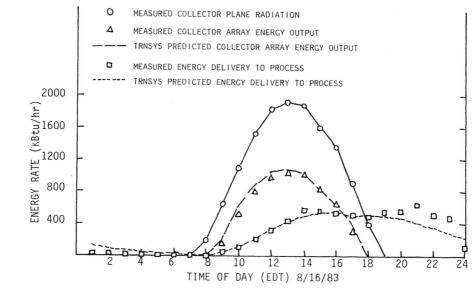

Figure 8.9. Actual and predicted system operation. August 16, 1983.

cessing in the U.S. using TRNSYS. The unit operations (and their respective temperatures) for this plant compatible with solar thermal energy are:

1. Boiler feed make-up water (100°C),
2. pasteurizer make-up water (21°C),
3. case washer and rinsing (49°C),
4. clean up (71°C),
5. high temperature short time clean up (79°C), and
6. bottle water (93°C).

They estimated that for a plant producing 170,000 Kg/week of milk and 98,000 kg/week of orange juice, a total of 621.3 GJ (or 80 percent) of energy demand was compatible with solar energy in summer and 724.9 GJ (or 93 percent) of energy demand was compatible in winter. Simulating a solar system similar to the one in Fig. 8.6 they found that a 4000 m² collector area could provide about 30–35% solar fraction for the milk processing plant in Madison, WI, Fresno, CA or Charleston, SC [46].

8.4 LONG-TERM PERFORMANCE OF SIPH SYSTEMS

In order to assess the economic viability of any solar process, its cumulative energy delivery over its economic life (in years or decades) must be known. It is very difficult to calculate this number accurately since (1) solar systems and their energy delivery are subject to the vagaries of local microclimate which can change on a time scale on the order of hours, and (2) future weather cannot be predicted at this level of detail. The standard approach used to estimate future performance of a solar system is to use a typical year of past weather data and assume that it will represent the future on the average, to engineering accuracy. Two common methods used to predict long-term performance of SIPH systems are the utilizability method and the TRNSYS [42]. These methods can also be used for the design of SIPH systems. The following sections describe both methods.

8.4.1 Critical Solar Intensity Ratio X

The instantaneous efficiency equation for many solar collectors has been shown to be of the form

$$\eta_c = F(\eta_o - U_c \Delta T^+/I_c), \quad (\eta_c > 0) \tag{8.1}$$

where ΔT^+ is the value of a collector to ambient temperature difference if positive and η_c is zero otherwise, and F is a heat exchanger factor (F', F_R), the expression for which depends on the definition of ΔT^+. It is technically correct but not always economical to operate the solar collector system if $\eta_c > 0$. In practice, $\eta_c \geq \eta_{min} > 0$ is usually the system turn-on criterion since it is not worthwhile to operate collector loops for cases where η_c is very small.

Equation (5.24) can be used to determine the solar intensity level above which useful energy collection can take place. Solving Eq. (5.24) for I_c,

$$I_c \geq U_c \Delta T^+ / (\eta_o - \eta_{min}/F). \tag{8.2}$$

A dimensionless critical intensity ratio X is generally used and since $\eta_{min} \ll 1$ and F is close to 1, for convenience the second term in the denominator above is dropped:

$$X \equiv U_c \Delta T^+ / \eta_o I_c \leq 1.0. \tag{8.3}$$

X is seen to be the ratio of collector heat loss to absorbed solar flux at $\eta_c = 0$, i.e., at the no-net-energy-delivery condition. In many cases the daily or monthly averaged daily critical intensity ratio $\overline{X}$ is of more interest and is defined as[1]

$$\overline{X} \equiv U_c \overline{\Delta T^+} \, \Delta t_c / \overline{\eta_o} \overline{I_c}, \tag{8.4}$$

where $\overline{\eta_o}$ is the daily averaged optical efficiency and $\overline{\Delta T^+}$ is the daily mean temperature difference *during collection*. These can also be expressed

$$\overline{\Delta T} = \frac{1}{\Delta t_c} \int_{t_o}^{t_o + \Delta t_c} (T_c - T_a) dt, \tag{8.5}$$

$$\overline{\eta_o} = \frac{\int_{t_o}^{t_o \Delta t} \eta_o I_c dt}{\int_{t_o}^{t_o + \Delta t_c} I_c dt}, \tag{8.6}$$

and

$$\overline{I_c} = \frac{1}{\Delta t_c} \int_{t_o}^{t_o + \Delta t_c} I_c dt. \tag{8.7}$$

The collector cut-in time t_o and cut-off time $t_o + \Delta t_c$ are described shortly. The time $t = [0,24]$ h and is related to the solar hour angle h_s by $t = (180 + h_s)/15$; Δt_c is the collection period in hours. In Eq. (5.29) T_c can be collector surface, average fluid, inlet fluid or outlet fluid temperature, depending upon the efficiency data basis.

8.4.2 The Utilizability Method

Utilizability, ϕ, has been used to describe the fraction of solar flux absorbed by a collector which is delivered to the working fluid. On a monthly time scale

$$\overline{\phi} = Q_u / F \overline{\eta_o} \overline{I_c} < 1.0, \tag{8.8}$$

---

[1]Note that U_c can be defined to include piping heat loss per collector array [3].

where the overbars denote monthly means and Q_u is the monthly averaged daily total useful energy delivery. ϕ is the fraction of the absorbed solar flux which is delivered to the fluid in a collector operating at a fixed temperature T_c. The ϕ concept does not apply to a system comprised of collectors, storage, and other components wherein the value of T_c varies continuously. The fixed temperature mode will occur if the collector is a boiler, if very high flow rates are used, if the fluid flow rate is modulated in response to flux variations to maintain a uniform T_c value, or if the collector provides only a minor fraction of the thermal demand. However, if the flow is modulated, note that the value of F (i.e., F', F_R) may not remain constant to engineering accuracy.

When T_c is not constant in time as in the case of a collector coupled to storage, the $\overline{\phi}$ concept cannot be applied directly. However, for most concentrators for CR >10, the value of U_c is small and the collector is relatively insensitive to a *small* range of operating temperatures. To check this assumption for a particular process, values of $\overline{\phi}$ at the extremes of the expected temperature excursion can be compared.

The value of $\overline{\phi}$ depends upon many system and climatic parameters. However, Collares-Pereira and Rabl [5] have shown that only three are of first order—the clearness index $\overline{K}_T$ (see Chapter 3), the critical intensity ratio $\overline{X}$ [Eq. (5.27)], and the ratio r_d/r_T (see Chapter 2). The first is related to insolation statistics, the second to collector parameters and operating conditions, and the last to collector tracking and solar geometry.

Empirical expressions for $\overline{\phi}$ have been developed for several collector types [5]. For nontracking collectors,

$$\overline{\phi} = \exp\{-[\overline{X} - (0.337 - 1.76\overline{K}_T + 0.55r_d/r_T)\overline{X}^2]\} \tag{8.9}$$

for $\overline{\phi} > 0.4$, $\overline{K}_T = [0.3, 0.5]$, and $\overline{X} = [0, 1.2]$.

Also,

$$\overline{\phi} = 1 - \overline{X} + (0.50 - 0.67\,\overline{K}_T + 0.25r_d/r_T)\overline{X}^2 \tag{8.10}$$

for $\overline{\phi} > 0.4$, $\overline{K}_T = [0.5, 0.75]$, and $\overline{X} = [0, 1.2]$.

The $\overline{\phi}$ expression for tracking collectors (CR >10) is

$$\overline{\phi} = 1.0 - (0.049 + 1.44\overline{K}_T)\overline{X} + 0.341\overline{K}_T\overline{X}^2 \tag{8.11}$$

for $\overline{\phi} > 0.4$, $\overline{K}_T = [0, 0.75]$, and $\overline{X} = [0, 1.2]$.

Also,

$$\overline{\phi} = 1.0 - \overline{X} \tag{8.12}$$

for $\overline{\phi} > 0.4$, $\overline{K}_T > 0.75$ (very sunny climate), and $\overline{X} = [0, 1.0]$ for any collector type.

Equations (5.32)–(5.35) were developed using curve-fitting techniques emphasizing large $\overline{\phi}$ values since this is the region of interest for most practical designs. Hence, they should be considered accurate to ± 5 percent only for $\overline{\phi} > 0.4$. Empirical equations for utilizability were also given by Klein and his coworkers, however these equations were restricted to flat plate collectors. Since SIPH systems may use concentrating solar collectors, equations given by Ref. (5) are used in this chapter.

8.4.3 Example Calculation

To illustrate the use of the long-term method an example will be worked in stepwise fashion. The several steps used are

1. Evaluate $\overline{K}_T$ from terrestrial $\overline{H}_h$ data and extraterrestrial $\overline{H}_{o,h}$ data.
2. Calculate r_d/r_T for the concentration ratio and tracking mode for the collector.
3. Calculate the critical intensity ratio $\overline{X}$ from Eq. (5.27) using a long-term optical efficiency value $\overline{\eta}_o$ and monthly average collector-plane insolation.

$$\overline{I}_c = (r_T - r_d \overline{D}_h / \overline{H}_h)\overline{H}_h. \tag{8.13}$$

The collection time Δt_c may need to be determined in some cases for non-tracking, low-concentration collectors by an iterative method as described in the next section.

Example 8.1. Find the energy delivery of a polar-mounted, parabolic trough collector operated for 8 h per day ($\Delta t_c = 8$) during March in Kabul, Afghanistan ($L = 34.5°N$). The collector has an optical efficiency $\overline{\eta}_o$ of 60 percent, a heat loss coefficient $U_c = 0.5$ W/m² °C, CR = 20, and heat removal factor $F_R = 0.95$. The collector is to be operated at 150°C. The mean, horizontal solar flux is 450 Ly/day (5.23 kW h/m² day) and the ambient temperature is 10°C.

Solution. Following the three-step procedure above, the clearness index is calculated:

$$\overline{H}_{o,h} = 8.15 \text{ kW h/m}^2 \text{ day} \quad \text{(Table A2.2b)} \quad \overline{K}_T = 5.23/8.15 = 0.64.$$

The geometric factors r_d and r_T are calculated from expressions in Table 3.11:

$$r_T = (ah_{\text{coll}} + b \sin h_{\text{coll}})/d \cos L,$$
$$r_d = (h_{\text{coll}}/d)(1/\cos L + \cos h_{\text{sr}}(\alpha = 0)/(\text{CR})) - \sin h_{\text{coll}}/d(\text{CR}),$$

where

$$h_{coll} = 60° = 1.047 \text{ rad}$$
$$(h_{coll} = \Delta t_c/2) \times 15°,$$

If the collection period is centered about solar noon),

$$h_{sr}(\alpha = 0) = 90° = 1.571 \text{ rad},$$
$$a = 0.409 + 0.5016 \sin 30° = 0.66,$$
$$b = 0.6609 - .4767 \sin 30° = 0.42,$$
$$d = \sin 90° - 1.571 \cos 90° = 1.0,$$

in which case

$$r_T = (0.66 \times 1.047 + 0.42 \times \sin 60°)/1.0 \cos 34.5° = 1.28,$$
$$r_d = (1.047/1.0)[1/\cos 34.5° + \cos 90°/20] - \sin 60°/1.0 \times 20 = 1.31.$$

Finally, the critical intensity ratio is

$$\overline{X} = U_c\overline{\Delta T^+} \Delta t_c/\overline{\eta}_o \overline{I}_c,$$

and the collector plane insolation $\overline{I}_c$ from Eq. (3.146) is

$$\overline{I}_c = (1.28 - 1.31 \times 0.34) \times 5.23 = 5.4 \text{ kW h/m}^2,$$

so

$$\overline{X} = 0.5 \times (150 - 10) \times 8/0.6 \times 5400 = 0.173.$$

The utilizability $\overline{\phi}$ from Eq. (5.35) is

$$\overline{\phi} = 1.0 - (0.049 + 1.44 \times 0.64)(0.173) + 0.341 \times 0.64 \times (0.173)^2 = 0.84.$$

Finally, the useful energy is

$$Q_u = F_R\overline{\eta}_o\overline{I}_c\overline{\phi} = 0.95 \times 0.6 \times 5.4 \times 0.84 = 2.58 \text{ kWh/m}^2 \text{ day}$$

for the month of March on the average.

8.4.4 Collection Period (Δt_c)

The collection period Δt_c can be dictated either by optical or thermal constraints. For example, with a fixed collector, the sun may pass beyond the acceptance limit or be blocked by another collector, and collection would then cease. Alternately, a high efficiency, solar-tracking concentrator operating at relatively low temperature might be able to collect from sunrise to sunset. A third scenario would be for a relatively low concentration device operating at high temperature to cease to have a positive effi- ciency during daylight at the time that heat losses are equal to absorbed flux. In this case, the cutoff time is dictated by thermal properties of the collector and the operating conditions.

Collares-Pereira and Rabl [5] have suggested a simple procedure to find the proper value of Δt_c. Useful collection Q_c is calculated using the optical time limit first, i.e., $\Delta t_c =$ 2 min $\{[h_{sr}(\alpha = 0), h_{sr}(i = 90)]/15\}$. Second, Q_u is calculated for a time period slightly shorter, say by one-half hour, than the optical limit. If this value of Q_u is larger than that for the first, optically limited case, the collection period is shorter than the optical limit. The time period is then further reduced until the maximum Q_u is reached.

The above method assumes that collection time is symmetric about solar noon. This is almost never the case in practice since the heat collected for an hour or so in the morning is required to warm the fluid and other masses to operating temperature. A symmetric phenomenon does not occur in the afternoon. If the time constant of the thermal mass in the collector loop is known, the collection period may be assumed to begin at t_o, $\Delta t_c/2$ h (from above symmetric calculation) before noon decreased by two or three time constants. Another asymmetry can occur if solar flux is obstructed during low sun angle periods in winter. It is suggested that r_T and r_d from Table 3.11 under asymmetric collection conditions be calculated from

$$r_T = [r_T(h_{s,stop}) + r_T(h_{s,start})]/2, \tag{8.14}$$

$$r_d = [r_d(h_{s,stop}) + r_d(h_{s,start})]/2, \tag{8.15}$$

where the collection starting and stopping hour angles account for transients, shading, etc., as described above:

$$h_{s,start} = 180 - 15t_o, \tag{8.16}$$

$$h_{s,stop} = 180 - 15(t_o + \Delta t_c). \tag{8.17}$$

Example 8.2. Calculations in Example 8.1 were based on $\Delta t_c = 8$ h. Repeat for 10 h to see the effect of collection time if a symmetric collection period about noon is used.

Solution. The values of r_T and r_d for $h_{coll} = 75° = 1.31$ rad are

$$r_T = (0.66 \times 1.31 + 0.42 \times \sin 75°)/1.0 \cos 34.5° = 1.98,$$

$$r_d = (1.31/1.0)(1/\cos 34.5°) - \sin 75°/20 = 1.54.$$

The collector plane insolation is then

$$\bar{I}_c = (1.98 - 1.54 \times 0.34) \times 5.23 = 7.6 \text{ kWh/m}^2,$$

and

$$\bar{X} = [0.5 \times (150 - 10) \times 10]/0.6 \times 7600 = 0.154.$$

Then $\overline{\phi}$ is 0.86 from Eq. (8.10) and the useful energy delivery is 3.7 kWh/m² day. Hence, it is worthwhile operating the collector for at least 10 h. The calculation can be repeated by the reader for an asymmetric case 4 h before noon and 6 h after to determine the effect of warm up.

8.4.5 Long-Term Performance of Collector Systems with Storage

The previous section of this chapter described a method of predicting long-term performance of a solar collector operated at a temporally constant temperature. This situation is a good approximation of the operating conditions experienced by several types of generic thermal systems. Other systems, however, do not operate at constant temperature and the $\overline{\phi}$ method cannot be used. Although there is no simplified performance method now extant for varying temperature systems, Klein and Beckman [31] have correlated some modeling results on collector-heat exchanger-storage subsystems coupled to a uniform, process-like load operating above some temperature T_{proc}. The method is called the $\overline{\phi}$, f chart and is described below. Although the method was developed for flat-plate collectors and uses a different $\overline{\phi}$ calculation method than used above, it can be applied equally well to concentrators [30].

The calculation method requires first the determination of the utilizability $\overline{\phi}$ from Eqs. (5.32)–(5.35) above. This represents the maximum energy deliverable to a load at T_{proc}. When storage is present and collected solar heat is greater than the demand, the temperature of storage, and hence the collector inlet temperature, will rise. (The $\overline{\phi}$, f method applies only to well-mixed, sensible heat storage with liquid heat transfer fluids and storage media.) Hence, the monthly averaged, daily useful energy collected Q_u will be less than $F\,\overline{\eta}_o\,\overline{I}_c\overline{\phi}$, but storage may permit a greater fraction of the demand to be met by solar since the maximum amount of heat $F\,\overline{\eta}_o\overline{I}_c\overline{\phi}$ collectable may be more than can be used, depending on the demand amount. The $\overline{\phi}$, f method can be employed to find Q_u in a system with storage.

The technical basis for the $\overline{\phi}$, f method lies in the nondimensionalization of governing energy equations for a solar thermal system. Dimensionless groups, so identified, are used to correlate monthly thermal energy delivery-to-load for various systems simulated in various climates by an hourly time-scale computer model. The two dimensionless groups identified for use in the $\overline{\phi}$, f method, in addition to $\overline{\phi}$ ($\overline{\phi}$ is defined in this context relative to the minimum temperature acceptable to the process T_{proc}, not relative to the collector temperature as in the previous section), are a solar parameter P_s, a measure of long-term solar gain by the collector receiver, and a collector heat-loss parameter P_L, a measure of long-term heat loss at a fixed collector-to-ambient temperature difference of 100°C. This 100°C value does not restrict the generality of the results, however. In equation form,

$$P_s = F_R\overline{\eta}_o\overline{I}_cA_cN_d/L, \tag{8.18}$$

$$P_L = F_RU_cA_cN_h\,100/L, \tag{8.19}$$

in which F_R is given in Eqs. (3.43)–(3.44). L is the monthly thermal demand, and N_d and N_h are the number of days and hours in a month. The $\overline{\phi}$, f chart predicts the monthly solar load fraction f_s $(\overline{\phi}, P_s, P_L) \equiv Q_u/L$ by an empirical equation [31]

$$f_s = \overline{\phi}P_s - a[e^{3.85f_s} - 1][1 - e^{-0.15P_L}],$$ (8.20)

where $a = 0.015 \left[\dfrac{Mc_p}{350 A_c}\right]^{-0.76}$ in which $m = $ (kg), $c_p = $ (kJ/kg°C), and $A_c = $ (m²).

Figure 8.10 is one $\overline{\phi}$, f chart for a standard storage size of 350 kJ/°C-m² and is entered with values of $\overline{\phi} P_s$ and P_L to give a monthly value of f_s. The calculation of $\overline{\phi} P_s$ and P_L is done once for each month of an average year, and the totals are added to give annual performance. An example below shows how the method is used. It is noted that $\overline{\phi} P_s$, the ordinate, is the ratio of maximum possible energy delivered by a collector operating at fixed T_{proc}, $(F\eta_0\overline{I}_c\phi A_c)$, to the monthly load L. At values of $f_s > 0.4$, the $\overline{\phi}$, f curves are not independent of X since at progressively higher load fractions the average storage and collector temperatures are higher and collected solar heat per unit area smaller because collector efficiency is lower at higher temperature.

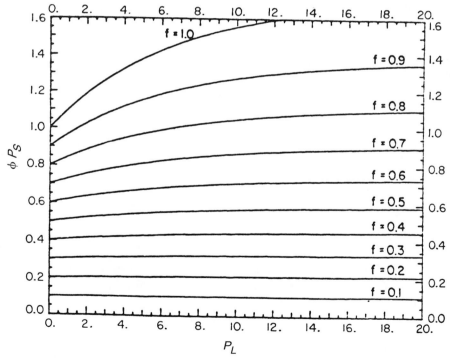

Figure 8.10. The $\overline{\phi}$, f chart used to calculate average, monthly solar fraction $f(f_s)$ of solar-thermal systems. From [31].

This $\bar{\phi}, f$ chart is based upon several limiting assumptions which should be noted in interpreting f_s values:

1. The load L is distributed uniformly over the month between the hours of 6:00 and 18:00.
2. Standard storage amount is fixed at 350 kJ/°C m² (about 2 gal of H_2O/ft_c^2 or 84 l/m_c^2). For non-standard storage, correction a (see Eq. 8.20) is applied.
3. No energy is rejected from storage; therefore, the vessel is assumed to be designed for the peak temperature and pressure expected.
4. Storage is well mixed and no storage-to-load heat exchanger is used.
5. The load device uses solar heat at temperature-independent efficiency to meet the load L. Therefore, the load device cannot be a turbine, for example.
6. No parasitic heat losses from storage occur.

Some of these restrictions can be relaxed using work on the $\bar{\phi}, f$ method conducted by Klein and his co-workers [31]. Users of the $\bar{\phi}$ methods must exercise caution in the proper choice of the $\bar{\phi}$ time scale. In the method presented here $\bar{\phi}$ and $\bar{I}_c$ are calculated over the collection period Δt_c, not over all daylight hours. The method developed by Klein uses $\bar{\phi}$ and $\bar{I}_c$ for all daylight hours. Although the Collares-Pereira $\bar{\phi}$ value can be used with the $\bar{\phi}, f$ chart, the two methods of finding $\bar{\phi}$ itself must not be confused.

Example 8.3. Repeat Example 8.1 from the previous section for Kabul, Afghanistan, for a monthly averaged load of 260 kWh/day using a collector of 100 m². From the previous example recall that $F_R\bar{\eta}_o = 0.57$, $F_RU_c = 0.475$ W/m² °C, $\bar{I}_c = 5.4$ kW h/m², and $\bar{\phi} = 0.84$. What is the effect of storage on energy delivery per unit collector area?

Solution. First calculate P_s and P_L, then use the $\bar{\phi}, f$ chart to find the solar fraction.

$$P_s = F_R\bar{\eta}_o\bar{I}_cA_cN_d/L = 0.57 \times 5.4 \times 100 \times 31/(260 \times 31) = 1.18,$$
$$P_L = F_RU_cA_cN_h\,100/L$$
$$= 0.475 \times 100 \times (31 \times 24) \times 100/(260 \times 31 \times 1000) = 0.44.$$

The value of f_s from the chart with $P_L = 0.44$ and $\bar{\phi}P_s = 0.99$ is $f_s = 0.97$. Therefore, the energy delivery per unit area is $0.97 \times 260/100 = 2.52$ kWh/m² day, nearly identical to the result using the $\bar{\phi}$ method. This is a result of the low value of U_c for the concentrator and its resulting insensitivity to temperature fluctuations above T_{proc}. The reader may repeat the calculations for a 200-m² collector with $U_c = 2.0$ W/m² °C to show that $\bar{\phi} = 0.43$, $P_L = 1.75$, $P_s = 2.37$, $f_s = 0.85$. The energy delivery per unit area is then 1.11 kWh/m² day compared with 1.32 kWh/m² day predicted by the $\bar{\phi}$ method. Hence, the effect of storage is to reduce the unit energy delivery by 16 percent for the more lossy collector.

8.5 TRNSYS-COMPUTER SIMULATION PROGRAM

TRNSYS (Transient System Simulation) is a sequential-modular transient simulation program developed at the Solar Energy Laboratory of the University of Wisconsin [29]. It is a widely used, detailed, design tool involving hourly simulations of a solar energy system over an entire year. TRNSYS computer program contains FORTRAN subroutines of almost all the components that are necessary to build a solar energy system. The component models which are either empirical or analytical, describe the component performance with algebraic and/or differential equations. A system simulation model is created by interconnecting the models of individual components. The resulting set of simultaneous algebraic and/or differential equations is solved by TRNSYS. Table 8.7 gives a list of standard TRNSYS component subroutines available in the TRNSYS library. This public domain software is constantly being upgraded and is backed by technical support.

8.6 THERMAL POWER

Electricity is fast becoming the energy form of choice all over the world even, unfortunately, for space and water heating applications. There are two basic approaches to solar electric power generation. One is by photovoltaic process, a direct energy conversion. The other approach is to convert sunlight to heat and then heat to mechanical energy by a thermodynamic power cycle and, finally, convert the mechanical energy to electricity. This indirect approach, called Solar Thermal Power is based on well established principles of thermal power. A vast majority of electricity in the world is produced by thermal power conversion. Most of the thermal power production in the world is based on Rankine Cycle and to a smaller extent Brayton Cycle. Both of these are applicable to solar thermal power conversion, with Rankine Cycle being the most popular. Stirling Cycle has also shown great potential and solar thermal power systems based on this cycle are under development.

8.6.1 Rankine Cycle

Most of the existing thermal power plants are based on the Rankine cycle. The basic ideal Rankine Cycle is shown in Fig. 8.11, which also shows a Temperature-Entropy (T-s) diagram for steam as a working fluid. The ideal cycle consists of the processes:

Process

1–2 Saturated liquid from the condenser at state 1 is pumped to the boiler at state 2 isentropically.

2–3 Liquid is heated in the boiler at constant pressure. The temperature of the liquid rises until it becomes a saturated liquid. Further addition of heat vaporizes the liquid at constant temperature until all of the liquid turns into saturated vapor. Any additional heat superheats the working fluid to state 3.

3–4 Steam expands isentropically through a turbine to state 4.

Table 8.7. List of the standard TRNSYS component subroutines.

Utility components	**Heat exchangers**
Data reader	Heat exchanger
Time-dependent forcing function	Waste heat recovery
Algebraic operator	
Solar radiation processor	**Building loads and structures**
Quantity integrator	Energy/degree-hour space heating/cooling load
Collector array shading	Pitched roof and attic
Psychometrics	Detailed zone (Transfer function)
Load profile sequencer	Overhand and wingwall shading
Convergence promoter	Window with variable insulation
Weather data generator	Thermal storage wall
	Attached sunspace
Solar collectors	Multi-zone building and BID
Linear efficiency data	
Detailed performance map	**Combine sub-systems**
Single/Biaxial incidence angle modifier	Liquid collector-storage subsystem
Theoretical flat-plate	Air collector-storage subsystem
Theoretical CPC	Domestic water heating subsystem
	Thermosyphon collector-storage subsystem
Thermal storage	
Stratified liquid storage	**Outputs**
Rock bed thermal storage	Printer
Algebraic tank (Plug-flow)	Plotter
Variable volume tank	Histogram plotter
	Simulation summary
Controllers	Economic analysis
On/off differential controller	
Four-stage room thermostat	**User contributed components**
Microprocessor controller	Electric storage battery
	Regulator/invertor
Auxiliary heating and cooling	Combined PV subsystem
On/off auxiliary heater	Flat-plate solar collector supplement
Absorption air-conditioner	
Dual-source heat pump	**Hydronics**
Cooling coil	Pump or fan
Conditioning equipment	Flow diverter/tee piece/mixing valve
Part load performance	Pipe or duct
Cooling tower	
Parallel chillers	

4–1 Steam exiting the turbine is condensed at constant pressure until it returns to state 1 as saturated liquid.

In an actual Rankine Cycle the pumping and the turbine expansion processes are not ideal. The actual processes are 1–2′ and 3–4′, respectively. For the above cycle

$$\text{Turbine Efficiency } \eta_{\text{turbine}} = \frac{h_3 - h_{4'}}{h_3 - h_4} \qquad (8.21)$$

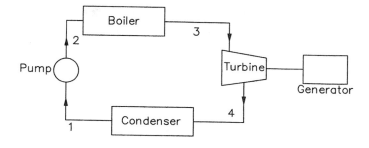

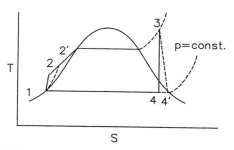

Figure 8.11. Basic Rankine Power Cycle.

$$\text{Pump Efficiency } \eta_{\text{pump}} = \frac{h_1 - h_2}{h_1 - h_{2'}} \tag{8.22}$$

$$\text{Net Work Output} = (h_3 - h_{4'}) - (h_{2'} - h_1) \tag{8.23}$$

$$\text{Heat Input} = h_3 - h_{2'} \tag{8.24}$$

$$\text{Pump Work} = h_{2'} - h_1 = \frac{v(P_2 - P_1)}{\eta_{\text{pump}}} \tag{8.25}$$

$$\text{Cycle Efficiency} = \frac{\text{Net Work Output}}{\text{Heat Input}} = \frac{(h_3 - h_{4'}) - (h_{2'} - h_1)}{h_3 - h_{2'}} \tag{8.26}$$

where h represents enthalpy and v is the specific volume at state 1.

Example 8.4. In a simple steam Rankine Cycle, steam exits the boiler at 7.0 MPa and 540°C. The condenser operates at 10 kPa and rejects heat to the atmosphere at 40°C. Find the Rankine Cycle efficiency and compare it to the Carnot cycle efficiency. Both pump and turbine operate at 85 percent efficiencies.

The cycle is shown in Fig. 8.11.

The Rankine Cycle Efficiency

$$\eta_R = \frac{(h_3 - h_{4'}) - (h_{2'} - h_1)}{h_3 - h_{2'}},$$

$$\eta_{turbine} = \frac{h_3 - h_{4'}}{h_3 - h_4}, \text{ and}$$

$$\eta_{pump} = \frac{h_2 - h_1}{h_{2'} - h_1} = \frac{v_1(P_2 - P_1)}{h_{2'} - h_1}.$$

The enthalpies at the state points are found as follows:

$$h_1 = h_f(10 \text{ kPa}) = 191.8 \text{ kJ/kg}.$$

Pump work from 1 to 2′

$$_1W_{2'} = \frac{v_1(P_2 - P_1)}{\eta_{pump}} = h_{2'} - h_1,$$

$$h_{2'} = h_1 + \frac{v_1(P_2 - P_1)}{\eta_{pump}}$$

$$= 191.8 \text{ kJ/kg} + \frac{(.00101 \text{ m}^3/\text{kg})(7000 - 10)\text{kPa}}{0.85}$$

$$= 200.1 \text{ kJ/kg}.$$

$$h_3(540°C, 7 \text{ MPa}) = 3506.9 \text{ kJ/kg},$$

states 3 and 4 have the same entropy.
    Therefore,

$$s_3 = 6.9193 \text{ kJ/kg K} = s_4.$$

Saturated vapor entropy at state 4 (10 kPa)

$$s_{4g} = 8.1502 \text{ kJ/kgK}.$$

Since s_4 is less than s_{4g}, 4 is a wet state

$$s_4 = s_f + Xs_{fg} \text{ or } s_g - Ms_{fg},$$

where X is the vapor quality and M is the moisture.

$$M_4 = \frac{s_g - s_4}{s_{fg}} = \frac{8.1502 - 6.9193}{7.5009} = 0.1641 \text{ or } 16·41\%,$$

$$h_4 = h_g - Mh_{fg} = 2584.7 - 0.1641\ (2392 \cdot 8) = 2192\ kJ/kg.$$

Therefore,

$$h_{4'} = h_3 - \eta_{\text{turbine}}(h_3 - h_4)$$
$$= 3507 - 0.85\ (3507 - 2192) = 2389\ kJ/kg.$$

Actual moisture at the turbine exhaust

$$h_{4'} = h_g - Mh_{fg},$$
$$M_{4'} = \frac{h_g - h_{4'}}{h_{fg}} = \frac{2584.7 - 2389}{2392}$$
$$= 0.0817\ \text{or}\ 8.17\%.$$

Net Work $= (h_3 - h_{4'}) - (h_{2'} - h_1) = 1109\ kJ/kg.$

$$\eta = \frac{\text{Net Work}}{h_3 - h_{2'}} = \frac{1109}{3507 - 200} = 0.3354\ \text{or}\ 33 \cdot 54\%.$$

$$\eta_{\text{Carnot}} = \frac{813\ K - 313\ K}{813\ K} = 0.615\ \text{or}\ 61.5\%$$

Several improvements can be made to the basic Rankine Cycle in order to improve the cycle efficiency. The efficiency of the Rankine Cycle may be increased by increasing the pressure in the boiler. However, that will result in increased moisture in the steam exiting the turbine. In order to avoid this problem, the steam is expanded to an intermediate pressure and reheated in the boiler. The reheated steam is expanded in the turbine until the exhaust pressure is reached. Figure 8.12 shows the Rankine Cycle with reheat.

The cycle efficiency of the Rankine Cycle with reheat

$$\eta = \frac{(h_3 - h_{4'}) + (h_5 - h_{6'}) - (h_{2'} - h_1)}{(h_3 - h_{2'}) + (h_5 - h_{4'})}. \tag{8.27}$$

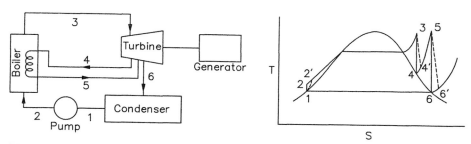

Figure 8.12. Rankine Cycle with reheat.

Example 8.5. A reheat Rankine cycle uses steam as a working fluid. The steam leaves the boiler and enters the turbine at 5 MPa and 350°C, and it leaves the condensor as a saturated liquid. After expansion in the turbine to $P = 1.4$ Mpa, the steam is reheated to 350°C and then expands in the low-pressure turbine to 20 kPa. If the efficiencies of the pump and turbine are 0.9 each, determine the cycle efficiency.

Solution. *Point 3.* $P_3 = 20$ kPa, $T_3 = 350°C$.

From Steam Tables

$$h_3 = 3068 \text{ kJ/kg},$$

$$s_3 = 6.449 \text{ kJ/kg K}.$$

Point 4. $s_4 = s_3 = 6.449$ kJ/kg K, $P_4 = 1.4$ MPa.

From Steam Tables

$$h_4 = 2781 \text{ kJ/kg}.$$

Using h_4 and turbine efficiency to find $h_{4'}$

$$\eta_T = \frac{h_3 - h_{4'}}{h_3 - h_4}$$

$$h_{4'} = h_3 - \eta_T(h_3 - h_4) = 3082 - 0.9(3082 - 2781) = 2811.1 \text{ kJ/kg}.$$

Point 5. $P_5 = 1.4$ MPa, $T_5 = 350°C$.

From Steam Tables

$$h_5 = 3149 \text{ kJ/kg},$$

$$s_5 = 7.136 \text{ kJ/kg K}.$$

Point 6.

$$P_6 = 20 \text{ kPa},$$

$$s_6 = s_5 = 7.136 \text{ kJ/kg K}.$$

$$x = \frac{7.136 - 0.8319}{7.908 - 0.8319} = 0.89,$$

$$h_6 = h_f + xh_{fg} = 251.4 + 0.89(2610 - 251.4) = 2352.$$

Using the turbine efficiency to find $h_{6'}$,

$$h_{6'} = h_5 - \eta_T(h_5 - h_6) = 3149 - 0.9(3149 - 2352) = 2432 \text{ kJ/kg}.$$

Point 1.

$$P_1 = 20 \text{ kPa},$$

Saturated liquid,

$$v_1 = 0.001017 \text{ m}^3/\text{kg},$$

$$h_1 = 251.4 \text{ kJ/kg}.$$

Point 2.

$$w = h_2 - h_1 = v(P_2 - P_1),$$

$$\eta_P = \frac{h_2 - h_1}{h_{2'} - h_1}$$

$$h_{2'} - h_1 = \frac{h_2 - h_1}{\eta_P}.$$

But $h_2 - h_1 = v(P_2 - P_1),$

$$h_{2'} - h_1 = \frac{v(P_2 - P_1)}{\eta_P} = \frac{0.001017(5000 - 20)}{0.9} = 5.6274, \text{ and}$$

$$h_{2'} = 257.03 \text{ kJ/kg}.$$

Computing the cycle efficiency

$$\eta = \frac{(h_3 - h_{4'}) + (h_5 - h_{6'}) - (h_{2'} - h_1)}{(h_3 - h_{2'}) + (h_5 - h_{4'})} = 0.30 = 30\%.$$

Another improvement to the basic Rankine Cycle is the regenerative cycle in which expanded steam is extracted at various points in the turbine and mixed with the condensed water to preheat it in the feedwater heaters. Figure 8.13 shows a schematic diagram and a T-s diagram of a Rankine Cycle with regeneration.

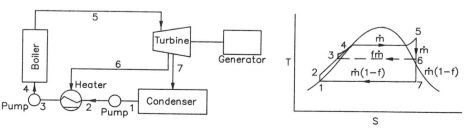

Figure 8.13. Rankine Cycle with regeneration.

If a fraction f of the steam in the turbine is bled at state 5 to mix with the feedwater. The efficiency of the Rankine Cycle with regeneration as shown in Fig. 8.13 is:

$$\eta = \frac{(h_5 - h_6) + (1 - f)(h_6 - h_7) - (1 - f)(h_2 - h_1) - (h_4 - h_3)}{h_5 - h_4}, \quad (8.28)$$

h_3 can be found from the energy balance:

$$\dot{m}h_3 = f\dot{m}h_5 + (1 - f)\dot{m}h_2,$$

or

$$h_3 = f(h_5 - h_2) + h_2. \quad (8.29)$$

Example 8.6. A regenerative Rankine cycle uses steam as the working fluid. Steam leaves the boiler and enters the turbine at $P = 5$ MPa, $T = 500°C$. After expansion to 400 kPa, a part of the steam is extracted from the turbine for the purpose of heating the feedwater in the feedwater heater. The pressure in the feedwater heater is 400 kPa and the water leaving it is saturated liquid at 400 kPa. The steam that is not extracted expands to 10 kPa. Assuming the efficiencies of the turbine and the two pumps are 100 percent each, determine the fraction of the steam extracted to the feedwater heater (f) and the cycle efficiency (η).

Solution. *Point 5.* $P_5 = 5$ MPa, $T_5 = 500°C$

From the steam table, steam is superheated.

$$s_5 = 6.976 \text{ kJ/kg K},$$
$$h_5 = 3434 \text{ kJ/kg}$$

Point 6. $s_6 = s_5 = 6.976$ kJ/kg K,

$$P_6 = 400 \text{ kPa.}$$

From the steam table, steam is superheated.

Using s_6, we can interpolate for h_6.

$$h_6 = 2773 \text{ kJ/kg.}$$

Point 7. $P_7 = 10$ kPa. Again $s_7 = s_5 = 6.976$ kJ/kg K, mixture.

Computing,

$$x = \frac{6.976 - 0.6492}{8.1501 - 0.6492} = 0.84,$$

$$h_7 = h_f + xh_{fg} = 191.81 + 0.84(2392.8) = 2210 \text{ kJ/kg}.$$

Point 1.

$$P_1 = 10 \text{ kPa},$$

Saturated liquid,

$$h_1 = 191.81 \text{ kJ/kg}.$$

Point 2.

$$P_2 = 400 \text{ kPa},$$

$$h_2 - h_1 = v_1(P_2 - P_1) = .00101(400 - 10) = 0.3939 \text{ kJ/kg},$$

$$h_2 = (191.81 + 0.3939) = 192.2 \text{ kJ/kg}.$$

Point 3.

$$P = 400 \text{ kPa},$$

Saturated liquid,

$$v_3 = 0.001084 \text{ m}^3/\text{kg},$$

$$h_3 = 604.73 \text{ kJ/kg}.$$

Using energy balance of feedwater heater, f may be computed as

$$h_3 = fh_6 + (1 - f)h_2$$

$$f = \frac{h_3 - h_2}{h_6 - h_2} = \frac{604.73 - 192.2}{2773 - 192.2} = 0.16.$$

Point 4

$$w_{p2} = h_4 - h_3 = v_3(P_4 - P_3) = .001084(5000 - 400) = 4.9864 \text{ kJ/kg}.$$

$$h_4 = h_3 + 4.9864 = 609.72 \text{ kJ/kg}.$$

Compute Efficiency (η) may be computed from equation (8.28) as:

$$\eta = \frac{(h_5 - h_6) - (1 - f)(h_6 - h_7) - (1 - f)(h_2 - h_1) - (h_4 - h_3)}{(h_5 - h_4)} = 0.399 = 39.9\%.$$

8.6.2 Components of a Rankine Power Plant

Major components of a Rankine power plant include boiler, turbine, condenser, pumps (condensate pump, feedwater booster, boiler feed pump), and heat exchangers (open heaters and closed heaters). All of the components of a solar thermal power plant are the same as those in a conventional thermal power plant except the boiler. The boiler in a solar thermal power plant includes a solar collection system, a storage system, an auxiliary fuel heater and heat exchangers. Figure 8.14 shows a schematic representation of a solar boiler. The maximum temperature from the solar system depends on the type of solar collection system (parabolic trough collectors, central receiver with heliostat field, parabolic dishes, etc.). These collection systems are described in detail in Chapter 3. If the temperature of the fluid from the solar system/storage is less than the required temperature for the turbine, the auxiliary fuel is used to boost the temperature. A fossil fuel or a biomass fuel may be used as the auxiliary fuel.

A condenser is a large heat exchanger that condenses the exhaust vapor from the turbine. Steam turbines employ surface type condensers, mainly shell and tube heat exchangers operating under vacuum. The vacuum in the condenser reduces the exhaust pressure at the turbine blade exit to maximize the work in the turbine. The cooling water from either a large body of water such as a river or a lake, or from cooling tower, circulates through the condenser tubes. The cooling water is cooled in the cooling tower by evaporation. The air flow in the cooling tower is either natural draft (hyperbolic towers) or forced draft (see Fig. 8.15). The condensate and feed water pumps are motor-driven centrifugal pumps, while the boiler feed pumps may be motor- or turbine-driven centrifugal pumps.

Steam turbines and generators are described in detail in a number of books [44,50,24,10,11] and will not be described here.

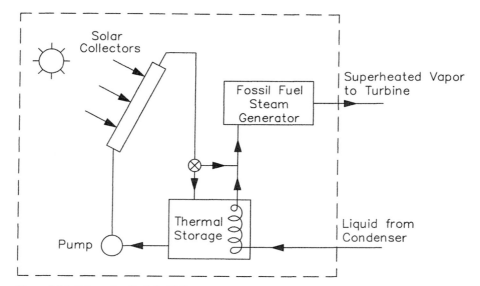

Figure 8.14. Schematic of a Solar Boiler.

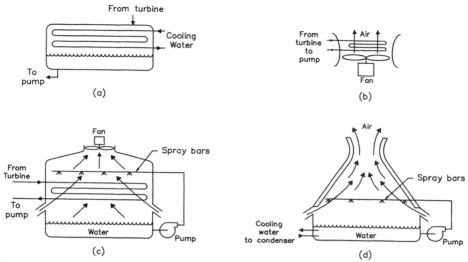

Figure 8.15. Types of condenser and/or heat rejection used in Rankine cycle solar power systems: (a) tube-and-shell condenser; (b) dry cooling tower; (c) wet cooling tower; and (d) natural-draft cooling tower. Adapted from [48].

8.6.3 Choice of Working Fluid

Working fluid in a solar Rankine cycle is chosen based on the temperature from the solar collection system. The working fluid must be such that it optimizes the cycle efficiency based on the expected temperature from the source. Steam is the most common working fluid in a Rankine cycle. Its critical temperature and pressure are 374°C and 22.1 MPa. Therefore, it can be used for systems operating at fairly high temperatures. Systems employing parabolic trough, parabolic dish or central receiver collection systems can use steam as a working fluid. Other major advantages of steam are that it is non-toxic, environmentally safe, and inexpensive. Its major disadvantage is its low molecular weight which requires very high turbine speeds in order to get high turbine efficiencies.

Steam is a wetting fluid, that is, as it is expanded in a turbine, once it reaches saturation, any further expansion increases the moisture content. In other words, steam becomes wetter as it expands as shown in Fig. 8.16a. On the other hand, a fluid that has a T-s (Temperature-Entropy) diagram similar to that shown in Fig. 8.16b is called a drying fluid. As seen from this figure, even though the working fluid passes through the two phase region, it may exit the turbine as superheated. Normally, the turbine speed is so high that, in such a case, there is no condensation in the turbine. Examples of drying fluids include hydrocarbons (toluene, methanol, isobutane, pentane, hexane), chlorofluorocarbons (CFC's such as R-11, R-113), and ammonia. Since a drying fluid does not get wetter on expansion from a saturated vapor condition in an ideal or real process, it does not have to be superheated. Therefore, a Rankine cycle using a drying fluid may be more efficient than the cycle using a wetting fluid. In fact, a drying fluid may be heated

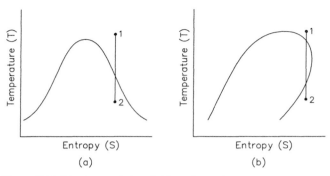

Figure 8.16. T-s characteristics of; (a) wetting and
(b) drying types of working fluids.

above its critical point so that that upon expansion it may pass through the two-phase
dome. Because of the T-s characteristics, the fluid may pass through the two-phase region and still exit from the turbine as superheated. These characteristics can be used to
increase the resource effectiveness of a cycle by as much as 8 percent [14].

Figure 8.17 and Table 8.8 give some characteristics of candidate working fluids
for Rankine cycles.

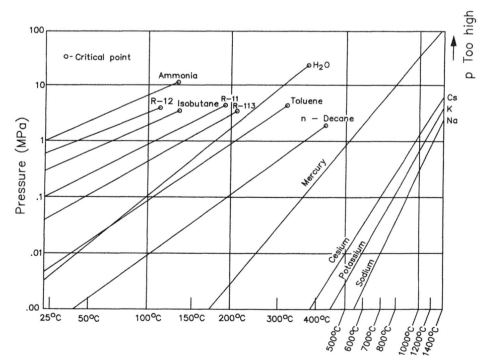

Figure 8.17. Saturation pressure-temperature relationships for potential Rankine cycle
working fluids. Adapted from [48].

Table 8.8. Physical and thermodynamic properties of prime candidate Rankine cycle working fluids.

Property	Water	Methanol	2-Methyl Pyridine, H₂O	Fluorinol 85	Toluene	R-113	Ammonia	Isobutane
Molecular weight	18	32	33	88	92	187	17	58
Boiling point (1 atm) (°C)	100	64	93	75	110	48	−330	−12
Liquid density (kg/m³)	999.5	749.6	934	1370	856.9	2565	682	594
Specific volume (saturated vapor at boiling point) (m³/kg)	1.69	0.80	0.87	0.31	0.34	0.14	1.124	0.35
Maximum stability temperature (°C)	—	175–230	370–400	290–330	400–425	175–230	300*	>200
Wetting-drying	W	W	W	W	D	D	D	D
Heat of vaporization at 1 atm (kJ/kg)	2256	1098	879	442	365	1370	1370	367
Isentropic enthalpy drop across turbine (kJ/kg)	348–1160	162–302	186–354	70–186	116–232	23–46	200–600	120–380

Adapted from [48].

*Anhydrous ammonia in the presence of iron. Small trace of water increases this limit.

8.7 EXAMPLES OF RANKINE SOLAR THERMAL POWER PLANTS

There are successful examples of solar thermal power plants working on Rankine cycle and using parabolic troughs, parabolic dishes, and central receiver towers. An example of each type is given below.

8.7.1 Parabolic Trough Based Power Plant

Luz Corporation developed components and commercialized parabolic trough collector-based solar thermal power by constructing a series of such power plants from 1984 to 1991. Starting with their first 14 MW_e Solar Electric Generating Station (SEGS I) in Southern California they added a series of SEGS power plants with total generating capacity of 354 MW_e. Figure 8.18 shows a schematic of the SEGS VIII and IX plants. Table 8.9 provides information about the solar field, power block, and other general information about SEGS I and SEGS IX plants. All of these plants use natural gas as the auxiliary fuel so that, on average, 75 percent of the energy is supplied from the sun and 25 percent from natural gas. With power plant electrical conversion efficiencies of the order of 40 percent and the solar field efficiencies of 40 to 50 percent, overall efficiencies for solar to electricity conversion of order of 15 percent are being achieved in these plants. The cost of electricity from these plants has been decreased from about 30¢/KWh to less than 10¢/KWh.

8.7.2 Central Receiver Systems

Central receiver tower-based solar thermal power plants have been actively pursued in the U.S., Germany, Switzerland, Spain, France, Italy, Russia, and Japan since the con-

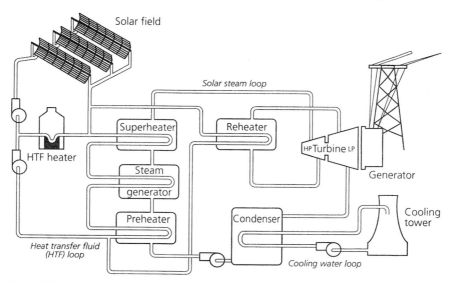

Figure 8.18. Flow of heat-transfer fluid through the SEGS VIII and IX plants (Adapted from [8]).

Table 8.9. Important characteristics of SEGS I to SEGS IX plants.

	Units	I	II	III	IV	V	VI	VII	VIII	IX
Power Block										
Turbine-generator output	gross MW$_e$	14.7	33	33	33	33	33	33	88	88
Output to utility	net MW$_e$	13.8	30	30	30	30	30	30	80	80
Turbine-generator set:										
Solar steam conditions:										
Inlet pressure	bar	35.3	27.2	43.5	43.5	43.5	100	100	100	100
Reheat pressure	bar	0	0	0	0	0	17.2	17.2	17.2	17.2
Inlet temperature	°C	415	360	327	327	327	371	371	371	371
Reheat temperature	°C	na	na	na	na	na	371	371	371	371
Gas mode steam conditions[a]										
Inlet pressure	bar	0	105	105	105	105	100	100	100	100
Reheat pressure	bar	0	0	0	0	0	17.2	17.2	17.2	17.2
Inlet temperature	°C	0	510	510	510	510	510	510	371	371
Reheat temperature	°C	na	na	na	na	na	371	371	371	371
Electrical conversion efficiency										
Solar mode[b]	percent	31.5	29.4	30.6	30.6	30.6	37.5	37.5	37.6	37.6
Gas mode[c]	percent	0	37.3	37.3	37.3	37.3	39.5	39.5	37.6	37.6

Table 8.9. (*Continued*)

	Units	I	II	III	IV	V	VI	VII	VIII	IX
Solar Field										
Solar collector assemblies										
LS 1 (128 m²)		560	536	0	0	0	0	0	0	0
LS 2 (235 m²)		48	518	980	980	992	800	400	0	0
LS 3 (545 m²)		0	0	0	0	32	0	184	852	888
Number of mirror segments		41,600	96,464	117,600	117,600	126,208	96,000	89,216	190,848	198,912
Field aperture area	m²	82,960	190,338	230,300	230,300	250,560	188,000	194,280	464,340	483,960
Field inlet temperature	°C	240	231	248	248	248	293	293	293	293
Field outlet temperature	°C	307	321	349	349	349	390	390	390	390
Annual thermal efficiency	percent	35	43	43	43	43	42	43	53	50
Peak optical efficiency	percent	71	71	73	73	73	76	76	80	80
System thermal losses	percent of peak	17	12	14	14	14	15	15	15	15
Heat transfer fluid										
Type		Esso 500	VP-1	VP-1	VP-1	VP-1	VP-1	VP-1	VP-1	VP-1
Inventory	m³	3,213	416	403	403	461	416	416	1,289	1,289
Thermal storage capacity	MWh_t	110	0	0	0	0	0	0	0	0
General										
Annual power outlet	net MWh/year	30,100	80,500	91,311	91,311	99,182	90,850	92,646	252,842	256,125
Annual gas power use	10^9 m³/year	4.76	9.46	9.63	9.63	10.53	8.1	8.1	24.8	25.2

[a] Gas superheating contributes 18 percent of turbine inlet energy.
[b] Generator gross electrical output divided by solar field thermal input.
[c] Generator gross electrical output divided by thermal input from gas-fired boiler or HTF-heater.

Table 8.10. Central-receiver pilot plants constructed in several countries [8].

Name	Location	Size MW_e	Receiver fluid	Start-up	Sponsors
Eurelios	Adrano, Sicily	1	Water/steam	1981	European Community
SSPS/CRS	Tabernas, Spain	0.5	Sodium	1981	Austria, Belgium, Italy, Greece, Spain, Sweden, Germany, Switzerland, United States
Sunshine	Nio, Japan	1	Water/steam	1981	Japan
Solar One	Daggett, California	10	Water/steam	1982	U.S. DOE, SCE[a], CEC[a], LADWP[a]
Themis	Targasonne, France	2.5	Hitec salt	1982	France
CESA-1	Tabernas, Spain	1	Water/steam	1983	Spain
MSEE	Albuquerque, New Mexico	0.75	Nitrate salt	1984	U.S. DOE, EPRI[a], U.S. Industry and Utilities
C3C-5	Crimea, former Soviet Union	5	Water/steam	1985	Former Soviet Union

[a]SCE = Southern California Edison Company, CEC = California Energy Commission, LADWP = Los Angeles Department of Water and Power, EPRI = Electric Power Research Institute.

cept was first proposed by Alvin Hildebrandt and Lorin Vant Hull in the early 1970s [8]. Table 8.10 summarizes the central receiver plants built in these countries. Solar One is a 10 MW_e power plant built in Southern California as a joint venture of the U.S. Department of Energy, Southern California Edison Company, and the State of California. Figure 8.19 shows a schematic and a photograph of the Solar One plant. The solar field generated superheated steam at 510°C and 10.3 MPa which ran the turbine. The system had a crushed rock and sand storage system which was charged by steam through a heat transfer oil. Solar One was operated successfully for six years and then shut down for a redesign. The redesigned plant is called Solar Two. A schematic of Solar Two is shown in Fig. 8.20. Solar Two uses a nitrate salt as the working fluid in the solar loop and as the storage medium.

8.7.3 Parabolic Dish Systems

A parabolic dish based solar thermal power plant was built in Shenandoah, Georgia. This plant was designed to operate at a maximum temperature of 382°C and to provide electricity (450 kW_e), air-conditioning, and process steam (at 173°C) for an industrial complex. The system consisted of 114 parabolic dish concentrators. Figure 8.21 shows a schematic of the system. The system was decommissioned in 1990.

(a)

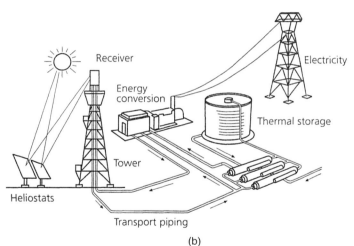

(b)

Figure 8.19. Solar One central receiver pilot plant in Barstow, California: (a) view of tower and heliostats, and (b) schematic of plant (Adapted from [8]).

8.8 STIRLING CYCLE

Stirling cycle is becoming very important in recent years because of its potential to operate at very high efficiency. In fact, theoretically, its efficiency is the same as Carnot efficiency. It was proposed by Rev. Robert Stirling, a Scottish minister, as a solar alternative to a steam engine (Fig. 8.22). In the 1930's Phillips Research Laboratory in Eindhoven, The Netherlands, developed the technology to commercialize it. More recently, interest in Stirling engines has increased with increased interest in solar and biomass energy sources [47]. The first solar application of a Stirling engine was con-

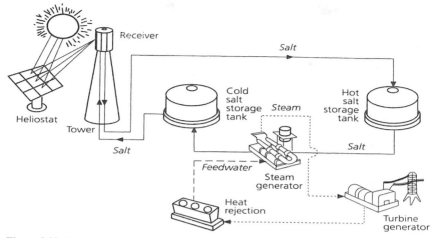

Figure 8.20. Schematic of Solar Two central-receiver plant configuration (Adapted from [8]).

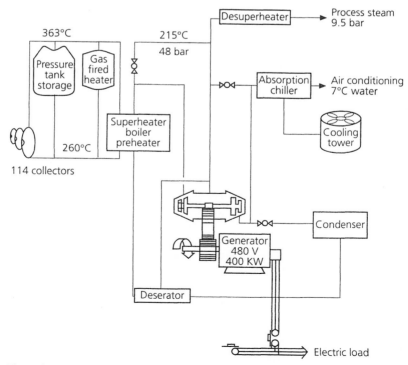

Figure 8.21. Schematic of the Shenandoah Solar Total Energy Project [8].

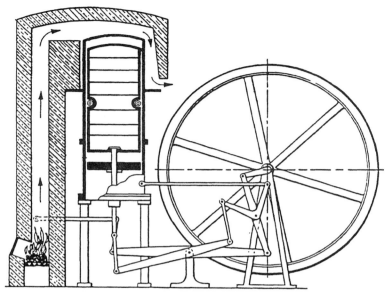

Figure 8.22. The original patented Stirling engine of Rev. Robert Stirling.

ducted by John Ericsson, the Swedish engineer who spent a long and productive career in the U.S. A Stirling cycle engine can use air as the working fluid, therefore it is also called a Hot Air engine. Since it is an external combustion engine it can use any fuel or concentrated sunlight. High performance Stirling engines operate at very high temperatures, typically 600°C to 800°C, resulting in conversion efficiencies of 30 to 40 percent [49]. Stirling engines are being developed in small power capacities, typically 10 to 100 kW.

8.8.1 Thermodynamics of Stirling Cycle

Figure 8.23 shows thermodynamic diagrams of an ideal Stirling cycle. A gas is used as the working fluid. The cycle consists of two isothermal and two constant volume processes. The working gas is compressed isothermally (constant temperature) from state 1 to 2. This process is accomplished by heat rejection at the low temperature of the cycle, T_L. It is then heated at constant volume to state 3. The gas is then expanded isothermally from 3 to 4. This process is accompanied by heat addition at the high temperature in the cycle, T_H. Finally, the gas is cooled at constant volume from temperature T_H to T_L (process 4–1). The hatched area under the process 2–3 represents the heat addition to the working gas while raising its temperature from T_L to T_H. And the hatched area under the process 4–1 represents heat rejection while the gas goes from T_H to T_L. The heat addition from 2 to 3 is equal to the heat rejection from 4 to 1, and they are between the same temperature limits

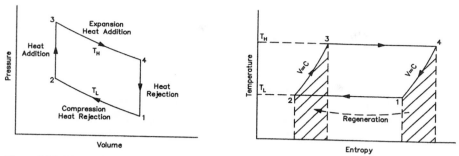

Figure 8.23. Thermodynamic diagrams of an ideal Stirling engine.

$$Q_{2-3} = mC_v(T_H - T_L), \tag{8.30}$$

and

$$Q_{4-1} = -mC_v(T_H - T_L). \tag{8.31}$$

Ideally, if the heat rejected in process 4–1 could be stored and transferred to the gas later in the process 2–3 (perfect regeneration), the only external heat addition in the cycle would be in the process 3–4

$$Q_{3-4} = -W_{3-4} = \int_3^4 pdV = mRT_H \ln\frac{V_4}{V_3}. \tag{8.32}$$

The work input for compression from 1–2 is

$$W_{1-2} = -\int_1^2 pdV = mRT_L \ln\frac{V_1}{V_2}$$

$$= -mRT_L \ln\frac{V_2}{V_1}, \tag{8.33}$$

since

$$\frac{V_2}{V_1} = \frac{V_3}{V_4}.$$

The net work output is

$$mR \ln\frac{V_3}{V_4}(T_H - T_L). \tag{8.34}$$

Therefore, the cycle efficiency

$$\eta = \frac{\text{Net Work Out}}{\text{Heat Input}} = \frac{T_H - T_L}{T_H}. \tag{8.35}$$

This efficiency, which is equal to the Carnot cycle efficiency, is based on the assumption that regeneration is perfect, which is not possible in practice. Therefore, the cycle efficiency would be lower than that indicated by the above equation. For a regeneration effectiveness e as defined below, the efficiency is given by [48]

$$\eta = \frac{T_H - T_L}{T_H + [(1 - e)/(k - 1)][(T_H - T_L)/\ln(V_1/V_2)]} \tag{8.36}$$

where $e = \dfrac{T_R - T_L}{T_H - T_L}$ and T_R = Regenerator temperature, and $k = C_p/C_v$ for the gas.

For perfect regeneration ($e = 1$) the above expression reduces to the Carnot efficiency. It is also seen from the above equation, that regeneration is not necessary for the cycle to work, because even for $e = 0$ the cycle efficiency is not zero.

Example 8.7. A Stirling engine with air as the working fluid operates at a source temperature of 400°C and a sink temperature of 80°C. The compression ratio is 5.
Assuming perfect regeneration, determine the following:

1. Expansion work.
2. Heat input.
3. Compression work.
4. Efficiency of the machine.

If the regenerator temperature is 230°C, determine:

5. The regenerator effectiveness.
6. Efficiency of the machine.
7. If the regeneration effectiveness is zero, what is the efficiency of the machine?

Solution.

(1) Expansion work per unit mass of the working fluid. Assuming air as an ideal gas,

$$w_{34} = -\int_3^4 P dv = RT_H \ln \frac{v_4}{v_3} = (0.287)(400 + 273) \ln 5 = -310.9 \text{ kJ/kg}.$$

Minus sign shows work output.

(2) Heat input per unit mass of the working fluid.

$$q_{34} = w_{34} = 310.9 \text{ kJ/kg.}$$

(3) Compression work per unit mass of the working fluid.

$$w_{12} = -\int_1^2 P dv = RT_L \ln \frac{v_2}{v_1} = -(0.287)(80 + 273) \ln\left(\frac{1}{5}\right) = 163.1 \text{ kJ/kg.}$$

(4) Efficiency of the machine.

$$\eta = \frac{T_H - T_L}{T_H} = \frac{400 - 80}{(400 + 273)} = 0.475 = 47.5\%.$$

(5) The regenerator effectiveness.

$$e = \frac{T_R - T_L}{T_H - T_L} = \frac{230 - 80}{400 - 80} = 0.469.$$

(6) Efficiency of the machine.

$$\eta = \frac{T_H - T_L}{T_H + \dfrac{(1 - e)(T_H - T_L)}{(k - 1)\ln(v_1/v_2)}} = \frac{400 - 800}{(400 + 273) + \dfrac{(1 - 0)}{(1.4 - 1)} \dfrac{(400 - 80)}{\ln(5)}} = 0.341,$$

$$\eta = 34.1\%.$$

(7) If the regeneration effectiveness is zero, the efficiency of the machine.

$$\eta = \frac{T_H - T_L}{T_H + \dfrac{(1 - e)(T_H - T_L)}{(k - 1)\ln(v_1/v_2)}} = \frac{400 - 80}{(400 + 273) + \dfrac{(1 - 0)}{(1.4 - 1)} \dfrac{(400 - 80)}{\ln(5)}} = 0.273,$$

$$\eta = 27.3\%.$$

In order to understand how the Stirling cycle shown in Fig. 8.23 may be achieved, in practice, the simple arrangement and sequence of processes shown in Fig. 8.24 is helpful. In the proposed arrangement two cylinders with pistons are connected via a porous media, which allows gas to pass through from one cylinder to the other. As the gas passes through the porous media it exchanges heat with the media. The porous media, therefore, serves as the regenerator. In practice, this arrangement can be realized in three ways as shown in Fig. 8.25, alpha, beta and gamma types.

The choice of a working fluid for Stirling engine depends mainly on the thermal conductivity of the gas in order to achieve high heat transfer rates. Air has traditionally been used as the working fluid. Since hydrogen has 40 percent higher thermal conduc-

State

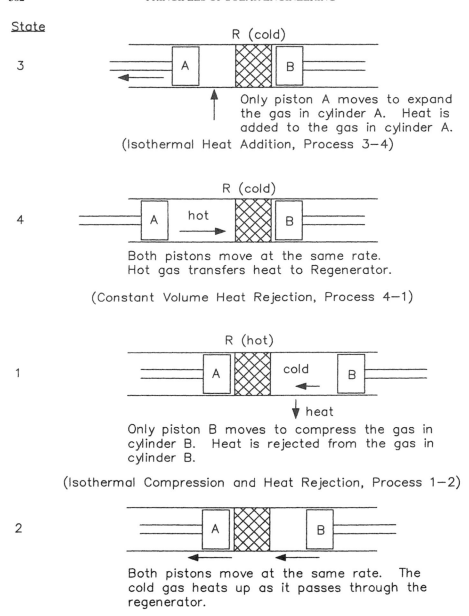

3

R (cold)

A B

Only piston A moves to expand
the gas in cylinder A. Heat is
added to the gas in cylinder A.

(Isothermal Heat Addition, Process 3—4)

4

R (cold)

A   hot   B

Both pistons move at the same rate.
Hot gas transfers heat to Regenerator.

(Constant Volume Heat Rejection, Process 4—1)

1

R (hot)

A   cold   B

heat

Only piston B moves to compress the gas in
cylinder B. Heat is rejected from the gas in
cylinder B.

(Isothermal Compression and Heat Rejection, Process 1—2)

2

A   B

Both pistons move at the same rate. The
cold gas heats up as it passes through the
regenerator.

(Constant Volume Heat Addition, Process 2—3)

Figure 8.24. Stirling cycle states and processes with reference to Fig. 8.23.

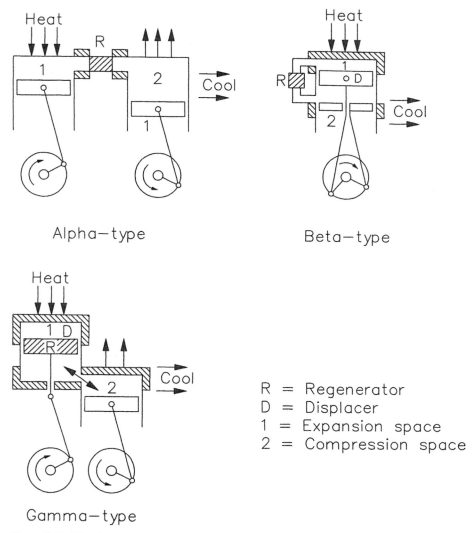

Figure 8.25. Three basic types of Stirling engine arrangements.

tivity at 500°C it is preferable over air. Helium has a higher ratio of specific heats (k) which lessens the impact of imperfect regeneration.

In the alpha configuration there are two cylinders and pistons on either side of a regenerator. Heat is supplied to one cylinder and cooling is provided to the other. The pistons move at the same speed to provide constant volume processes. When all the gas has moved to one cylinder the piston of that cylinder moves with the other remaining fixed to provide expansion or compression. Compression is done in the cold cylin-

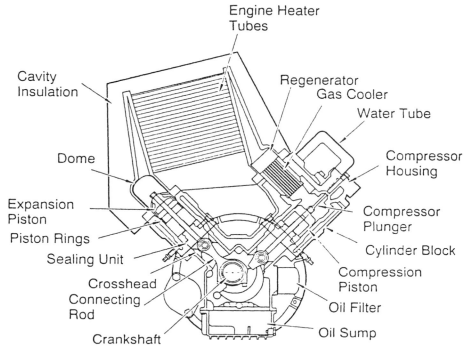

Figure 8.26. Stirling Power Systems / Solo Kleinmotoren V-160 alpha-configuration Stirling engine. Adapted from [49].

der and expansion in the hot cylinder. The Stirling Power Systems V-160 engine (Fig. 8.26) is based on an alpha configuration.

The beta configuration has a power piston and a displacer piston which divides the cylinder into hot and cold sections. The power piston is located on the cold side and compresses the gas when the gas is in the cold side and expands it when it is in the hot side. The original patent of Robert Stirling was based on beta configuration, as are free piston engines.

The gamma configuration also uses a displacer and a power piston. In this case the displacer is also the regenerator which moves gas between the hot and cold ends. In this configuration the power piston is in a separate cylinder.

8.8.2 Piston/Displacer Drives

The power and displacer pistons are designed to move according to a simple harmonic motion to approximate the Stirling cycle. This is done by a crankshaft or a bouncing spring/mass second order mechanical system [49].

8.8.3 Kinematic or Free Piston Engines

Stirling engines are designed as kinematic or free piston depending on how power is removed. In a kinematic engine the power piston is connected to the output shaft by a connecting rod crankshaft arrangement.

Free piston arrangement is an innovative way to realize the Stirling cycle. In this arrangement the power piston is not connected physically to an output shaft. The piston bounces between the working gas space and a spring (usually a gas spring). The displacer is also usually free to bounce. This configuration is called the Beale free piston Stirling engine after its inventor, William Beale [49]. Since a free piston Stirling engine has only two moving parts, it offers the potential of simplicity, low cost and reliability. Moreover, if the power piston is made magnetic it can generate current in the stationary conducting coil around the engine as it moves. This is the principle of the free piston/linear alternator in which the output from the engine is electricity. Figure 8.27 shows a schematic of a free piston Stirling engine with a linear alternator.

8.8.3 Examples of Solar Stirling Power Systems

Stirling engines can provide very high efficiencies with high concentration solar collectors. Since practical considerations limit the Stirling engines to small sizes (5 to 100 kW), a Stirling engine fixed at the focal point of a parabolic dish provides an optimum match. Therefore, all of the commercial developments to date have been in Parabolic Dish-Stirling Engine combination. The differences in the commercial systems have been in the construction of the dish and the type of Stirling engine.

Advanco Corporation developed and tested a 25 kW$_e$ dish Stirling system (called Vanguard I) in Rancho Mirage, CA in 1984 and achieved net conversion efficiency of about 30 percent [20]. The system consisted of an 11 m diameter parabolic dish with a concentration ratio of 2100, and a United Stirling Mark II engine operating at 800°C. Despite the technical success, the effort had to be abandoned by Advanco because of an exclusive agreement between McDonnell Douglas Corporation and United Stirling AB of Sweden.

McDonnell Douglas and United Stirling jointly developed a 25 kW$_e$ dish Stirling system and tested it in California. The concentrator was made from glass mirror facets on a structure to approximate a parabolic dish. Figure 8.28 shows a picture of the system. Tests showed the system efficiency to be in the same range as the Vanguard I system. McDonnell Douglas built six prototypes, but decided to abandon the commercialization efforts.

Recently, Cummins Power Generation and Stirling Technology Company have independently developed 25 kW$_e$ free piston, linear alternator engines. Cummins Power also developed a 5 kW$_e$ dish Stirling power system for remote power applications with a view to mass produce the system (see Fig. 8.29). The company, however, abandoned the effort in 1996.

Schlaich Bergermann and Partner has also developed a 7.7 kW$_e$ dish Stirling power system using a V-160 two-cylinder Stirling engine developed by Stirling Power Systems and a stretched membrane dish.

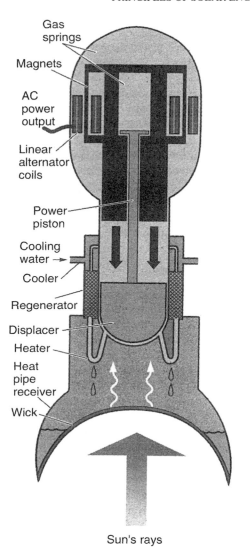

Gas springs

Magnets

AC power output

Linear alternator coils

Power piston

Cooling water →

Cooler

Regenerator

Displacer

Heater

Heat pipe receiver

Wick

Sun's rays

Free Piston Stirling Engine

Figure 8.27. Basic components of a Beale free-piston Stirling converter incorporating a sodium heat pipe receiver for heating with concentrated solar energy.

Figure 8.28. The McDonnell Douglas/United Stirling dish-Stirling 25 kW$_e$ module.

Stirling engine development in Japan includes a 3 kW$_e$ engine developed by Toshiba Corp. for residential heat pump applications and a 2.4 kW$_e$ engine developed for space testing by Aisin Seiki company [25,40].

Even though commercial efforts in solar Stirling engine systems have not been successful so far, this technology holds a great deal of promise for remote applications and developing countries because of potential high efficiency and modularity.

8.8.4 Recent Developments in Solar Thermal Power Cycles

Solar thermal power technologies were successfully demonstrated in the 1980's, as noted in the earlier sections in this chapter. In fact, SEGS thermal power plants are producing over 350 MW$_e$ commercially. At the same time, learning experience has been gained which has brought the price of electricity from the SEGS plants to below 10¢/kWh. However, the capital cost of $3500/kW for a solar power plant is four times higher than the natural gas combustion turbines and the plant efficiency of 12 to 14

Figure 8.29. The Cummins Power Generation dish-Stirling engine power system.

percent has the potential to improve by as much as three times [17]. This potential has drawn the interest of researchers to fundamental thermodynamic improvements in power cycles for solar energy.

8.8.4.1 Hybrid and combined cycles. The SEGS plants and the CPG dish/Stirling power plants were designed to use natural gas as a supplemental fuel. Most recently, it has also been suggested to use natural gas as a supplemental fuel for the Solar Two project [4]. If the supplemental natural gas flows in the power system are large enough, it makes sense to use the natural gas in a combined-cycle mode. With the present-day technology of natural gas combustion turbines, a combined combustion turbine/steam Rankine cycle power plant can achieve a cycle efficiency as high as 58 percent. Figure 8.30 shows a schematic of a proposed combined-cycle solar/natural gas power plant [58]. With the present cost of the commercially available, aeroderivative combustion

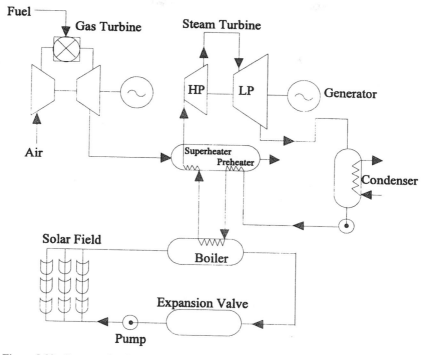

Figure 8.30. Concept of an integrated solar combination-cycle power system [58].

turbines as low as $500/kW, and the cost of a SEGS-type solar power plant being approximately $3500/kW, the capital cost of a combined-cycle solar/natural gas power plant could be as low as $1000–$3000/kW, depending on the natural gas fraction. This idea seems very attractive, especially where natural gas is to be used as a backup fuel anyway. However, this idea does not in any way improve the thermodynamic performance or reduce the cost of the solar part of the plant. The capital cost of the solar part being seven times the cost of the natural gas part of the power plant, the solar part is indeed a burden on the overall cost and, therefore, unattractive to the investors.

8.8.4.2 Kalina cycle. Kalina [26] proposed a novel bottoming cycle for a combined-cycle power plant that improves the overall performance of the cycle. This novel cycle, now known as the Kalina cycle, uses a mixture of ammonia and water as the working fluid for the bottoming cycle expansion turbine. Using a two-component working fluid and multipressure boiling, one can reduce heat transfer related irreversibilities and therefore improve the resource effectiveness (Fig. 8.31). Also, the ammonia-water mixture can be used as a working fluid for a lower temperature (250°C) solar energy system employing lower cost parabolic trough technology. According to Kalina and Tribus [27], it is possible to improve the efficiency of the bottoming cycle by more than 45 percent over the steam Rankine cycle for a lower temperature resource (~250°C). Therefore, if the Kalina cycle is used for solar thermal power in place of the

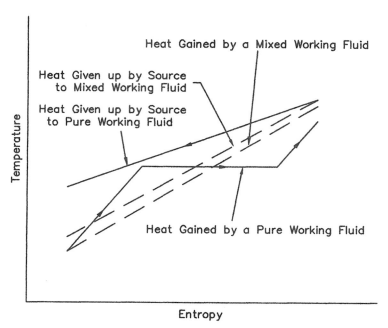

Figure 8.31. Heat transfer between a heat source and a working fluid.

Rankine cycle, the capital cost could be reduced by at least 45 percent because of the higher efficiency and the possibility of using cheaper concentrating collectors due to the lower temperature requirement. These improvements may be enough to make solar thermal power economically competitive by itself and not a burden on natural gas for combined-cycle power plants. This offers an opportunity for research in solar thermal power utilizing the Kalina cycle.

Although use of mixed working fluids such as ammonia/water provide a big advantage in utilizing sensible heat sources in a boiler, they present a disadvantage in the condenser, since condensation must also take place with change in temperature. This disadvantage could be overcome if the normal condensation process is replaced by the absorption condensation process as proposed by Rogdakis and Antononpoulos [43] and Kouremenous et al, [32] as a modification of the Kalina cycle and, more recently, by Goswami [18] as an ammonia-based combined power/cooling cycle.

8.8.4.3 Aqua-ammonia combined power/absorption cycle. Kouremenous et al. [32] showed an analysis of an aqua-ammonia power cycle that is similar to the Kalina cycle; however, it replaces the conventional condensation process of the aqua-ammonia mixture coming out of the low-pressure turbine with an absorption condensation process. They showed that with an ammonia concentration of 72.86 percent in the mixed working fluid in the turbine, cycle efficiencies of the order of 25–45 percent could be achieved for a source temperature of 500°C and pressure ratios from 25 to 100.

8.8.4.4 Combined power/cooling cycle. Goswami [18] has proposed a new thermodynamic cycle that improves the cycle efficiency and, therefore, resource utilization by producing power and refrigeration in the same cycle. The proposed new cycle uses ammonia as the working fluid in an innovative combination of an ammonia-based Rankine cycle and an ammonia-water absorption refrigeration cycle. Figure 8.32 shows a schematic of the cycle.

In a simulation of the cycle [19] shown in Tables 8.11 and 8.12, with the working fluid entering the turbine at 410K and 30 bar pressure and exiting at a 2 bar pressure, a first law efficiency of 23.54 percent is achieved. By contrast, the Carnot cycle efficiency for the same source temperature (410K) and a sink temperature of 280K is 31.7 percent. The cycle efficiency of a conventional Steam Rankine cycle between the same

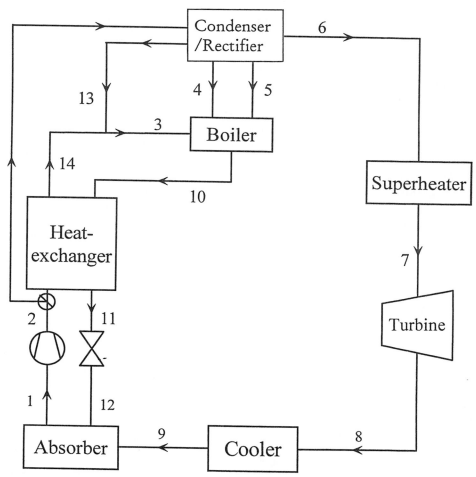

Figure 8.32. A modified ammonia-based combined power/cooling cycle.

Table 8.11. Typical operating conditions [19].

State	T (K)	P (bar)	h (kJ/kg)	s (kJ/kg K)	x	Flow rate m/m$_1$
1	280.0	2.0	−214.1	−0.1060	0.5300	1.0000
2	280.0	30.0	−211.4	−0.1083	0.5300	1.0000
3	378.1	30.0	246.3	1.2907	0.5300	1.0000
4	400.0	30.0	1547.2	4.6102	0.9432	0.2363
5	360.0	30.0	205.8	1.1185	0.6763	0.0366
6	360.0	30.0	1373.2	4.1520	0.9921	0.1997
7	410.0	30.0	1529.7	4.5556	0.9921	0.1997
8	257.0	2.0	1148.9	4.5556	0.9921	0.1997
9	280.0	2.0	1278.7	5.0461	0.9921	0.1997
10	400.0	30.0	348.2	1.5544	0.4147	0.8003
11	300.0	30.0	−119.0	0.2125	0.4147	0.8003
12	300.0	2.0	−104.5	0.2718	0.4147	0.8003

source and sink temperatures will be much lower than both the Carnot cycle and the combined power/cooling cycle.

In addition, a system designed to produce 2 MW of electrical power will produce more than 50 tons of refrigeration. By boiling a mixture (ammonia/water), this cycle reduces the irreversibilities associated with the heat transfer from a sensible heat source. Also, by expanding almost pure ammonia in the last-stage turbine, it allows the working fluid to go down to a much lower temperature, providing refrigeration. The working fluid is condensed by absorption in the water. The net effect is lowering the sink temperature of the cycle. The cycle can use source temperatures as low as 100°C, thereby making it a useful power cycle for low-cost solar thermal collectors, geothermal resources, and waste heat from existing power plants. Since it can utilize low-cost solar collectors, it has the potential to make solar thermal power plants cost competitive with fossil-fuel-fired power plants.

Table 8.12. Results from Table 8.11 state conditions. (All energy units are kJ/kg basic solution.)

Boiler heat input	390.4
Superheat input	31.3
Condenser heat rejection	−83.8
Absorber heat rejection	−385.8
Refrigeration output	25.9
Turbine work output	76.0
Turbine liquid fraction	0.0692
Turbine vapor fraction	0.9308
Pump work, input	2.7
Total heat input	421.6
Net power and refrigeration output	99.23
Thermal efficiency	23.54%
Carnot efficiency (between 410 K and 280 K)	31.7%

8.9 SOLAR DISTILLATION OF SALINE WATER

Solar distillation for the production of potable water from saline water has been practiced for many years. A solar distillation plant, covering 4,740 m² of land, was built in Las Salinas, Chile, in 1872, to provide fresh water from salt water for use at a nitrate mine [54]. Single-glass-covered flat-plate collectors with salt water flowing downward over slanting roofs were used to vaporize some of the water, which was then condensed on the air-cooled underside of the roof. This plant ran effectively for over forty years and produced up to 23,000 liters of fresh water per day until the nitrate mine was exhausted.

The stills used to date are called shallow basin-type stills and Fig. 8.33 illustrates their method of operation. The still is irradiated by direct and diffuse solar radiation I_s as well as some infrared radiation I_i from the surroundings. The long-wavelength radiation is absorbed by the glass, but $\tau_{s,g}I_s$ of the solar radiation reaches the saline water; $\tau_{s,w}\tau_{s,g}I_s$ reaches the bottom of the trough where $\alpha_{s,t}\tau_{s,g}\tau_{s,w}I_s$ is absorbed per unit still area. A value of 0.8 has been suggested as a reasonable approximation for $\alpha_{s,t}\tau_{s,w}\tau_{s,g}$ if good glass, a blackened tray surface, and a 0.2 m thick layer of water are used [9]. Typical shallow basin stills are 1.5 m wide, 2–25 m long, and approximately 0.2 m deep; the glass roof should have a slope of 10–15°.

Of the energy absorbed at the bottom, one part $q_{k,s}$ is lost through the insulation by conduction, while the other part will be transferred to the saline water in the still tray above. Of the latter portion, some will heat the water, if the water temperature is less than the tray temperature, while the rest is transferred from the surface of the water by free convection, radiation and evaporation to the underside of the glass cover. Some of this heat passes through the glass by conduction and is transferred from the outer surface of the glass cover by convection and radiation to the surrounding atmosphere. The thermal circuit for this system is shown in Fig. 8.34.

To obtain the greatest yield, the rate of evaporation of water, which is proportional to q_e, should be as large as possible. An inspection of the thermal circuit shows that to achieve large values of q_e, $\tau_{s,g}\tau_{s,w}\alpha_{s,t}I_s$ should be as large as possible, while $q_{c,s}$, $q_{r,s}$, and

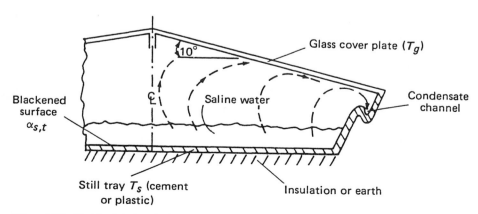

Figure 8.33. Sketch of shallow basic-type solar still.

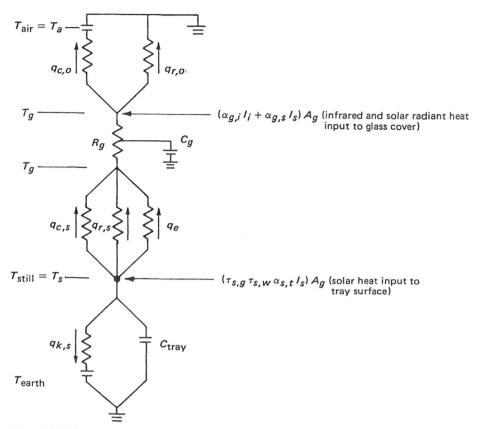

Figure 8.34. Thermal circuit for solar still.

$q_{k,s}$ should be as small as possible. To achieve these results the transmittance of the glass cover and the absorptance of the tray surface must be large, and the bottom of the tray should be well insulated. Obviously, the amount of solar irradiation should be as large as possible.

In order to analyze the thermal performance, heat balances must be written for the glass cover and the tray, and the resulting equations must be solved simultaneously to obtain the tray and glass cover temperatures. The objective is to obtain the thermal efficiency, defined as the ratio of the energy used in evaporating saline water to the incident solar radiation.

If steady-state conditions prevail, the glass resistance R_g is neglected, and the water temperature is assumed equal to the tray temperature, a heat balance on the glass cover gives

$$q_e + q_{r,s} + q_{c,s} + \alpha_{g,i}IA_g + \alpha_{g,s}I_sA_g = q_{c,o} + q_{r,o}, \qquad (8.37)$$

where $q_{r,s}$ = rate of heat transfer by radiation from the water in the tray to the inner surface of the glass = $A\bar{h}_{r,s}(T_s - T_g)$,

$$\bar{h}_{r,s} = \sigma(T_s^2 + T_g^2)(T_s + T_g)/(1/\epsilon_w + 1/\epsilon_g - 1),$$

$q_{c,s}$ = rate of heat transfer by convection from water to inner surface of the glass
$$= A\bar{h}_{c,s}(T_s - T_g),$$

$q_{c,o}$ = rate of convection heat loss from the glass to the ambient air
$$= A\bar{h}_{c,o}(T_g - T_a),$$

$q_{r,o}$ = rate of radiation heat loss from the glass = $A\bar{h}_{r,o}(T_g - T_a)$, and
$$\bar{h}_{r,o} \cong \epsilon_g\sigma(T_g^2 + T_a^2)(T_g + T_a).$$

Similarly, a heat balance on the saline water gives

$$\alpha_{s,l}\tau_{s,g}\tau_{s,w}I_s = q_k + q_{c,s} + q_{r,s} + q_e. \tag{8.38}$$

In the enclosed space between the water surface and the glass cover, heat and mass transfer occur simultaneously as brine is evaporated, condenses on the lower surface of the cooler glass, and finally flows into the condensate channel. Since the glass cover is nearly horizontal, the correlations developed in Chapter 3 for free convection between parallel plates can be used to calculate the Nusselt number and heat-transfer coefficient. However, because heat and mass transfer occur simultaneously in the still, the buoyancy term in the Grashof number must be modified to take account of the density gradient resulting from composition as well as temperature. As shown in [36], in horizontal enclosed air spaces in the range $3 \times 10^5 < Gr_L < 10^7$, the relationship

$$Nu' = \frac{h'_c L}{k} = 0.075\left(\frac{L^3\rho^2 g\beta_T\Delta T'}{\mu^2}\right)^{1/3}, \tag{8.39}$$

can be used to calculate the heat-transfer coefficient if $\Delta T'$ is considered an equivalent temperature difference between the water and the glass cover and includes the molecular weight difference resulting from the change in vapor concentration in evaluating the buoyancy. For an air-water system $\Delta T'$ is given by [9] in English units.

$$\Delta T' = (T_s - T_g) + \left(\frac{p_{w,s} - p_{w,g}}{39 - p_{w,s}}\right)T_s, \tag{8.40}$$

where

$p_{w,s}$ = partial pressure of water (in psia) at the temperature of the water surface T_s, and
$p_{w,g}$ = vapor pressure of water in psia at the temperature of the glass cover T_g.

Since the size of the still does not affect the heat-transfer coefficient in the Grashof number range of interest, $\bar{h}_c$ in Eq. (8.19) can be approximated by evaluating all physical constants at an average still temperature or

$$h_c' = 0.13\left[(T_s - T_g) + \left(\frac{p_{w,s} - p_{w,g}}{39 - p_{w,s}}\right)T_s\right]^{1/3}. \tag{8.41}$$

The rate of heat transfer by convection between the water and the glass is then

$$q_c = \bar{h}_c' A_s (T_s - T_g), \tag{8.42}$$

and, using the analogy between heat and mass transfer, the rate of mass transfer q_m (in lb/hr) is

$$q_m = 0.2\bar{h}_c' A_s (p_{w,s} - p_{w,g}). \tag{8.43}$$

The heat-transfer rate resulting from evaporation q_e (in Btu/hr) equals the rate of mass transfer times $h_{f,g}$, the latent heat of evaporation at T_g, or

$$q_e = 0.2\bar{h}_c' A_s (p_{w,s} - p_{w,g}) h_{f,g}. \tag{8.44}$$

If the tray of the still is resting on the ground, the conduction loss is difficult to estimate because it depends on the conductivity of the earth, which varies considerably with moisture content. If the bottom of the tray is insulated and raised above the ground, the preferable arrangement, unless too expensive, the bottom heat loss is given by

$$q_k = U_b A_s (T_s - T_a). \tag{8.45}$$

where the bottom conductance U_b is composed of the tray insulation and the free convection elements in series and can be calculated by the methods given in Chapter 3.

Since $p_{w,s}$ and $p_{w,g}$ in Eqs. (8.23) and (8.24) are functions of T_s and T_g, respectively, Eqs. (8.17) and (8.18) can be solved simultaneously for T_s and T_g, provided I_s and all the physical properties are known. However, this approach is more complex than assuming a still temperature and solving for T_g and I_s.

Example 8.5. For a solar still operating at a brine temperature of 155°F (410 K) in ambient air at 76°F (298 K) with a bottom conductance $U_b = 0.3$ Btu/hr ft^2 °F (1.7 W/m^2 · °C), and $(1/\epsilon_g + 1/\epsilon_w - 1) = 0.9$, calculate the rates of heat transfer by convection q_c, radiation q_r, and conduction q_k, and the rate of evaporation per unit still area q_e. Then calculate the still efficiency.

Solution. First the temperature of the glass cover is calculated from Eq. (8.37) [the partial pressure of water vapor was obtained from [28]. This temperature is found to be 135°F (390.5 K) and the various heat fluxes are shown below:

$q_{c,s}/A_s$, convection flux from brine = 10 Btu/hr ft²(31.5 W/m²),

q_e/A_s, rate of evaporation per unit area = 158 Btu/hr ft² (497 W/m²),

$q_{r,s}/A_s$, radiation flux from brine = 28 Btu/hr ft² (88 W/m²),

$q_{k,s}/A_s$, conduction back loss flux = 23 Btu/hr ft² (72 W/m²),

$\alpha_{s,t}\tau_{s,g}\tau_{s,w}I_s$ = 219 Btu/hr ft² (689 W/m²), and

I_s = 219/0.8 = 274 Btu/hr ft² (274 W/m²).

The thermal efficiency is then

$$\eta = \frac{q_e}{I_s} = \frac{158}{274} = 58 \text{ percent.}$$

A potentially important factor omitted in the preceding analysis is the effect of air leakage. It increases with increasing temperature and can also be accelerated by high wind. Leakage should be minimized for good performance.

The maximum performance of a still is limited by the heat of vaporization of water and the solar insolation. Under favorable conditions with I_s = 250 Btu/hr ft² (789 W/m²), η_s = 40 percent, and $h_{f,g}$ = 1000 Btu/lb (2326 kJ/kg), one square foot of still can provide 0.1 lb of water/hour, or about 1 lb/day. Thus, approximately 10 ft² (1 m²) is required for a production rate of 1 gal/day ($\sim$ 3.8 ℓ/day) and 27 ft² ($\sim$ 2.5 m²) of still would be required to produce 1000 gal/yr ($\sim$ 3800 ℓ/yr).

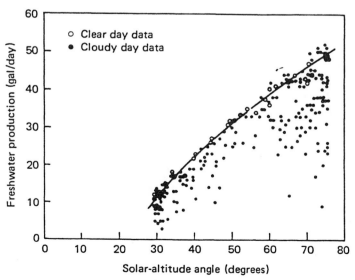

Figure 8.35. Production rates for 1958 University of California solar still No. 16.

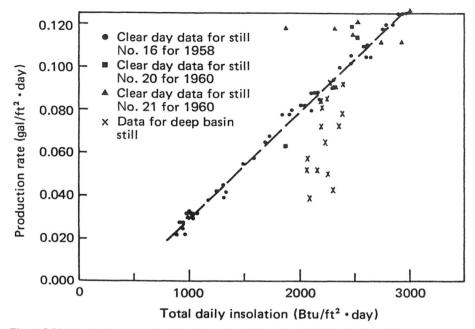

Figure 8.36. Production versus insolation for various University of California solar stills.

8.10 NONCONVECTING SOLAR PONDS

The nonconvecting solar pond is a horizontal-surfaced solar collector using the absorption of solar radiation at the bottom of a 1- or 2-m-deep body of water to generate low-temperature heat. Since heat storage is an integral part of ponds, they have promise in some parts of the world for continuous energy delivery to process or space-conditioning systems. Modern solar ponds were first studied scientifically in Israel by Bloch, Tabor et al. [52,53]. This section describes ponds in which a temperature substantially above ambient—by 50 K or more—may be achieved.

8.10.1 Introduction

When solar energy enters a pond, the infrared component is absorbed within a few centimeters, near the surface, since water is opaque to long-wave radiation. The visible and ultraviolet components of sunlight can penetrate clear water to a depth of several meters. These radiation components can be absorbed at the bottom of the pond by a dark-colored surface. The lowest layer of water is then the hottest and would tend to rise because of its relatively lower density if measures were not taken to prevent it.

In nonconvecting solar ponds, the water at the bottom is made heavier than that at the top by dissolving salt in the water as shown in Fig. 8.37. The concentration of salt is decreased from bottom to top so that the natural tendency of ponds to mix by the

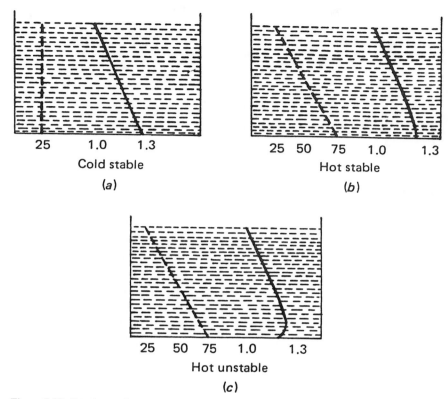

Figure 8.37. Density gradients in a solar pond. If the gradient is positive when cold (a), it becomes less positive when heated (b); if the lowest layers become too hot, the density profile may reverse (c) and the onset of convection will follow. Dashed lines represent temperature profiles and the solid lines represent density profiles (units are °C and g/cm³).

creation of convection currents is effectively eliminated if the density gradient is adequate.

Since stationary water is quite an effective insulator, it is possible for the lowest layers of a well-designed solar pond to boil. Boiling, of course, must be avoided, for it destroys the stable density gradient. Therefore, the design of a solar pond for heat generation must involve a mechanism for useful heat removal in sufficient quantity to avoid boiling (Fig. 8.38).

Since solar ponds are usually envisioned to be on the order of hectares in size a heat-exchanger pipe network using a separate working fluid is impractical. However, hydrodynamic principles predict that a layer of fluid could be removed slowly from the bottom of the pond without disturbing the main body of water. This is evidenced on a large scale by the ability of ocean currents—Gulf Stream, Benguela Current, etc.—to retain their identity over thousands of miles. The fluid in the removed layer of a pond is then passed through a heat exchanger for useful heat removal and returned to the bottom of the pond. Since the returned fluid is cooler than the extracted fluid, the re-

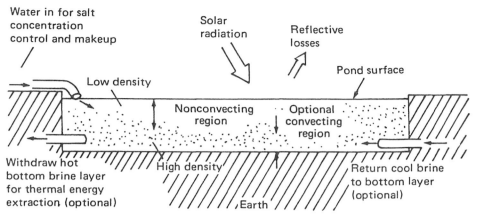

Figure 8.38. Schematic diagram of a nonconvecting solar pond showing conduits for heat withdrawal; surface washing; and an optional convecting zone near the bottom.

quired density gradient is maintained. In practice, the horizontal flows can disturb the hydrostatic equilibrium somewhat, depending on the Reynolds and Froude numbers, and a mixed layer can exist at the pond's bottom. In this layer, convection can occur, although it can be confined to a zone on the order of centimeters. Nielsen and Rabl [38] have described a pond in which a convective zone is an intentional feature of the design proposed for space heating.

Several practical difficulties arise during long-term use of solar ponds in the field. Since a salinity gradient exists, diffusion of salt from regions of high concentration to low concentration occurs naturally. Hence, the density gradient required for hydrostatic stability tends to destroy itself. This diffusion can amount to 60 tons/km$^2 \cdot$ day [53]. To maintain the gradient, it is necessary to supply salt to the bottom of the pond and to wash the surface with a weak brine solution. The supply of salt to a pond is a major maintenance cost for this type of solar collector. Tabor [53] has suggested an idea whereby the bottom layer of a pond is decanted, partially evaporated, and returned to the pond bottom. Simultaneously, water is added to the top of the pond to replace that evaporated from the layer. As a result, the net flow of water is downward, and the *relative* flow (diffusion) of salt is upward. By proper matching of these rates, the salt can remain stationary in a fixed frame of reference.

The replacement of water that evaporates naturally from the surface of a pond is a second practical consideration. If water is not replaced at the top, a local reversal of the stable salinity gradient will occur and a convective layer will form at the surface. This is to be avoided, since it destroys part of the insulation effect of the stagnant pond. Nielsen and Rabl [38] have proposed the use of a plastic cover over the pond to help maintain the proper gradient. Since reflection of solar radiation from plastic is greater than that for water (see Chapter 3), they also suggest maintaining a thin water layer over the plastic cover.

Other practical matters that must be considered are the effects of wind, waves, rain and leaks on the pond gradient (halocline). Waves can destroy the halocline and therefore must be controlled. An economical and effective method for this task has yet to be developed. Very severe rainstorms and wind can destroy the surface gradient to a depth of 35–50 cm [39]. A plastic cover could presumably avoid the problem if it could stay intact during severe storms. The problem of dirt in a pond depends on the location of the particles. Either on the surface or suspended within the pond, they reduce solar transmittance and must be removed by withdrawing, filtering and replacing a layer of the pond. Dirt on the bottom can likewise be removed but may not cause a major problem if left in place. Leaves in a pond should be avoided, since they may float, may lose their color to the water, and may plug fluid circulation machinery and conduits. Algae growth in ponds should be readily controllable by chemical means.

A final feature of a solar pond determines it geographic limits. Since the ponds are horizontal and the sun is low in the sky north or south of the mid-latitudes in winter, ponds must be used near the equator if winter yields are not to be curtailed sharply. Other climatic effects in the tropic—monsoons or storms—must also be considered in finding the best sites for large-scale pond usage.

8.10.2 Solar Pond Stability Criteria

One of the principal costs of a solar pond is the dissolved salt required to establish the stable density gradient. For economical application it is essential that no more salt be used than necessary to assure hydrostatic stability. Figure 8.39 shows the solubility of several common salts in water as a function of temperature. Salts like NH_4 NO_3 or

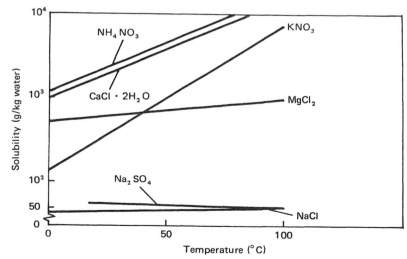

Figure 8.39. Solubility of some common inorganic salts usable in solar ponds.
Source: [34].

KNO_3, whose solutions are as transparent as water, are best for solar ponds, since solubility increases significantly with temperature thereby ensuring a significant density gradient. Salts such as Na_2SO_4 with the opposite solubility characteristic would not be suitable for ponds.

Weinberger [59] has established the criteria for the hydrostatic and hydrodynamic stability of large solar ponds. To avoid convection, the fluid density must decrease upward from the bottom, that is,

$$\frac{d\rho}{dz} = \frac{\partial\rho}{\partial s}\frac{ds}{dz} + \frac{\partial\rho}{\partial T}\frac{dT}{dz} \geq 0, \tag{8.46}$$

where s is the salt concentration, ρ is the density, T is the temperature, and z is the vertical coordinate, increasing downward.

However, to avoid the growth of oscillatory motion, a stronger condition is required [38]:

$$\left(\frac{v + K_s}{\rho}\right)\frac{\partial\rho}{\partial s}\frac{ds}{dz} + \left(\frac{v + K_T}{\rho}\right)\frac{\partial\rho}{\partial T}\frac{dT}{dz} \geq 0, \tag{8.47}$$

where v, K_s and K_T are momentum, salt and thermal diffusivities, respectively. Based on this condition, a convective layer of thickness h will grow if [38]

$$q_{net} > 0.89h(\rho c)\sqrt{\frac{K_T}{t}}\left(-\frac{dT}{dz} - \frac{(v + K_s)}{(v + K_T)}\frac{(\partial\rho/\partial s)}{(\partial\rho/\partial T)}\frac{ds}{dz}\right), \tag{8.48}$$

where t is the time during which heat is absorbed at rate q_{net} in a region of heat capacity (ρc) per unit volume.

If the net heat rate is above that required for convective layer growth, the layer grows in thickness as [34]

$$h^2 = Cq_{net} + h^2(t = 0), \tag{8.49}$$

where

$$C = \frac{2}{(\rho c)[(ds/dz)(\partial\rho/\partial s)/(\partial\rho/\partial T) + dT/dz}. \tag{8.50}$$

If the heat rate q_{net} is less than the critical value from Eq. (8.48), the convective layer may not decrease, as is the case in many other hydrodynamic phenomena. This is a result of the much greater numerical value of thermal diffusivity compared with salt diffusivity. Nielsen and Rabl [36] have shown that the salinity gradient to stabilize a convective zone is about five times that required to maintain a stagnant zone. A density difference of 25 percent has successfully stabilized ponds in practice.

Table 8.13. The Spectral absorption of sunlight in water[a]

Wavelength (μm)	0	Layer depth			
		1 cm	10 cm	1 m	10 m
0.2–0.6	23.7	23.7	23.6	22.9	17.2
0.6–0.9	36.0	35.3	36.0	12.9	0.9
0.9–1.2	17.9	12.3	0.8	0.0	0.0
1.2 and over	22.4	1.7	0.0	0.0	0.0
Total	100.0	73.0	54.9	35.8	18.1

[a]Numbers in the table give the percentage of sunlight in the wavelength band passing through water of the indicated thickness.

8.10.3 Thermal Performance of Solar Ponds

In a solar pond, solar radiation is partially reflected at the surface, partially absorbed in the water, and partially absorbed at the bottom. The absorption of solar energy by water does not follow a simple Bouger's law since the absorption phenomena differ widely with wavelength. Table 8.13 is a summary of absorption of sunlight in water as a function of wavelength. Absorption in solutions of inorganic salts used in solar ponds is expected to be nearly the same.

The data from Table 8.13 are plotted in Fig. 8.40. An adequate curve fit of these data is represented by the sum of several exponential terms. If $\tau(x)$ is the transmittance of water of depth x, it can be related to x by [38]

$$\tau(x) = \sum_{i=1}^{4} a_i e^{-b_i x} \tag{8.51}$$

where the regression coefficients a_i and b_i are (see Table 8.14).

Equation (8.51) does not include the infrared spectrum ($\lambda > 1.2$ μm), since it is of no interest in solar pond analysis. As shown in Refs. [22,42,58], a detailed analysis of heat transfer in a solar pond is very complex and must include effects of volumetric absorption, variation of density, and conductivity with salinity.

Figure 8.40 shows that about 30 percent of the incident radiation reaches the absorbing bottom surface of a 2-m solar pond. This represents the upper limit of collection efficiency of a pond. But since the bottom is an imperfect absorber and heat losses occur from the top and bottom surfaces, the thermal efficiency is an inadequate index of solar system viability. Economics must be considered. Since solar ponds are presently not in commercial use, their economic analyses are not reliable. However, Rabl and Nielsen [42] have predicted that solar ponds for building heating may be

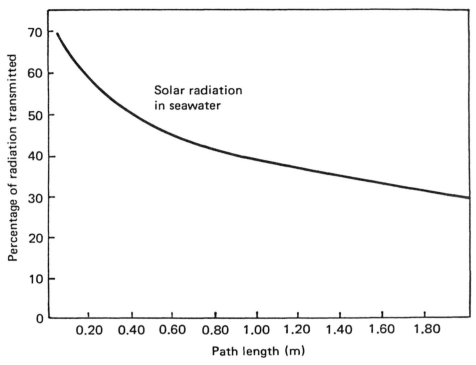

Figure 8.40. Transmittance of water to solar radiation as a function of the thickness of the water layer. See Table 8.13 for numerical values.

more economical than fuel oil heating in some areas. Styris et al. [51] have analyzed solar ponds for process heat as well.

An additional consequence of the data shown in Fig. 8.40 should be noted. The insulating effect of the nonconvecting layer in a solar pond increases linearly with depth. In most practical cases [22,42,58] this insulating effect increases more rapidly than does the attenuation of sunlight because of increased depth. For example, a 1-m-deep pond that can achieve a no-load temperature of 100°C, could achieve a temperature of about 180°C if it were 2-m-deep. Of course, brine costs increase with pond depth and pond depth should therefore be selected to match the task for which the heat is required.

Table 8.14.

Wavelength (μm)	a_i	b_i ($\times 10^{-3}$ cm^{-1})
0.2–0.6	0.237	0.32
0.6–0.75	0.193	4.5
0.75–0.90	0.167	30.0
0.90–1.20	0.179	350.0

PROBLEMS

8.1. A reheat Ranking cycle uses steam as a working fluid. The steam leaves the condenser as saturated liquid. The steam leaves the boiler and enters the turbine at 5 MPa, 350°C. After expansion in the turbine to $P = 1.2$ MPa, the steam is reheated to 350°C and then expands in the low-pressure turbine to $P = 10$ kPa. If the efficiencies of the pump and turbine are 0.9, determine:

a. Work output from high-pressure turbine per unit mass of working fluid;

b. Work output from low-pressure turbine per unit mass of working fluid;

c. Work input to the pump per unit mass of working fluid;

d. Heat added to the boiler;

e. Quality of vapor at the exit of high pressure turbine and low-pressure turbine; and

f. Cycle efficiency.

8.2. A regenerative Rankine cycle uses steam as the working fluid. Steam leaves the boiler and enters the turbine at $P = 5$ MPa, $T = 500$°C. After expansion to 400 kPa, a part of steam is extracted from the turbine for the purpose of heating the feed water in the feed water heater. The pressure in the feed water heater is 400 kPa and the water leaving it is saturated liquid at 400 kPa. The steam not extracted expands to 10 kPa. Assuming the efficiencies of the turbine and two pumps are 90 percent, determine:

a. The fraction of steam extracted to the feed water heater (f).

b. Heat added to the boiler per unit mass of working fluid.

c. Work obtained from the turbine.

d. Work for both pumps per unit mass of working fluid.

e. The cycle efficiency (η).

8.3. Consider a Stirling cycle from Example 8.7 with imperfect regenerator ($T_R = 230$°C). Assume ideal gas.

a. Compute the efficiency from

$$\eta = \frac{\text{work}_{\text{net}}}{\text{Heat}_{\text{input}}},$$

where

$$\text{Heat}_{\text{input}} = q_{34} + mC_v(T_H - T_R).$$

Verify that the efficiency you get is the same as the efficiency obtained by Eq. (8.16).

b. Start from the efficiency equation above. Derive the efficiency in the Eq. (8.16).

8.4. A solar electric engine operating between 5 and 95°C has an efficiency equal to one-half the Carnot efficiency. This engine is to drive a 4 kW pump. If the collector has an efficiency of 50 percent, calculate the area needed for operation in Egypt where the insolation averages about 2800 kJ/m² hr during the day. State any additional assumptions.

8.5. The graph below shows the Rankine cycle efficiency as a function of maximum cycle temperature for different working fluids. Prepare a thermal analysis matching a line focusing collector, a paraboloid dish collector, and a *CPC* collector with a concentration ratio of 3:1 to a suitable working fluid in Washington, D.C., and Albuquerque, NM. Comment on storage needs in both locations. Low-temperature organic and steam efficiencies were computed with the following assumptions: expander efficiency, 80 percent; pump efficiency, 50 percent; mechanical efficiency, 95 percent; condensing temperature, 95°F; regeneration efficiency, 80 percent; high side pressure loss, 5 percent; low side pressure loss, 8 percent.

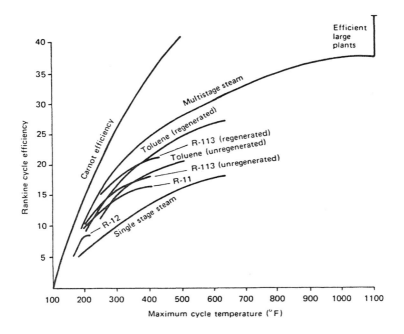

8.6. The schematic diagram shown below for a solar-driven irrigation pump was developed by Battelle Memorial Institute and uses tracking parabolic trough collectors. Solar energy is to heat the water in the collectors to 423 K, which then vaporizes an organic working fluid that powers the pump-turbine. Calculate the surface area needed to power a 50-hp pump capable of delivering up to 38,000 liters/min of water at noon in Albuquerque, NM (Answer: 510 m².)

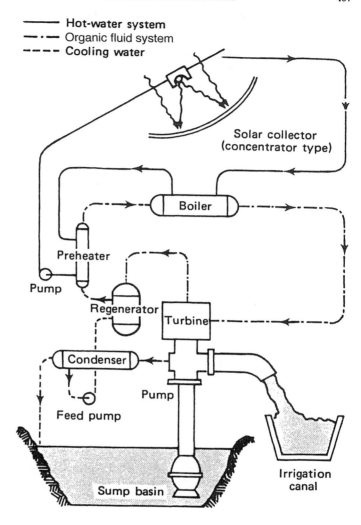

Hot-water system
Organic fluid system
Cooling water

Solar collector
(concentrator type)

Boiler

Preheater

Pump

Regenerator

Turbine

Condenser

Pump

Feed pump

Irrigation
canal

Sump basin

8.7. Write a closed-form expression for the work output of a solar-powered heat engine if the energy delivery of the solar collector at high temperature is given approximately by the expression

$$q_u = (\tau\alpha)_{\text{eff}} I_c - \frac{\sigma\epsilon \overline{T}_f^4}{CR},$$

(convection and conduction losses are neglected) where $\overline{T}_f$ is the average fluid temperature and CR is the concentration ratio. Two heat engines are to be evaluated:

(1) Carnot cycle:

$$\text{Cycle efficiency } \eta_c = 1 - \frac{T_\infty}{T_f},$$

and (2) Brayton cycle:

$$\text{Cycle efficiency } \eta_B = 1 - \frac{C_B T_\infty}{T_f} \quad C_B \geq 1,$$

where $C_B = (r_p)^{(k-1)/k}$

$$r_p = \text{the compressor pressure ratio}$$

$$k = \text{the specific heat ratio of the working fluid}$$

Write an equation with $\overline{T}_f$ as the independent variable that, when solved, will specify the value of $\overline{T}_f$ to be used for maximum work output as a function of concentration ratio, surface emittance and $(\tau\alpha)_{eff}$ product, insolation level, and C_B. Optional: solve the equation derived above for $CR = 100$, $\epsilon = 0.5$, and $(\tau\alpha)_{eff} = 0.70$, at an insolation level of 1 kW/m². What is the efficiency of a solar-powered Carnot cycle and a Brayton cycle for which $C_B = 2$?

REFERENCES AND SUGGESTED READINGS

1. Aoki, T., Takahashi, H., and Horigome, T. 1970. *The Development of Solar Systems for Industrial Processes.* Toshima-Ku, Tokyo, Japan: NEDO.
2. Becker, M., and M. Böhmer. 1987. Achievements of high and low flux receiver development. Proc. of the ISES Solar World Congress, Hamburg. Oxford: Pergamon Press. 1744–1783.
3. Beckman W. 1978. Simulations of the performance of open cycle dessicant systems using solar energy. *Sol. Energy* 21: 531.
4. Bohn, M.S., T. Williams, and H.W. Price. 1995. Combined cycle power tower. *Solar Engineering.* N.Y., ASME, 597–606.
5. Collares-Pereira M. and A. Ralb. 1979. Simple procedure for predicting long term average performance of nonconcentrating and of concentrating solar collectors. *Sol. Energy* 23: 235–253.
6. Daniels, F. 1964. *Direct Use of the Sun's Energy.* New Haven: Yale University Press.
7. Daniels, F. 1949. Solar Energy. *Science* 109: 51–57.
8. De Laquil, P., D. Kearney, M. Geyer, and R. Diver. 1993. Solar-thermal electric technology. In *Renewable energy.* Washington D. C.: Island Press.
9. Dunkle, R.V. 1961. Solar water distillation: The roof type still and a multiple effect diffusion still. *Proc. 1961 International Heat Transfer Conf.,* New York: American Society of Mechanical Engineers.
10. Fitzgerald, A.E., C.F. Kingsley, and A. Kusko. 1971. *Electric machinery,* 3rd ed. New York: McGraw-Hill.
11. Frank Kreith, ed. 1997. *The CRC handbook of mechanical engineering.* Boca Raton, FL: CRC Press.
12. Fraser, M.D. InterTechnology Corporation assesses industrial heat potential. *Sol. Engineering* 1, pp. 11–12, September 1976.
13. Goswami, B.C., and J. Langley. 1987. A review of the potential of solar energy in the textile industry. In *Progress in solar engineering,* Washington, D.C.: Hemisphere Publishing Corporation.
14. Goswami, D.Y., et al. 1991. A laser-based technique for particle sizing to study two-phase expansion in turbines. *Jour. Sol. Energy Engineering, Trans ASME* 113(3): 211–218.

15. Goswami, D.Y., and D.E. Kelett. 1982. Solar industrial process heat for textile industries. Paper presented at the 1982 ASME Textile Industries Division Conference at Raleigh, NC.
16. Goswami, D.Y., Klett, D.E., Raiford, M.T., and E.K. Stefankos. 1979. *Solar radiation design data of North Carolina.* Raleigh, NC: Energy Division, NC Department of Commerce.
17. Goswami, D.Y. 1998. Solar thermal power technology: Present status and ideas for the future. *Energy Sources* 20: 137–145.
18. Goswami, D.Y. 1995. Solar thermal power: Status of technologies and opportunities for research. *Proc. of ASME Heat and Mass Transfer Conference,* Surathkal, India, December.
19. Goswami, D.Y., and F. Xu. 1998. Analysis of a new thermodynamic cycle for combined power and cooling using low and mid temperature solar collectors. *Solar Engineering International Solar Energy Conference 1998,* ASMEm Fairfield, NJ. 111–120.
20. Gupta, B.P. 1985. Status and progress in solar thermal research and technology. In *Progress in solar engineering.* Washington, D.C.: Hemisphere Publishing Corporation.
21. Hebrank, W.H. 1975. Options for reducing fuel usage in textile finishing tenter dryers. *American Dyestuff Reporter* 63: 34.
22. Hipser, M.S. and R.F. Roehm. 1976. Heat transfer considerations of a non-convecting solar pond heat exchanger. ASME Paper 76-WA/Sol-4.
23. Intertechnology Corporation. 1977. Analysis of the economic potential of solar thermal energy to provide industrial process heat. ERDA Rep. No. COO/2829–1, 3 vols.
24. Japikse, D. and C.B. Nicholas. 1994. *Introduction to turbomachinery concepts.* Norwich, VT: ETI.
25. Kagawa, N., M. Okuda, S. Nagamoto, S. Ichikawa, and M. Sakamoto. 1989. Mechanical analysis and durability for a 3KW Stirling engine. *Proc. of the 24th IECEC Conference,* August 6–11. N.Y., IEEE, Vol. 5, pp. 2369–2374.
26. Kalina, A.I. 1984. Combined cycle system with novel bottoming cycle. *Jour. of Engineering for Gas Turbines and Power* 106: 737–742.
27. Kalina, A.I., and M. Tribus. 1992. Advances in Kalina cycle technology (1980–91), Part I: Development of a practical cycle, energy for the transition age. In *Proc. of FLOWERS'92,* 97–109. Nova Science Publishers.
28. Keenan, J.H. and F.G. Keys. 1956. *Thermodynamic properties of steam.* New York: John Wiley & Sons.
29. Klein, S.A., et al. 1979. TRNSYS—A transient system simulation. User's Manual. Madison: University of Wisconsin. Engineering Experiment Station Report 38–10.
30. Klein S.A. 1978. Private Communication.
31. Klein S.A. and W.A. Beckman. 1979. General design method for closed loop solar energy systems. *Sol. Energy* 22: 269.
32. Kouremenos, D.A., E.D. Rogdakis, and K.A. Antonopoulos. 1994. Cogeneration with combined gas and aqua-ammonia absorption cycles. *Thermodynamics and the Design, Analysis, and Improvement of Energy Systems.* New York: American Society of Mechanical Engineers, Advanced Energy Systems Division(Publication) AES 33 1994, ASME, 231–238.
33. Kreider, J.F. 1979. *Medium and High Temperature solar processes.* New York: Academic Press.
34. Leshuk, J.P., R.J. Zaworski, D.L. Styris, and O.K. Harling. 1976. Solar pond stability experiments. In *Sharing the Sun,* vol. 5: 188–202, Winnipeg: ISES.
35. Lowery, J.F. et. al. 1977. Energy conservation in the textile industry. *Technical Reports of Department of Energy Project No. Ey-76S-05-5099.* Atlanta, GA: School of Textile Engineering, Georgia Institute of Technology, 1977, 1978, and 1979.
36. McAdams, W.H. 1954. *Heat transmission,* 3rd ed. New York: McGraw-Hill Book Co.
37. Nielsen, C.E., and A. Rabl. 1976. Salt requirement and stability of solar ponds. In *Sharing the Sun,* vol. 5: 183–187. Winnipeg: ISES.
38. Nielsen, C.E. 1976. Experience with a prototype solar pond for space heating. In *Sharing the Sun,* vol. 5. Winnipeg: ISES, 169–182.
39. Nogawa, M., T. Hamagima, H. Ishikawa, Y. Momose, and N. Tanatsugu. 1989. Development of solar Stirling engine alternator for space experiments. Proc. of the 24th IECEC Conference, August 6–11. N.Y., IEEE, Vol. 5, pp. 2375–2378.
40. Proctor, D. and R.N. Morse. 1975. *Solar energy for Australian food processing industry.* East Melbourne, Victoria, Australia: CSIRO Solar Energy Studies.

41. Rabl, A., and C.E. Nielsen. 1975. Solar ponds for space heating. *Sol. Energy* 17, 1–12.
42. Reddy, T.A. 1987. The design and sizing of active solar thermal systems. Oxford: Oxford University Press.
43. Rogdakis, E.D. and K.A. Antonopoulos. 1991. High efficiency NH_3/H_2O absorption power cycle. *Heat Recovery Syst. CHP* 11(4): 263–275.
44. Salisbury, J.K. 1950. *Steam Turbines and Their Cycles.* Malabar, FL: Robert K. Krieger Publishing. Reprint 1974.
45. Singh, R.K., D.B. Lund, F.H. Buelow, and J.A. Duffie. 1980. Compatibility of solar energy with fluid milk processing energy demands. *Trans. ASAE* 23(3): 762.
46. Singh, R.K., D.B. Lund, and F.H. Buelow. 1986. Compatibility of solar energy with food processing energy demands. In *Alternative energy in agriculture* vol. I. Boca Raton, FL: CRC Press.
47. Stine, W.B., and R.E. Diver. 1994. A compendium of solar dish/Stiring technology. SAND93-7026, Albuquerque: Sandia National Laboratory.
48. Stine, W.B., and R.W. Harrigan. 1985. *Solar energy fundamentals and design: With computer applications.* New York: Wiley.
49. Stine, W.B., Stiring engines. 1998. In *The CRC handbook of mechanical engineering.* Boca Raton, FL: CRC Press.
50. Stodola, A. and L.C. Loewenstein. 1927. *Steam and Gas Turbine,* New York: Peter Smith. Reprint 1945.
51. Styris, D.L., O.K. Harling, R.J. Zaworski, and J. Leshuk. 1976. The nonconvecting solar pond applied to building and process heating. *Sol. Energy* 18: 245–251.
52. Tabor, H. 1966. Solar Ponds. *Sci. J.* 66–71.
53. Tabor, H. Solar ponds. 1963. *Sol. Energy* 7: 189–194.
54. Talbert, S.G., J.A. Eibling, and G.O.G. Löf. 1970. Manual on solar distillation of saline water. *Office of Saline Water, Research and Development Progress Rept. 546.* Washington, D.C.: U.S. Department of Interior.
55. 1981. *TRNSYS: A Transient Simulation Program,* Version 11.1. *Engineering Experiment Station, Report 38.* Madison, WI: Solar Energy Laboratory, University of Wisconsin.
56. Trombe, F. 1956. High temperature furnaces. *Proc. World Symposium on Applied Solar Energy,* Phoenix, Arizona, 1955. Menlo Park, CA: Stanford Research Institute.
57. Wagner, R. 1977. Energy conservation in dyeing and finishing. *Textile Chemists and Colorists* 9: 52.
58. Washam, B.M., M. Willrich, J.C. Schaefer, and D. Kearney. 1993. *Integrated Solar Combined Cycle Systems (ISCCS) Utilizing Solar Parabolic Trough Technology—Golden Opportunities For the '90s.* Diablo, CA: Spencer Management Associates.
59. Weinberger, H. 1964. The physics of the solar pond. *Sol. Energy* 8: 45–56.

PHOTOVOLTAICS

Photovoltaic conversion is the direct conversion of sunlight into electricity with no intervening heat engine. Photovoltaic devices are solid state; therefore, they are rugged and simple in design and require very little maintenance. Perhaps the biggest advantage of solar photovoltaic devices is that they can be constructed as stand-alone systems to give outputs from microwatts to megawatts. That is why they have been used as the power sources for calculators, watches, water pumping, remote buildings, communications, satellites and space vehicles, and even megawatt scale power plants. Photovoltaic panels can be made to form components of building skin, such as roof shingles and wall panels. With such a vast array of applications, the demand for photovoltaics is increasing every year. In 1995, 80 MW_p of photovoltaic panels were sold for the terrestrial markets.

In the early days of solar cells in the 1960s and 1970s, more energy was required to produce a cell than it could ever deliver during its lifetime. Since then, dramatic improvements have taken place in the efficiencies and manufacturing methods. In 1996, the energy payback periods were reduced to about two and a half to five years, depending on the location of use [11], while panel lifetimes were increased to over 25 years. The costs of photovoltaic panels have come down to $5–10 per peak watt over the last two decades and are targeted to reduce to around $1.00 per peak watt.

Historically, the photoelectric effect was first noted by Becquerel in 1839 when light was incident on an electrode in an electrolyte solution [1]. Adams and Day first observed the effect in solids in 1877 while working with selenium. Early work was done with selenium and copper oxide by pioneers such as Schottkey, Lange and Grandahl. In 1954 researchers at RCA and Bell Laboratories reported achieving efficiencies of about 6 percent by using devices made of p and n types of semiconductors. The space race between the USA and the Soviet Union resulted in dramatic improvements in the photovoltaic devices. Ref. [3] gives a review of the early developments in photovoltaic conversion.

9.1 SEMICONDUCTORS

A basic understanding of the atomic structure is quite helpful in understanding the behavior of semiconductors and their use as the photovoltaic energy conversion devices. Any fundamental book on physics or chemistry generally gives adequate background for basic understanding. Ref. [2] presents an in-depth treatment of a number of topics in semiconductor physics.

For any atom, the electrons arrange themselves in orbitals around the nucleus so as to result in the minimum amount of energy. Table 9.1 shows the distribution of the electrons in various shells and subshells in light elements. In elements that have electrons in multiple shells, the innermost electrons have the minimum energy and therefore, require the maximum amount of externally imparted energy to overcome the attraction of the nucleus and become free. Electrons in the outermost band of subshells are the only ones that participate in the interaction of an atom with its neighboring atoms. If these electrons are very loosely attached to the atom, they may attach themselves with a neighboring atom to give that atom a negative charge, leaving the original atom as a positive charged ion. The positive and negatively charged ions become attached by the force of attraction of the charges thus forming *ionic bonds*. If the electrons in the outermost band do not fill the band completely but are not loosely attached either, they arrange themselves so that neighboring atoms can share them to make the outermost bands full. The bonds thus formed between the neighboring atoms are called *covalent bonds*.

Since electrons in the outermost band of an atom determine how an atom will react or join with a neighboring atom, the outermost band is called the **valence band**. Some electrons in the valence band may be so energetic that they jump into a still higher band and are so far removed from the nucleus that a small amount of impressed force would cause them to move away from the atom. Such electrons are responsible for the conduction of heat and electricity, and this remote band is called a **conduction band**. The difference in the energy of an electron in the valence band and the innermost subshell of the conduction band is called the **Band Gap**, or the forbidden gap.

Materials whose valence bands are full have very high band gaps (>3eV). Such materials are called *insulators*. Materials, on the other hand, that have relatively empty valence bands and may have some electrons in the conduction band are good *conductors*. Metals fall in this category. Materials with valence bands partly filled have intermediate band gaps (≤3eV). Such materials are called *semiconductors* (Fig. 9.1). Pure semiconductors are called *intrinsic semiconductors*, while semiconductors doped with very small amounts of impurities are called *extrinsic semiconductors*. If the dopant material has more electrons in the valence band than the semiconductor, the doped material is called an *n*-type of semiconductor. Such a material seems to have excess electrons available for conduction even though the material is electronically neutral. For example, silicon has four electrons in the valence band. Atoms of pure silicon arrange themselves in such a way that, to form a stable structure, each atom shares two electrons with each neighboring atom with covalent bands. If phosphorous, which has five valence electrons (one more than Si) is introduced as an impurity in silicon, the doped material seems to have excess electrons even though it is electrically neutral. Such a

Table 9.1[†]. Electronic structure of atoms

Principal quantum number n				1	2	2	3	3	3	4	5
Azimuthal quantum number l				0	0	1	0	1	2	0	1
Letter designation of state				1s	2s	2p	3s	3p	3d	4s	4p
Z	Symbol	Element	V_i volts								
1	H	Hydrogen	13.60	1							
2	He	Helium	24.58	2							
3	Li	Lithium	5.39		1						
4	Be	Beryllium	9.32		2						
5	B	Boron	8.30		2	1					
6	C	Carbon	11.26		2	2					
7	N	Nitrogen	14.54		2	3					
8	O	Oxygen	13.61		2	4					
9	F	Fluorine	17.42		2	5					
10	Ne	Neon	21.56		2	6					
11	Na	Sodium	5.14				1				
12	Mg	Magnesium	7.64				2				
13	Al	Aluminum	5.98				2	1			
14	Si	Silicon	8.15				2	2			
15	P	Phosphorus	10.55				2	3			
16	S	Sulfur	10.36				2	4			
17	Cl	Chlorine	13.01				2	5			
18	A	Argon	15.76				2	6			
19	K	Potassium	4.34							1	
20	Ca	Calcium	6.11							2	
21	Sc	Scandium	6.56						1	2	
22	Ti	Titanium	6.83						2	2	
23	V	Vanadium	6.74						3	2	
24	Cr	Chromium	6.76						5	1	
25	Mn	Manganese	7.43						5	2	
26	Fe	Iron	7.90						6	2	
27	Co	Cobalt	7.86						7	2	
28	Ni	Nickel	7.63						8	2	
29	Cu	Copper	7.72						10	1	
30	Zn	Zinc	9.39						10	2	
31	Ga	Gallium	6.00						10	2	1
32	Ge	Germanium	7.88						10	2	2
33	As	Arsenic	9.81						10	2	3
34	Se	Selenium	9.75						10	2	4
35	Br	Bromine	11.84						10	2	5
36	Kr	Krypton	14.00						10	2	6

(Helium core for Z = 3–10; Neon core for Z = 11–18; Argon Core for Z = 19–36.)

[†]From Charlotte E. Moore, *Atomic energy levels*, Vol. II. Washington, D.C.: National Bureau of Standards Circular 467, 1952.

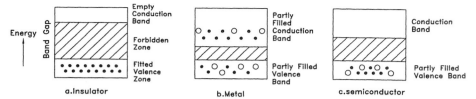

Figure 9.1. Electrical conduction is described in terms of allowed and forbidden energy bands. Band gap for insulators (a) is the highest, followed by semiconductors (c) and metals (b), respectively.

doped material is called *n*-type silicon. If on the other hand, silicon is doped with Boron which has three valence electrons (one less than Si), there seems to be a positive hole (missing electrons) in the structure, even though the doped material is electrically neutral. Such material is called *p*-type silicon. Thus, *n*- and *p*-type semiconductors make it easier for the electrons and holes respectively to move in the semiconductors.

9.1.1 *p-n* Junction

As explained earlier, an *n*-type material has some impurity atoms with more electrons than the rest of the semiconductor atoms. If those excess electrons are removed, the impurity atoms will fit more uniformly in the structure formed by the main semiconductor atoms; however, the atoms will be left with positive charges. On the other hand, a *p*-type material has some impurity atoms with fewer electrons than the rest of the semiconductor atoms. Therefore, these atoms seem to have holes that could accommodate excess electrons even though the atoms are electrically neutral (Fig. 9.2). If additional electrons could be brought to fill the holes, the impurity atoms would fit more uniformly in the structure formed by the main semiconductor atoms, however, the atoms will be negatively charged.

The above scenario occurs at the junction when a *p* and an *n* type of material are joined together as shown in Fig. 9.3. As soon as the two materials are joined, "excess"

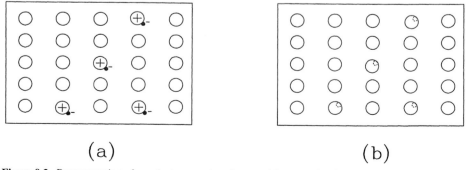

Figure 9.2. Representation of *n* and *p* type semiconductors: (a) *n*-type showing "excess" electrons as dots; (b) *p*-type showing "excess" positive holes as ⊙.

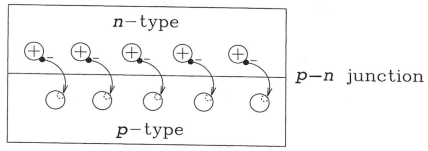

Figure 9.3. "Excess" electrons from *n*-material jump to fill "excess" holes on the *p*-side of a *p-n* junction, leaving the *n*-side of the junction positively charged and the *p*-side negatively charged.

electrons from the *n* layer jump to fill the "holes" in the *p* layer. Therefore, close to the junction, the material has positive charges on the *n* side and negative charges on the *p* side. The negative charges on the *p* side restrict the movement of additional electrons from the *n* side to the *p* side, while the movement of additional electrons from the *p* side to the *n* side is made easier because of the positive charges at the junction on the *n* side. This restriction makes the *p-n* junction behave like a diode. This diode character of a *p-n* junction is made use of in solar photovoltaic cells as explained below.

9.1.2 The Photovoltaic Effect

When a photon of light is absorbed by a valence electron of an atom, the energy of the electron is increased by the amount of energy of the photon. If the energy of the photon is equal to or more than the band gap of the semiconductor, the electron with excess energy will jump into the conduction band where it can move freely. If, however, the photon energy is less than the band gap, the electron will not have sufficient energy to jump into the conduction band. In this case, the excess energy of the electrons is converted to excess kinetic energy of the electrons which manifests in increased temperature. If the absorbed photon had more energy than the band gap, the excess energy over the band gap simply increases the kinetic energy of the electron. It must be noted that a photon can free up only one electron even if the photon energy is a lot higher than the band gap. This fact is a big reason for the low conversion efficiency of photovoltaic devices. The key to using the photovoltaic effect for generating useful power is to channel the free electrons through an external resistance before they recombine with the holes. This is achieved with the help of the *p-n* junction.

Figure 9.4 shows a schematic of a photovoltaic device. As free electrons are generated in the *n* layer by the action of photons, they can either pass through an external circuit, recombine with positive holes in the lateral direction or move toward the *p* layer. The negative charges in the *p* layer at the *p-n* junction restrict their movement in that direction. If the *n* layer is made extremely thin, the movement of the electrons and, therefore, the probability of recombination within the *n* layer is greatly reduced

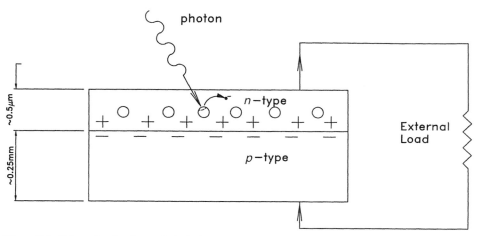

Figure 9.4. Schematic of a photovoltaic device.

unless the external circuit is open. If the external circuit is open, the electrons generated by the action of photons eventually recombine with the holes, resulting in an increase in the temperature of the device.

In a typical crystalline silicon cell, the n layer is about 0.5 μm thick and the p layer
is about 0.25 mm thick. As explained in Chapter 2, energy contained in a photon E_p is
given by

$$E_p = h\upsilon, \tag{9.1}$$

where h is the Planck's constant (6.625×10^{-34} J-sec) and υ is the frequency which is
related to the wavelength λ and the speed of light c as

$$\upsilon = \frac{c}{\lambda}.$$

Therefore,

$$E_p = hc/\lambda. \tag{9.2}$$

For silicon, which has a band gap of 1.11 eV, the following example shows that
photons of solar radiation of wavelength 1.12 μm or less are useful in creating electron-hole pairs. This spectrum represents a major part of the solar radiation. Table 9.2
lists some candidate semiconductor materials for photovoltaic cells along with their
band gaps.

Example 9.1. Calculate the wavelength of light capable of forming an electron-
hole pair in silicon.

Table 9.2. Energy gap for some candidate materials for photovoltaic cells [7].

Material	Band gap (eV)	Material	Band gap (eV)
Si	1.11	$CuInTe_2$	0.90
SiC	2.60	InP	1.27
$CdAs_2$	1.00	In_2Te_3	1.20
CdTe	1.44	In_2O_3	2.80
CdSe	1.74	Zn_3P_2	1.60
CdS	2.42	ZnTe	2.20
$CdSnO_4$	2.90	ZnSe	2.60
GaAs	1.40	AlP	2.43
GaP	2.24	AlSb	1.63
Cu_2S	1.80	As_2Se_3	1.60
CuO	2.00	Sb_2Se_3	1.20
Cu_2Se	1.40	Ge	0.67
$CuInS_2$	1.50	Se	1.60
$CuInSe_2$	1.01		

Solution. The band gap energy of silicon is 1.11 eV. From Eq. (9.2) we can write

$$\lambda = \frac{hc}{E}.$$

For $c = 3 \times 10^8$ m/sec, $h = 6.625 \times 10^{-34}$ J · sec, and 1 eV $= 1.6 \times 10^{-19}$ J, the above equation gives the required wavelength as:

$$\lambda = \frac{(6.625 \times 10^{-34} \text{ J} \cdot \text{sec})(3 \times 10^8 \text{ m/sec})}{(1.11)(1.6 \times 10^{-19} \text{ J})} = 1.12 \text{ } \mu\text{m}.$$

Example 9.2. A monochromatic red laser beam emitting 1 mW at a wavelength of 638 nm is incident on a silicon solar cell. Find:

a. the number of photons per second incident on the cell; and
b. the maximum possible efficiency of conversion of this laser beam to electricity.

Solution

a. The intensity of light in the laser beam (I_p) is equal to the energy of all the photons in it. If the number of photons is N_{ph}, then

$$I_p = N_{ph} \cdot E_p \tag{9.3}$$

$$1 \times 10^{-3}\text{W} = N_{ph} \cdot E_p \quad (1 \text{ W} = 1 \text{ J/sec})$$

$$E_p = hc/\lambda$$

$$= \frac{(6.625 \times 10^{-34} \text{ J} - \text{sec}) \cdot 3 \times 10^8 \text{ m/sec}}{638 \times 10^{-9} \text{ m}}$$

$$= 3.12 \times 10^{-19} \text{ J}$$

$$\therefore N_{ph} = \frac{1 \times 10^{-3} \text{ J/sec}}{3.12 \times 10^{-19} \text{ J}} = 3.21 \times 10^{15} \text{ photons/sec}$$

b. Assuming that each photon is able to generate an electron, a total number of N_{ph} electrons will be generated. Therefore, the electrical output will be equal to $N_{ph} \cdot$ (Band Gap). Therefore, the maximum possible efficiency is

$$\eta_{max} = \frac{(N_{ph}) \cdot (B.G.)}{(N_{ph}) \cdot E_p} = \frac{B.G.}{E_p} \tag{9.4}$$

$$= \frac{1.11 \times 1.6 \times 10^{-19} \text{ J}}{3.12 \times 10^{-19} \text{ J}} = .569 \text{ or } 56.9\%.$$

From the above examples, it is clear that for a silicon solar cell, none of the photons of the sunlight over 1.12 μm wavelength will produce any electricity. However,

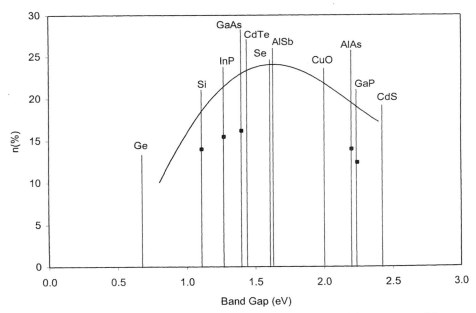

Figure 9.5. The maximum solar energy conversion efficiency as a function of the energy gap of the semiconductor. The measured efficiency for various materials is denoted by a solid square energy gap of the material. The curve has been calculated for an ideal junction outside the atmosphere.

photons of sunlight at a wavelength of 1.12 μm may be converted to electricity at a maximum efficiency of 100 percent, while photons at lower wavelengths will be converted at lower efficiencies. The overall maximum efficiency of a cell can be found by integrating the efficiency at each wavelength over the entire solar spectrum:

$$\eta = \frac{\int \eta_\lambda I_\lambda d\lambda}{\int I_\lambda d\lambda} \tag{9.5}$$

In addition, other factors such as probability of electron-hole recombination reduce the theoretical maximum achievable efficiency of a solar cell. Figure 9.5 shows a comparison of the maximum energy conversion of cells using different materials.

Figure 9.5 shows that the optimum band gap for terrestrial solar cells is around 1.5 eV.

9.2 ANALYSIS OF PHOTOVOLTAIC CELLS

This section presents an electrical analysis of a photovoltaic cell which will be useful in the design of photovoltaic devices for various applications. The physics leading to the expressions for the number density of electrons and holes in n and p materials at a temperature T will not be presented here. For such details, the reader is referred to books such as Refs. [1] and [2]. It would suffice to point out here that at the p-n junction a current is generated called the *junction current*. The junction current J_j is the net current due to the current J_o from the p side to the n side (called the **dark current** or the **reverse saturation current**) and a **light induced recombination current** J_r from the n side to the p side. Based on the temperature T, a certain number of electrons in the p material exist in the conduction band. These electrons can easily move to the n side to fill the holes created at the p-n junction, generating a current J_o. Normally, the electrons occupying the conduction band due to the temperature in the n material do not have enough potential energy to cross the p-n junction to the p side. However, if a foward bias voltage V is applied, which in a photovoltaic cell is due to the action of the photons of light, some of the electrons thus generated have enough energy to cross over and recombine with the holes in the p region. This gives rise to a light-induced recombination current J_r, which is proportional to J_o and is given by

$$J_r = J_o \exp(e_o V / kT), \tag{9.6}$$

where e_o = charge of an electron
$= 1.602 \times 10^{-19}$ Coulombs or J/V,
k = Boltzman's Constant $= 1.381 \times 10^{-23}$ J/K
The junction current J_j is the net current due to J_r and J_o.

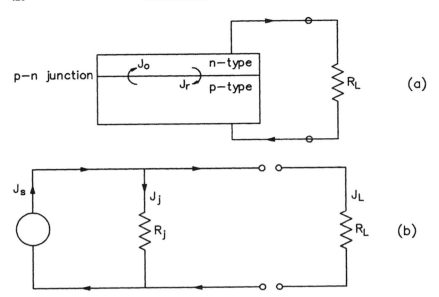

Figure 9.6. Equivalent circuit of a photovoltaic cell.

$$J_j = J_r - J_o$$
$$= J_o [\exp(e_o V/kT) - 1].$$

Referring to Fig. (9.6), it is clear that the current generated in the cell has two parallel paths, one through the junction and the other through the external resistance R_L. Figure (9.6) shows an equivalent circuit of a photovoltaic cell. It must be pointed out here that the current generated in a photovoltaic cell, including the junction current, is proportional to the area of the cell. Therefore, it is appropriate to analyze in terms of the current density J (current per unit area) instead of the current I. The relationship between the two is:

$$I = J \cdot A. \tag{9.8}$$

Referring to Fig. (9.6), we can write:

$$J_L = J_s - J_j \tag{9.9}$$
$$= J_s - J_o [\exp(e_o V/kT) - 1],$$

where J_s is the **short circuit current**.

For short circuit, $V = 0$ and $J_L = J_s$.
For open circuit, $J_L = 0$ and $V = V_{oc}$,

which gives

$$0 = J_s - J_o \left[\exp(e_o V_{oc}/kT) - 1\right], \tag{9.10}$$

or

$$V_{oc} = \frac{kT}{e_o}\ln\left(\frac{J_s}{J_o} + 1\right).$$

Figure (9.7) shows a typical performance curve (I-V) of a solar cell. The power output is the product of the load current and voltage and is a function of the load resistance

$$P_L = AJ_L V = I_L V \tag{9.11}$$
$$= I_L^2 R_L,$$

where A is the area of the cell.

The power output exhibits a maximum. To find the condition for the maximum power output (P_{max}), differentiate P with respect to V and equate it to zero:

$$\exp(e_o V_m/kT)\left(1 + \frac{e_o V_m}{kT}\right) = 1 + \frac{J_s}{J_o}, \tag{9.12}$$

where V_m stands for voltage at maximum power. The current at max power condition $J_{L,m}$ and the maximum power P_{max} can be found from Eqs. (9.9) and (9.11), respectively.

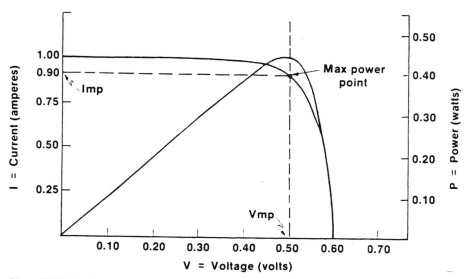

Figure 9.7. Typical current, voltage and power characteristics of a solar cell Adapted from [6].

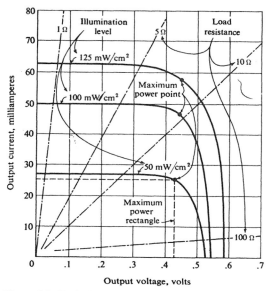

Figure 9.8. Typical current-voltage characteristics of a
silicon cell showing the effects of illumination level and
load resistance.

$$J_{L,m} = J_s - J_o [\exp(e_o V_m/kT) - 1]. \tag{9.13}$$

Combining Eqs. (9.12) and (9.13), $J_{L,m}$ is found to be:

$$J_{L,m} = \frac{e_o V_m/kT}{1 + (e_o V_m/kT)} (J_s + J_o), \tag{9.14}$$

$$P_{max} = \frac{e_o V_m^2/kT}{1 + (e_o V_m/kT)} (J_s + J_o) \cdot A. \tag{9.15}$$

Figure (9.8) shows the effect of illumination intensity and the load resistance on
the performance of a silicon cell. Temperature also affects the performance in such a
way that the voltage and thus the power output decreases with increasing temperature.

Example 9.3. The dark current density for a silicon solar cell at 40°C is 1.8 ×
$10^{-8} A/m^2$ and the short circuit current density is 200 A/m^2. Calculate:

a. Open circuit voltage
b. Voltage at maximum power
c. Current density at maximum power
d. Maximum power

e. Maximum efficiency
f. The cell area required for an output of 25 W when exposed to solar radiation of 900 W/m².

Solution. Given:

$$J_o = 1.8 \times 10^{-8} A/m^2$$

$$J_s = 200 \, A/m^2,$$

and

$$T = 40°C = 313 \, K.$$

a. Using Eq. (9.10),

$$V_{oc} = \frac{kT}{e_o} \ln\left(\frac{J_s}{J_o} + 1\right).$$

Since e_o/kT will be needed for other parts also it can be evaluated separately as

$$e_o/kT = \frac{1.602 \times 10^{-19} \, J/V}{(1.381 \times 10^{-23} \, J/K)(313K)} = 37.06 V^{-1}.$$

Therefore,

$$V_{oc} = \frac{1}{37.06} \ln\left(\frac{200}{1.8 \times 10^{-8}} + 1\right).$$

$$= 0.624V$$

b. Voltage at maximum power condition can be found from Eq. (9.12) by an iterative or trial and error solution:

$$\exp(37.06V_m)(1 + 37.06V_m) = 1 + \frac{200}{1.8 \times 10^{-8}}$$

$$\text{or } V_m = .542V.$$

c. Current density at maximum power can be found from Eq. (9.14):

$$J_{L,m} = \frac{e_o V_m/kT}{1 + (e_o V_m/kT)}(J_s + J_o)$$

$$= \frac{(37.06)\cdot(0.542)}{1 + (37.06)\cdot(0.542)}(200 + 1.8 \times 10^{-8})A/m^2$$

$$= 190.5A/m^2.$$

d.
$$P_{max} = V_m \cdot J_m \cdot A$$

$$\frac{P_{max}}{A} = (0.542V) \cdot (190.5A/m^2)$$

$$= 103.25 \text{ W}/m^2.$$

e.
$$\eta_{max} = \frac{103.25 \text{ W}/m^2}{900 \text{ W}/m^2} = 11.5\%.$$

f. Cell area required:

$$A = \frac{P_{out}}{P_{max}/A} = \frac{25W}{103.25 \text{ W}/m^2}$$

$$= 24.2 \text{ cm}^2.$$

9.2.1 Efficiency of Solar Cells

Theoretical limitation on the efficiency of a solar cell can be calculated using Eq. (9.5). These efficiency limitations and the practical efficiencies of some of the cells are shown in Figure 9.5. Some of the reasons for the actual efficiency being lower than the theoretical limitation are:

1. Reflection of light from the surface of the cell. This can be minimized by anti-reflection (AR) coating. For example, AR coatings can reduce the reflection from a Si cell to 3% from more than 30% from an untreated cell.
2. Shading of the cell due to current collecting electrical contacts. This can be minimized by reducing the area of the contacts and/or making them transparent, however, both of these methods will increase the resistance of the cell to current flow.
3. Internal electrical resistance of the cell.
4. Recombination of electrons and holes before they can contribute to the current. This effect can be reduced in polycrystalline and amorphous cells by using hydrogen alloys.

9.2.2 Multijunction Solar Cells

The limits imposed on solar cells due to band gap can be partially overcome by using multiple layers of solar cells stacked on top of each other, each layer with a band gap higher than the layer below it. For example (Fig. 9.9), if the top layer is made from a cell of material A (Band gap corresponding to λ_A), solar radiation with wavelengths less than λ_A would be absorbed to give an output equal to the hatched area A. The solar radiation with wavelength greater than λ_A would pass through A and be converted by the bottom layer cell B (Band gap corresponding to λ_B) to give an output equal to the hatched area B. The total output and therefore the efficiency of this tandem cell would

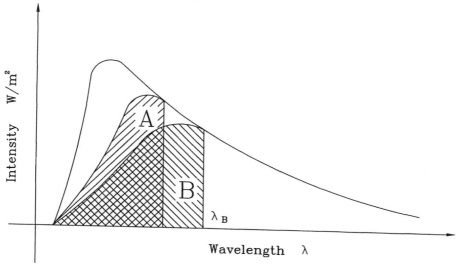

Figure 9.9. Energy conversion from a two-layered stacked cell.

be higher than the output and the efficiency of each single cell individually. The efficiency would increase with the number of layers. For this concept to work, each layer must be as thin as possible, which puts a very difficult if not an insurmountable constraint on crystalline and polycrystalline cells to be made multijunction. As a result, this concept is being investigated mainly for thin film amorphous solar cells.

At present, a triple junction a-Si solar cell is under development. This cell consists of layers of cells made from a-Si,C:H (an amorphous Silicon, Carbon and hydrogen alloy) with a band gap of 2.0 eV, a-Si:H (an amorphous Silicon and hydrogen alloy) with a band gap of 1.75 eV, and a-Si, Ge:H (an amorphous Silicon, Germanium and hydrogen alloy) with a band gap of 2.3 eV. The efficiency of a multijunction cell can be about 50 percent higher than a corresponding single cell. Figure 9.10 shows a conceptual device structure of a triple junction a-Si cell.

9.2.3 Design of a Photovoltaic System

Solar Cells may be connected in series, parallel, or both to obtain the required voltage and current. When similar cells or devices are connected in series, the output voltages and current are as shown in Figure 9.11. A parallel connection results in the addition of currents as shown in Figure 9.12. If the cells or devices 1 and 2 have dissimilar characteristics, the output characteristics will be as shown in Figure 9.13. Cells are connected to form modules, modules are connected to form panels and panels are connected to form arrays. Principles shown in Figures 9.11 and 9.12 apply to all of these connections.

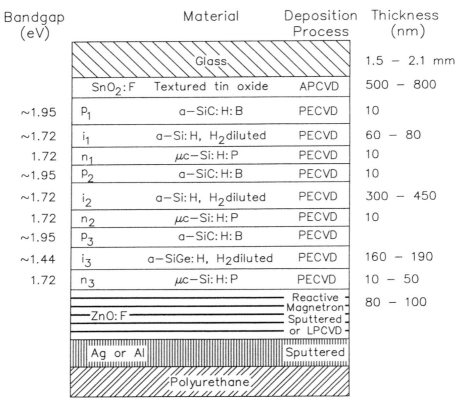

Bandgap (eV)		Material	Deposition Process	Thickness (nm)
		Glass		1.5 − 2.1 mm
	SnO_2:F	Textured tin oxide	APCVD	500 − 800
~1.95	P_1	a−SiC: H: B	PECVD	10
~1.72	i_1	a−Si: H, H_2 diluted	PECVD	60 − 80
1.72	n_1	μc−Si: H: P	PECVD	10
~1.95	P_2	a−SiC: H: B	PECVD	10
~1.72	i_2	a−Si: H, H_2 diluted	PECVD	300 − 450
1.72	n_2	μc−Si: H: P	PECVD	10
~1.95	P_3	a−SiC: H: B	PECVD	
~1.44	i_3	a−SiGe: H, H_2 diluted	PECVD	160 − 190
1.72	n_3	μc−Si: H: P	PECVD	10 − 50
		ZnO: F	Reactive Magnetron Sputtered or LPCVD	80 − 100
		Ag or Al	Sputtered	
		Polyurethane		

Figure 9.10. Typical triple-junction *a*-Si cell structure. Source: [4].

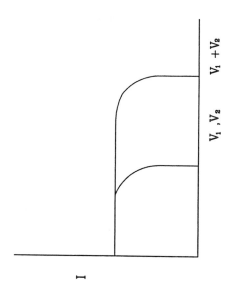

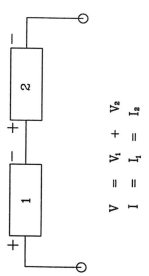

$$V = V_1 + V_2$$
$$I = I_1 = I_2$$

Figure 9.11. Characteristics of two similar cells connected in series.

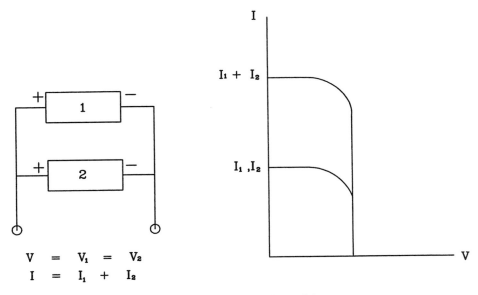

$$V = V_1 = V_2$$
$$I = I_1 + I_2$$

Figure 9.12. Characteristics of two similar cells connected in parallel

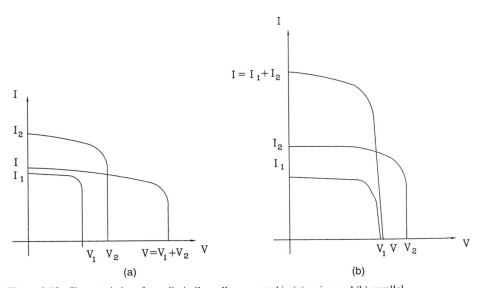

(a)   (b)

Figure 9.13. Characteristics of two dissimilar cells connected in (a) series, and (b) parallel.

Example 9.4. An application requires 300 Watts at 28V. Design a PV panel using solar cells from Example 9.3 each with an area of 6cm².

Solution. Assuming that the cells will be operated at maximum power conditions, the voltage and current from each cell are

$$V_m = 0.542\text{V}, \quad I_m = \left(190.5\,\frac{\text{A}}{\text{m}^2}\right)(6 \times 10^{-4}\,\text{m}^2)$$

$$= 0.1143\text{A},$$

$$\text{Power/Cell} = 0.542 \times 0.1143 = 0.062\text{W},$$

$$\text{Number of cells required} = \frac{300\text{W}}{0.062\text{W/cell}} = 4840,$$

$$\text{Number of cells in series} = \frac{\text{System Voltage}}{\text{Voltage/cell}} = \frac{28\text{V}}{0.542\text{V}} = 52,$$

$$\text{Number of rows of 52 cells connected in parallel} = \frac{4840}{52} \approx 93.1.$$

Since the number of rows must be a whole number we may increase the number to 94 rows which will give 303 W output (Fig. 9.14).

A blocking diode is used in series with a module or an array to prevent the current from flowing backward, for example, from the battery to the cells under dark conditions. A bypass diode is used in parallel with a module in an array to bypass the module if it is shaded. A photovoltaic system may be connected to a DC or an AC load as shown in Figure 9.15.

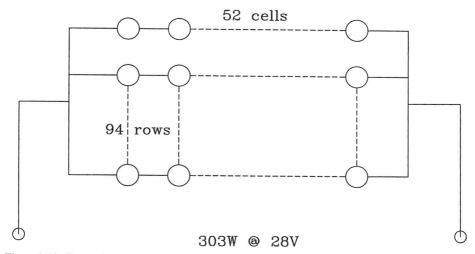

Figure 9.14. Connection of cells in rows and columns for Problem 9.4.

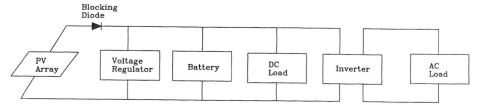

Figure 9.15. Schematic of a photovoltaic system.

9.3 MANUFACTURE OF SOLAR CELLS AND PANELS

Manufacture of crystalline silicon solar cells is an outgrowth of the manufacturing methods used for microprocessors. A major difference is that silicon used in microprocessors is ultra pure, which is not needed for photovoltaic cells. Therefore, a major source of feedstock for silicon solar cells has been the waste material from the microelectronics industry. Solar cells are also manufactured as polycrystalline and thin films. Below are some of the common methods of manufacture of silicon solar cells.

9.3.1 Single Crystal and Polycrystalline Cells

Single crystal silicon cells are produced by a series of processes: (1) growing crystalline ingots of *p*-silicon, (2) slicing wafers from the ingots, (3) polishing and cleaning the surface, (4) doping with *n* material to form the *p-n* junction, (5) deposition of electrical contacts, (6) application of anti-reflection coating and (7) encapsulation. Figure 9.16 illustrates the process.

Czochralski Method (Fig. 9.17(a)) is the most common method of growing single crystal ingots. A seed crystal is dipped in molten silicon doped with a *p*-material (Boron) and drawn upward under tightly controlled conditions of linear and rotational speed, and temperature. This process produces cylindrical ingots of typically 10 cm diameter, although ingots of 20 cm diameter and more than 1 m long can be produced for other applications. An alternative method is called the **float zone** method (Fig. 9.17(b)). In this method a polycrystalline ingot is placed on top of a seed crystal and the interface is melted by a heating coil around it. The ingot is moved linearly and ro-

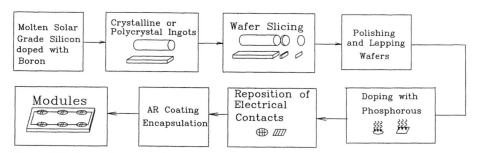

Figure 9.16. Series of processes for the manufacture of crystalline L polycrystalline cells.

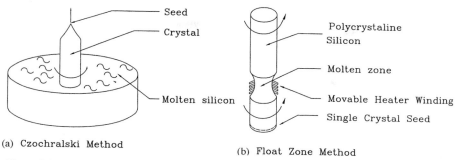

(a) Czochralski Method

(b) Float Zone Method

Figure 9.17. Crystalline silicon ingot production methods.

tationally, under controlled conditions. This process has the potential to reduce the cell cost.

Polycrystalline ingots are produced by casting silicon in a mold of preferred shape (rectangular) as shown in Figure 9.18. Molten silicon is cooled slowly in a mold along one direction in order to orient the crystal structures and grain boundaries in a preferred direction. In order to achieve efficiencies of greater than 10 percent, grain sizes greater than 0.5 mm are needed and the grain boundaries must be oriented perpendicular to the wafer. Ingots as large as 400 cm × 40 cm × 40 cm can be produced by this method.

Ingots are sliced into wafers by internal diameter (ID) saws or multiwire saws impregnated with diamond abrasive particles. Both of these methods result in high wastage of valuable crystalline silicon.

Alternative methods that reduce wastage are those that grow polycrystalline **thin films**. Some of the thin film production methods include Dendric Web Growth (Fig. 9.19), Edge-Defined Film-fed Growth (EFG) (Fig. 9.20), Ribbon against drop (RAD)

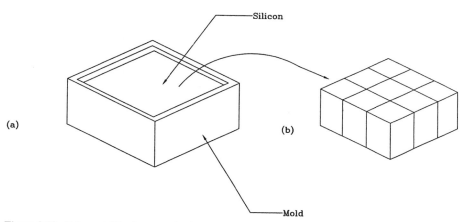

Figure 9.18. Polycrystalline ingot production.

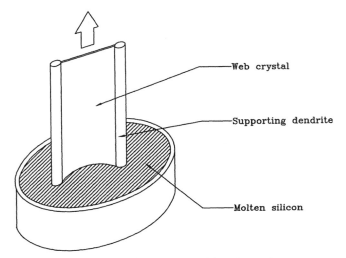

Figure 9.19. Thin film production by dendritic web growth.

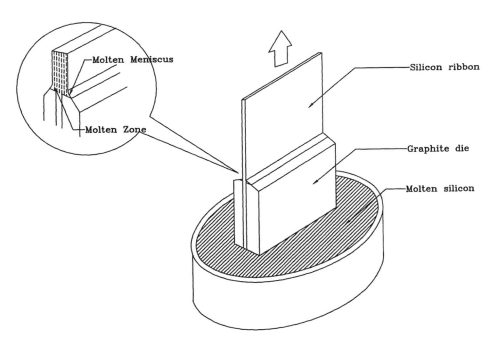

Figure 9.20. Thin film production by edge defined film-fed growth (EFG).

method, Supported Web method, and Ramp assisted foil casting technique (RAFT) (Fig. 9.21).

A p-n junction is formed in the cell by diffusing a small amount of n material (phosphorous) in the top layer of a p-silicon wafer. The most common method is diffusion of phosphorous in the vapor phase. In this case, the back side of the wafer must be covered to prevent the diffusion of vapors from that side. An alternate method is to deposit a solid layer of the dopant material on the top surfaces followed by high temperature (800–900°) diffusion.

Electrical contacts are attached to the top surface of the cell in a grid pattern to cover no more than 10 percent of the cell surface and a solid metallic sheet is attached to the back surface. The front grid pattern is made by either vacuum metal vapor deposition through a mask or by screen printing. Figure 9.22 shows how cells are connected to form modules.

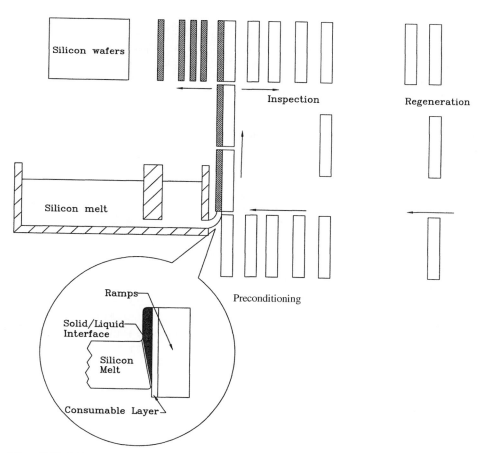

Figure 9.21. Schematic of RAFT processing.

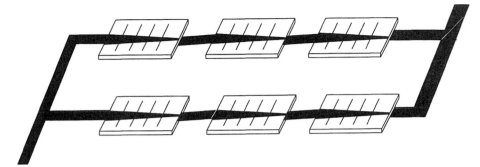

Figure 9.22. Assembly of solar cells to form a module.

Anti-reflection (AR) coatings of materials such as silicon dioxide (SiO_2), titanium dioxide (TiO_2) and tantalum pentaoxide (Ta_2O_5) are deposited on the cell surface to reduce reflection from more than 30 percent for untreated Si to less than 3 percent. AR coatings are deposited by vacuum vapor deposition, sputtering or chemical spraying. Finally, the cells are encapsulated in a transparent material to protect them from the environment. Encapsulants usually consist of a layer of either polyvinyl butyryl (PVB) or ethylene vinyl acetate (EVA) and a top layer of low iron glass.

9.3.2 Amorphous Silicon and Multijunction Thin Film Fabrication

Amorphous Silicon (*a*-Si) cells are made as thin films of *a* Si:H alloy doped with phosphorous and boron to make *n* and *p* layers respectively. The atomic structure of an *a* Si cell does not have any preferred orientation. The cells are manufactured by depositing a thin layer of *a*-Si on a substrate (glass, metal or plastic) from glow discharge, sputtering or chemical vapor deposition (CVD) methods. The most common method is by an RF glow discharge decomposition of silane (SiH_4) on a substrate heated to a temperature of 200–300°C. To produce *p*-silicon, diborane (B_2H_6) vapor is introduced with the silane vapor. Similarly phosphene (PH_3) is used to produce *n*-Silicon. The cell consists of an *n*-layer, an intermediate undoped *a*-Si layer, and a *p*-layer on a substrate. The cell thickness is about 1 μm. The manufacturing process can be automated to produce rolls of solar cells from rolls of substrate. Figure 9.23 shows an example of roll-to-roll *a*-Si cell manufacturing equipment using a plasma CVD method. This machine can be used to make multijunction or tandem cells by introducing the appropriate materials at different points in the machine.

9.4 DESIGN FOR REMOTE PHOTOVOLTAIC APPLICATIONS

Photovoltaic power may be ideal for a remote application requiring a few watts to hundreds of KW of electrical power. Even where a conventional electrical grid is available, for some applications, where uninterruptible or emergency standby power is nec-

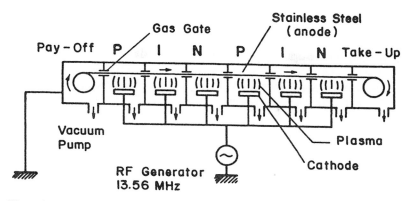

Figure 9.23. A schematic diagram of a roll-to-roll plasma CVD machine [8].

essary, photovoltaic power would be appropriate. Some examples of remote PV applications include water pumping for potable water supply and irrigation, power for remote houses, street lighting, battery charging, telephone and radio communication relay stations, and weather stations. Examples of some other applications include, electrical utility switching stations, peak electrical utility power where environmental quality is a concern, data acquisition systems and speciality applications such as ventilation fans and vaccine refrigeration.

The design of a PV system is based on some basic considerations for the application.

1. Which is more important, the daily energy output or the power (average or peak)?
2. Is a back-up energy source needed and/or available?
3. Is energy storage important? What type—Battery, pumped water etc.?
4. Is the power needed as AC or DC? What voltage?

There are three basic steps in the design of a PV system:

1. Estimation of load and load profile;
2. Estimation of available solar radiation; and
3. Design of PV system, including area of PV panels, selection of other components, and electrical system schematic.

Each of these steps will be explained in the following examples. These examples are based on Refs. [12] and [13].

9.4.1 Estimation of Loads and Load Profiles

Precise estimation of loads and their timings (load profile) are important for PV systems since the system is sized as the minimum required to satisfy the demand over a day. For example if power is needed for five different appliances requiring 200W, 300W, 500W, 1000W and 1500W respectively so that only one appliance is on at any one time and each appliance is on for an average of 1 hour a day, the PV system would be sized based on 1500W peak power and 3500Watt hours (Wh) of daily energy requirement. The multiple loads on a PV system are intentionally staggered to use the smallest possible system, since the capital costs of a PV system are the most important as opposed to the energy costs in a conventional fuel based system.

Example 9.5. *Daily load calculations.* How much energy per day is used by a remote weather station given the following load characteristics?

Load	Load Power (watts)	Run-time (hours/day)
Charge Controller	2.0	8
Data gathering	4.0	3
Modem (standby)	1.5	22.5
Modem (send/receive)	30.0	1.5

Solution.

Daily energy $= (2.0W)(8\text{ hr}) + (4.0W)(3\text{ hr}) + (1.5W)(22.5\text{ hr}) + (30W)(2.5\text{ hr})$
$\qquad\qquad = 106.75\text{ Wh}$
Daily Energy use is about 107 Wh per day.

Example 9.6. *Load calculations.* An owner of a remote cabin wants to install a PV power system. The loads in the home are described below. Assume that all lights and electronics are powered by AC. Find the daily and weekly peak and average energy use estimates. The system used is a 24-volt DC system with an inverter.

Lights	4, 23-watt compact fluorescent bulbs	On at night for 5 hours
Lights	6, 13-watt compact fluorescent bulbs	2 hours each (daytime)
Stereo	110 watts (amplifier), 15 watts (other)	On for 8 hours per week
Water pump	55 watts (3.75 amp start current)	Runs for 2 hours per day
Computer	250 watts (monitor included)	On for 1½ hours daily (weekend nights only)
Bathroom fan	40 watts (3.5 amp start current)	On for 1 hour per day
Microwave	550 watts (AC)–1000 W surge	On for 30 minutes per day

Solution. Loads need to be broken down according to: (1) run-time, (2) peak power, (3) night or day use, and (4) AC or DC loads. The load profile is as follows:

Load name	Power (W)		Run time (hours)		Energy (Wh)	
description	Average	Peak	Day	Week	Day	Week
Lights (AC)	(4)(23)	(4)(23)	5.0	35	460	3220
Lights (AC)	(6)(13)	(6)(13)	2.0	14	156	1092
Stereo (AC)	(1)(110)	(1)(110)	—	8	—	880
Pump (DC)	(1)(55)	(3.75A)(24V)	2.0	14	110	770
Computer (AC)	(1)(250)	(1)(250)	1.5	3	—	750
Fan (DC)	(1)(40)	(3.5A)(24V)	1.0	7	40	280
Microwave (AC)	(1)(550)	(1)(1000)	0.5	3.5	275	1925

Average DC load: $[770 + 280]/7 =$ **150 Wh/day**.
Average AC load: $[3220 + 1092 + 880 + 750 + 1925] / 7 =$ **1124 Wh/day**.
Peak DC load: max $[\{(3.5)(24) + 40\} :: \{(3.75)(24) + 55\}] =$ **145 watts**[1].
Peak AC load: $(1000) +$ max $[(4)(23) :: (6)(13)] + 250 + 100 =$ **1442 watts**[2].

9.4.2 Estimation of Available Solar Radiation

Methods of estimation of available solar radiation are described in Chapter 2. If long term measured solar radiation values are available at a location, Eqs. (2.56) to (2.64) can be used to estimate the average solar radiation per day. Otherwise, data for clear day can be used along with percent sunshine data (if available). For designing a PV system, a decision is made whether the PV panel will be operated as tracking the sun or will be fixed at a certain tilt and azimuth angle. For fixed panels, a tilt angle of latitude $+15°$ works best for winter and latitude $-15°$ for summer. To keep the panel fixed year round an angle equal to the latitude provides the maximum yearly energy (see Figure 9.24).

9.4.3 PV System Sizing

If meeting the load at all times is not critical, PV systems are usually sized based on the average values of energy and power needed, available solar radiation, and component efficiencies. This is known as the **heuristic approach**. It is important to note that a system designed by this approach will not give the best design but may provide a good start for a detailed design. A detailed design accounts for the changes in the efficiencies of the components depending on the load and the solar radiation availability and whether the system is operating in a PV-to-load, PV-to-storage, or storage-to-load mode.

Example 9.7. *Heuristic approach to PV system sizing.* A PV system using 50-watt, 12-volt panels with 6-volt, 125 amp-hour batteries is needed to power a home in Farmington, NM, with a daily load of 1700 watt-hours. System voltage is 24 volts.

---

[1]It can be assumed that the pump and fan will not start precisely at the same instant.
[2]It is assumed that the night and day lighting loads will not be on simultaneously.

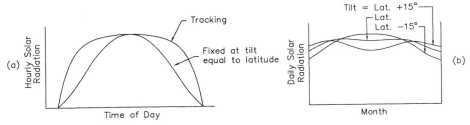

Figure 9.24. Solar radiation on panels at different tilt angles.

There are an average of 5 daylight hours in the winter. Specify the collector and storage values for the system using the heuristic approach.

Solution.

Load   = 1700 Wh/day,
Daylight hours   = 5 hrs/day,
Average Panel output = 50W,

$$\text{Number of panels} = \frac{1700 \text{ Wh/Day}}{(5 \text{ hr/day})(50\text{W/panel})} = 6.8, \text{ round off to 7 panels.}$$

Since the system voltage is 24V, but each panel produces only 12 volts an even number of panels will be needed. Therefore number of panels = 8.

Farmington, NM, is a very sunny location, so 3 days of storage is plenty. Assuming a battery efficiency of 75 percent and a maximum depth of discharge 70 percent,

Storage   = (1700)(3)/(0.75 × 0.7) = 9714 Wh,
Number of Batteries = (9714/Wh)/(125Ah × 6V)
                     = 13 (Rounded off to next whole number).

Since the system voltage is 24V, and each battery provides 6V, the number of batteries is increased to 16. In a detailed design, the efficiencies of battery storage, inverter, and the balance of system (BOS) must be accounted for. The following example shows how these efficiencies increase the energy requirements of the PV panel.

Example 9.8. *System operating efficiency.* Using the cabin electrical system from Example 9.6, calculate the overall system efficiency for each operating mode possible for the system. Estimate the amount of energy required per day for the system. When load timing (day or night), assume half of the load runs during the day and half runs at night. The inverter used has a component efficiency of 91 percent, the battery efficiency is 76 percent, and the distribution system efficiency is 96 percent.

Solution. From the example, the loads are:

Average DC load: 150Wh/day,
Average AC load: 1124 Wh/day.

The various system efficiencies are:

PV to load (DC): 0.96 (day, DC),
Battery to load (DC): (0.76)(0.96) = 0.73 (night, DC),
PV to load (AC): (0.96)(0.91) = 0.874 (day, AC),
Battery to load (AC): (0.76)(0.91)(0.96) = 0.664 (night/AC).

Expected day and night loads are:

Day (DC): (0.5)(110) + (0.5)(40) = 75 Wh/day,
Night (DC): (0.5)(110) + (0.5)(40) = 75 Wh/day,
Day (AC): (156) + (0.5)(880 + 750) /7 + (0.5)(275) = 409.9 Wh/day,
Night (AC): (460) + (0.5)(880 + 750)/7 + (0.5)(275) = 713.9 Wh/day.

Without considering system efficiency, the daily energy requirement is:

$$E_{day} = (150) + (1124) = 1274 \text{ Wh/day.}$$

The expected daily energy requirement is:

$$E_{day} = (75)/(0.96) + (75)/(0.73) + (409.9)/(0.874) + (713.9)/(0.664)$$

$$E_{day} = 1725 \text{ Wh/day}$$

The actual energy requirement is 35 percent higher than simple calculation.

9.4.4 Water Pumping Applications

Water pumping for drinking water or irrigation at remote locations is an important application of PV. For a simple schematic shown in Figure 9.25 the power needed to pump water at a volumetric rate $\dot{V}$ is given by $P = \rho \dot{V} g H / \eta_p$ where ρ is the density of water, g is the acceleration due to gravity, and η_p is the pump efficiency. The static head H_s is (A + B). In case the water level is drawn down, the static head would be (A + B + C). The pump must work against the total head H which includes the dynamic head H_d also,

$$H_d = H_f + \frac{v^2}{2g}$$

where H_f is the frictional head loss in the pipe and the bends and v is the velocity of the water at the pipe outlet. The pump efficiency η_p is a function of the load (head and flowrate) and is available as a characteristic curve from the manufacturer. For general

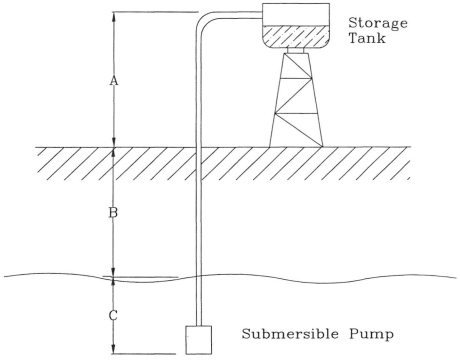

Figure 9.25. Water pumping using a submersible pump.

design purposes typical values given in Table 9.3 may be used. The table lists two basic types of pumps, centrifugal and positive displacement. These pumps can be driven by AC or DC motors. DC motors are preferable for the PV applications, because they can be directly coupled to the PV array output. Centrifugal pumps with submersible motors are the optimum for PV applications because of their efficiency, reliability and economy. However, for deep wells Jack pumps may be necessary. Jack pumps are the piston type of positive displacement pumps that move chunks of water

Table 9.3. Typical range of pump performance parameters.

Head (m)	Type pump	Wire-to-water efficiency (%)
0–5	Centrifugal	15–25
6–20	Centrifugal with Jet	10–20
	Submersible	20–30
21–100	Submersible	30–40
	Jack pump	30–45
>100	Jack pump	35–50

with each stroke. They require very large currents, therefore they are connected through batteries.

Example 9.9. A PV system is designed to pump water for livestock in the vicinity of El Paso, TX. The following information is available:

Site:	Near El Paso, TX,
	32° 20′ N 106° 40′ W 1670m,
Ambient Temp:	−5 to 45°C
Water Source:	Cased Borehole 15 cm
Static Head:	106 m
Maximum Draw-down:	8m
Water Required:	8325 l/day, June to Aug.

Solution. Since the required head is very high, a deep well jack pump with DC motor is needed. Assuming a friction factor of 0.05 the total head = 106 + 8 + (106) (0.05) = 122 m. We select a 75V DC Jack pump with an average efficiency of 0.45.

Daily Energy Required $= \rho \dot{V} gH/\eta_p$

$$= (8325l/day)\left(\frac{1kg}{l}\right)\left(9.81 \; \frac{m}{sec^2}\right)(122m)\frac{1}{0.45}$$

$$= 22.14 \times 10^6 J/day$$

$$= 6,150 \; Wh/day.$$

Since the system will be used from June to August, a tilt angle of lat. −15° would be optimum. Daily solar radiation at El Paso for 1 axis tracking and tilt angle of 17° is:

June	11.51	kWh/m²-day
July	10.58	KWh/m²-day
August	10.02	KWh/m²-day

Since August has the minimum insolation, the panel area will be based on insolation for this month. We select PV panels with the following specifications (this information may be obtained from a manufacturer):

Voltage	$V_{oc} = 21.7$ V	V_m @ 25° = 17.4V
Current	$I_s = 3.5$A	$I_{L.m} = 3.1$A

Temperature Correction Factor for voltage $= 1 - 0.0031 (T_c - T_{ref})$

Assuming that the panels will operate at a maximum temperature of 45° + 15° = 60°C, the voltage at the highest expected temperature

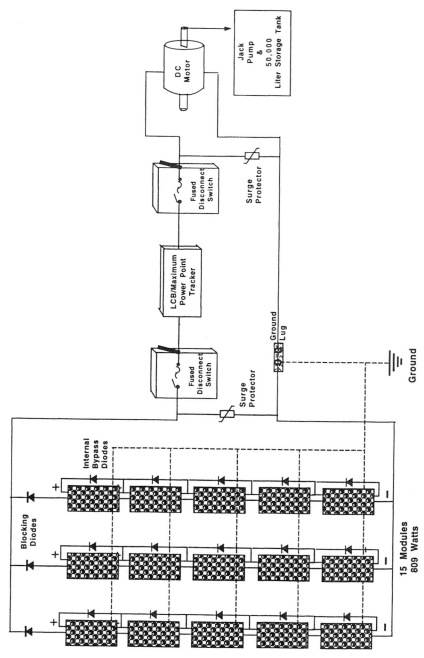

Figure 9.26. A system schematic for Problem 9.9.

442

$$= 17.4 [1 - 0.00331 (60 - 25)]$$

$$= 15.5V$$

Voltage Required $= 75V$.

Therefore, number of panels in series $= 75/15.5 \approx 5$.

Power output/panel $= 3.1$ A $\times$ 15.5 V $= 48.05$ W at a standard insolation of 1000 W/m²

$$\text{Power output/panel/day} = 48.05W \times \frac{10.02KWh/m^2 - day}{1KW/m^2}$$

$$= 481.5 \text{ Wh/day}.$$

Assuming an overall efficiency of 90 percent due to insolation times, wiring etc.

$$\text{Number of panels required} = \frac{6,150 \text{ Wh/day}}{0.9 \times 481.5 \text{ Wh/day}} \approx 15$$

The array will consist of 3 parallel rows of 5 panels each in series.

As a check, the daily water pumping rate for August

$$= \frac{15 \times 481.5 \times 0.9 \text{ Wh/day} \cdot 3600 \text{ sec/hr}}{\left(9.81 \dfrac{m}{sec^2}\right)(1 \text{ Kg}/l)(122 \text{ m})} \times 0.45$$

$$= 8798 \; l/\text{day}$$

Therefore, the system will meet the requirement. A system schematic is in Fig. 9.26.

PROBLEMS

9.1 Find the wavelength of radiation whose photons have energy equal to the band gap of GaAs.

9.2. What is the theoretical maximum efficiency of conversion if blue light of wavelength 0.45 μm is incident on a GaAs solar cell?

9.3. Find the theoretical maximum overall efficiency of GaAs solar cells in space.

9.4. The reverse saturation current I_o of a silicon cell at 40°C is 1.8×10^{-7} amp. The short circuit current when exposed to sunlight is 5 amp. From this information compute

a. open circuit voltage,

b. max. power output of the cell, and

  c. the number of 4 cm × 4 cm cells needed to supply 100 W at 12 V. How must
     the cells be arranged?

9.5. At what efficiency is a photovoltaic array running if insolation on the collector
     is 650 W/m², the total collector area is 10 m², the voltage across the array is 50
     volts, and the current being delivered is 15 amps?

9.6. If a PV array has a maximum power output of 10 watts under an insolation level
     of 600 W/m², what must the insolation be to achieve a power output of 17 watts?
     Would you expect the open circuit voltage to increase or decrease? Would you
     expect the short-circuit current to increase or decrease?

9.7. A PV battery system has an end-to-end efficiency of 77 percent. The system is
     used to run an all-AC load that is run only at night. The charge controller effi-
     ciency is 96 percent and the inverter efficiency is 85 percent. How much energy
     will need to be gathered by the PV array if the load is 120 watts running for 4
     hours per night?

9.8. If the average output of the PV system in problem 9.7 is 200 watts, the load is
     changed to run during the day, how much PV output energy is needed for the
     same load conditions? Assume that the battery bank is at 100 percent charge and
     that input efficiency is equal to output efficiency.

9.9. For the system in problem 9.7, how many hours of sunlight are needed to ensure
     that the battery bank is at 100 percent charge at the end of the day assuming the
     same load?

Problems 9.10–9.14.   The owner of a small cabin would like to convert her home to
PV power. She has the following equipment and associated run times:

Household equipment	Power (watts)	Run time (day) (hours)	Run time (night) (hours)
Lighting (DC)	25	2	4
Stereo (AC)	40	3	2
Refrigerator (DC)	125	3*	3*
Water Pump (DC)	400	1.5	0.5
Alarm Clock (DC)	8	12	12
Computer (AC)	250	3	0
Printer	175	0.25	0
Outdoor Safety Lights	48	0	8
Answering Machine (AC)	7	12	12
Coffee Pot (AC)	1200	0.25	0

   *The refrigerator is assumed to run 25 percent of the time

9.10. a. What is the homeowner's daily energy requirement as measured from the
         load?
      b. If she replaces her alarm clock with a wind-up clock, how much energy per
         day will she avoid using?
      c. What would you suggest she do to cut back her daily load?

9.11. How many 50-watt panels will the owner require assuming battery storage is 75
      percent efficient and all loads are DC (no inverter)?

a. for a stationary system "seeing" 5 hours of sunlight per day?

b. for a tracking system "seeing" 8 hours of sunlight per day?

9.12. For the loads listed:

a. What size inverter (peak watts) should she purchase?

b. If the inverter is 88 percent efficient, how much more daily energy is required from the PV array as compared to an all-DC system?

9.13. The homeowner decides to hire you to design a system for her. She has cut a deal with a local solar supplier for the following equipment. Specify the system and provide a line diagram showing system connection.

PV panels:	42-watts, nominal 12-volt,
Batteries:	125 Ah, 6-volt, end-to-end efficiency = 72 percent,
Charge Controller:	95 percent efficient, 12-volt, and
Inverter:	90 percent efficient, sizes of 500, 1000, 2000 and 4000 watts available, 12-volt input.

9.14. Through your connections, you have the following equipment available. Re-design the system

PV panels:	51-watts, nominal 12-volt,
Batteries:	200 Ah, 12-volt, end-to-end efficiency = 78 percent,
Charge Controller:	97 percent efficient, 24-volt, and
Inverter:	91 percent efficient, sizes of 500, 1200, 2500 and 5000 watts available, 24 − volt input.

9.15. A flashing beacon is mounted on a navigation buoy in the shipping channel at a port at 30° N latitude. The load consists of a single lamp operating 1.0 second on and 3.6 seconds off during the hours of darkness. Hours of darkness vary from 9.8 hours in July to 13.0 hours in December. The lamp draws 2 amps at 12 V when lighted. A flasher controls the lamp and draws 0.22 amps when the lamp is on. There is a surge current of 0.39 amps each time the flasher turns on. This current flows approximately 1/10 of the time the flasher is on. The design has 14 days of battery capacity. Provision has to be made to disconnect the load if the battery voltage drops below 11 V.

The available module has rated voltage of 17.2 V at 25°C (15 V @ 55°C) and 2.3 amps @ 1 KW/m². The available battery has a rated capacity of 105 Ah at 12 V. Assume maximum depth of discharge 30 percent. Design the PV system.

9.16. Design a PV system for the following application: A refrigerator/freezer unit for vaccine storage in a remote island of Roatan, Honduras (16° N lat., 86° W long., Temp. range 15–30°C).

Two compressors—one each for refrigerator and freezer.

Each compressor draws 5 amps @ 12 V.

Compressors remain on for	Summer	Winter
Refrigerator	9 hrs/day	5 hrs/day
Freezer	7 hrs/day	4 hrs/day

Design the PV system using the panels and the batteries described in problem 9.15.

9.17. A village in Antigua (17°N, 61°W, 15–30°C Temp. range), West Indies, requires 5000 gallons of water per day for community water supply. Assuming a year round average insolation of 8 kWh/day, design the system using the following components:

PV Panels: Solarex panels 17.5 V and 3.6 A at 1000 W/m^2 and 25°C

Pump: Grundfos multistage pump input 105 V DC, 9 Amps, 30 percent efficiency

REFERENCES AND SUGGESTED READINGS

1. Angrist, S.W. 1976. *Direct Energy Conversion.* Boston, MA: Allyn and Bacon, Inc., 3rd Edition.
2. Böer, K.W. 1990. *Survey of Semiconductor Physics.* New York: Van Nostrand Reinhold.
3. Bube, R.H. 1960. *Photoconductivity of Solids.* New York: John Wiley & Sons, Inc.
4. Crandall, R., and W. Luft. 1995. The future of amorphous silicon photovoltaic technology. NREL/TP-441-8019. Golden, CO: National Renewable Energy Laboratory.
5. Crossley, P.A., G.T. Noel, and M. Wolf. 1968. Review and evaluation of past solar cell development efforts. *RCA Astro-Electronics Division Report,* AED R-3346, Contract NASW-1427.
6. Florida Solar Energy Center. 1991. Photovoltaic system design. *FSEC-GP-31–86.* Cocoa Beach, FL: Florida Solar Energy Center.
7. Garg, H.P. 1987. *Advances in Solar Energy Technology,* vol. 3. Dordrecht, Holland: D. Reidel Publishing Company.
8. Lasnier, F., and T.G. Ang. 1990. *Photovoltaic Engineering Handbook.* New York: A. Hilger Publishing.
9. Loferski, J.J. 1963. Recent research on Photovoltaic Solar Energy Converters. *Proc. of the IEEE* 51: 667–674.
10. Moore, C.E. 1952. *Atomic Energy Levels.* vol. 2, National Bureau of Standards Circular 467.
11. Nijs, J., et. al. 1997. Energy payback time of crystalline silicon solar modules. In *Adv. in Sol. Energy,* vol. 11, Karl W. Böer, ed. Chapter 6, 291–327.
12. Post, H.N., and V.V. Risser. 1995. *Stand-Alone Photovoltaic Systems—A Handbook of Recommended Design Practices.* Albuquerque: Sandia National Lab. Report SAND-87-7023.
13. Taylor, M., and J. Kreider. 1998. *Solar Energy Applications,* 3rd ed. Boulder, CO: Univ. of Colorado Bookstore.

SOLAR PHOTOCHEMICAL APPLICATIONS

The use of solar energy for the production of food, fiber, and heat has been known to mankind for a long time. Research over the last five decades has also made it possible to produce mechanical and electrical power with solar energy. Although the potential of solar radiation for disinfection and environmental mitigation has been known for years, only recently has this technology been scientifically recognized and researched.

For anyone who has observed colors fading over long exposure to sunlight, certain materials deteriorating in the sunlight, or skin getting sunburned, it is not hard to imagine sunlight causing reactions that could be used beneficially to break up toxic chemicals. Older civilization considered it essential that human dwellings be designed to allow sunshine in. This may have been for disinfection or to keep the growth of microorganisms in check.

Research over the last three decades has not only confirmed the capability of sunlight for detoxification and disinfection, but also accelerated the natural process by the use of catalysts. When sunlight is used to cause a chemical reaction by direct absorption, the process is called **photolysis**. If the objective is achieved by the use of catalysts, it is known as **photocatalysis**. Recent research has concentrated mainly on photocatalytic reactions for detoxification, disinfection and production of hydrogen. Hydrogen energy systems, especially fuel cells, are being considered as a clean solution to transportation needs. Hydrogen can also provide solar energy storage that can be transported over long distances using pipelines and supertankers, and regionally in special containers via road and rail [67].

10.1 PHOTOCATALYTIC REACTIONS

The fundamentals of how photons affect molecules when absorbed has been described in Chapter 9 on photovoltaics. Basically, if a photon has more energy than the bandgap of a material, it will free up an electron when absorbed. The bandgap of a material represents the difference in the energy of the electrons in the valence band of the atom and the conduction band. As an electron moves up from the valence band to the conduction band and becomes free, it leaves a positive hole behind. The positive hole and the negative electron may recombine with the release of thermal energy, unless they interact with neighboring atoms of other materials to cause chemical reactions. Such reactions are known as **photoreactions**, since they are initiated by photons. If the reaction involves atoms or molecules that act as catalysts, the reaction is known as photocatalytic. Sunlight may be used in both photolytic and photocatalytic reactions that could result in useful applications, such as the oxidation of toxic organic chemicals or production of hydrogen.

As noted in Chapter 9, the energy ϵ of a photon is

$$\epsilon = h\upsilon = \frac{hc}{\lambda} \tag{10.1}$$

where h is the Planck's constant (6.625×10^{-34} J-s), υ is the frequency, λ the wavelength and c is the speed of light (3×10^8 m/s).

Titanium Dioxide (TiO_2) has a bandgap of 3.2eV; therefore the wavelength λ of a photon with energy equal to the bandgap of TiO_2 is

$$\lambda = \frac{hc}{\epsilon}$$

$$= \frac{(6.625 \times 10^{-34} \text{ J} - \text{s}) \times (3 \times 10^8 \text{ m/s})}{3.2\text{eV}} \times \frac{1\text{eV}}{(1.6 \times 10^{-19} \text{ J})}$$

$$= 0.388 \times 10^{-6}\text{m or } 0.388\mu\text{m or } 388 \text{ nm}.$$

Therefore, a photon of sunlight with a wavelength of 388nm or less (i.e., energy 3.2eV or higher) will excite an electron from a valence band (vb) to a conduction band (cb) when absorbed, resulting in a free electron (e^-) and a positive hole (hole+) (6,34).

$$TiO_2 + h\upsilon(\text{photon}) \rightarrow \text{hole}_{vb}^+ (TiO_2) + e_{cb}^- (TiO_2). \tag{10.2}$$

The holes (hole+) and the electrons (e^-) are both highly energetic and mobile. They may recombine and release heat or migrate to the surface. On the surface, they may react with adsorbed molecules of other species and cause a reduction or oxidation of that species.

Since recombination, in the bulk or near the surface, is the most common reaction, the **quantum yields** (molecules reacted/photons absorbed) of most photolytic reactions are low. Separation of the electron-hole pairs is aided by formation of a potential gradient near the surface of the semiconductor. This *space charge region* results from the different electrical potential of the solid semiconductor and the liquid phase of the ambient solution. For TiO_2 this potential drives valance band holes toward the particle surface and conduction band electrons away from the surface. The electrons and holes

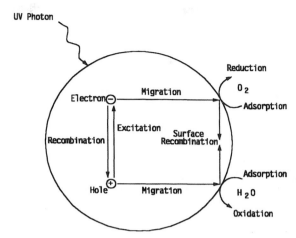

Figure 10.1. Oxidation-Reduction reaction occuring at the TiO_2 surface upon irradiation.

at the surface become active sites for oxidation and reduction of adsorbed molecules, as illustrated in Figure 10.1. The positive holes cause oxidation of the surface adsorbed species while the electrons cause reduction. Both reactions must take place in order to maintain electroneutrality. Thus, if the objective is the oxidation of organics, the electrons must be consumed in a reduction reaction such as absorption by oxygen molecules to form a superoxide, in order to keep the holes available for oxidation. On the other hand, if the objective is the reduction and recovery of metals, all other reducible species, such as oxygen, must be eliminated or kept away. The following equations describe the oxidation and reduction reactions [4,11,59]:

$$TiO_2 + h\nu \rightarrow TiO_2(e_{cb}^- + hole_{vb}^+), \tag{10.3}$$

$$TiO_2(e_{cb}^- + hole_{vb}^+) \overset{recomb}{\rightarrow} TiO_2 + (heat). \tag{10.4}$$

If the objective is oxidation of an organic pollutant, the electron is consumed by an adsorbed compound/molecule:

$$O_2 + e_{cb}^- \Rightarrow O_2^{-\cdot}. \tag{10.5}$$

If a water molecule is adsorbed on the surface of TiO_2, it may exist as hydroxyl and hydrogen ions.

$$H_2O \Rightarrow OH^- + H^+ \tag{10.6}$$

The negatively charged hydroxyl ion adsorbed on the surface of TiO_2 gives up its negative charge (electron) to the positive hole to regenerate the neutral TiO_2 and it in turn becomes a neutral hydroxyl radical ($OH^\cdot$).

$$TiO_2(hole_{vb}^+) + OH_{ads}^- \Rightarrow OH^{\cdot} + TiO_2 \qquad (10.7)$$

$$2O_2^{-\cdot} + 2H_{aq}^+ \rightarrow H_2O_2 + O_2 \qquad (10.8)$$

H_2O_2 may yield additional hydroxyl radicals by any of the following reactions:

$$H_2O_2 + e_{cb}^- \rightarrow OH^{\cdot} + OH^-, \qquad (10.9)$$

$$H_2O_2 + O_2^{-\cdot} \rightarrow OH^{\cdot} + OH^- + O_2 \qquad (10.10)$$

$$H_2O_2 + h\upsilon \rightarrow 2OH^{\cdot} \qquad (10.11)$$

The hydroxyl radical is a very potent oxidizing agent which can oxidize a pollutant organic molecule $C_lH_mX_n$, into CO_2 and H_2O, directly or through intermediate compounds as:

$$C_lH_mX_n + yOH^{\cdot} + zO_2 \rightarrow lCO_2 + mH_2O + nHX \qquad (10.12)$$

Table 10.1 shows the oxidation power of various species relative to that of chlorine. The oxidation power of the hydroxyl radical is ranked second among these known strong oxidizing agents. This demonstrates the hydroxyl radical's potential for oxidizing normally hard to destroy pollutants like halogenated organics, surfactants, herbicides and pesticides.

If the objective is to reduce and precipitate a metal from a metal compound, the electron is consumed by the metal compound as:

$$ne_{cb}^- + M^{n+} \rightarrow M^{\circ} \qquad (10.13)$$

Table 10.1. Oxidation power for various species relative to chlorine [21].

Species	Relative oxidation power
Flourine	2.23
Hydroxyl radical	2.06
Atomic oxygen (singlet)	1.78
Hydrogen peroxide	1.31
Perhydroxyl radical	1.25
Permanganate	1.24
Hypobromous Acid	1.17
Chlorine dioxide	1.15
Hypochlorous Acid	1.10
Chlorine	1.00
Bromine	.80
Iodine	.54

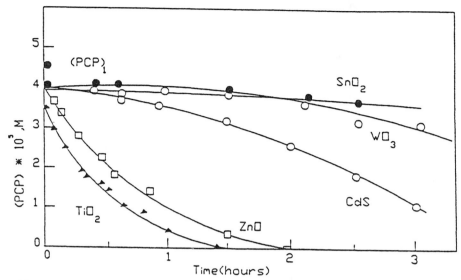

Figure 10.2. Photodegradation of pentachlorophenol (PCP) in the presence of various semiconductors [5].

Although the above photocatalytic reactions have been explained based on TiO_2, almost all of the semiconductors may be used with varying degrees of usefulness. Figure 10.2 shows a comparison of some of the catalysts for photodegradation of a hazardous organic compound.

10.2 SOLAR PHOTOCATALYTIC DETOXIFICATION

Solar photocatalytic detoxification refers to the destruction of hazardous pollutants from the environment by solar photocatalytic oxidation or reduction reactions. Solar detoxification has shown great promise for the treatment of groundwater, industrial wastewater, and contaminated air and soil. In recent years the process has also shown great potential for disinfection of air and water, making possible a number of applications. Research studies on the photocatalytic oxidation process have been conducted over at least the last three decades [8]. Blake [8] published a bibliography listing 660 publications, and even this long list is incomplete. He compiled a list of organic and inorganic compounds with references which shows over 300 compounds, including about 100 on the US EPA priority pollutant list, that can be treated by the photocatalytic process. Blake also lists 42 review articles that cover various aspects of photocatalytic chemistry and technology including: Blake et al. [10]; Kamat [39]; Legrini et al. [41]; Ollis, Pellizetti and Serpone [56]; and Venkatadri and Peters [73].

TiO_2 has been the most commonly used photocatalyst. The use of TiO_2 in water detoxification was first demonstrated by Carey et al. in 1976. They showed that polychlorinated biphenyls, PCBs, were dechlorinated in aqueous suspensions of TiO_2. Other semiconductors have also been investigated as alternatives to TiO_2, however, TiO_2 has generally been shown to be the most active [5]. Figure 10.2 shows a typical result from such a comparative study. Only ZnO has an activity similar to TiO_2. However, zinc oxides dissolve in acidic solutions which make them inappropriate for technical applications. Titanium dioxide, on the other hand, is insoluble under most conditions, photostable and nontoxic.

The energy needed to activate TiO_2 is 3.2 eV or more, which corresponds to near UV radiation of a wavelength of 388 nm or less. As 4–6 percent of sunlight reaching the earth's surface is characterized by these wavelengths, the sun can be used as the illumination source [30,31]. However, since UV radiation does form just 4–6 percent of the usable solar spectrum, recent research has been aimed at improving the catalyst's performance by improving the reaction kinetics, increasing the useful wavelength range to utilize larger portions of the solar spectrum, developing appropriate reactors, and finding new engineering applications of the process for practical problems.

The following sections describe the engineering aspects of solar detoxification applications, including reactor design, modeling of kinetic reactions, system design, and industrial and commercial applications.

10.3 SOLAR REACTORS

Designs of solar photocatalytic reactors have followed the well-known designs of solar thermal collectors including concentrating and nonconcentrating designs. The key differences are: (1) the fluid to be treated in the reactors must be exposed to UV solar radiation; therefore, the absorber must be transparent to UV solar radiation; and, (2) no insulation is needed, since temperature does not play a significant role in the photoreaction. As a result, the first engineering scale outdoor reactor developed was a simple conversion of a parabolic trough solar thermal collector. The conversion replaced absorber/glazing tube combination of the thermal collector with a simple pyrex glass tube through which contaminated water can flow (Figure 10.3). Pacheco and Tyner [58] used this reactor to treat water contaminated with Trichloroethylene (TCE). The catalyst, TiO_2 powder, was mixed with contaminated water to form a slurry which was passed through the pyrex glass tube (reactor tube) located at the focal line of the parabolic trough. Since that time, a number of reactor concepts and designs have been advanced by researchers all over the world. Basically all these reactors fall into the following categories:

A. Reactor Configuration:
   1. Concentrating Reactors,
   2. Nonconcentrating Reactors.
B. Catalyst Deployment:
   1. Fixed Catalyst,
   2. Slurry,
   3. Neutral Density Large Particles.

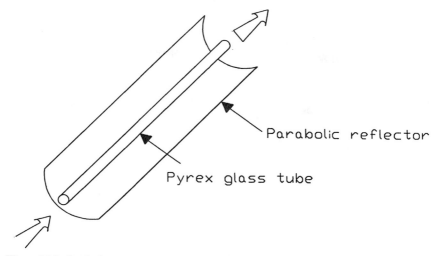

Figure 10.3. Parabolic trough solar photocatalytic reactor.

10.3.1 Concentrator Reactors

As mentioned before, Pacheco and Tyner [58] developed a simple modification of the parabolic trough solar thermal collector which worked successfully. They used this reactor in experiments with the catalyst deployed as a slurry. Since then, another configuration of this reactor has been developed and used by researchers at the Sandia National Laboratory (SNL) and at the National Renewable Energy Laboratory (NREL), in which the catalyst is fixed on a loosely woven fiberglass matrix which is inserted into the reactor tube. Both of these configurations have worked well. Parabolic concentrating types of reactors have been used for applications such as ground water remediation [9] and metal removal from water [59].

A disadvantage of concentrating reactors is their inability to use diffuse solar radiation. For solar thermal applications, this limitation is not a major problem since diffuse radiation forms a small fraction of the total solar radiation. However, solar photocatalytic detoxification with TiO_2 as a catalyst uses only the UV portion of the solar radiation, about 4–6 percent of the total spectrum. As much as 50 percent or more of the UV radiation can be diffuse, especially at locations with high humidity or during cloudy or partly cloudy periods. Therefore, concentrating reactors would be more useful at dry, high direct insolation locations. Another disadvantage of the concentrating reactors is low quantum efficiency [4,43], resulting in a reaction rate constant dependence on the intensity of UV radiation as:

$$k \propto I^{1/2} \tag{10.14}$$

These disadvantages tend to favor the use of nonconcentrating reactors. However, a big advantage of the concentrating reactors is that the reactor tube area is very small, which allows the use of expensive but high quality UV transmitting materials such as pyrex glass. Such materials may be used without increasing the cost too much, while also improving the lifetime of use of these reactors. Ultimately, the choice of these types of reactors or any other types will be based on the overall economics of the solar detoxification system. As explained later in this chapter, under certain conditions the reactor capital cost may be a small portion of the overall costs. In that case the balance of the system costs, the catalyst costs, and the operation and maintenance costs dictate the economics of the solar detoxification process.

Other concentrating reactors that have been constructed and tested include very high concentration, high temperature solar furnaces (Fig. 10.4), and a heliostat concentration system with a vertical falling film reactor [72]. The high concentration, high temperature solar furnaces combine solar photolytic and thermal effects and have been demonstrated to work well for gas phase detoxification as well as other applications.

The solar catalytic steam reforming furnace shown in Figure 10.4, uses a parabolic dish to concentrate the sunlight into a reactor through a quartz window, where it heats the reactor absorber to 700–1,000°C. The toxic organic waste is destroyed by a steam reforming process over a rhodium catalyst supported by a porous ceramic absorber. A version of this direct catalytic absorption receiver (DCAR) reactor has been tested at the SNL solar furnace for engineering scale experiments to demonstrate various steam reforming reactions including CO_2/methane, steam/methane, steam/propanol, and steam/TCE (68). Another version of this system was built in Ger-

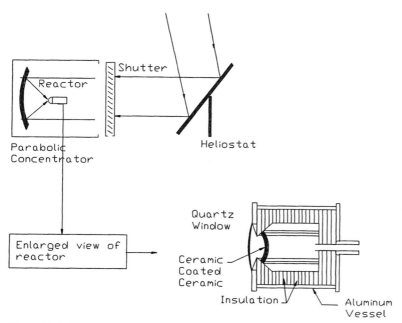

Figure 10.4. Direct catalytic absorption steam reforming furnace. A furnace based on a similar concept was also built at NREL [24,25].

many by a joint collaboration involving the SNL of USA and the *Deutsche Forsch-ungsanstalt für Luft- und Raumfahrt* (DLR) of Germany. A concentrating furnace has been built at NREL which has been used for detoxification tests for various hazardous chemicals [24–26]. Although the high concentration, high temperature solar furnaces have worked well, the major impediment to their widespread use is the high capital cost.

10.3.2 Non-Concentrating Reactors

Non-concentrating reactors have a major advantage over concentrating reactors because they can utilize the diffuse part of solar UV radiation in addition to the beam part. While diffuse solar radiation forms only about 10–15 percent of the total solar radiation over the whole solar spectrum, that is not the case for UV solar radiation. Since UV is not absorbed by water vapor, it can be as much as 50 percent of the diffuse solar radiation at certain locations, especially locations with high atmospheric humidities [66]. Moreover, the quantum efficiency of the photocatalytic process due to nonconcentrated incident radiation is also higher than that due to a concentrated beam. In addition, nonconcentrating reactors have the potential to be simple in design and low in cost. A disadvantage of the nonconcentrating reactors is the requirement of a much larger reactor area (although the total solar aperture would be less) than the concentrating reactors. Researchers have proposed a number of different designs of nonconcentrating solar reactors. They are:

1. Flat Plate
2. Tubular
3. Falling Film
4. Shallow Solar Pond

Researchers at the University of Florida have developed a number of nonconcentrating solar reactors including trickle-down flat plate, pressurized flat plate, tubular (rigid and inflatable), free falling film, and shallow solar pond.

10.3.3 Flat-Plate Reactors

Figure 10.5 shows a trickle-down flat plate reactor with a thin film glazing [75]. This reactor consists of a back plate on which water trickles down from a spray bar at the top. A woven mesh (fiberglass) covers the back plate in order to damp out surface waves as the water trickles down and to even out the flow. This allows a thin even film of water. A UV transparent glazing prevents any evaporation from the flow. This design allows contaminated water, mixed with the catalyst particles as a slurry, to be treated as the water trickles down when exposed to the sun. The reactor can be operated in a fixed catalyst configuration by replacing the plain woven mesh with a mesh with the catalyst fixed on it.

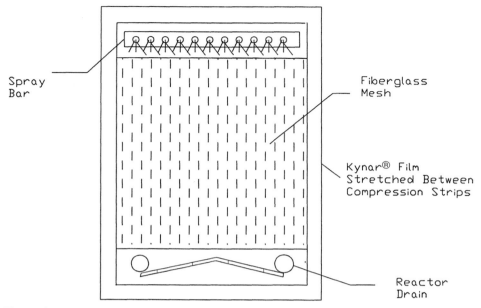

Spray
Bar

Fiberglass
Mesh

Kynar® Film
Stretched Between
Compression Strips

Reactor
Drain

Figure 10.5. Trickle-down flat plate reactor [75].

10.3.4 Tubular Reactors

Because of the weight of water, the glazing of the pressurized flat plate requires additional structural support. A simple concept of such a reactor is shown in Figure 10.6 [27]. In this system the contaminated water flows through the small channels between the lower and the upper headers. A variation of this concept is to substitute rigid transparent tubes in place of rectangular channels (Figure 10.7). Such tubular reactors were tested at the University of Florida to treat water contaminated by volatile organic compounds [53]. These reactors were also tested in the field at Tyndall Air Force Base to treat groundwater contaminated with fuel, oil and lubricants [30,31]. Figure 10.8 shows the performance of this system.

A simple reactor concept consists of transparent inflatable tubes connected in parallel between two headers [27]. As the water flows through the reactors under pressure, the tubes inflate. Large areas of this reactor can be rolled or folded into a small volume for portability and transported for on-site use.

10.3.5 Shallow Solar Ponds

Shallow pond-type reactors, developed at the University of Florida, may be constructed on-site especially for industrial wastewater treatment [7]. Since industries already use holding ponds for microbiological treatment of wastewater, shallow solar ponds can be used for the front end or the back end of a combined solar/microbiologi-

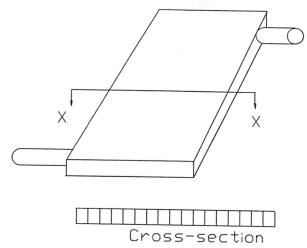

Cross-section

Figure 10.6. Pressurized flat plate reactor [27].

cal treatment of wastewater. These reactors would be ideal for wastewater treatment in industries such as pulp and paper, textiles, pharmaceuticals, and chemicals.

The reactor concept is simple, as shown in Figure 10.9. The reactor can be operated in a slurry or fixed catalyst configuration. If TiO_2 is used as a catalyst in a slurry configuration, it settles down to the bottom, if it is not continuously mixed. While the disadvantage of this configuration is that continuous mechanical mixing is needed, the advantage is that after the catalyst settles down, the treated water can be removed from the top without filtration. Bedford et al. tested shallow pond reactors in both slurry and

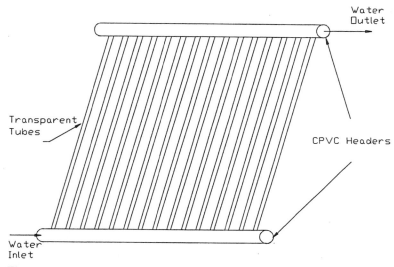

Figure 10.7. Flat plate reactor with rigid transparent tubes.

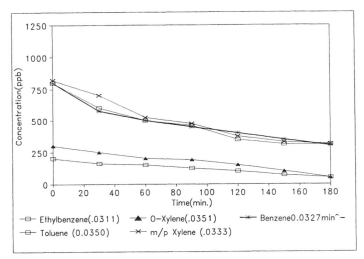

Figure 10.8. Destruction of BTEX in groundwater at Tyndall AFB using nonconcentrating tubular reactors [30].

fixed configuration, and various area-to-depth ratios, for treating contaminated water. The reactors worked extremely well under various insolation conditions (sunny, partly cloudy and cloudy).

10.3.6 Falling-Film

Some of the other nonconcentrating reactors developed include a vertically free-falling film reactor developed by Goswami [27]. The vertical film of this reactor is open to the atmosphere on both sides, which allows it to make maximum use of the diffuse atmo-

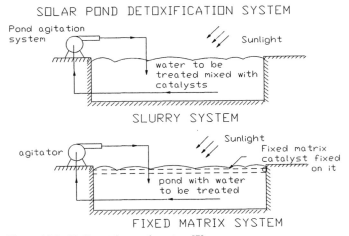

Figure 10.9. Shallow solar pond reactors [7].

spheric radiation. The integrity of the film is maintained by means of vertical strings appropriately spaced apart.

Finally, solar detoxification would work just as well, on-site if the application involves treatment of large bodies of contaminated water, such as spills in a lake, or the sea. Heller and Brock [35] developed neutral density spheres coated with TiO_2 which can be floated in large bodies of water to treat the contamination in situ in the presence of the sun [36].

10.4 KINETIC MODELS

An understanding of reaction rates and how the reaction rate is influenced by different parameters is important for the design and optimization of an industrial system. The rate of photolytic degradation depends on several factors including illumination intensity, catalyst type, oxygen concentration, pH, presence of inorganic ions, and concentration of organic reactant. As is typical of many photoassisted reactions, the effect of temperature is small. Figure 10.10 shows a typical change in pollutant concentration C with time t.

The destruction rates of organics in photocatalytic oxidation have been modeled by different kinetic models. Langmuir-Hinshelwood (L-H) kinetics seem to describe many of the reactions fairly well. The rate of destruction is given by:

$$-\frac{dC}{dt} = \frac{K_1 K_2 C}{(1 + K_2 C)} \tag{10.15}$$

In the ideal case for which the L-H model is derived, C is the bulk solute concentration, K_1 the reaction rate constant, K_2 the equilibrium adsorption constant, and t represents time. In the photocatalytic system, these definitions are less clear due to the importance of reactive radical species.

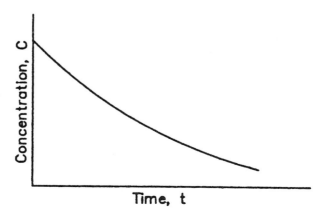

Figure 10.10. Typical Concentration-time history.

The L-H reaction rate constants are useful for comparing the reaction rates under different experimental conditions. Once the reaction constants K_1 and K_2 have been evaluated, the disappearance of reactant can be estimated if all other factors are held constant.

For low solute concentrations, C is almost equal to zero, which makes the denominator in the L-H expression equal to one. This reduces the L-H expression to a pseudo-first-order expression:

$$-\frac{dC}{dt} = K_1 K_2 C = kC. \tag{10.16}$$

This equation has been shown to apply to many photocatalyzed reactions [4]. Industrial pollutant levels are typically on the order of ppm, low enough for the reaction rate to follow pseudo first-order kinetics.

For the design of a system, the reaction rate constant k can be determined from experimental data as described here. Figure 10.11 shows a simplified schematic of a batch type solar photocatalytic detoxification facility. The contaminated water is stored in a containment tank and pumped through a reactor which is illuminated by the sun or another light source. The partially treated water coming out of the reactor is mixed with the water in the containment tank. The process continues until the mixed concentration in the containment tank reaches an acceptable level. Laboratory treatment studies normally employ such a facility to determine reaction rate constants that can be used in the design of treatment systems.

If the reaction rate constant is determined from (10.15) or (10.16) by simply using the concentration-time history in the tank, the results would be in error, unless the tank volume (V_T) is negligible as compared to the reactor volume (V_R), which is usually not

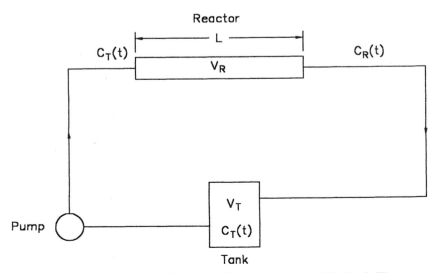

Figure 10.11. Simplified schematic diagram of a batch-type solar detoxification facility.

the case. Wulfrum and Turchi [74] suggested that an apparent reaction rate constant k_{app} may be calculated using (10.16) and the concentration-time history in the tank, from which actual reaction rate constants may be approximated by

$$k = k_{app} \frac{1 + \gamma}{\gamma}, \text{ where } \gamma = \frac{V_R}{V_T}. \tag{10.17}$$

From a rigorous analysis, Klausner et al. determined that the approximation given by Eq. (10.17), gives reaction rate constants that are accurate to within 5 percent for

$$\frac{\gamma + 0.79}{\kappa^{0.62}} > 3.30, \text{ where } \kappa = \frac{V_R k}{Q}, \tag{10.18}$$

and Q is the volume flow rate through the reactor.

For $\kappa < 0.1$ Eq. (10.16) is in error by less than 5 percent, regardless of γ.

Example 10.1. Assuming pseudo-first order kinetics can model the destruction of benzene as shown by the experimental concentration history data in Fig. 10.8, determine the apparent and actual reaction rate constants if the reactor volume is 20 gallons and the tank volume is 500 gallons (neglect volume in pipes). If the solution flows at 30 gal/min, comment on the accuracy of the calculated actual reaction rate. What happens when the tank volume becomes much smaller than the reactor volume?

Solution. Using Eq. (10.16),

$$\frac{-dC}{dt} = k_{app} C$$

$$\int_{c_i}^{c_f} \frac{dC}{C} = - k_{app} \int_0^{\Delta t} dt$$

$$k_{app} = \frac{\ln C_i/C_f}{\Delta t}.$$

From Fig. 10.8 for benzene; $C_i = 800$ ppb, $C_f = 350$ ppb, $\Delta t = 180$ min. Plugging into the above equation, we get

$$k_{app} = 0.00459 \text{ min}^{-1}.$$

From Eq. (10.17),

$$k = k_{app} \frac{1 + \gamma}{\gamma} = k_{app} \left(\frac{V_T}{V_R} + 1 \right),$$

or,

$$k = 0.119 \text{ min}^{-1}.$$

Note that the actual reaction rate is faster than the apparent one, since no reaction occurs in the large tank where the concentration is measured.

From Eq. (10.18),

$$\frac{\gamma + 0.79}{K^{0.62}} = \frac{\dfrac{v_R}{v_T} + 0.79}{\left(\dfrac{v_R k}{Q}\right)^{0.62}} = 3.99 > 3.30.$$

The inequality of Eq. (10.18) holds, indicating the actual reaction rate constant is accurate to within 5 percent.

As the tank volume decreases,

$$V_T \ll V_R \quad \text{or} \quad \gamma \gg 1.$$

With $\gamma \gg 1$, Eq. (10.17) reduces to

$$k = k_{app}.$$

This indicates all the solution is in the reactor; hence the reaction rate measured is the actual reaction rate.

It should be noted here that the reaction rate constant k determined herein is not the traditional rate constant used in reactor engineering. Due to the nature of the photocatalytic reaction, it is a function of external system parameters such as UV irradiation intensity, pH, catalyst loading, and possible degree of mixing in the fluid. It may also depend on the geometry of the photoreactor. Of these parameters, the intensity of the solar UV radiation cannot be controlled, therefore the experimentally determined reaction rate constant, k_o, must be adjusted for the actual UV intensity according to the following equations:

Concentrating reactors [43]:

$$\frac{k}{k_o} = a\left[\frac{I}{I_o}\right]^x, \tag{10.19}$$

where a is a constant given as A/V [43] and x is an intensity-dependent exponent which varies between 0.5 and 1. A value of 0.5 is recommended for x [4].

Nonconcentrating tubular or channel-type reactors [30]:

$$\frac{k}{k_o} = a\left[\frac{I}{I_o}\right], \tag{10.20}$$

where $a = 0.88$ for 0.1% TiO_2 and 1.14 for 0.01% TiO_2 [75].

Trickle film flat plate reactor [75]:

$$\frac{k}{k_o} = a\left(\frac{I}{I_o}\right)\left(\frac{\delta_o}{\delta}\right).$$
(10.21)

Here δ represents the film thickness and the value of a is 0.97 [75]. δ_o is the reference film thickness corresponding to k_o and I_o.

Shallow-pond reactor [7]:

$$\frac{k}{k_o} = m\left[\frac{I(A/V)}{I_o(A/V)_o}\right]^n$$
(10.22)

Values of m and n are given in Table 10.2 [7].

It can be seen from the above equations that the liquid film thickness (δ) is an important parameter for a trickle bed reactor, while the ratio of the area to volume (A/V) is critical for the shallow ponds.

Example 10.2. Three reactors, a flat-plate, a parabolic trough concentrator and a shallow solar pond, were tested for effectiveness in treating a water stream contaminated with 4CP. Using 0.1% TiO_2 solution, all three gave a reaction rate constant k_o of 0.0172 min^{-1} for an incident UV radiation of 31 W/m^2.

1. Determine the reaction rate for a 0.1% TiO_2 solution of the contaminated water passing through each reactor experiencing a steady UV insolation of 43 W/m^2.
2. Determine the depth of a rectangular shallow pond reactor to give (a) the reaction rate found in part (1), and (b) double the reaction rate found in part (1).

Table 10.2. Empirical constants m and n for destruction of 4-Chlorophenol in a shallow pond reactor [7]

% TiO_2	$C_o \times 10^{-4}$ (M)	m	n
0.01	1	0.28	0.50
0.1	1	0.66	0.70
0.3	1	0.95	1.00
0.01	4	0.52	0.44
0.1	4	0.95	0.65
0.3	4	1.23	0.66

Use the following reference values:

$(A/V)_0 = 19.7$ m^{-1} (for shallow pond reactor)
$a = 1.0$ (for concentrating reactor).

Solution.

(1) From Eq. (10.20), for a flat plate reactor,

$$k = k_o a\left[\frac{I}{I_o}\right] = (0.0172 \text{ min}^{-1})(0.88)\left[\frac{43 \text{ W/m}^2}{31 \text{ W/m}^2}\right]$$

$$k = 0.021 \text{ min}^{-1}.$$

For a concentrating reactor, Eq. (10.19) yields

$$k = k_o a\left[\frac{I}{I_o}\right]^x, \text{ where } x = 0.5$$

$$k = (0.0172)(1.0)(43/31)^{0.5} = 0.0203 \text{ min}^{-1}.$$

For a shallow solar pond, Eq. (10.22) yields:

$$k = k_o \cdot m\left\{\frac{I(A/V)}{I(A/V)_o}\right\}^n.$$

For the same A/V ratio

$$k = (0.0172)(0.66)(43/31)^{0.7} = 0.0143 \text{ min}^{-1}.$$

(2) For the same reaction rate constant it was assumed that

$$(A/V) = (A/V_o) = 19.7 \text{ m}^{-1}$$

For a rectangular pond $V = A \cdot D,$ where D is the depth of the pond

$$\therefore \quad 1/D = 19.7 \text{ m}^{-1},$$
$$\text{or} \quad D = 0.051 \text{ m or } 5.1 \text{ cm}$$

For double the reaction rate constant $k = 0.0143 \times 2 = 0.0286$ min^{-1} for the same insolation, from Eq. (10.22)

$$(A/V) = (A/V)_o \cdot (I_o/I)(k/mk_o)^{1/n}$$
$$= (19.7 \text{ m}^{-1})(1)(2/0.66)^{1/0.7}$$

$$= 96 \text{ m}^{-1},$$

$$\therefore D = 1/96 = 0.0104 \text{ m or } 1.04 \text{ cm}.$$

10.5 USEFUL INSOLATION

Photocatalytic detoxification using TiO_2 as catalyst requires photons of ultraviolet (UV) radiation with wavelength less than 388 nm. The extraterrestrial radiation supplied by the sun contains about 10 percent UV radiation when it first reaches the earth's atmosphere, which is reduced to approximately 4–6 percent as the solar radiation travels to the surface (ground-level). Typically, the maximum value of UV radiation is about 50 W/m² for a south-facing surface in the northern hemisphere or a north-facing surface in the southern hemisphere tilted at an angle equal to the latitude of the location. Typical solar radiation data available for most locations in the world represent the insolation values for the entire spectrum. However, utilization of this data for modeling detoxification processes requires the estimation of the ultraviolet component. Riordan et al. [63] developed correlations that related measured UV to measured total values. Based on their study, global horizontal and direct normal UV radiation can be calculated as:

$$\frac{I_{uv,h}}{I_t} = 0.14315K_t^2 - 0.20445K_t + 0.135544, \qquad (10.23)$$

$$\frac{I_{uv,b}}{I_{t,b}} = 0.0688e^{-0.575m} = 0.688 \exp\left(\frac{-0.575}{\sin \alpha}\right), \qquad (10.24)$$

where K_t is the cloudiness index and m is the air mass. These correlations were based on limited measurements at only a few locations (Cape Canaveral, FL; San Ramon, CA; and Denver, CO) with an uncertainty greater than 20 percent.

The functional form of irradiance on the reactor aperture varies according to the reactor type. For a flat plate reactor ultraviolet irradiance can be calculated by using the following equation [66] which is a modified form of an equation given by Rabl [61] for total solar radiation:

$$I_{uv} = I_{uv,b} K(i) \max \{\cos i,0\} + 0.5K(i)[I_{uv,d}(1 + \cos \beta)$$
$$+ \rho_{uv,gr}I_{uv,h}(1 - \cos \beta)] \qquad (10.25)$$

where i is the incident angle of the reactor aperture (calculated at the midpoint of each hour), $K(i)$ is the incident angle modifier, β is the reactor tilt, and $\rho_{uv,gr}$ is the ultraviolet reflectance of the ground. Ground reflectance values in the ultraviolet region are determined for various ground covers using a data fit developed by Green [33]. Ground cover may be assumed to be farmland, where reflectance is approximately 4 percent, unless covered by snow, where reflectance is approximately 29 percent [33]. The incident angle modifier $K(i)$, defined as the ratio of the UV transmittance-absorptance product for the reactor at an angle i to the normal, is dependent on the materials used for re-actors. Actual values of $K(i)$ for the reactor under consideration must be used when-ever possible. In the absence of actual values, the following expression for a solar ther-mal collector may be used, although reactor receiver covers and the useful solar spectrum are generally not the same as those for solar thermal collectors.

$$K(i) = 1 - 0.10\left(\frac{1}{\cos i} - 1\right). \tag{10.26}$$

March et al. [46] demonstrated the use of the above described reaction kinetic mod-els and solar insolation models to simulate the performance of solar detoxification systems.

10.6 CATALYST DEVELOPMENT

At present, TiO_2 in anatase form is the most common catalyst used in solar photocat-alytic detoxification. As explained earlier, this limits the useful range of the solar spec-trum to wavelengths of less than about 388 nm. Only about 4–6 percent of the solar ra-diation is available in this wavelength range. Therefore, the catalyst improvement research is concentrated on:

1. Physical and chemical modification of TiO_2 to improve the catalyst performance.
2. Dye sensitization to increase the useful wavelength range of the solar radiation.
3. Development of homogenous catalysts.

Figure 10.12 shows the results of the comparative tests conducted on modified cata-lysts. It can be seen that several methods, by themselves or in combination, can be used to improve the performance of the TiO_2 catalyst.

According to Magrini et al. [44], thermal treatment at 550–600°C can double the reaction rates. Manipulating the surface area and particle size can also improve the cat-alyst performance. Metallization of TiO_2 particles with Pt can improve the perfor-mance five times, while Pd, Ag and WO_3 can improve it by factors of two to three[1].

Another improvement being investigated by researchers is the use of photosensi-tizer (dye) molecules on the surface of a semiconductor. In dye sensitization, a dye molecule absorbs visible light, gets excited to a higher energy state, and interacts with the semiconductor molecule (TiO_2), water, and oxygen to produce hydroxyl radicals. The result is an extension of the photoresponse of the semiconductor photocatalyst, making it capable of using a broader spectrum of solar radiation than the semiconduc-tor alone [19,39]. Dye sensitization of TiO_2 photocatalyst has the potential to reduce the overall cost of treatment. Methylene blue and Rose Bengal have been the most common dyes investigated so far. The results have been inconclusive.

---

[1]For detailed data on the catalyst improvement research, see Ref. [17, 42, 44, and 45].

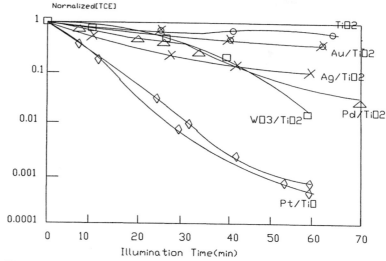

Figure 10.12. Concentration vs. Illumination time data for metallized TiO_2. The metal-containing catalyst contained 1–2 wt% metal, and the WO_3/TiO_2 contained 3 wt% WO_3. All reactions were conducted in a 0.51 circulating batch reactor with 0.05 wt% catalyst at 22° C, 25 ppm initial TCE, and illuminated with a 1,000 W Xe source [44].

10.7 SYSTEM DESIGN METHODOLOGY

The effectiveness of solar detoxification systems has been demonstrated commercially [28]. Goswami et al. [29] have developed a methodology which can be used for the design of simple solar detoxification systems.

The procedure for designing a solar detoxification system requires the selection of the reactor, reactor operational mode (slurry or fixed matrix), reactor field configuration (series or parallel), treatment system mode (once through or batch), flow rate, pressure drop, pretreatment, catalyst and oxidant loading, pH control and catalyst reuse system. For the treatment of groundwater or industrial wastewater, the following must be known or determined a priori:

1. Complete analysis of the water to be treated to determine pretreatments.
2. Target chemicals, their initial, and final desired concentrations.
3. Amount of water to be treated daily.

The following steps describe a design procedure.

10.7.1 Laboratory Treatment Study

A laboratory treatment study must be conducted to determine the following parameters for optimum treatment:

1. Reaction rate constant
2. Catalyst loading
3. Oxidant (H_2O, O_2, O_3, etc.) loading, if any
4. Required pH
5. Pre- and post-treatment

Indoor reactors, with simulated radiation similar to the expected solar radiation and reactor geometry similar to the chosen outdoor reactor, may be used for the laboratory
treatment study.

Experiments are conducted to find the concentration-time history of the pollutant
chemical under consideration from which the reaction rate constant k may be calculated as explained earlier. If more than one target chemical is present, reaction rate
constants for all the chemicals must be determined. In this case, the lowest value of k
should be used for the reactor design.

10.7.2 Treatment Facility Operational Mode

A solar detoxification facility may be operated in the following modes:

1. Batch mode (Figure 10.13a): In this mode, the effluent is stored in a tank and is
   continuously recirculated though the reactors until the desired destruction is
   achieved. Operation in this mode requires one or more storage tanks. If the desired

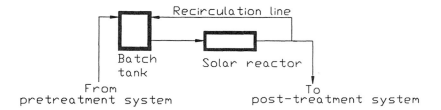

(a) Batch mode operation

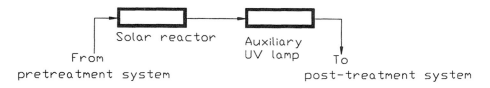

(b) Single pass mode operation

Figure 10.13. Schematics of treatment facility operational modes [10]: (a) Batch mode;
(b) Single pass mode.

destruction of the contaminated water is not achieved in a single day, the system is operated the following day until the desired destruction is obtained. Shallow solar ponds that combine reactor and storage fall in this category.
2. Single-Pass Mode (once through) (Figure 10.13b): In this mode, the reactor area and the flow rates are designed so that the desired destruction is achieved in a single-pass. The flow rates in this case are normally lower. The solar insolation varies throughout the day and hence, the flow rate through the reactors should be varied with the intensity of the solar insolation. As the UV intensity decreases, the flow rate through the reactor should be decreased in order to maintain the same final concentration.

10.7.3 Residence Time

For the first order reaction kinetics, the required residence time t is calculated by using equation (10.16) as:

$$\left(\frac{C_f}{C_i}\right) = e^{-kt} \tag{10.27}$$

or

$$t = \frac{\ln(C_i/C_f)}{k} \tag{10.28}$$

where

C_f = Final concentration of the contaminant,
C_i = Initial concentration of the contaminant, and
k = Rate constant (min^{-1}).

The rate constant k is obtained from k_o (which is the measured rate constant for the intensity I_o) by adjusting it for the actual intensity, I, according to Equations (10.19–10.22). Since the intensity I changes throughout the day, k would require continuous adjustment. An approximate and simplified value can be used by adjusting k_o for an average value of I for the useful part of the day.

The time needed for the detoxification of the wastewater depends on the desired amount of destruction of the contaminant. As can be seen from the above equation, the time increases exponentially for destruction approaching 100 percent.

10.7.4 Reactor Area

Given the amount of wastewater to be treated per day, q, the average useful time of operation per day, T, and the reactor geometry, the fluid velocity in the reactor can be calculated. For a reactor with N parallel tubes/channels which are connected between two headers, the flow velocity can be calculated as (Figure 10.14):

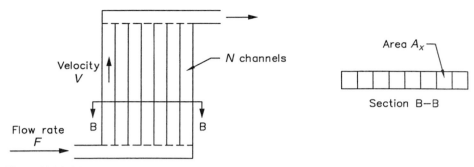

Figure 10.14. Flow through a reactor with N channels.

$$\text{Total flow rate } F = \frac{q}{T} \tag{10.29}$$

$$\text{Velocity in one channel } v = \frac{F}{(NA_x)} \tag{10.30}$$

where A_x = cross-sectional area of a channel.

From the required residence time t as calculated from Eq. (10.28), the flow velocity v is calculated from Eq. (10.30), and the length of one reactor L and the total number of reactors for a once-through system can be calculated as follows:

$$\text{Residence time for one reactor } t_r = \frac{L}{v} \tag{10.31}$$

Total number of reactors = t/t_r (rounded off to the next integer number).

The following is given as an example of the reactor design using the experimental data from field tests conducted at the Tyndall Air Force Base [29–31].

Example 10.3. Using the experimental data for treatment of ground water at Tyndall Air Force Base in Florida (Table 10.3), find (1) the residence time for 99.75 percent destruction of the contaminents, and (2) the number of reactors required to treat 36,000 liters of contaminated groundwater per day, using tubular reactors.

Each reactor is 2.44m × 1.83m (nominal) and contains 132 tubes that are 2.44m long and 6.4mm diameter. The volume capacity of each reactor is 25l. Assume a yearly average UV radiation value of 28 W/m² for an average of 6 hours each day.

Solution. From Table 10.3, the minimum k value is selected for the residence time and the design of the reactor. For example, for Test 1, minimum k value is 0.0311 (min⁻¹). The solar UV intensity for this test data is 45.3W/m². Since k value depends on the UV intensity, it needs to be adjusted for the design UV intensity. In this case we assume the design for yearly average conditions when the average UV intensity is 28W/m². Thus, the adjusted k value is given by the following equation:

Table 10.3. Experimental parameters and results for TyAFB Field Tests [30]

Test #	UV (W/m²)	% TiO₂	pH	H₂O₂	Benzene	Toluene	Ethyl Benzene	m,p-xylene	o-oxylene
							Rate constants, K (min⁻¹)		
1	45.3	0.1	6	100	0.0327	0.0350	0.0311	0.0333	0.0351
2	47.7	0.05	6	100	0.0263	0.321	0.0304	0.0337	0.0309
3	47.2	0.1	5	100	0.364	0.364	0.364	0.364	0.364

Note: Test 3 consisted of city water spiked with BTEX.

$$k = k_o\left(\frac{I}{I_o}\right),\tag{10.32}$$

or

$$k = 0.0311 \times \left(\frac{28}{45.3}\right) = 0.0192 (\text{min}^{-1})\tag{10.33}$$

Assuming first order reaction kinetics, for a 99.75% destruction ($C_f/C_i = 0.0025$) the residence time can be found from Eq. (10.28) as

$$t = \frac{\ln(C_i/C_f)}{k}, \text{ or } t = \frac{\ln(1/(0.0025))}{0.0192} = 312 \text{ min},$$

or, $t = 312$ minutes (approximately 5.2 hours).

The time that is needed for detoxification increases exponentially for destruction rates approaching 100 percent. For example, for the above kinetics, the residence time increases from 72.2 minutes for 75 percent destruction, to 156 minutes for 95 percent destruction; and to 312 minutes for 99.75 percent destruction rate.

For a single pass (once-through) system, the number of reactors depends upon several parameters, such as: amount of the material to be treated, residence time required, and the size and configuration of reactors. In this example, we have to treat 36,000 liters of wastewater per day.

Assuming a length of a day as six hours, the volume of water to be treated per minute would equal 36,000/(6 × 60) = 100 lpm. Knowing the flow rate, we can now determine the residence time of wastewater in one reactor as follows:

Total flow rate = 100 lpm,
Flow in one tube = 100/132 = 0.76 lpm = 760 cm³/min, and
Cross-sectional area of one tube = (π/4) d² = 0.32 cm².

Hence, velocity = 760/0.32 = 2,375 cm/min = 0.39 m/s, and residence time for one reactor = length of tube/velocity = 2.44m / 0.39m/s = 6.26s.

Therefore, the number of reactors required = total residence time / (residence time in one reactor) = (5.2 hr × 3,600) / 6.26
= 2,990 reactors.

The number of reactors strongly depends upon the reaction rate constant k. For example, the k value for the city water—spiked with the same amount of BTEX as found in the groundwater—was determined to be 0.361 in comparison to an average of 0.031 for the groundwater. The residence time required for the k value of 0.361 will be only 27 minutes, as compared to 5.2 hrs for a k value of 0.031. Following the above procedure in this case, the number of reactors required will be only 262 as compared to 2,990 for the groundwater.

10.7.5 Catalyst Life

Catalyst life and the fraction of the catalyst useful after each run are important parameters in the process economics. Catalyst may be poisoned by contaminants and particles or washed away in the discharge water, reducing the catalyst's life and requiring additional catalyst for each run. Useful catalyst life can be estimated by conducting tests with actual contaminated water a number of times, and determining the reduction in reaction rate constant each time [31]. In the field tests conducted at Tyndall Air Force Base, Goswami, et al. [30] found that approximately 10 percent of the catalyst was lost in each run.

10.8 GAS PHASE PHOTOCATALYTIC DETOXIFICATION

Since air contains approximately 21% oxygen and contaminant levels are often in the range of 0.1 to 100 ppm, it follows that contaminated air has a great deal of excess oxidant available over that needed for total oxidation of impurities [54]. Published literature shows that gas phase photocatalytic oxidation can be used successfully to oxidize paraffins, olefins, and alcohol [18], trichloroethylene (TCE) [2,18,52], trans-dichloroethylene (trans-DCE), and cis-dichloroethylene (cis-DCE) [51], toluene [37], airborne nitroglycerin (NG) [62], and 3-chlorosalicylic acid [64,69].

10.8.1 Photoreactors

Many photoreactors have been successfully studied for liquid phase photocatalysis, but reactor design information for gas phase photocatalytic processes is scarce. There are two major differences between gas-phase systems and aqueous systems that affect the performance of the reactor. First, the concentration of oxygen in a gas-phase reactor, which is 20 percent by volume, provides a sufficient supply of electron acceptors. Second, the concentration of water vapor present in gas-phase reactors can vary considerably unlike concentrations in aqueous photoreactors in which water molecules are always the predominant species in contact with the catalyst [2].

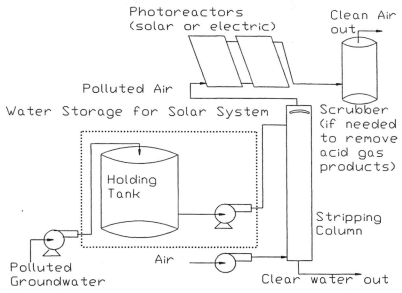

Figure 10.15. Schematic of an air stripper with photocatalytic off-gas treatment [71].

Because of intense commercial interest in using photocatalysis to clean indoor air and industrial emissions, very little information is available in the open literature on gas phase reactors.

Figure 10.15 shows an off-gas photocatalytic treatment system based on a commercial system [48,71]. This system is designed to transfer volatile contaminants from groundwater to air in an air stripper tower. The contaminants, transferred to air, are then treated by solar radiation or UV lamps in photoreactors.

10.9 COMMERCIAL/INDUSTRIAL APPLICATIONS

Solar photocatalytic detoxification research to date has shown great potential for the application of this technology to treatment of groundwater and soils contaminated with toxic organic chemicals, as well as treatment of certain industrial wastewaters. Recent research is also showing the potential of this technology for gas phase detoxification and for disinfection of water and air. However, to date, there has been very little commercial/industrial use of this technology. Published literature shows only three engineering scale applications for groundwater treatment in the USA and one industrial wastewater treatment in Spain. Field demonstrations of this technology for groundwater remediation took place at Lawrence Livermore National Laboratory (LLNL) and Tyndall Air Force Base (TyAFB). Engineering scale field experiments were conducted by NREL at LLNL to treat groundwater contaminated with trichloroethylene (TCE). The field system consisted of 158 m² of parabolic trough reactors, as described earlier, and used Degussa P25 TiO_2 particles as catalyst in a slurry flow configuration. With

0.1% TiO$_2$ concentration in the slurry, the TCE concentration was reduced from 200 ppb to less than 5 ppb. The field experiment is detailed in Ref. [9].

Engineering scale demonstration of the nonconcentrating solar reactor technology was conducted at the TyAFB in 1992 [30]. A "one-sun" solar detoxification facility was constructed to treat groundwater contaminated with fuel, oil, and lubricants leaking from the underground storage tanks. The contaminants of interest included benzene, toluene, ethylbenzene and xylenes (BTEX) although there were other chemicals also present in the groundwater. The facility shown in Fig. 10.16 consisted of three rows of five-each series connected 2.4m × 3.1m (nominal) photoreactors developed by the University of Florida, an equipment skid from NREL containing storage tanks and metering pumps for the addition of acid, base, H$_2$O$_2$ and TiO$_2$ to the contaminated water.

The field tests were successful in destroying the BTEX in the groundwater as seen from the results shown earlier in Figure 10.8. However, in a laboratory test using the same reactors in similar sunlight conditions and city water spiked with the same amounts of BTEX as in the groundwater, the reaction rates were an order of magnitude faster. These results suggest a careful site treatability study and establishment of appropriate pretreatment methods are extremely important in the successful field deployment of solar photocatalytic processes.

Reference [12] has described an engineering scale field demonstration of treatment of industrial wastewater from a resins factory containing organic contaminants such as phenols, fornol, phthalic acid, fumeric acid, maleic acid, glycols, xylene, toluene, methanol, butanol and phenylethylene, amounting to 600 ppm TOC. They used 12 two-axis tracking parabolic troughs of total aperture area of 384m², with borosilicate glass tube absorber reactors. 100 mg/l of TiO$_2$ was used as the catalyst and 0.007 molar concentration of sodium peroxydisulphate (Na$_2$S$_2$O$_8$) as an oxidizing ad-

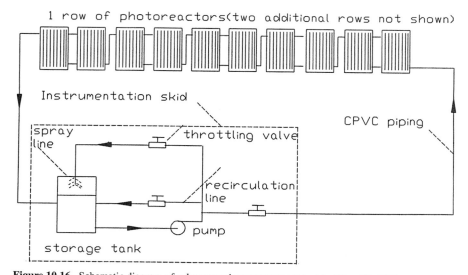

Figure 10.16. Schematic diagram of solar groundwater treatment test facility at TyAFB.

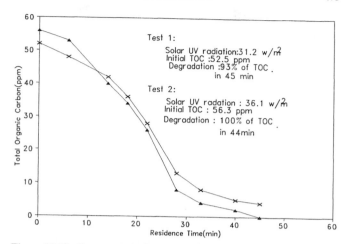

Figure 10.17. Treatment of industrial wastewater containing organic contaminants [12].

ditive. Figure 10.17 shows 100 percent degradation of TOC in 44 minutes, amounting to 1.25mg/l/min-degradation rate of the mixed contaminants. Based on this rate, they estimated treatment costs of $19.3/m³ of industrial wastewater treating 10m³/day. Other investigations of treatment of real industrial wastewater, although not to engineering scale, do show the potential of the solar treatment process where conventional treatment methods have been unsuccessful. Anheden, Goswami and Svedburg [3] demonstrated the potential for solar photocatalytic oxidation for decolorization and COD reduction of wastewater from 5-fluorouracil (a cancer drug) manufacturing plant. They showed that using 0.1% TiO_2 as the catalyst and 2400 ppm H_2O_2 as an oxidizing additive, the color of the wastewater is reduced by 80 percent in one hour and the COD of the wastewater was reduced by 70 percent in 16 hours. All of the conventional (non-solar) treatments tried by the manufacturers for the reduction of color and COD failed. Zaidi [76] showed the potential of the solar photocatalytic technology to reduce color and COD of distillery wastewater pretreated by anaerobic microbiological methods. Figure 10.18 shows that both color and COD can be reduced successfully by the solar process where conventional methods have failed to work. Both of the above studies report that the solar treatment works much better if the wastewater is first diluted with clean water to about 10–20% of the initial concentration. Although

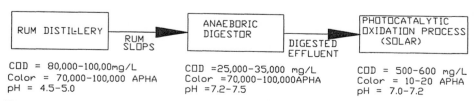

Figure 10.18. Treatment of distillery effluent by anaerobic digestion and post-treatment by solar photocatalytic oxidation process [76].

this method increases the volume of water to be pumped, a continuous system can be designed by recirculating a fraction of the treated water for dilution.

In laboratory study, Turchi et al. [70] showed the potential of photocatalytic treatment of the wastewater from pulp and paper mills. Although no studies have been published on the potential use of the solar process for the clean-up of textile mill wastewater, preliminary studies indicate that the solar process can be very successful in treating textile plant wastewater, namely dyehouse wastewater. Since many manufacturing plants, such as textiles, pulp and paper, and chemicals, are located in areas where a large part of the solar UV is diffuse and where large open areas with solar access are available, the shallow solar pond reactors may be used.

10.10 SOLAR DISINFECTION OF WATER AND AIR

UV disinfection has been widely used in the past to destroy biological contaminants by using UV radiation from germicidal lamps primarily at 254 nm wavelength. Solar UV which is primarily at 290–400 nm wavelength, is much less active as a germicide. However, a review of the literature has shown that photocatalytic disinfection has been demonstrated in Japan by Matsunaga [47] and later by others [38,49,50,57]. In a recent study, Block and Goswami [13] have studied the antibacterial effect of solar photocatalytic reaction and the conditions that affect it. Their study showed that several common bacteria (Serratia Marcescens, E. Coli and Streptococcus Aureus) were killed in just a few minutes on solar exposure in the presence of TiO_2, whereas without TiO_2 it took over an hour to destroy them (Figure 10.19). A concentration of 0.01% TiO_2 was most effective in killing bacteria and even 0.001% was quite effective. However, bactericidal activity went down at 0.1% and higher concentrations of TiO_2.

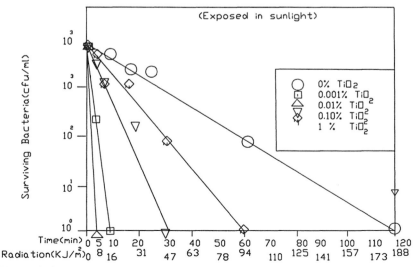

Figure 10.19. Effect of concentration of TiO_2 on survival of *Serratica Marcescens* bacteria [13].

Recently, some studies have appeared in the literature that report disinfection and deodorization of air. Goswami et al. [32] demonstrated the antibacterial effect of titanium dioxide in indoor air. They used a TiO_2 coated fiberglass matrix reactor to kill *Serratia Marcescens* bacteria in air. Results indicated that the photocatalytic process can be a viable technique for controlling indoor air quality.

10.11 SUMMARY

Solar photocatalytic detoxification and disinfection processes have shown great promise for the treatment of contaminated groundwater, industrial wastewater, air, and soil. Development of these processes have reached a point where the solar technology can be competitive with the conventional treatment methods. In some cases requiring decolorization and reduction of COD, the photocatalytic technology may be the only effective treatment technology. The next logical step is the commercial demonstration of the technology. Although only a few commercial demonstrations have been reported to date, it is expected and hoped that activity will pick up considerably in the future.

PROBLEMS

10.1. 1. A laboratory test with TiO_2 as the catalyst produced a concentration time plot of O-xylene destruction identical to that shown in Fig. 10.8. The reactor used was of 40 gallons while the storage tank contained 100 gallons with minimal volume in piping. Determine the reaction rate constant corrected for the finite volumes of the system.

2. A single reactor is to be designed to decontaminate water with the same initial concentration of O-xylene as above to acceptable levels of 10 ppb. If the reaction rate is the same as found above, what is the required residence time for this process? What is the residence time for 99.9 percent destruction? 100 percent destruction?

3. Each of 250 nonconcentrating channel-type reactors connected in series is composed of 140 channels 3 m long and 1 cm wide. The system is designed to treat 10,000 liters of O-xylene contaminated water per day. If the system operates for 8 hours each day, what must be the channel depth of each reactor? At what velocity does the solution flow? Use the residence time for destruction to acceptable levels of contaminant.

4. A similar reactor tested with 35 W/m² insolation in a controlled experiment yielded a reaction rate constant of 0.0183 min⁻¹. What must be the insolation in the field trials if a 0.01% TiO_2 catalyst solution is used? Can the field reactor site dictate which catalyst to use?

10.2. Discuss the effects of the following factors on the photocatalytic destruction of organic contaminants.
   a. illumination intensity
   b. catalyst type

    c. addition of inorganic and organic particles
    d. dissolved oxygen concentration
    e. pH of the solution
    f. temperature
    g. metallization of the catalyst
    h. dye sensitization
    How are the concentration-time histories of the contaminant affected?

REFERENCES AND SUGGESTED READINGS

1. Acher, A. J. 1984. Sunlight photo-oxidation of organic pollutants in wastewater. *Water Science and Technology* 17: 623–632.

2. Anderson, M., S. Nishiida, and S. Cervera-March. 1993. Photodegradation of trichlorethylene in the gas phase using TiO_2 porous ceramic membrane. In *Photocatalytic Purification of Water and Air*. Vol. 1, 405–420.

3. Anheden, M., D.Y. Goswami, and G. Svedberg. 1995. Photocatalytic treatment of wastewater from 5-fluorouracil manufacturing. *Solar Engineering 1995, Proc. of the 1995 ASME International Solar Energy Conference*.

4. Bahnemann, D., D. Bockelmann, and R. Goslich. 1991. Mechanistic studies of water detoxification in iluminated TiO_2 suspensions. *Sol. Energy Materials*, 24: 564–583.

5. Barbeni, M., E. Pramauro, E. Pelizzetti, E. Borgarello, and N. Serpone. 1985. Photodegradation of pentachlorophenol catalyzed by semiconductor particles, *Chromosphere*, 14 (2): 195–208.

6. Bard, A.J. 1979. Photoelectrochemistry and heterogeneous photocatalysis at semiconductors. *J. Photochem.* 10(1): 59–75.

7. Bedford, J., J.F. Klausner, D.Y. Goswami, and K.S. Schanze. 1994. Performance of non-concentrating solar photocatalytic oxidation reactors, Part II: shallow pond configuration. *J. of Solar Energy Engineering* 116(1): 8–13.

8. Blake, D.M. 1994. Bibliography of work on the photocatalytic removal of hazardous compounds from water and air. *NREL Report #NREL/TP-430-6084*. Available from NTIS, Springfield, VA.

9. Blake, D.M. 1994. Solar processes for the destruction of hazardous chemicals. In *Alternative Fuels and the Environment*, F.S. Sterrett, ed. Boca Raton, FL: Lewis Publishers.

10. Blake, D.M., H.F. Link, and K. Eber. 1992. Solar photocatalytic detoxification of water. In *Advances in solar energy*, Vol. 7, K.W. Boer, ed. 167–210. Boulder, CO: American Solar Energy Society.

11. Blake, D.M., J. Webb, C. Turchi, and K. Magrini. 1991. "Kinetic and mechanistic overview of TiO_2-photyocatalyzed oxidation reactions in aqueous solution." *Sol. Energy Materials* 24, pp. 584–593.

12. Blanco, J., and S. Malato. 1994. Solar photocatalytic mineralization of real hazardous waste water at pre-industrial level. *Proc. of the ASME International Solar energy conf.* D.E. Klett, R.E. Hogan, and T. Tanaka, eds., 103–109.

13. Block, Seymour S. and D.Y. Goswami. 1995. Chemically enhanced sunlight for killing bacteria. *Solar Engineering, Proc. of the ASME International Solar Energy Conference,* Hawaii, March 1995, 431–438.

14. Bolton, J. 1994. Private Communication, Seminar at the University of Florida, April.

15. Carey, J.H., J. Lawrence, and H.M. Tosine. 1976. Photodechlorination of polychlorinated biphenyls in the presence of titanium dioxide in aqueous solution. *Bull. Environ. Contam. Toxicol.* 16 (6): 697–701.

16. Cooper, J.H., and A.M. Ratcliffe. 1992. *Apparatus for Photocatalytic Treatment of Liquids*. U.S. Patent #5,174,877, Dec. 29.

17. Crittendon, J.C., Y. Zhang, D.W. Hand, and D.L. Perram. 1995. Destruction of organic compounds in water using fixed bed photocatalysts. *Solar Engineering 1995. Proc. of the 1995 ASME International Solar Engineering Conference,* 449–457.

18. Dibble, L.A. and G.B. Raupp. 1990. Kinetics of the gas-solid heterogeneous photocatalytic oxidation of trichlorethylene by near UV illuminated titanium dioxide. *Catalysis Letters* 4: 345–354.

19. Dieckmann, M.S., K.A. Gray, and R.G. Zepp. 1994. The sensitized photocatalysis of azo dyes in a solid system: A feasibility study, *Chemosphere* 28(4): 1021–1034.
20. Duffie, J.A., and W.A. Beckman. 1980. *Solar Engineering of Thermal Processes*, 263. New York: Wiley-Interscience.
21. Elizardo, K. 1991. Fighting pollution with hydrogen peroxide, *Pollution Engineering*, Sept 1991: 106–109.
22. Fujishima, A. 1994. New application of photocatalysis under room light illumination. *Proc. of the First International Conference on Advanced Oxidation Technologies for Water and Air Remediation*. London, Ontario, June, 98–99.
23. Gergov, M., M. Priha, E. Talka, and O. Valtilla. 1988. Chlorinated organic compounds in effluent treatment at Kraft Mills. *Environment Conference, TAPPI Proc.* Atlanta, pp. 443–455.
24. Glatzmaier, G.C., M.S. Mehos, and R.G. Nix. 1990a. Solar destruction of hazardous chemicals. *J. of Environmental Science and Health*, A25(5), 571–581.
25. Glatzmaier, G.C., M.S. Mehos, and R.G. Nix. 1990b. Solar destruction of hazardous chemicals. *Solar Engineering 1990, Proc. of the ASME International Solar Energy Conference*, 153–158.
26. Glaztmaier, G.C., and M.S. Bohn. 1993. Solar assisted combustion of 1–2 dichlorobenzene. *Proc. of the AIChE/ASME National Heat Transfer Conference*, August.
27. Goswami, D.Y. 1995. Engineering of solar photocatalytic detoxification and disinfection processes. *Advances in Solar Energy*, 13: 208.
28. Goswami, D.Y. and D. Blake. 1996. Cleaning up with sunshine. *Mechanical Engineering* 118(8):56–59.
29. Goswami, D.Y., G.D. Mathur and C.K. Jotshi. 1994. Methodology of design of nonconcentrating solar detoxification systems. EDSA 1994, Proc. of the ASME Engineering Systems and Design Analysis Conf, PD-vol. 64-3, New York: ASME. 117–122.
30. Goswami, D.Y., J. Klausner, G.D. Mathur, A. Martin, K. Schanze, P. Wyness, C. Turchi, and E. Marchand. 1993a. Solar photocatalytic treatment of groundwater at Tyndall AFB, field test results," *Solar 1993*, Proc. of the American Solar Energy Society Annual Conference, 235–239.
31. Goswami, D.Y., J.F. Klausner, P. Wyness, A. Martin, and G.D. Mathur. 1993b. Solar photocatalytic treatment of groundwater at Tyndall AFB, field test results. University of Florida, Department of Mechanical Engineering, *Rep. #UFME/SEECL-9302*, January.
32. Goswami, D.Y., D. Trivedi, and S.S. Block. 1995. Photocatalytic disinfection of indoor air. *Solar Engineering, Proc. of the ASME International Solar Energy Conference*, Hawaii, March, 421–430.
33. Green, A.E.S. 1983. The penetration of ultraviolet radiation to the ground. *Phys. Plant* 58, 351.
34. Heller, A. 1981. Conversion of sunlight into electrical power and photoassisted electrolysis of water in photoelectrochemical cells. *Acc. Chem. Res.*, 14: 154–162.
35. Heller, A. and J.R. Brock. 1991. Materials and methods for photocatalyzing oxidation of organic compounds in water. United States Patent 4.997.576, March 5.
36. Heller, A., M. Nair, L. Davidson, Z. Luo, J. Schwatzgebel, J. Norrell, J.R. Brock, S.E. Lindquiest, and J.G. Eskerdt. 1992. Photoassisted oxidation of oil and organic spill on water. *Proc. of the First International Conference on TiO₂ Photocatalytic Purification and Treatment of Water and Air*, London, Ontario, pp. 39–155.
37. Ibusuki, T., and K. Takeuchi. 1986. Toluene oxidation on UV-irradiated titanium dioxide with and without O₂, NO₂, or H₂O at ambient temperature. *Atmospheric Environment* 20: 1711–1715.
38. Ireland, J.S., P. Klostermann, E.W. Rice, and R.M. Clark. 1993. Inactivation of *Escherichi coli* by titanium dioxide photocatalytic oxidation, *Appl. Environmental Microbiol.* 59: 1668–1670.
39. Kamat, P.V. 1993. Photochemistry on nonreactive and reactive (semiconductor) surfaces. *Chemical Reviews* 93: 267–300.
40. Klausner, J.F., A.R. Martin, D.Y. Goswami, and K.S. Schanze. 1994. On the accurate determination of reaction rate constants in batch-type solar photocatalytic oxidation facilities. *J. of Solar Energy Engineering, ASME* 116(1): 19–24.
41. Legrini, O., E. Oliveros, and A.M. Braun. 1993. Photochemical processes for water treatment. *Chem. Rev.* 93, pp. 671–698.
42. Linder, M., D.W. Bahnemann, B. Hirthe, and W.D. Griebler. 1995. Solar water detoxification: Novel TiO₂ powders as highly active photocatalysts. *Solar Engineering 1995. Proc. of the ASME International Solar Engineering Conference*, 399–408.

43. Link, H., and C.S. Turchi. 1991. Cost and performance projections for solar water detoxification system. *Solar Engineering 1991. Proc. of the ASME International Solar Energy Conference,* 289–294.

44. Magrini, K.A., R.M. Goggin, A.S. Watt, A.M. Taylor and A.L. Baker. 1994. Improving catalyst performance for the solar based photocatalytic oxidation of organics. *Solar Engineering 1994. Proc. of the 1994 ASME International Solar Engineering Conference,* 163–170.

45. Magrini, K.A., A. Watt, and B. Rinehart. 1995. Photocatalyst evaluation for solar based aqueous organic oxidation, *Solar Engineering 1995. Proc. of the 1995 ASME International Solar Engineering Conference,* 415–420.

46. March, M., A. Martin, and C. Saltiel. 1995. Performance modelling of nonconcentrating solar detoxification systems. *Sol. Energy* 54 (3): 143–151.

47. Matsunaga, T., 1985. Sterilization with particulate photosemiconductor. *J. Antibact. Antifung. Agents.* 13, pp. 211–220.

48. Miller, R.A., and R. Fox. 1993. Treatment of organic contaminants in air by photocatalytic oxidation: A commercialization perspective. In *Photocatalytic Purification and Treatment of Water and Air,* D.F. Ollis and H. El-Akabi eds., 573–578. Elsevier Publishers.

49. Morioka, T., T. Saio, Y. Nara, and K. Onoda. 1988. Antibacterial action of powdered semiconductor on a serotype g *Streptococcus mutans. Caries Res.* 22: 230–231.

50. Nagame, S., T. Oku, M. Kambara, and K. Konishi. 1989. Antibacterial effect of the powdered semiconductor TiO_2 on the viability of oral microorganisms. *J. Dent. Res.,* 68: 1696–1697.

51. Nimlos, M., W. Jacoby, D. Blake, and T. Milne, 1993. Direct mass spectrometric studies of the destruction of hazardous wastes: Gas-phase photocatalytic oxidation of trichloroethylene over TiO_2. *Env. Science & Tech.* 27: 732–740.

52. Nishida, S., K. Nagano, S. Phillips, and M. Anderson. 1993. Photocatalytic degradation of trichlorethylene in the gas-phase using titanium dioxide pellets. *J. of Photochem. Photobiol. A: Chem.,* vol. 70. 95–99.

53. Oberg, V., D.Y. Goswami, and G. Svedberg. 1993. On photocatalytic detoxification of water containing volatile organic compounds. *Solar Engineering 1993, Proc. of the ASME International Solar Engineering Conference,* 147–153.

54. Ollis, David F. 1994. *Photoreactors for Purification and Decontamination of Air.* Department of Chemical Engineering, North Carolina State University. Raleigh, NC.

55. Ollis, D.F., E. Pelizzetti, and N. Serpone. 1989. Heterogeneous photocatalysis in the environment: Application to water purification. In *Photocatalysis Fundamentals and Applications,* N. Serpone and E. Pelizzetti eds., NY: Wiley-Interscience, 603–637.

56. Ollis, D.F., E. Pelizzetti, and N. Serpone. 1989. Chapter 18 In *Photocatalysis fundamentals and applications.* NY: Wiley-Interscience.

57. Onoda, K. J. Watanabe, Y. Nakagawa, and I. Izumi. 1988. Photocatalytic bactericidal effect of powdered TiO_2 on *streptococcus mutans. Denki Kagaku* 56 (12): 1108–1109.

58. Pacheco, J.E., C.E. Tyner. 1990. Enhancement of processes for solar photocatalytic detoxification of water. *Solar Engineering 1990, Proc. of the ASME Solar Energy Conference,* 163–166.

59. Prairie, M.R., J.E. Pacheco, and L.R. Evans. 1992. Solar detoxification of water containing chlorinated solvents and heavy metals via TiO_2 photocatalysis. *Solar Engineering 1992, Proc. of the ASME International Solar Energy Conference* 1, 1–8.

60. Pacheco, J., M. Prairie, and L. Yellowhorse. 1991. Photocatalytic destruction of chlorinated solvents with solar energy. *Solar Engineering 1991, Proc. of the 1991 ASME/JSME/JSES International Solar Energy Conference,* 275–282.

61. Rabl, A. 1981. Yearly average performance of the principal solar collector types. *Sol. Energy,* 27: 215.

62. Raissi, A. and N. Muradov. 1993. Flow reactor studies of TiO_2 photocatalytic treatment of airborne nitroglycerin. *Photocatalytic Purification and Treatment of Water and Air,* 1: 435–454.

63. Riordan, C.J., R.L. Hustrom, and D.R. Myers. 1990. Influences of atmospheric conditions and air mass on the ratio of ultraviolet to total solar radiation. *SERI Rep., SERI/TP-215-3895.*

64. Sabate, J., M. Anderson, H. Kikkawa, M. Edwards, and C. Hill, 1991. A kinetic study of the photocatalytic degradation of 3-chlorosalicylic acid over TiO_2 membranes supported on glass. *J. of Catalysis* 127: 167–177.

65. Saito, T., T. Iwase, J. Horie, and T. Morioka. 1991. Mode of photocatalytic bactericidal action of powdered semiconductor TiO$_2$ on *mutans streptococci*. *J. Photochem. Photobiol. B: Biol.* 14: 369–379.

66. Saltiel, C., A. Martin, and D.Y. Goswami. 1992. Performance analysis of solar water detoxification systems by detailed simulation. *ASME International Solar Energy Conference,* Maui, Hawaii.

67. Sherif, S.A., F. Barbir, and T.N. Veziroglu. 1999. Hydrogen energy system. In the *Wiley Encyclopedia of Electrical and Electronics Engineering*, John G. Webster, ed. New York: John Wiley & Sons, Inc.

68. Skocypec, R.D., and R.E. Hogan. 1990. Investigation of a direct catalytic absorption reactor for hazardous waste destruction. *Solar Engineering 1990, Proc. of the ASME International Solar Energy Conference*, 167–173.

69. Tunesi, S., and M. Anderson. 1991. Influence of chemisorption on the photodecomposition of salicylic acid and related compounds using suspended TiO$_2$ ceramic membranes. *J. of Physical Chemistry*, 95: 3399–3405.

70. Turchi, C.S., L. Edmundson, and D.F. Ollis. 1989. Application of heterogeneous photocatalysis for the destruction of organic contaminants from a paper mill alkali extraction process. Paper presented at *TAPPI 5th International Symposium on Wood and Pulping Chemistry*, Raleigh, NC.

71. Turchi, C.S., E.J. Wolfrum, and R.A. Miller. 1994. Off-gas treatment by photocatalytic oxidation: Concepts and economics. *Proc. of the First International Conference on Advanced Oxidation Technologies for Water and Air Remediation, Book of Abstracts,* London, Ontario, Canada, 125–128.

72. Tyner, C.E., J.E. Pacheco, C.A. Haslund, and J.T. Holmes. 1989. Rapid destruction of organic chemicals in groundwater using sunlight. Paper presented at the *Hazardous Materials Management Conference/HASMAT CENTRAL.*

73. Venkatadri, R., and R.W. Peters. 1993. Chemical oxidation technologies: Ultraviolet light/hydrogen peroxide, Fenton's reagent and titanium dioxide-assisted photocatalysis. *Hazardous Waste and Hazardous Materials* 10, pp. 107–149.

74. Wolfrum, E.J. and C.S. Turchi. 1992. Comments on reactor dynamics in the evaluation of photocatalytic oxidation kinetics. *J. Catal.* 136: 626–628.

75. Wyness, P., J.F. Klausner, D.Y. Goswami, and K.S. Schanze. 1994. Performance of non-concentrating solar photocatalytic oxidation reactors, Part I: flat plate configuration. *J. of Solar Energy Engineering, ASME*, 116 (1): 2–7.

76. Zaidi, A., 1993. Solar photocatalytic post-treatment of anaerobically digested distillery effluent. M.S. thesis, University of Florida, Gainesville, FL.

CAPTURING SOLAR ENERGY THROUGH BIOMASS

INTRODUCTION

Biomass is any material of recent biological origin. It may be grown as a crop, but the vast majority of biomass in the world occurs as forests, prairies, marshes, and fisheries. The energy to build these chemical bonds comes from sunlight. Solar energy collected by green plants is converted into energetic chemical bonds to produce proteins, oils, and carbohydrates. This stored chemical energy represents fuel that can be converted into heat, power, or transportation fuels.

Wood was the primary biomass fuel for most pre-twentieth century societies, although there are traditions of using grass for steel-making in Africa and "lightly-processed" grass (buffalo dung) for cooking fires in the American West. Biomass energy's preeminence began to wane when the great forests of the northern hemisphere were depleted by the voracious fuel demand of early manufacturing processes and steam engines. Biomass can once again become an important energy resource if more efficient conversion processes are developed and biomass reserves are properly managed.

Because biomass is a solid fuel, it is at an enormous disadvantage to petroleum and natural gas. The goal of many biomass conversion processes is to convert solid fuel into more useful forms: gaseous or liquid fuels. Examples of conversion of biomass to gaseous fuels include anaerobic digestion of wet biomass to produce methane gas and high temperature gasification of dry biomass to produce flammable gas mixtures of hydrogen, carbon monoxide, and methane. Examples of converting biomass to liquid fuels include fermentation of sugars to ethanol, thermochemical conversion of biomass to pyrolysis oils or methanol, and processing of vegetable oils to biodiesel. The resulting liquid and gaseous fuels can then be used in machinery to produce heat and power. Biomass can also be converted to heat and power by burning it, as occurs in boilers and steam power plants. This chapter examines the properties of various biomass feedstocks and the conversion processes for producing useful energy from them.

11.1 BIOMASS FEEDSTOCKS

Feedstocks are the raw materials required for an industrial process. Biomass feedstocks have historically been classified as either waste materials or energy crops, although these distinctions disappear when waste streams are viewed as useful coproducts. For example, corn processing plants often generate enormous quantities of agricultural residues in the form of corn cobs and husks that are currently viewed as wastes. If these materials could be economically converted into process heat, electricity, or liquid fuels, they would be viewed as one of several valuable coproducts from the processing of corn.

11.1.1 Waste Materials

Waste materials, by the traditional classification scheme, are any materials of recent biological origin discarded because they have no apparent value or they represent nuisances or even pollutants to the local environment. Categories of waste materials that qualify as biomass feedstocks include agricultural residues, yard waste, municipal solid waste, food processing waste, and even sewage. Agricultural residues are simply that part of a crop discarded after harvest such as corn stover (husks and stalks), rice hulls, wheat straw, bagasse (fibrous material remaining after the milling of sugarcane), grapevine prunings, and almond shells, to name a few. Yard waste is urban biomass crops: grass clippings, leaves, and tree trimmings. Municipal solid waste (MSW) is whatever is thrown out in the garbage, not all of it suitable as biomass feedstock. In some communities, yard waste may constitute up to 18 percent of MSW, although a growing number of communities have ordinances against disposal of yardwaste with garbage in an effort to conserve landfill space. In communities where yard waste is excluded from MSW, the important components are paper (50 percent), plastics and other fossil fuel-derived materials (20 percent), and food wastes (10 percent). Nonflammable materials (glass and metal) represent 20 percent of MSW. Food processing waste is the effluent from a wide variety of industries ranging from breakfast cereal manufacturers to alcohol breweries. These wastes may be dry solids or watery liquid streams. Sewage represents a source of chemical energy and is often converted into electric power at municipal wastewater treatment plants. The recent concentration of animals into giant livestock facilities has led to calls to treat animal wastes in a manner similar to that for human wastes. Consequently, many strategies for manure management integrate waste treatment with heat and power generation.

Waste materials share few common traits other than the difficulty of characterizing them because of their variable and complex composition. Municipal solid waste is the leavings of thousands of households and industries that yield a feedstock that may be easy to process one day and difficult the next. Yard wastes show seasonal variations in quantity and composition: the spring brings high moisture grass clippings that are displaced by dry leaves in the autumn. Waste streams from food processing plants, on the other hand, may be relatively invariant in composition but contain a wide assortment of complex organic compounds that are not amenable to a single conversion process. Thus, waste biomass presents special problems to engineers who are tasked with converting this sometimes unpredictable feedstock into reliable power or high quality fuel.

The major virtue of waste materials are their low cost. By definition, waste materials have little apparent economic value and often can be acquired for little more than the cost of transporting the material from its point of origin to a conversion plant. Increasing costs for solid waste disposal and sewer discharges and restrictions on land-filling certain kinds of wastes allow some biomass wastes to be acquired at negative cost; that is, a biomass processing company is paid by a company seeking to dispose of a waste stream. For this reason, many of the most economically attractive opportunities in the biomass industry involve biomass waste feedstocks. For example, the seed corn industry, which sells seed grown specifically for planting new crops, has an annual waste disposal problem. Seed for which germination cannot be guaranteed after a certain period of storage cannot be marketed nor can it be sold for animal feed or even landfilled, because the seed is treated with fungicide. Seed corn companies often pay brokers to accept this obsolete seed who, in turn, sell it as an inexpensive fuel for boilers and cement kilns.

Clearly, a waste material that can be used as feedstock for an energy conversion process is no longer a waste material. As demand for these new-found feedstocks increases, those who generate it come to view themselves as suppliers and may demand payment for the one-time waste. Such a situation developed in the California biomass power industry during the 1980s. Concerns about air pollution in California led to restrictions on open-field burning of agricultural residues, a practice designed to control infestations of pests. With no means for getting rid of these residues, an enormous reserve of biomass feedstocks materialized. These feedstocks were so inexpensive that independent power producers recognized that even small, inefficient power plants using these materials as fuel would be profitable. A number of plants were constructed and operated on agricultural residues. Eventually, the feedstock producers had plant operators bidding up the cost of their once valueless waste material. In the end, many of these plants were closed because of the escalating cost of fuel.

11.1.2 Energy Crops

Energy crops are defined as plants grown specifically as an energy resource. It is important to note that firewood obtained from cutting down an old-growth forest does not constitute an energy crop. An energy crop is planted and harvested periodically. Harvesting may occur on an annual basis, as with sugar beets or switchgrass, or on a 5–7 year cycle, as with certain strains of fast-growing trees such as hybrid poplar or willow. The cycle of planting and harvesting over a relatively short time period assures that the resource is used in a sustainable fashion; that is, the resource will be available for future generations.

Energy crops can fulfill one or more market niches. In some instances, the whole plant is used as feedstock for production of electricity or liquid fuels, or both. Such is the case when trees are grown and harvested specifically as boiler fuel for steam power plants. Another possibility is that energy as well as food, feed, and fiber coproducts are coaxed from a single crop. For example, alfalfa is being evaluated for its potential to yield both energy and feed from a single crop. The high protein leaves would be removed after harvesting and processed into animal feed while the fibrous stems would be used as fuel in a gasification power plant. The least desirable and most wasteful scenario for energy crops is extracting the highest-value portion of the crop for conversion into an energy product and discarding the rest of the plant as waste. Milling sugar cane to extract sugar for fermentation to ethanol and discarding the rest of the plant material (known as bagasse) is an example of a conversion process that is wasteful of biomass resource. A better strategy for utilizing this resource would be to extract the sugar as food and convert the bagasse to electric power.

Energy crops contain significant quantities of one or more of four important energy-rich components: oils, sugars, starches, and lignocellulose (fiber). Crops rich in the first three have historically been grown for food and feed: oils from soybeans and nuts; sugars from sugar beets, sorghum, and sugar cane; and starches from corn and cereal crops. Oil, sugars, and starches are easily metabolized. On the other hand, lignocellulose is indigestible by humans although certain domesticated animals with specialized digestive tracts are able to break down the polymeric structure of lignocellulose and use it as an energy source. From this discussion it might appear that the best strategy for developing biomass resources is to grow crops rich in oils, sugars, and starches. However, even for oil crops or starch crops the largest single constituent is invariably lignocellulose, which is the structural (fibrous) material of the plant—stems, leaves, and roots. If oils, sugars, and starches are harvested and the lignocellulose is left behind as an agricultural residue rather than used as fuel, the greatest portion of the biomass crop is wasted.

Not only should lignocellulose be valued, there is good reason to maximize its production at the expense of lipids and simple carbohydrates if energy production is the primary purpose for growing the crop. Research has shown that energy yields (kilojoules per hectare per year) are usually greatest for plants that are mostly roots and stems. In other words, there will be more energy available when plant resources are directed toward the manufacture of lignocellulose rather than oils, sugars, and starches. As a result, there has been a bias toward development of energy crops that focus on lignocellulosic biomass, which is reflected in the discussion that follows.

Energy crops are conveniently divided into woody energy crops and herbaceous energy crops. Woody crops grown on a sustainable basis are harvested on a rotation of 5–7 years; thus, they are often referred to as short-rotation woody crops (SRWC). Hardwoods are more promising than softwoods because of their higher productivity potentials, lower costs, and the ability to resprout from stumps (coppice). Promising hardwoods include hybrid poplar, willow, and black locust. Research in recent years has increased SRWC yields from 4.5 ton/ha to 11–16 ton/ha. Herbaceous energy crops (HEC) include both annual crops and perennial crops. Annual crops die at the end of a growing season and must be replanted in the spring. Examples include corn and sweet sorghum. Perennial herbaceous crops die-back each year in temperate climates but reestablish themselves each spring from rootstock. Examples include switchgrass,

Indian grass, and other native prairie grasses. Both annual and perennial HECs are harvested on at least an annual basis, if not more frequently, with annual yields of 11–16 ton/ha.

11.1.3 Important Properties of Biomass

Evaluation of biomass resources as potential energy feedstocks generally requires information about their composition, heating value, production yields (in the case of energy crops), and bulk density. Compositional information can be reported in terms of biochemical analysis, proximate analysis, or ultimate analysis.

Biochemical analysis reports kinds and amounts of plant chemicals as proteins, oils, sugars, starches, and lignocellulose (fiber). In the case of energy crops, engineers are particularly interested in the lignocellulosic component and how it is partitioned among cellulose, hemicellulose, and lignin. Tables 11.1a and 11.1b provide the biochemical composition of important categories of energy crops. This information is particularly useful in designing biological processes that convert plant chemicals into liquid fuels.

Proximate analysis is important in developing thermochemical conversion processes for biomass. Proximate analysis reports the yields (% mass basis) of various products obtained upon heating the material under controlled conditions. These products include moisture, volatile matter, fixed carbon, and ash. Since the moisture content of biomass is so variable and can be easily determined by gravimetric methods (weighting, heating at 100°C, and reweighing), the proximate analysis of biomass is commonly reported on a dry basis. Volatile matter is that fraction of biomass that decomposes and escapes as gases upon heating a sample at moderate temperatures (about 400°C) in an inert (nonoxidizing) environment. Knowledge of volatile matter is important in designing burners and gasifiers for biomass. The remaining fraction is a mixture of solid carbon (fixed carbon) and mineral matter (ash), which can be distinguished by further heating the sample in the presence of oxygen. The carbon is converted to carbon dioxide leaving only the ash. Table 11.2 contains the proximate analysis (dry basis) of a wide range of biomass materials as well as some fossil fuels for reference. Note that the relatively high volatile content of biomass (50–75%) com-

Table 11.1a. Biochemical composition of cellulosic biomass (dry basis)

Feedstock	Cellulose	Hemicellulose	Lignin	Other*
Bagasse	35	25	20	20
Corn Stover	53	15	16	16
Corn Cobs	32	44	13	11
Wheat Straw	38	36	16	10
Short Rotation Woody Crops	50	23	22	5
Herbaceous Energy Crops	45	30	15	10
Waste Paper	76	13	11	0

*Includes both high-value plant chemicals such as proteins and oils and mineral matter such as silica and alkali [5, 18]

Table 11.1b. Biochemical composition of starch & sugar biomass (dry basis)

Feedstock	Protein	Oil	Starch	Sugar	Fiber
Corn grain	10	5	72	<1	13
Wheat grain	14	<1	80	<1	5
Jerusalem Artichoke	<1	<1	<1	75	25
Sugar Cane	<1	<1	<1	50	50
Sweet Sorghum	<1	<1	<1	50	50

Ref [18].

pared to coal (typically less than 25%), makes biomass very suitable for gasification. Also note the relatively low ash content of biomass compared to coal, which eases ash disposal problems.

Ultimate analysis is simply the (major) elemental composition of the biomass on a gravimetric basis: carbon, hydrogen, oxygen, nitrogen, sulfur, and chlorine along with moisture and ash. Table 2 contains the proximate analysis of several biomass materials on a dry basis. Sometimes this information is presented on a dry, ash-free (daf) basis. This information is very important in performing mass balances on biomass conversion processes, as will be seen in 11.2 of this chapter. Evident from Table 11.2 is the relatively high oxygen content of biomass compared to coal.

Heating value is the net energy released upon reacting a particular fuel with oxygen under isothermal conditions (the starting and ending temperatures are the same). If water vapor formed during reaction condenses at the end of the process, the latent heat of condensation contributes to what is known as the higher heating value (HHV). Otherwise, the latent heat does not contribute and the lower heating value (LHV) prevails. These measurements are typically performed in a bomb calorimeter and yield the higher heating value for the fuel. Table 2 reports higher heating values for several biomass materials. Obviously, heating values are important in performing energy balances on biomass conversion processes.

The inorganic constituents of biomass are important to different extents, depending on the conversion process under consideration. No comprehensive information on inorganic constituents will be provided here, although such information can be found in the literature on biomass. However, knowledge of the alkali metal content of biomass (that is, potassium and sodium) can be very important if the fuel is to be used in combustors. Experience in burning biomass reveals that excessive alkali salts in biomass, which are particularly concentrated in fast-growing biomass, can lead to ash fouling of boiler tubes, as described in 11.5 of this chapter. Table 11.3 contains information on alkali in ash for selected biomass materials that is useful in designing biomass combustion systems.

Planning a biomass conversion facility requires estimates of the total amount of land that must be put into production of biomass crops and of how far crops must be transported to a facility. Thus, the annual yield of biomass crops (kilograms per hectare) is important information for an engineer working on such a project. Unfortunately, yield information does not lend itself to tabulation, since it depends on so many

Table 11.2. Thermochemical properties of selected biomass

Biomass	HHV (dry) (MJ/kg)	Proximate Analysis (% wt., dry)			Ultimate Analysis (% wt., dry basis)						
		Volatile	Ash	Fixed C	C	H	O	N	S	Cl	Ash
Alfalfa seed straw	18.45	72.60	7.25	20.15	46.76	5.40	40.72	1.00	0.02	0.03	6.07
Almond shells	19.38	73.45	4.81	21.74	44.98	5.97	42.27	1.16	0.02		5.60
Black locust	19.71	80.94	0.80	18.26	50.73	5.71	41.93	0.57	0.01	0.08	0.97
Black oak	18.65	85.60	1.40	13.00	48.97	6.04	43.48	0.15	0.02		1.34
Cedar (western red)	20.56	86.50	0.30	13.20							
Corn cobs	18.77	80.10	1.36	18.54	46.58	5.87	45.46	0.47	0.01		1.40
Corn stover	17.65	75.17	5.58	19.25	43.65	5.56	43.31	0.61	0.01	0.21	6.26
Corn grain	17.20	86.57	1.27	12.16	44.00	6.11	47.24	1.24	0.14	0.60	1.27
Douglas fir	20.37	87.30	0.10	12.60	50.64	6.18	43.00	0.06	0.02		0.01
Food waste	7.59				17.93	2.55	12.85	1.13	0.06	0.38	5.10
Grape vines		80.10	2.20	17.70							
Hemlock (western)	19.89	87.00	0.30	12.70	50.40	5.80	41.40	0.10	0.10		2.20
Maize straw					47.09	5.54	39.79	0.81	0.12		5.77
Manure (beef, aged)	15.12				33.00	4.90	17.50	0.70	0.80		41.60
Manure (beef, fresh)	17.36				45.40	5.40	31.00	1.00	0.30		15.90
Municipal Solid Waste	19.87	76.30	12.00	11.70	47.60	6.00	32.90	1.20	0.30		12.00
Oak bark	19.47				49.70	5.40	39.30	0.20	0.10		5.30
Orchard prunings	19.05	83.30	2.10	14.60	49.20	6.00	43.20	0.25	0.04		1.38
Ponderosa pine	20.02	82.54	0.29	17.17	49.25	5.99	44.36	0.06	0.03	0.01	0.30

Table 11.2. (*Continued*)

Biomass	HHV (dry) (MJ/kg)	Proximate Analysis (% wt., dry)			Ultimate Analysis (% wt., dry basis)						
		Volatile	Ash	Fixed C	C	H	O	N	S	Cl	Ash
Poplar	19.38	82.32	1.33	16.35	48.45	5.85	43.69	0.47	0.01	0.10	1.43
Redwood (combined)	20.72	79.72	0.36	19.92	50.64	5.98	42.88	0.05	0.03	0.02	0.40
Refuse-derived fuel	17.40				42.50	5.84	27.57	0.77	0.48	0.57	22.17
Rice hulls	16.14	65.47	17.86	16.67	40.96	4.30	35.86	0.40	0.02	0.12	18.34
Rice straw (fresh)	16.28	69.33	13.42	17.25	41.78	4.63	36.57	0.70	0.08	0.34	15.90
Rice straw (weathered)	14.56	62.31	24.36	13.33	34.60	3.93	35.38	0.93	0.16		25.00
Sorghum stalks	15.40				40.00	5.20	40.70	1.40	0.20		12.50
Sudan grass	17.39	72.75	8.65	18.60	44.58	5.35	39.18	1.21	0.08	0.13	9.47
Sugarcane bagasse	17.33	73.78	11.27	14.95	44.80	5.35	39.55	0.38	0.01	0.12	9.79
Switchgrass	18.64	81.36	3.61	15.03	47.45	5.75	42.37	0.74	0.08	0.03	
Vineyard prunings	16.82				48.00	5.70	39.60	0.86	0.08		
Walnut shells	20.18	78.28	0.56	21.16	49.98	5.71	43.35	0.21	0.01	0.03	1.41
Water hyacinth	16.02		22.40		41.10	5.29		1.96	0.41		0.71
Wheat straw	17.51	71.30	8.90	19.80	43.20	5.00	39.40	0.61	0.11	0.28	11.40
White fir	19.95	83.17	0.25	16.58	49.00	5.98	44.75	0.05	0.01	0.01	0.20
Yard waste	9.32				23.29	2.93	17.54	0.89	0.15	0.13	10.07

Ref.: Various sources

Table 11.3. Alkali content of biomass

	Heating value (MJ/kg)	Ash in fuel (%)	Alkali in ash (%)
Hybrid Poplar	19.0	1.9	19.8
Pine Chips	19.9	0.7	3.0
Tree Trimmings	18.9	3.6	16.5
Urban Wood Waste	19.0	6.0	6.2
White Oak	19.0	0.4	31.8
Almond Shells	17.6	3.5	21.1
Bagasse - washed	19.1	1.7	12.3
Rice Straw	15.1	18.7	13.3
Switch Grass	18.0	10.1	15.1
Wheat Straw	18.5	5.1	31.5

Ref. [11]

variables: plant variety, soil type, landscape, climate, weather, and water drainage, to name a few. Table 11.4 has been included to give an idea of the kinds of yields that might be expected in various geographical locations for some energy crops that have been widely studied. Site specific information will require discussions with state extension agents and local agronomists in combination with field trials in advance of detailed plant design.

Bulk density is determined by weighing a known volume of biomass that is packed or baled in the form anticipated for its transportation or use. Clearly, solid logs will have higher bulk density than the same wood chipped. Bulk density will be an important determinant of transportation costs and the size of fuel storage and handling equipment. Volumetric energy content is also important in transportation and storage issues. Volumetric energy content, which is simply the energy content of fuel per unit

Table 11.4. Nominal annual yields of biomass crops

Biomass crop	Geographical location	Annual yield (kg/ha)
Corn: grain	North America	7,000
Corn: cobs	North America	1,300
Corn: stover	North America	8,400
Jerusalem artichoke: Tuber	North America	45,000
Jerusalem artichoke: Sugar	North America	6,400
Sugar Cane: Crop	Hawaii	55,000
Sugar Cane: Sugar	Hawaii	7,200
Sugar Cane: Bagasse (dry)	Hawaii	7,200
Sweet sorghum: Crop	Midwest U.S.	38,000
Sweet sorghum: Sugar	Midwest U.S.	5,300
Sweet sorghum: Fiber (dry)	Midwest U.S.	4,900
Switchgrass	North America	14,000
Hybrid poplar	North America	14,000
Wheat: grain	Canada	2,200
Wheat: straw	Canada	6,000

Ref. [18]

Table 11.5. Density and volumetric energy content of various solid and liquid fuels

Fuel	Density (kg/m³)	Volumetric Energy Content (GJ/m³)
Ethanol	790	23.5
Methanol	790	17.6
Biodiesel	900	35.6
Pyrolysis oil	1280	10.6
Gasoline	740	35.7
Diesel fuel	850	39.1
Agricultural residues	50–200	0.8–3.6
Wood	160–400	3–9
Coal	600—900	11–33

Ref. [7].

volume, is calculated by multiplying the higher heating value of a fuel by its bulk density. Table 11.5 compares bulk densities and volumetric energy contents of various liquid and solid fuels. The cost of collecting large quantities of biomass can be significant. Wood or other biomass resources must generally be produced within no more than a 50-mile radius of the power plant to be economical, given the high transportation costs and low densities of biomass.

11.2 THERMODYNAMIC CALCULATIONS FOR BIOMASS ENERGY

Mass and energy balances on process streams are critical to understanding the efficiency and economics of biomass conversion. Accounting for these balances is more complicated than for energy conversion processes that do not include chemical reaction because chemical constituents change and energy is released from the rearrangement of chemical bonds.

11.2.1 Mass Balances

In the absence of chemical reaction, the change in mass of a particular constituent within a control volume is equal to the difference in net mass flow of the constituent entering and exiting the control volume. Figure 11.1 illustrates mass balance for a system consisting of five inlets and five exits. In general, the mass balance for a given chemical constituent can be written in the form:

$$\frac{dm_{cv}}{dt} = \sum_i \dot{m}_i - \sum_e \dot{m}_e, \tag{11.1}$$

where m_{CV} is the amount of mass contained within the control volume, $\dot{m}_i$ and $\dot{m}_e$ are, respectively, the rates at which mass enters at i and exists at e, where we allow for the

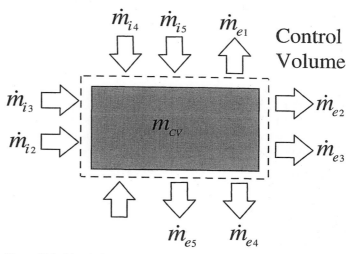

Figure 11.1. Mass balance on steady-flow control volume with five inlets and five exits.

possibility of several inlets and exits. At steady flow conditions, the net quantity of mass in the control volume is unchanging with time and Eq. (11.1) can be written as:

$$\sum_i \dot{m}_i = \sum_e \dot{m}_e. \qquad (11.2)$$

However, when chemical reaction occurs, chemical compounds are not conserved as they flow through the system. For example, methane (CH_4) and oxygen (O_2) entering a combustor are consumed and replaced by carbon dioxide (CO_2) and water (H_2O):

$$CH_4 + 2O_2 \rightarrow CO_2 + 2H_2O \qquad (11.3)$$

Accordingly, mass balance cannot be written for methane and oxygen using either Eq. (11.1) (unsteady flow) or Eq. (11.2) (steady flow). Although chemical compounds are not conserved, the chemical elements making up these compounds are conserved; thus, elemental mass balances can be written.

In the case of the reaction of CH_4 with O_2, mass balances can be written for the chemical elements carbon (C), hydrogen (H), and oxygen (O). However, because chemical compounds react in distinct molar proportions, it is usually more convenient to write molar balances on the elements. Recall that a mole of any substance is the amount of mass of that substance that contains as many individual entities (whether atoms, molecules, or other particles) as there are atoms in 12 mass units of carbon-12. For engineering systems, it is usually more convenient to work with kilograms as the unit of mass; thus, for this measure kilomole (kmol) will be employed instead of the gram-mole (gmol) that often appears in chemistry books. The number of kilomoles, n, of a substance is related to the number of kilograms, m, of a substance by its molecular weight, M:

$$n = \frac{m}{M}.$$

Working on a molar basis, it is straightforward to account for mass changes that occur during chemical reactions: an overall chemical reaction is written that is supported by molar balances on the elements appearing in the reactant and product chemical compounds.

Example 11.1. One kilogram of methane reacts with air. (a) If all of the methane is to be consumed, how many kilograms of air will be required? (b) How many kilograms of carbon dioxide, water, and nitrogen will appear in the products?

Solution. One kilogram of methane, with a molecular weight of 16, is calculated to be 1/16 kmol using Eq. 11.4. Air is approximated at 79% nitrogen and 21% oxygen on a molar basis. The overall chemical reaction can be written as:

$$(\tfrac{1}{16})CH_4 + a\left(O_2 + \frac{0.79}{0.21}N_2\right) \rightarrow xCO_2 + yH_2O + zN_2,$$

where a is the number of moles of oxygen required to consume 1/16 moles of CH_4 and x, y, and z are the moles of CO_2, H_2O, and N_2, respectively, in the products. The unknowns in this equation can be found from molar balances on the elements C, H, O, and N:

carbon:

$$\tfrac{1}{16} = x \text{ (kmoles)}, \therefore m_{co_2} = n_{co_2} \times M_{co_2} = \tfrac{1}{16} \times 44 = 2.75 \text{ kg};$$

hydrogen:

$$\tfrac{1}{16} \times 4 = 2y$$

$$y = \tfrac{1}{8} \text{ (kmoles)}, \therefore m_{H_2O} = n_{H_2O} \times M_{H_2O} = \tfrac{1}{8} \times 18 = 2.25 \text{ kg};$$

oxygen:

$$2a = 2x + y = 2 \times \tfrac{1}{16} + \tfrac{1}{8} = \tfrac{1}{4}$$

$$a = \tfrac{1}{8} \text{ (kmoles)}, \therefore m_{O_2} = n_{O_2} \times M_{O_2} = \tfrac{1}{8} \times 32 = 4 \text{ kg};$$

nitrogen:

$$\frac{0.79}{0.21}a = \frac{0.79}{0.21} \times \frac{1}{8} = 0.47 = z \text{ (kmoles)},$$

$$\therefore m_{N_2} = n_{N_2} \times M_{N_2} = 0.47 \times 28 = 13.2 \text{ kg}.$$

A check shows that 18.2 kg of methane and air are converted into 18.2 kg of products in the form of carbon dioxide, water, and nitrogen.

In some processes, like combustion and gasification, it is useful to compare the actual oxygen provided to the fuel to the amount theoretically required for complete oxidation (the stoichiometric requirement). There are several useful comparisons, described below.

The fuel-oxygen ratio, F/O, is defined as the mass of fuel per mass of oxygen consumed (a molar fuel-oxygen ratio is also sometimes defined). Another frequently used ratio is the equivalence ratio, ϕ:

$$\phi = (F/O)_{actual}/(F/O)_{stoichiometric}. \tag{11.5}$$

This ratio is less than unity for fuel-lean conditions and greater than unity for fuel-rich conditions. Two other ratios, theoretical air and excess air, can be calculated on either mass or molar bases:

$$\text{Theoretical air (\%)} = (\text{actual air/stoichiometric air}) \times 100, \tag{11.6}$$

$$\text{Excess air (\%)} = (\text{actual air} - \text{stoichiometric air}) \times 100/\text{stoichiometric air}. \tag{11.7}$$

11.2.2 Energy Balances

In the absence of chemical reaction, the net change in stored energy within a control volume is given by the net flow of energy into the control volume. This is in the form of heat and work as well as kinetic energy, potential energy, and enthalpy associated with mass flowing into and out of the control volume. Figure 11.2 illustrates the energy balance for a control with two inlets and one outlet with work transferred in and heat transferred out. More generally, a system undergoing steady flow processes can be described by an energy balance of the form:

$$\frac{dE_{CV}}{dt} = \dot{Q}_{CV} - \dot{W}_{CV} + \sum_i \dot{m}_i \left(h_i + \frac{1}{2}V_i^2 + gz_i \right) - \sum_e \dot{m}_e \left(h_e + \frac{1}{2}V_e^2 + gz_e \right), \tag{11.8}$$

where E_{CV} is the stored energy in the control volume. $\dot{Q}_{CV}$ and $\dot{W}_{CV}$ are the rates at which heat and work cross the control volume boundary, h is enthalpy, V is velocity, and z is elevation with respect to an arbitrary datum for the mass flows at the inlet, i,

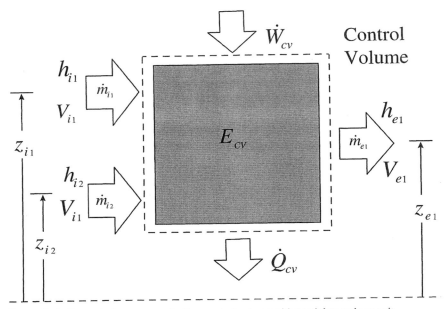

Figure 11.2. Energy balance on steady-flow control volume with two inlets and one exit.

and outlet, e. In steady flow, with a single inlet and single outlet and no velocity or elevation changes in the system, Eq. 11.8 simplifies to:

$$\dot{Q}_{CV} - \dot{W}_{CV} = \dot{m}(h_e - h_i). \tag{11.9}$$

However, for a chemically reacting system, this formulation of an energy balance does not take into account changes in the chemical composition of the system nor the chemical energy absorbed or released during these reactions. Like mass conservation, it is more convenient to present energy conservation in a molar formulation rather than a mass formulation when chemical reaction occurs. The intensive property enthalpy, h, with units of kJ/kg, is replaced by the intensive property molar enthalpy, $\bar{h}$, with units of kJ/kmol. Table 11.6 is an abbreviated collection of molar enthalpies of selected gases as a function of temperature. More extensive collections are available in thermodynamics textbooks [12].

Energy conservation on a molar basis for a steady flow system consisting of one inlet for reactants r and one outlet for products p (and neglecting velocity or elevation changes) is of the form:

$$\dot{Q}_{CV} - \dot{W}_{CV} = \sum_p \dot{n}_p \bar{h}_p - \sum_r \dot{n}_r \bar{h}_r, \tag{11.10}$$

Table 11.6. Thermodynamic properties for selected gases

	Molar Enthalpies (kJ/kmol)					
T (K)	N_2	O_2	H_2O	CO	CO_2	H_2
0	0	0	0	0	0	0
298	8,669	8,682	9,904	8,669	9,364	8,468
500	14,581	14,770	16,828	14,600	17,678	14,350
1000	30,129	31,389	35,882	30,355	42,769	29,154
1500	47,073	49,292	57,999	47,517	71,078	44,738
2000	64,810	67,881	82,593	65,408	100,804	61,400
2500	82,981	87,057	108,868	83,692	131,290	78,960
3000	101,407	106,780	136,264	102,210	162,226	97,211
3250	110,690	116,827	150,272	111,534	177,822	106,545

Substance	Formula	$\bar{h}_f^o$ (kJ/kmol)
Carbon	C(s)	0
Hydrogen	H_2	0
Nitrogen	N_2	0
Oxygen	O_2	0
Carbon monoxide	CO	−110,530
Carbon dioxide	CO_2	−393,520
Water liquid	H_2O (l)	−285,830
Water vapor	H_2O (g)	−241,820
Methane	CH_4	−74,850
Methyl alcohol liquid	CH_3OH (l)	−238,810
Methyl alcohol vapor	CH_3OH (g)	−200,890
Ethyl alcohol liquid	C_2H_5OH (l)	−277,690
Ethyl alcohol vapor	C_2H_5OH (g)	−235,310

Ref. [12].

where $\dot{n}$ specifies the molar flow rate of a chemical constituent and the summation is over all the products p at the exit or all the reactants r at the inlet. Integrated over a finite time interval, this equation takes the form:

$$Q_{CV} - W_{CV} = \sum_p n_p \bar{h}_p - \sum_r n_r \bar{h}_r = H_p - H_r, \qquad (11.11)$$

where Q_{CV} and W_{CV} are the amounts of heat and work done over a designated time interval and n_r and n_p are the moles of reactants and products, respectively, crossing the control surface in the time interval. A convenient short-hand is to designate H_p and H_r as the mixture enthalpies (kJ) of the products and reactants, respectively.

An enthalpy-temperature diagram, shown in Figure 11.3, is useful in understanding the change in enthalpy that occurs in the presence of a chemical reaction. For a given mixture of reactants, say methane and oxygen, there is a unique enthalpy-

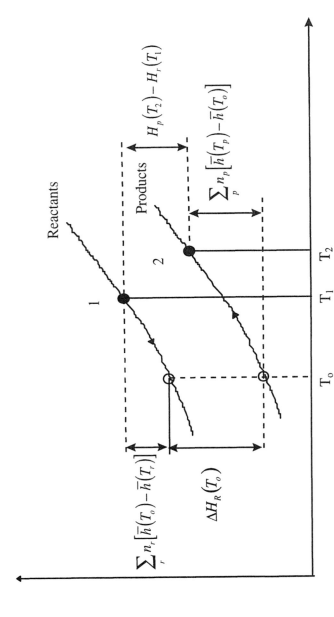

Figure 11.3. Relationship between mixture enthalpy and temperature for a chemically-reacting system.

temperature relationship. Similarly, a unique enthalpy relationship exists for the products of methane oxidation (carbon dioxide and water). It is also easy to understand the enthalpy change that occurs for a constant temperature chemical reaction: reactants at T_o are converted to products at T_o with a release or absorption of energy known as the enthalpy of reaction $\Delta H_R(T_o)$ at temperature T_o. Reactions that release energy (exothermic reactions) have negative enthalpies of reaction whereas reactions that absorb energy (endothermic reactions) have positive enthalpies of reaction. This situation becomes obvious by inspecting Eq. 11.11 and recalling the convention that Q_{CV} is negative for heat flow out of a system.

However, the typical chemical reaction is not isothermal; indeed, many combustion reactions are accompanied by temperature increases of over 1000 K. Thus, enthalpy changes must account for sensible enthalpy changes of the reactants, sensible enthalpy changes for the products, and the release or absorption of heat as a result of the chemical reaction. One way to handle this potentially complicated situation is to visualize the reaction as following the reaction pathway illustrated in Figure 11.3: reactants initially at temperature T_1 are cooled to temperature T_o at which point the reactants undergo isothermal chemical reaction to form products that are then heated to the final temperature T_2. Thus, the enthalpy change for nonisothermal chemical reactions can be calculated from the relationship:

$$H_p(T_2) - H_r(T_1) = \sum_p n_p[\bar{h}(T_2) - \bar{h}(T_o)]_p + \Delta H_R(T_o) - \sum_r n_r[\bar{h}(T_1) - \bar{h}(T_o)]_r.$$

$$(11.12)$$

Sensible enthalpies, $\bar{h}(T)$, for a variety of chemical substances are available as tabulations of thermodynamic properties of substances. Likewise, enthalpies of reaction at a specified reference temperature, T_o have been compiled for a number of chemical reactions. In the SI system, T_o is chosen as 298 K for the purpose of tabulating data. Using Eq. 11.12, tabulations of sensible enthalpies and enthalpies of reaction can be used to calculate enthalpy changes for reactions under a wide variety of conditions.

Example 11.2. One mole of biogas produced by anaerobic digestion of animal waste consists of 60% methane and 40% carbon dioxide by volume (that is, molar basis). The biogas reacts with 1.2 moles of oxygen to form carbon dioxide and water. The enthalpy of reaction for methane is $-890,330$ kJ/kmol at 298 K. Assume the reactants are initially at 298 K. Calculate the enthalpy change if the reactants are at 298 K and the products are at 1500 K.

Solution. The complete reaction is

$$(0.6CH_4 + 0.4CO_2) + 1.2O_2 \rightarrow CO_2 + 1.2H_2O$$

From Table 11.6 the following sensible enthapies are found:

Temperature (K)	$h_{CH_4}(T)$ (kJ/kmol)	$h_{O_2}(T)$ (kJ/kmol)	$h_{CO_2}(T)$ (kJ/kmol)	$h_{H_2O}(T)$ (kJ/kmol)
298	—	8,682	9,364	9,904
1500	—	—	71,078	57,999

Substituting values from Eq. 11.12:

$$[1(71,078 - 9,364) + 1.2(57,999 - 9,904)]$$
$$+ 0.6(-890,330) - [0.6(0) + 0.4(0) + 1.2(0)] = -414,770 \text{ kJ/kmol.}$$

Thus, 414.77 kJ is released by the combustion of 1 mole of biogas under these conditions.

For some well-characterized fuels, such as hydrogen, methane, and ethanol, enthalpies of reaction can be calculated from tabulations of specific enthalpies of formation, $\Delta\bar{h}_f^o$ of chemical compounds from their elements at a standard state:

$$\Delta H_R = \sum_p n_p (\Delta h_f^o)_p - \sum_r n_r (\Delta h_f^o)_r, \tag{11.13}$$

where n_r and n_p are the stoichiometric coefficients for reactants and products of a chemical reaction. However, many biomass fuels are not well characterized in terms of their chemical constituents because of variations in composition or the presence of complicated chemical compounds. As a result, it is often simpler to perform calorimetric tests on biomass fuels to determine enthalpy of reaction.

Example 11.3. Use standard enthalpies of formation to calculate the enthalpy of reaction of liquid ethanol (C_2H_5OH) with oxygen to form carbon dioxide and water vapor.

Solution. The stoichiometric reaction is expressed by:

$$C_2H_5OH(\ell) + 3O_2 \rightarrow 2CO_2 + 3H_2O(g).$$

From Table 11.6:

Compound	$C_2H_5OH_{(\ell)}$	O_2	CO_2	$H_2O(g)$
Δh_f^o (kJ/kmol)	-277,690	0	-393,520	-241,820
Stoichiometric Coefficient	1	2	1	2

Substituting into Eq. 11.13 yields:

$$\Delta H_R = 2(-393,520) + 3(-241,820) - (-277,690) - 3(0)$$
$$= -1,234,800 kJ/kmol.$$

11.2.3 Thermodynamic Efficiency

The conversion of chemical energy stored in biomass into more useful forms, such as gaseous and liquid fuels or electrical power, is accompanied by loss of energy to forms that are not easily recovered or utilized. There are many reasons for such losses. Separation processes can inadvertently reject valuable fractions of a feedstock to waste streams. Heat losses can reduce the amount of energy available to energy conversion processes. Entropy production, inherent in even ideal processes, limits the amount of energy that can be converted into useful forms. Every energy conversion process can be characterized by its thermodynamic efficiency defined as:

$$\eta = \frac{E_{out}}{E_{in}}, \tag{11.14}$$

where E_{in} = all forms of energy entering the conversion process, and
E_{out} = useful energy leaving the conversion process.

These energy flows represent net energy entering and leaving the process. Energy entering the process includes the chemical energy content of the feedstock, thermal energy needed for chemical reactors and distillation units, and power to drive mechanical devices such as tub grinders, pumps, and blowers. The useful energy leaving the process includes chemical energy of gaseous and liquid fuels, electric power, and heat that can be used external to the process, such as district heating of buildings. Obviously, energy must be conserved and the fact that most conversion processes have efficiencies substantially less than unity indicates that some energy leaves the process as waste energy (i.e., unconverted feedstock or waste heat).

In the electric power industry, the efficiency of converting chemical energy into electric power is frequently expressed as the heat rate, defined as:

$$HR = \frac{E_{in}}{E_{out}} \; (\text{Btu/kW} - \text{h}), \tag{11.15}$$

where E_{in} is the rate of chemical energy entering as boiler fuel measured in English units (Btu/h) and E_{out} is the electrical power output measured in SI units (kW). Heat rate is closely related to thermodynamic efficiency except that it is expressed in a mixed systems of units. Notice that low heat rates correspond to high thermodynamic efficiencies.

11.3 CONVERSION OF BIOMASS TO GASEOUS FUELS

11.3.1 Anaerobic Digestion

Anaerobic digestion is the decomposition of organic waste to gaseous fuel by bacteria in an oxygen-free environment. The process occurs in stages to successively break down the organic matter into simpler organic compounds. The desired product, known as biogas, is a mixture of methane (CH_4), carbon dioxide (CO_2), and some trace gases. Most digestion systems produce biogas that is between 55 percent and 75 percent methane by volume. Biogas, once treated to remove sulfur compounds, can substitute for natural gas in many applications, including stationary power generation. Biogas is a suitable fuel for engine generator sets, small gas turbines, and some kinds of fuel cells.

The biological processes within an anaerobic digester that lead to biogas are relatively complicated. The process consists of three basic steps: hydrolysis and fermentation, transitional acetogenic dehydrogenation, and methanogenesis. Hydrolysis and fermentation involve hydrolytic and fermentative bacteria that break down proteins, carbohydrates, and fats into simpler acids, alcohols, and neutral compounds. Hydrogen and carbon dioxide are also produced. This step is followed by transitional digestion through the action of acid-forming bacteria. Products of fermentation that are too complex for methane-forming bacteria to consume are further degraded in this step to acetate, hydrogen, and carbon dioxide. Traces of oxygen in the feedstock are consumed in this step, which benefits oxygen-sensitive, methane-forming bacteria. The final step, methanogenesis, converts acetate to methane by the action of methane-forming bacteria.

Basic anaerobic digestion systems are relatively simple, as illustrated in Figure 11.4. For some feedstocks, removal of scum may be desirable in advance of the anaerobic digestor. Whether batch or steady flow processing, two effluent streams result: biogas and sludge. The biogas may have to be scrubbed in advance of combustion to remove hydrogen sulfide that would otherwise appear as the pollutant sulfur dioxide in the exhaust gas.

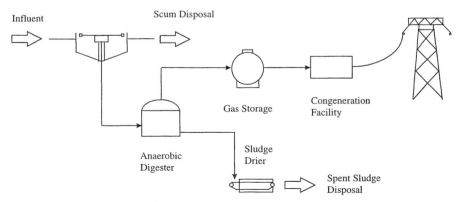

Figure 11.4. Basic anaerobic digestive system.

Examples of anaerobic reactor designs include the simple batch reactor, plug-flow reactors, continuously stirred tank reactors (CSTR), upflow reactors, and two-tank reactor systems. The batch reactor is a single vessel design in which waste is added and removed in batches, and all steps of the digestion process take place in the same environment. More advanced reactor designs aim to improve waste contact with active bacteria, to separate and control the environments for acid-forming and methane-forming bacteria or both. Relatively new two-tank reactor technology physically separates the acid formation and methane formation phases of digestion, so that each takes place under optimal conditions. Benefits of this technology are numerous: hydrolysis and acidification occur quicker than in conventional systems; the common problems of foaming in single-tank systems are reduced by the destruction of biochemical foaming agents before they reach the methane forming reactor; and the biogas produced is typically rich in methane.

The biological conversion process of anaerobic digestion can be used for a wide range of feedstocks of medium-to-high moisture content. Sewage sludge is most commonly used as feedstock for anaerobic digesters, but municipal solid waste (MSW), food processing wastes, agricultural wastes, Napier grass, kelp, bagasse, and water hyacinth can also be digested. Digesters are designed to maintain optimal conditions for a specific type of waste. Waste pretreatment, heating, mixing, nutrient addition, specialized bacteria addition, and pH adjustment can be manipulated to control digester performance.

Typical thermodynamic efficiency in converting the chemical energy of dry matter to methane is about 60 percent with methane yields ranging between 8,000–9,000 cubic feet/ton of volatile solids added to the digester. However, thermodynamic efficiencies approaching 90 percent with methane yields of 11,000–13,000 cubic feet/ton volatile solids have been achieved in research programs.

11.3.2 Thermal Gasification

Thermal gasification is the conversion of solid, carbonaceous fuels into flammable gas mixtures, sometimes known as producer gas, consisting of carbon monoxide (CO), hydrogen (H_2), methane (CH_4), nitrogen (N_2), carbon dioxide (CO_2), and smaller quantities of higher hydrocarbons. This gas can be burned directly in a furnace to generate process heat or it can fuel internal combustion engines, gas turbines, or fuel cells. It can also serve as feedstock for production of liquid fuels, as described in a subsequent section. Because of this flexibility of application, gasification has been proposed as the basis for energy refineries that would provide a variety of energy and chemical products, including electricity and transportation fuels. Gasification consists of two processes acting in parallel: combustion and pyrolysis. Combustion is the rapid oxidation of fuel accompanied by the release of large quantities of heat (that is, a fast, exothermic reaction). Pyrolysis is the thermal degradation of solid fuel into a variety of simple gases and organic vapors and liquids. Depending on the reaction conditions, these pyrolysis products further react to form the final products of gasification. Pyrolysis is an endothermic process, requiring an external source of heat for reaction to progress. Combustion is encouraged, by the addition of air or oxygen to the gasifier, only to the extent that heat is required to drive pyrolysis.

Gasification includes several physical and chemical steps. The process begins with the addition of heat to raise the temperature of the fuel particles. First moisture is driven off at temperatures around $100°C$. Depending on the kind of reactor, the fuel composition, and the size of fuel, this may require several minutes to accomplish or may occur almost instantaneously.

At about $400°C$, the complex structure of biomass begins to breakdown with the release of gases, vapors, and liquids. Many of these released components are flammable and contribute significantly to the heating value of the product gas from the gasifier. Of the original solid there remains only char, a porous solid consisting mostly of elemental carbon and ash. Pyrolysis reaction times and product yields also depend on fuel properties and reaction conditions. Reaction times range from milliseconds to minutes. Reaction yields range from mostly liquids to exclusively low-molecular weight gases. However, the total pyrolysis yield of pyrolysis products and the amount of char residue can be estimated from the proximate analysis of the fuel, described in section 11.1 of this chapter. The fuel's volatile matter roughly corresponds to the pyrolysis yield while the combination of fixed carbon and ash content can be used to estimate the char yield.

As the temperature approaches $700°C$, the char begins to react with oxygen, carbon dioxide, and water vapor to produce additional flammable gas. Some of these reactions are exothermic and drive the remaining endothermic reactions:

Carbon-oxygen reaction:

$$2C + O_2 = 2CO \tag{11.16}$$

Boudouard reaction:

$$C + CO_2 = 2CO \tag{11.17}$$

Carbon-water reaction:

$$C + H_2O = H_2 + CO \tag{11.18}$$

Water-shift reaction:

$$CO + H_2O = H_2 + CO_2 \tag{11.19}$$

Hydrogenation reaction:

$$C + 2H_2 = CH_4 \tag{11.20}$$

Unlike combustion, where important reactions go essentially to completion, the above gasification reactions are reversible, resulting in a complex equilibrium among the gaseous products, which depends on temperature and the proportions of fuel, oxidant, and steam added to the gasifier. For example, methane formation is favored by low temperatures, high pressures, low oxygen concentrations, and high water vapor pressures in the gasifier. The thermodynamic calculations necessary to predict product yields is beyond the scope of this chapter. However, typical gas composition for several kinds of gasifiers is given in Table 11.7.

Gasifiers are generally classified according to the method of contacting fuel and gas with the four main types being counterflow, coflow, entrained flow, and fluidized bed. Biomass gasifiers commonly operate near atmospheric pressure, although some recent designs operate at ten atmospheres of pressure or more.

Counterflow gasifiers, also called updraft gasifiers, are the simplest as well as the first type of gasifier developed. They were a natural evolution from charcoal kilns, which yielded smoky yet flammable gas as a waste product, and blast furnaces, which generated product gas that reduced ore to metallic iron. Counterflow gasifiers are little more than grate furnaces with fuel admitted from above and insufficient air for complete combustion entering from below. Just above the grate, where air first contacts the fuel, combustion occurs and very high temperatures are produced. Although the gas flow is depleted of oxygen higher in the fuel bed, hot carbon dioxide and water vapor from combustion near the grate reduces char to hydrogen and carbon monoxide. These reactions cool the gas, but temperatures are still high enough to heat, dry, and pyrolyze the fuel moving downward to the grate. Of course, pryolysis releases both condensible and noncondensible gases, and the producer gas leaving a counterflow gasifier contains large quantities of tars. The product gas can be passed through a catalytic reactor to crack the tar, but counterflow gasifiers are generally not strong candidates for biomass energy applications.

In cocurrent, or downdraft, gasifiers fuel and gas move in the same direction. Cocurrent gasifiers appear to have been developed near the end of the nineteenth century after the introduction of induced draft fans allowed air to be drawn downward through a gasifer in the same direction as the gravity-fed fuel. Contemporary designs usually include an arrangement of tuyeres that admits air or oxygen directly into the

Table 11.7. Producer gas composition from various kinds of gasifiers

Gasifier Type	Gaseous constituents (vol.% dry)						Gas quality	
	H_2	CO	CO_2	CH_4	N_2	HHV (MJ/m^3)	Tars	Dust
Air-blown fluidized bed	9	14	20	7	50	5.4	medium	high
Air-blown downdraft	17	21	13	1	48	5.7	low	medium
Oxygen-blown downdraft	32	48	15	2	3	10.4	low	low
Indirectly-heated fluidized bed	31	48	0	21	0	17.4	medium	high

Ref. [4].

region of combustion. This design assures that condensable gases released during pyrolysis are forced to flow through a bed of hot char, where tars are cracked. The producer gas is relatively free of tars, making it a satisfactory fuel for engines. A disadvantage of this type of reactor is that slagging or sintering of ash may occur due to the concentrated oxidation zone. Rotating ash grates or similar mechanisms can solve this problem. Higher gas outlet temperature and lower conversion of char to gas result in a lower efficiency compared to the countercurrent reactor.

In the entrained flow gasifier, finely pulverized fuel is injected into the reactor with a stream of gas. This reactor was developed for steam-oxygen gasification of coal at temperatures of 1100–1400°C. The technology is attractive for advanced coal power plants but is unlikely to be used for commercial biomass gasification. Processing of biomass material into finely divided particles appropriate to entrained flow gasifiers is prohibitively expensive. Furthermore, the low energy density of biomass makes it difficult to achieve the high temperatures characteristic of entrained flow gasifiers.

In a fluidized bed gasifier, a gas stream passes vertically upward through a bed of inert particulate material to form a turbulent mixture of gas and solid. Fuel is added at such a rate that it is only a few percent by weight of the bed inventory. Unlike the counterflow and concurrent flow gasifiers, no segregated regions of combustion, pyrolysis, and tar cracking exist. The violent stirring action makes the bed uniform in temperature and composition with the result that gasification occurs simultaneously at all locations in the bed. In a sufficiently deep bed, tar cracking eliminates most tar from the producer gas. Fluidized beds are gaining favor as the reactor of choice for biomass gasification. The high thermal mass of the bed imparts a high degree of flexibility in the kinds of fuels, including those of high moisture content, that can be gasified. Disadvantages include relatively high exit gas temperatures, which complicates efficient energy recovery, and relatively high particulate burdens in the gas due to the abrasive forces acting within the fluidized bed.

Depending on the kind of gasifier, product gas can be contaminated by tar and particulate matter. Tar is a black, viscous liquid formed from high molecular weight organic compounds released during pyrolysis. Tar cannot be tolerated in the operation of many heat and power machines, including internal combustion engines, fuel cells, and methanol synthesis reactors. A well-designed gasifier has a high temperature zone through which pyrolysis products pass, allowing tars to thermally decompose ("crack") into the low molecular gases predicted by equilibrium theory. Particulate matter is a combination of ash and unreacted char from the biomass. Particulate loadings vary considerably from one type of gasifier to another.

Gas cleaning usually involves separate processes for removing solid particles and tar vapor or droplets. A gas cyclone operated above the condensation temperature for the least volatile tar constituent can remove most ash and char particles larger than about 5 microns in diameter. There are two approaches to removing tars from product gas. The gas can be passed through a packed bed reactor containing granulated catalyst capable of cracking the tar to low-molecular weight hydrocarbons. Steam or oxygen are sometimes used to promote this reaction at temperatures on the order of 900°C. Catalytic reactors can be very efficient in removing tars but are relatively expensive. An approach that removes both tar and particles scrubs the gas stream with a mist of water or oil. Gas scrubbers are a well developed technology in the chemical industry and relatively inexpensive. However, removal of tar after gasification reduces overall energy conversion efficiency and increases problems with waste disposal. Methanol production requires even further cleaning of producer gas to remove sulfur and certain hydrocarbons. Hot-gas clean-up, as it is known, is a subject of current research. Possibilities include ceramic filters and moving bed filters.

The thermodynamic efficiency of gasifiers is strongly dependent on the kind of gasifier and how the product gas is employed. Some high temperature, high pressure gasifiers are able to convert 90 percent of the chemical energy of solid fuels into chemical and sensible heat of the product gas. However, these high efficiencies come at high capital and operating costs. Most biomass gasifiers have conversion efficiencies ranging between 70 and 80 percent. In some applications, such as process heaters or driers, both the chemical and sensible heat of the product gas can be utilized. In many power applications, though, the hot product gas must be cooled before it is utilized; thus, the sensible heat of the gas is lost. In this case, what is known as cold gas efficiency can be as low as 50–60 percent. Whether the heat removed from the product gas can be recovered for other applications, like steam raising or fuel drying, ultimately determines which of these conversion efficiencies is most meaningful. Gas cleaning to remove tar and particulate matter also has a small negative impact on gasifier efficiency since it removes flammable constituents from the gas (tar and char particles) and generally requires a small amount of energy to run pumps.

11.4 LIQUID FUELS FROM BIOMASS

11.4.1 Ethanol Fermentation

Ethanol, or ethyl alcohol (C_2H_5OH), is a liquid fuel that can be burned in modified internal combustion engines. However, it is most commonly used as a 5 percent blend with gasoline known as gasohol, which can be burned in unmodified automotive engines. Ethanol can also serve as feedstock for production of ethyl tertiary butyl ether (ETBE), a high-value, octane-enhancing gasoline additive. Ethanol, whether for fuel or alcoholic beverages, is produced by fermentation, which is the decomposition of complex organic molecules into simpler compounds by the action of microorganisms.

A variety of carbohydrates (sugars, starches, hemicellulose, and cellulose) can serve as feedstock in ethanol fermentation as long as they can be broken down to sugars that are susceptible to biological action. Plant materials high in sugar, such as sugar cane, sugar beets, and sweet sorghum, are the simplest to convert to ethanol since these sugars can be directly fermented. Sugar beets, for example, are the basis of Brazil's fuel ethanol industry.

The process becomes more difficult as the carbohydrate becomes more complicated. Starch, a storage carbohydrate for plants, requires an additional processing step before it can be used in fermentation. Starch is essentially long chains of maltose sugar connected by hydrogen bonds, which can be decomposed into simple sugars by breaking these bonds. The process, known as hydrolysis, is accomplished by cooking milled starch granules in a dilute acid solution or by reacting the starch with thermophyllic (heat tolerant) enzymes. The cost of hydrolyzing starch is a very small proportion of the cost of producing ethanol. The cost of starch crops dominates process economics. Starch crops suitable as feedstock include corn, grains, and potatoes. Corn starch has been the basis for the United States fuel ethanol industry for over two decades.

Inulin is another storage carbohydrate suitable for fermentation to ethanol. Like starch, inulin is essentially chains of sugars that can be hydrolyzed to simple sugars by acid or enzymes. Jerusalem artichokes and dahlias are notable for their high inulin content.

Cellulose and hemicellulose found in the plant cell walls of woody and herbaceous energy crops can also be converted to fermentable sugars, but the process is relatively difficult. Plant cell walls are actually a composite structure of cellulose fibers imbedded in a cross-linked matrix of lignin and hemicellulose complexes. The resulting material, known as lignocellulose, is both strong and durable and not easily depolymerized to fermentable sugars either by biological or chemical action. However, ruminant animals, such as cattle, have developed strategies for breaking down lignocellulose to digestable materials, processes that can be mimicked in a biomass conversion plant. The key process is disruption of the lignin-hemicellulose sheath, which increases pore volume and surface area accessible to the imbedded cellulose fibers. Acid or cellulose-hydrolyzing enzymes (cellulase) can then penetrate to and react with the cellulose to yield glucose. This disruption can be accomplished physically (milling or heat treatment) or chemically (basic or acid reactions). A number of processes have been developed; commercial development has not advanced sufficiently to know which method will prove the most cost-effective. However, many studies suggest that lignocellulose-to-ethanol processes that employ enzyme hydrolysis of cellulose (rather than acid hydrolysis) will ultimately prove the most economical.

Conversion of lignocellulose to ethanol is flow charted in Figure 11.5. The feedstock is first milled fine enough to be susceptible to a pretreatment process designed to disrupt the lignin-hemicellulose sheath. Depending on the pretreatment, hemicellulose may be hydrolyzed to simple sugars, lignin dissolved into solution (solubilized), or a combination of hydrolysis and solubilization may occur. The end result is the separation of lignocellulose into its three main components: cellulose, hemicellulose, and lignin. The hemicellulose, as the result of hydrolysis during pretreatment, appears as xylose, a five-carbon (C5) sugar. The cellulose, although liberated from the lignin-hemicellulose sheath, is relatively resistant to hydrolysis and must be subjected to an additional step of acid or enzymatic hydrolysis to produce glucose, a six-carbon (C_6) sugar. Lignin can, in principle, be processed into adhesives or octane enhancers for gasoline but economic studies to date suggest that lignin is best used as boiler fuel, the resulting steam being used in subsequent ethanol distillation.

Glucose is readily fermented to ethanol by a variety of yeast employed by the fermentation industry. Xylose, on the other hand, is more difficult to ferment. Only in recent years have strains of microorganisms been identified that can yield ethanol from xylose. Batch fermentation limits yields to about 10% alcohol solution. Continuous fermentation results in higher yields before the yeast is rendered ineffective. A distillation step separates the alcohol and water to yield 94% ethyl alcohol. Production of an-

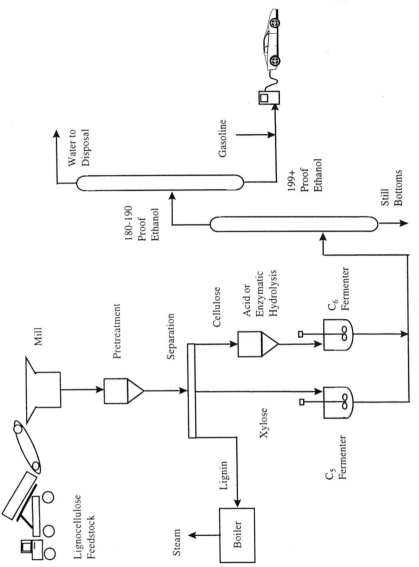

Figure 11.5. Conversion of lignocellulose to ethanol as fuel additive.

hydrous ethanol (99.8 percent purity) from this mixture requires an additional, energy-intensive step.

Although thermodynamic efficiencies for conversions of carbohydrates to ethanol can be calculated, it is more typical to report the volumetric yield of ethanol per unit mass of feedstock. The yield of ethanol from energy crops varies considerably. Among sugar crops, sweet sorghum yields 80 liters/ton, sugar beets yield 90–100 liters/ton, and sugar cane yields 75 liters/ton. Among starch and inulin crops, the ethanol yield is 350–400 liters/ton of corn, 400 liters/ton of wheat, 90 liter/ton of Jerusalem artichoke. Among lignocellulosic crops, the potential ethanol yield is 400 liters/ton of hybrid poplar, 450 liters/ton for corn stover, 510 liters/ton for corn cobs, and 490 liters/ton for wheat straw.

11.4.2 Chemical Synthesis of Methanol

Methanol, also known as methyl alcohol, is a liquid fuel that can be burned in modified internal combustion engines. It can be blended with gasoline, most commonly as an 85% methanol and 15% gasoline mixture known as M85. Methanol is also used as a feedstock for production of methyl tertiary butyl ether (MTBE), an octane-enhancing gasoline additive. Methanol (CH_3OH) is produced by the reaction of CO and H_2:

$$CO + 2H_2 \rightarrow CH_3OH \quad \Delta H = -90.77 \ kJ/mol. \qquad (11.21)$$

The CO and H_2 required for this process can be produced by gasifying biomass fuels. Gasification often produces less hydrogen than the 2:1 ratio of H_2 to CO indicated in Eq. 11.21 for methanol synthesis; thus, the gas mixture (syn gas) is often reacted with steam in the presence of a catalyst to promote a shift to higher hydrogen content:

$$CO + H_2O \rightarrow CO + H_2 \quad \Delta H = -40.5 \ kJ/mol. \qquad (11.22)$$

CO_2 and H_2S in the syn gas are removed prior to the methanol reactor where the gas reacts with a catalyst at elevated temperatures and pressures to produce methanol in a highly exothermic reaction.

The two most common commercial reactor systems for methanol synthesis are the Imperial Chemical Industries (ICI) reactor and the Lurgi reactor, which operate at temperatures between 200°C and 300°C and pressures between 500 and 100 bar. Both employ zinc-copper catalyst in gas-phase synthesis and differ mainly in their design and heat-removal ability. The ICI process uses packed beds of catalyst with cold synthesis gas injected between them to control reactor temperature. The Lurgi process uses a catalyst packed in the tubes of a shell and a tube heat exchanger with heat removed by converting water to steam.

Yields of methanol from woody biomass are expected to be in the range of 480–565 liter/ton. Variations in methanol yields from different biomass feedstocks are expected to be relatively small. The heating values of biomass materials are remarkably similar and the process of thermal decomposition by gasification is essentially the same for all biomass feedstocks.

11.4.3 Pyrolysis Oils

Pyrolysis is the thermal decomposition of organic compounds in the absence of oxygen. The resulting product streams depend on the rate and duration of heating. Liquid yields exceeding 70 percent are produced under fast pyrolysis conditions, which is characterized as having short residence times (<0.5 s), moderate temperatures (450–600°C), and rapid quenching at the end of the process. Rapid quenching is essential if high molecular weight liquids are to be condensed rather than further decomposed to low molecular weight gases.

Pyrolysis liquid from flash pyrolysis is a low viscosity, dark-brown fluid with up to 15 to 20% water, which contrasts with the black, tarry liquid resulting from slow pyrolysis. Fast pyrolysis liquid is a complicated mixture of hydrocarbons arising from the uncontrolled degradation of lignin in lignocellulosic biomass. The liquid is highly oxygenated, approximating the elemental composition of the feedstock, which makes it highly unstable. It contains many different compounds including phenols, sugars, and both aliphatic and aromatic hydrocarbons. The liquid, despite its high water content, shows no appreciable phase separation. The low pH of pyrolysis liquids, which arise from organic acids, makes the liquids highly corrosive. The liquid can also contain particulate char. The higher heating value of pyrolysis liquids range between 17 MJ/kg and 20 MJ/kg with liquid densities of about 1280 kg/m³. Assuming conversion of 72% of the biomass feedstock to liquid on a weight basis, yields of pyrolysis oil will be about 135 gal/ton.

Pyrolysis liquids can be used directly as a substitute for heating oil. In some circumstances they are also suitable as fuel for combustion turbines or modified diesel engines. Recovery of high-value chemicals is another possibility, suggesting an integrated approach to production of both chemicals and fuel.

Production of pyrolysis oils and its co-products is illustrated in Figure 11.6. Lignocellulosic feedstock, such as wood or agricultural residues, are milled to a fine powder to promote rapid reaction. The particles are entrained in steam, which transports the material to the pyrolysis reactor. Within the reactor the particles are rapidly heated

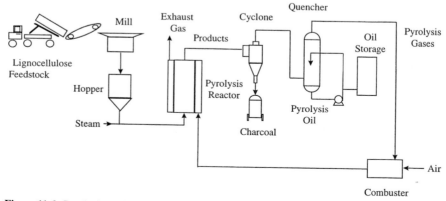

Figure 11.6. Pyrolysis production process.

and converted into condensable vapors, noncondensable gases, and solid charcoal. These products are transported out of the reactor into a cyclone operating above the condensation point of pyrolysis vapors where the charcoal is removed. Vapors and gases are transported to a quench vessel where a spray of pyrolysis liquid cools vapors sufficiently for them to condense. The noncondensable gases, which includes flammable carbon monoxide, hydrogen, and methane, are burned in air to provide heat for the pyrolysis reactor. A number of schemes have been developed for indirectly heating the reactor, including transport of solids into fluidized beds or cyclonic configurations to bring the particles in contact with hot surfaces.

There are several problems with pyrolysis liquids. Phase-separation and polymerization of the liquids and corrosion of containers make storage of these liquids difficult. The high oxygen and water content of pyrolysis liquids makes them incompatible with conventional hydrocarbon fuels. Furthermore, pyrolysis liquids have much lower quality than even Bunker C heavy fuel oil. Thus, upgrading pyrolysis liquids to more conventional hydrocarbon fuels such as gasoline is highly desirable.

Either traditional hydrotreatment or emerging zeolite technology can accomplish upgrading. Hydrotreating, common in the petrochemical industry, involves addition of hydrogen at high pressure in the presence of a catalyst. Hydrotreating has the advantage of yielding high quality products at maximum liquid yield. Disadvantages of hydrotreating include high hydrogen consumption and the need to operate at elevated pressures (13–17 MPa), which adversely affect the economics of the process. Upgrading can also be accomplished at atmospheric pressure without reducing gases by dehydration and decarboxylation over acidic zeolites. A yield of 17% of C_5-C_{10} hydrocarbons has been reported in a study of upgrading of pyrolytic liquids from poplar wood.

11.4.4 Vegetable Oils

A variety of grains and seeds contain relatively high concentrations of vegetable oils that are of high caloric value and can be used as liquid fuels. However, research has shown that the high viscosity of these raw vegetable oils (20 times that of diesel) would lead to coking of the injectors and rings. On the other hand, chemical modification of vegetable oils to methyl or ethyl esters yields excellent diesel-engine fuel without the viscosity problems associated with raw vegetable oil. **Biodiesel** is the generic name given to these vegetable oil esters. Suitable feedstocks include soybean, sunflower, cottonseed, corn, groundnut (peanut), safflower, rapeseed, waste cooking oils, and animal fats. Waste oils or tallow (white or yellow grease) can also be converted to biodiesel.

The production of biodiesel from oil seed is illustrated in Figure 11.7. Oil is squeezed from the seed in an oil press. By-product meal can be sold as a feed additive. Transesterification describes the process by which triglycerides in vegetable oil are reacted with methanol or ethanol to produce esters and glycerol.

One triglyceride molecule ($Gl(FA)_3$) reacts with three methanol molecules (MeOH) to produce three ester molecules (MeFA) and one glycerol molecule ($Gl(OH)_3$):

$$Gl(FA)_3 + 3\,MeOH \rightarrow 3\,MeFA + Gl(OH)_3, \tag{11.23}$$

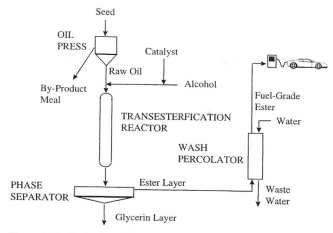

Figure 11.7. Production of fuel-grade biodiesel from seed.

where FA represents the fatty acid component of the ester molecule.

The reaction is catalyzed by lye (NaOH or KOH) dissolved in 50% excess of methanol. The reaction proceeds rapidly to completion at room temperature in about an hour. Small amounts of soap are also produced by the reaction of lye with fatty acids. Upon completion, the glycerol and soap are removed in a phase separator.

The heat of combustion of biodiesel is within 95 percent by weight of conventional diesel. Because they burn more efficiently, they have essentially the same fuel value as diesel. Biodiesel can be used in unmodified diesel engines with no excess wear or operational problems. In over 100,000 km of tests using soybean oil methyl esters in light and heavy trucks, few significant problems arose other than the need for more frequent oil changes because of the build-up of ester fuel in engine crankcases.

11.5 CONVERSION OF BIOMASS TO ELECTRICITY

11.5.1 Steam Power

Steam power involves the direct combustion of fuel to raise steam that is expanded through a turbine to produce electricity. Steam power plants are the basis of most of the electric power generation capacity in the United States. Although most of these plants are fired by coal or natural gas, in principle, biomass could substitute for these fossil fuels.

Combustion is the rapid oxidation of fuel to obtain energy in the form of heat. Since biomass fuels are composed primarily of carbon, hydrogen, and oxygen, the main oxidation products are carbon dioxide and water. Depending on the heating value and moisture content of the fuel, the amount of air used to burn the fuel, and the construction of the furnace, flames can exceed 3000°F.

Solid-fuel combustors can be categorized as grate-fired systems, suspension burners, or fluidized beds. Grate-fired systems were the first burner systems to be developed, evolving during the late nineteenth and early twentieth centuries into a variety of automated systems. The most common system is the spreader-stoker, consisting of a fuel feeder that mechanically or pneumatically flings fuel onto a moving grate where the fuel burns.

Suspension burners suspend the fuel as fine powder in a stream of vertically-rising air. The fuel burns in a fireball and radiates heat to tubes that contain water to be converted into steam. Suspension burners, also know as pulverized coal (PC) boilers, have dominated the U. S. power industry since World War II because of their high volumetric heat release rates and their ability to achieve combustion efficiencies often exceeding 99 percent.

Fluidized bed combustors are a recent innovation in boiler design. Air injected into the bottom of the boiler suspends a bed of sand or other granular refractory material. The turbulent mixture of air and sand is heated to high temperature, allowing a variety of fuels to be efficiently burned. The large thermal mass of the bed allows the unit to be operated at relatively low temperatures, around 1600°F, which lowers the emission of nitrogen compounds.

A biomass-fired steam power plant based on suspension burners is illustrated in Figure 11.8. Biomass is fed from a bunker through pulverizers designed to reduce fuel particle size enough to burn in suspension. The fuel particles are suspended in the primary air flow and feed to the furnace section of the boiler through burner ports where it burns as a rising fire ball. Secondary air injected into the boiler helps complete the combustion process. Heat is absorbed by steam tubes arrayed in banks of heat exchangers (waterwall, superheaters, and economizer) before exiting through a baghouse designed to capture ash released from the fuel. Steam produced in the boiler is part of a Rankine power cycle.

As an alternative to completely replacing coal with biomass fuel in a boiler, mixtures of biomass and coal can be burned together in a process known a cofiring. Cofiring offers several advantages for industrial boilers. Industries that generate large quantities of biomass wastes, such as lumber mills or pulp and paper companies, can use cofiring as an alternative to costly landfilling of wastes. Federal regulations also make cofiring attractive. The New Source Performance Standards, which limits particulate emissions from large coal-fired industrial boilers to 0.05 lb/MMBtu, doubles this allowance in cofired boilers in which the capacity factor for biomass exceeds 10 percent. Adopting cofiring is a good option for companies that are slightly out of compliance with their coal-fired boilers. Similarly, a relatively inexpensive method for a company to reduce sulfur emissions from its boilers is to cofire with biomass, which contains much less sulfur than coal. Cofiring capability also provides fuel flexibility, which is important during times of unstable fuel pricing.

Alkali in biomass fuels present a difficult problem for direct combustion systems. Compounds of alkali metals, such as potassium and sodium salts, are common in rapidly growing plant tissues. Annual biomass crops contain large quantities of alkali while the old-growth parts of perennial biomass contain relatively small quantities of alkali. These alkali compounds appear as oxides in the residue left after combustion of volatiles and char. They would not be troublesome except that they act as a fluxing agent, allowing other minerals in the fuel to melt. These sticky compounds bind ash particles to fuel

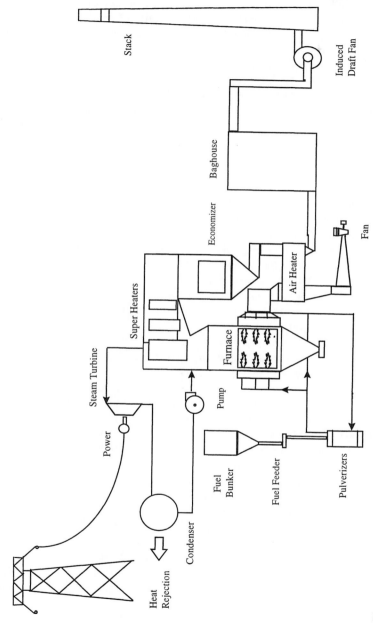

Figure 11.8. Biomass-fired steam power plant with suspension burners.

515

grates and heat exchanger surfaces. Boiler performance degrades as air flow and heat transfer are restricted by ash deposits. One solution to ash fouling is cofiring, the process by which biomass and coal are burned together. Some studies suggest that limiting biomass to 5–15 percent of the fuel requirement of a boiler can prevent ash fouling.

The best wood-fired power plants, which are typically 20–100 MW in capacity, have heat rates exceeding 12,500 Btu/kWh. In contrast, large, coal-fired power plants have heat rates of only 10,250 Btu/kWh. The relatively low thermodynamic efficiency of steam power plants at the sizes of relevance to biomass power systems may ultimately limit the use of direct combustion to convert biomass fuels to useful energy.

11.5.2 Gas Turbine Cycles

Gas turbines generate mechanical power by expanding a stream of hot, high pressure gas through an array of vanes attached to a rotating hub. In a conventional gas turbine cycle, compressing air, mixing it with fuel, and burning the mixture at high pressure produces this energetic gas stream. This approach works very well with fuels such as natural gas and kerosene, which are clean-burning and do not generate particulate matter or corrosive vapors that can damage turbine vanes. Gas turbine cycles are increasingly attractive for electric power generation because of the relative ease of plant construction, cost-effectiveness in a wide range of sizes (from tens of kilowatts to hundreds of megawatts), and potential for very high thermodynamic efficiencies when employed in advanced cycles.

Gas turbines cannot be directly fired with biomass because ash particles and alkali released from the burning biomass would damage turbine blades. The biomass must first be converted to a clean-burning fuel such as ethanol, pyrolysis oils, biogas, or producer gas before it can be used in gas turbines. Economics currently favor producer gas as a biomass-derived fuel for gas turbines, especially when employed in advanced cycles.

Figure 11.9 illustrates one of the most promising advanced cycles: integrated gasification/combined cycle (IGCC). Air is compressed and enters an oxygen plant, which separates oxygen from the air. The oxygen is used to gasify biomass in a pressurized gasifier to produce medium heating-value producer gas. The producer gas passes through cyclones and a gas clean-up system to remove particulate matter, tar, and other contaminants that may adversely affect gas turbine performance (alkali and chloride most prominent among these). These clean-up operations are best performed at high temperature and pressure to achieve high cycle efficiency. The clean gas is then burned in air and expanded through a gas turbine operating in a topping cycle. The gas exits the turbine at temperatures ranging between 400°C and 600°C. A heat recovery steam generator produces steam for a bottoming cycle that employs a steam turbine. Electric power is produced at two locations in this plant, yielding thermodynamic efficiencies approaching 47 percent.

Integrated gasifier/combined cycle (IGCC) systems based on gas turbines are attractive for several reasons. These reasons include their relative commercial readiness, the ability to construct small generating capacity units without being strongly influenced by economies of scale, and the expectation that they can generate electricity at the lowest cost of all possible biomass power options.

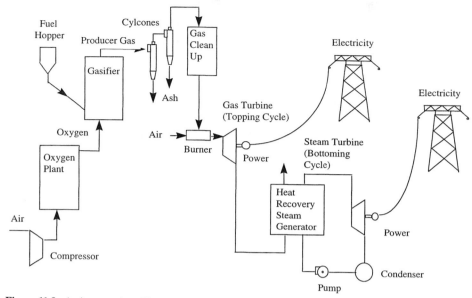

Figure 11.9. An integrated gasification/combined cycle (IGCC) power plant.

An alternative to IGCC is to generate steam for injection into the gas turbine combustor in order to increase mass flow and power output from the turbine. This variation, called a steam-injected gas-turbine (STIG) cycle, clearly is less capital intensive than IGCC, which employs a steam turbine. The STIG cycle is commercially developed for natural gas; lower flammability limits for producer gas make steam injection more problematic for biomass-derived producer gas.

11.5.3 Fuel Cell Cycles

Fuel cells convert chemical energy directly into electricity without combustion. They are similar to batteries except that they are continuously supplied with fuel so they never become discharged. The operation of a fuel cell is illustrated schematically in Figure 11.10. The fuel cell consists of two gas-permeable electrodes separated by an electrolyte, which is a transport medium for electrically charged ions. Hydrogen gas, the ultimate fuel in all current designs of fuel cells, enters the fuel cell through the anode while oxygen is admitted through the cathode. Depending on the fuel cell design, either positively charged hydrogen ions form at the anode or negatively-charged ions containing oxygen form at the cathode (represented by the oxygen ion, O^{-2}, in Figure 11.10, but hydroxyl and carbonate ions may also be employed in other fuel cells). In either case, the resulting ions migrate through the electrolyte to the opposite electrode from which they are formed. Hydrogen ions migrate to the cathode where they react with oxygen to form water. Oxygen-bearing ions migrate to the anode where they react with hydrogen to form water. Both ionic processes release chemical energy in the form

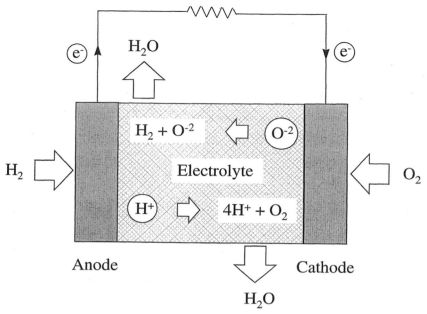

Figure 11.10. Schematic diagram of a fuel cell.

of electrons at the anode, which flow to the cathode through an external electric circuit. The flow of electrons from anode to cathode represents the direct generation of electric power from flameless oxidation of fuel. The inherently high thermodynamic efficiency of fuel cells make them attractive for biomass power where fuel costs are relatively expensive compared to many fossil fuels.

Several types of fuel cells have been developed or are under development. These fuel cells are classified according to the kind of electrolyte employed: phosphoric acid, polymeric, molten carbonate, and solid oxide. Despite differences in materials and operating conditions, all these fuel cells are based on the electrochemical reaction of hydrogen and oxygen. For biomass power applications, the molten carbonate and solid oxide systems are of particular interest. These types of fuel cells operate at elevated temperatures, which presents opportunities for heat recovery and integration into combined cycles.

The molten carbonate fuel cell operates at about 650°C. Although hydrogen is the ultimate energy carrier in the electrochemical reactions of this fuel cell, it has been designed to operate on a variety of hydrogen-rich fuels, including methane, kerosene, diesel fuel, ethanol, and producer gas. Within the fuel cell is a reformer that converts these fuels into mixtures of hydrogen, carbon monoxide, carbon dioxide, and water along with varying amounts of unreformed fuel. These decomposition reactions are strongly endothermic: not only does the reformer produce hydrogen and carbon dioxide required as electrochemical feedstocks, it also removes waste heat generated during the electrochemical oxidation of hydrogen in the fuel cell. The reformed fuel flows past the anode of the fuel cell, which absorbs hydrogen and releases water vapor and carbon dioxide. The exhaust gas stream, which still contains methane and hydrogen, is mixed with air and burned external to the fuel cell to produce a hot mixture of oxygen, carbon dioxide, and water vapor. This mixture passes over the cathode of the fuel cell to supply carbon dioxide and oxygen required for the proper operation of the system. Overall efficiency of converting methane into electricity is 50 percent, or a heat rate of 6850 Btu/kWh. Unlike many other power conversion devices, fuel cells can maintain high efficiencies even when operated at half their rated capacities. Molten carbonate fuel cells have completed extensive demonstration trials and are close to market entry.

The solid oxide system is the next generation fuel cell. The electrolyte is a solid, nonporous metal oxide, usually based upon yttrium and zirconium. The cell operates at 650–1000°C where ionic conduction by oxygen ions takes place. The solid electrolyte provides for a simpler, less expensive design and longer expected life than current fuel cell systems. The higher operating temperature compared to the molten carbonate system also enhance its attractiveness for combined cycle operation. However, solid oxide systems are still in the early stages of development.

Obviously, fuel cells designed to oxidize hydrogen electrochemically require some modification before they can convert the chemical energy stored in solid biomass into electricity. Several options for doing this are possible. Anaerobic digestion produces a mixture of methane and carbon dioxide that is ideal for a carbonate fuel cell. Ethanol can also be reformed in carbonate fuel cells to a suitable gaseous fuel. However, biogas and ethanol are relatively expensive fuels compared to producer gas; thus, much of the interest in using biomass with fuel cells has focused on integrating them with gasifiers.

Figure 11.11 illustrates an integrated gasifier/combined cycle (IGCC) power plant based on a molten carbonate fuel cell. Biomass is gasified in oxygen to yield producer gas. Gasification occurs at elevated pressure to improve the yield of methane, which is important for proper thermal balance of the fuel cell. Hot-gas clean-up to remove par-

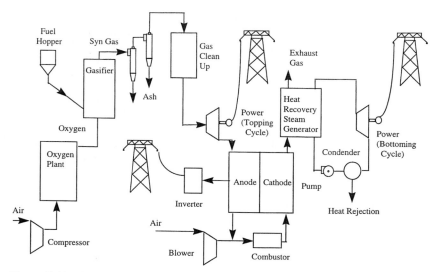

Figure 11.11. IGCC power plant with molten carbonate fuel cell.

ticulate matter, tar, and other contaminants is followed by expansion through a gas turbine as part of a topping power cycle. The pressure and temperature of the producer gas is sufficiently reduced after this to admit it into the fuel cell. High temperature exhaust gas exiting the cathode of the fuel cell enters a heat recovery steam generator, which is part of a bottoming cycle in the integrated plant. Thus, electricity is generated at three locations in the plant for an overall thermodynamic efficiency reaching 60 percent or more.

11.6 ECONOMIC ASSESSMENTS

Capital costs and operating costs presented here should be considered as approximations of actual costs since many of these technologies have either not been built at commercial scales or have not been operated with biomass feedstock. More detailed information can be found in the published literature.

Capital costs, even when presented on the basis of output (for example, dollars per gallon per year or dollars per kilowatt) are strongly dependent on the size of the plant and the year for which construction is assumed. Thus, corrections must be made for economies of scale, in the first instance, and rate of inflation, in the second instance.

The cost of most capital facilities is not linearly related to the size of the facility. The principle of economies of scale recognizes that capital cost per unit of production decreases as a plant gets larger. In many instances, this relationship is well approximated by the simple scaling law:

$$C_s = C_b(S_s/S_b)^n, \tag{11.24}$$

where C_s = predicted cost of the specified plant (dollars),
$\quad C_b$ = known cost of the baseline plant (dollars),
$\quad S_s$ = size of the specified plant (kW, tons/day, liters/yr, etc.),
$\quad S_b$ = size of the baseline plant (kW, tons/day, liters/yr, etc.), and
$\quad n$ = economy of scale factor (less than unity).

The value for the economy of scale factor depends on the kind of equipment being specified and sometimes depends on the size range of the equipment. This factor is a reasonably well known from industrial practice for a variety of parts and equipment and can be used to estimate overall costs of systems made up of such parts and equipment. This scaling law can also be applied to overall systems, but with diminishing accuracy as the system becomes more complicated. In the absence of more system-specific values, a reasonable estimate for n is 0.6 for many energy and chemical process plants.

Capital costs must also be adjusted for inflation since all published data is historical and costs change with time. These corrections are easy to make if tabulations of inflation indices as a function of year are available:

$$C_2 = C_1(I_2/I_1), \tag{11.25}$$

where C_2 = cost at date of evaluation (dollars),
$\quad C_1$ = cost at reference date (dollars),
$\quad I_2$ = inflation index at date of evaluation (dimensionless), and
$\quad I_1$ = inflation index at reference date (dimensionless).

Separate indices are available for cost of equipment and energy. Useful sources are *Chemical Engineering Magazine* for Marshall and Swift equipment indices, Ref [17] energy indices, and Ref [13].

11.6.1 Gaseous Fuels

Capital costs for anaerobic digestion facilities to produce biogas range from $63,000 to $122,000 for each ton/day of dry matter processed. These costs include feed preparation, digestion, gas cleanup, and residue processing. Gas costs of $5.10–$6.10/GJ in 1990 dollars have been realized with advanced conversion systems sized to deliver 1.14 trillion kJ/year by using water hyacinth and napier grass feedstocks. In comparison, the cost of natural gas in the United States, which shows large seasonal and geographical variations, ranges between $1.90–$2.40/GJ. In niche markets, where the feedstock is inexpensive and natural gas is not available, biogas can be a viable alternative energy resource. Anaerobic digestion is also a mature technology and is widely used as a waste reduction process with energy production a secondary consideration.

The capital cost for a gasification plant, including fuel feeding and gas clean-up, is dependent on both the size and the operating pressure of the system. A 50 MW_t atmospheric pressure gasifier would cost about $15 million. A comparably sized pressurized gasifier would cost nearly $60 million because of the added complexity of lock-hopper and pressure vessel design.

11.6.2 Liquid Fuels

The cost of producing ethanol from biomass varies tremendously depending on the feedstock employed, the size and management of the facility, and the market value of coproducts generated as part of some conversion process. Industrial-scale ethanol plants resemble petrochemical plants in that they are capital intensive and benefit from economies of scale.

Cost information for ethanol plants to be built in the United States is most reliable for those using corn starch, the basis of the U.S. ethanol industry. A 5,000 barrel per day plant (about 18.5 million liters per year) will have a capital cost of about $0.53/liter of annual capacity if built from the ground up. Smaller facilities can have capital costs as high as $0.79/liter of annual capacity, and poorly designed facilities of any size may cost $1.06/liter of annual capacity. On the other hand, ethanol plants that are built from existing facilities, such as refineries or chemical plants, or ethanol plants integrated into a larger industrial facility can have substantially lower capital costs, often in the range of $0.26 to $0.40/liter of annual capacity.

Low-end production costs are about $0.26/liter. However, the heating value of ethanol is only 70 percent that of gasoline. This production cost is equivalent to gasoline selling for $0.29/liter before tax, transportation, or profit. In contrast, refinery price for gasoline in 1990 dollars was about $0.20/liter. Currently, the economics of fermentation are such that the commercial viability of ethanol is entirely dependent on government incentives in the form of a tax credit, currently $0.16 for each liter of ethanol used for fuel blending. Also, a strong market for fermentation by-products is a key factor in the economic viability of ethanol-from-corn.

Detailed economic information on production of ethanol from lignocellulose is not available since this process is still under development. Some estimates place the price at $0.36/liter with technology improvements expected to reduce the cost of ethanol by this process to $0.16/liter in the first decade of the twenty-first century. However, some reports suggest that ethanol from cellulose will have to cost as little as $0.08 to 0.11/liter to be competitive with gasoline prices anticipated early in the twenty-first century.

Capital investment for a 5,000 barrel/day plant to produce methanol from biomass would be about $380 million in 1990 dollars. The cost of methanol from $1.90/GJ wood is projected to be about $0.57/liter in 1990 dollars. Since the heating value of methanol is only 57 percent that of gasoline, the production cost from this large plant is still equivalent to gasoline selling for $0.66/liter. Methanol from natural gas can be produced at significantly lower cost, but this assumes much larger plant capacities to capture economies of scale. Such large plants are not feasible for widely dispersed biomass feedstocks. New methanol synthesis technologies may be able to significantly reduce this price. The U.S. Department of Energy's methanol from biomass program has a goal of $0.15/liter ($7.90/GJ) based on feedstock cost of $1.90/GJ.

Capital investment for a 5000 barrel per day plant to produce pyrolysis liquids would be $60 million in 1990 dollars. This size plant could produce pyrolysis liquids for $0.17/liter from $1.90/GJ biomass feedstock.

Feedstock costs for production of biodiesel range from $0.16–$0.26/liter for waste fats to $0.53–$0.79/liter for vegetable oils. A 12,000 barrel/year facility could, under the best scenario, produce biodiesel from soybeans at a cost of $0.33/liter.

11.6.3 Electricity

The capital and operating costs for steam power plants fired with biomass are relatively well known because of significant operating experience with these systems. The capital cost for a new plant ranges between $1,400 and $1,800 per kilowatt capacity. Accordingly, a 50-MW biomass power plant based on direct-combustion would cost approximately $80 million. On the basis of a target price of $1.90/GJ for biomass, the cost of production for direct-fired biomass power is about $0.06/kWh in 1990 dollars.

A 50-MW gasification/gas-turbine power plant would have total capital costs of between $62 million and $98 million. Production costs would range from $0.068/kWh to $0.091/kWh if fuel is available at an optimistic $1.42/GJ.

Capital costs for molten carbonate fuel cells are expected to be $1500/kW at the time of market entry but decrease to about $1000/kW for a commercially mature unit.

The cost of electricity from a mature unit operating on natural gas is projected between $0.049 and $0.085 per kWh. More attractive economics result if less expensive fuel is available. The cost of electricity generated from landfill gas using mature fuel cell technology is expected to be comparable to that for an internal combustion engine/electric generator set, i.e., about $0.05/kWh.

PROBLEMS

1. A 50-MW power plant, expected to operate at 42 percent thermodynamic efficiency, will use biomass for fuel. Determine how many hectares of land will be required to run it for the following biomass fuels:
   (a) Hybrid poplar grown in North America.
   (b) Bagasse grown in Hawaii.
2. The large volume of biomass that must be transported to a central processing facility may severely limit the practical size of such facilities. Calculate the maximum capacity (liters per year) of a fuel ethanol plant that uses wheat straw for feedstock if no more than 25 trucks per hour can enter the facility. Assume the trucks can carry 20 cubic meters of wheat straw, which has a bulk density of 50 kg/m³.
3. Fermentation of starch and sugar crops yields about 10% molar volume of ethanol in water. Distillation of this mixture to remove the water and produce pure ethanol is an expensive and energy-intensive process. To reduce fuel costs, some researchers have proposed partial distillation to a 50% mixture of ethanol and water, which could be used to power a fuel cell. Calculate the enthalpy change in oxidizing this mixture in air if the reactants are at 298 K and the products are at 1000 K.
4. A 2-MW carbonate fuel cell power plant is reported to have a thermodynamic efficiency of 55 percent. Assume products of reaction exit at 298 K with water in the liquid phase.
   (a) What is the corresponding heat rate of this fuel cell?
   (b) How much methane (kg/h) is required to power this plant?
5. The manager of a coal-fired power plant proposes to cofire switchgrass with coal as a means to reduce sulfur emissions from the plant. However, the plant engineer warns that alkali in the switchgrass can lead to ash fouling in the boiler. The fuel blend should have no more than 0.34 kg of alkali per GJ of fuel heating value. The raw coal has heating value of 28 MJ/kg and a sulfur content of 4%.
   (a) What is the maximum weight-percent of switchgrass that can be blended with the coal without producing ash fouling?
   (b) What is the expected sulfur emission rate (kg/GJ) for the fuel blend?
6. A chemical plant for the production of 5000 barrels per day of pyrolysis oils would cost $60 million. Estimate the cost of a plant with twice this capacity.
7. A 2000 barrel per day ethanol plant using Jerusalem artichokes as a source of sugar for fermentation is proposed.
   (a) Determine the daily feedstock requirement of raw Jerusalem artichokes for the plant.

(b) Determine the number of hectacres that must be planted annually to provide feedstock to the plant.

8. Calculate the annual fuel costs for 10 MW power plants based on the following cycles. The fuel is hybrid poplar at $48/ton.

(a) Conventional steam power plant with a heat rate of 14,000 Btu/kWh.

(b) Integrated gasification/combined cycle (IGCC) based on gas turbines with overall thermodynamic efficiency of 50 percent based on gaseous fuel. However, the gasifier is only 85 percent energy efficient in converting the solid biomass fuel into gaseous fuel.

9. Some policy analysts have suggested a tax on carbon dioxide emissions (the so-called carbon tax) as an incentive for energy consumers to switch from fossil fuels to renewable biomass fuels. This idea is based on the argument that there is no net emission of carbon dioxide into the atmosphere from biomass, since the same amount of carbon dioxide is absorbed by biomass in the process of growing it. Assume that coal is available for $1/GJ while biomass in the form of switchgrass can be grown and delivered to a power plant for $2.50/GJ. Assume that the coal has heating value of 28 MJ/kg and contains 70% carbon by weight. How large of tax would have to be placed on carbon dioxide ($/ton of carbon dioxide emitted) before plant operators would consider burning biomass instead of coal?

10. Fermentation technologies that use lignocellulose rather than starch as feedstock represent a more efficient use of agricultural crops. Calculate the energy efficiency of converting switchgrass to ethanol and compare it to the conversion of corn grain to ethanol. Note that essentially the whole switchgrass crop is ligno-cellulose whereas only a fraction of the corn plant is corn grain.

REFERENCES AND SUGGESTED READINGS

1. Beenackers, A.A.C.M. and W.P.M. Van Swaaij. 1984. Methanol from wood: Process principles and technologies for producing methanol from biomass. *Int. J. Solar Energy* 2: 349–367.

2. Boehm, R.F. 1987. *Design Analysis of Thermal Systems*. New York: Wiley.

3. Borman, G.L. and K.W. Ragland. 1998. *Combustion Engineering*. New York: McGraw-Hill.

4. Bridgwater, A.V. 1995. The technical and economic feasibility of biomass gasification for power generation. *Fuel* 74:631–653.

5. Bull, S.R. 1991. The U.S. Department of Energy biofuels research program. *Energy Sources* 13:433–442.

6. Cliburn, J.K. et al. 1993. *Biogas Energy Systems: A Great Lakes Casebook*. Chicago, IL: Great Lakes Regional Biomass Energy Program, Council of Great Lakes Governors.

7. Grohmann, K., C.E. Wyman, and M.E. Himmel. 1992. Potential for fuels from biomass and wastes. In *Emerging Technologies for Material and Chemicals from Biomass*, R.M. Rowell, T.P. Schultz, and R. Narayan, eds. ACS Symp. Series 476. Washington, D.C.: American Chemical Society.

8. Hirschenhoofer, J.H., D.B. Stauffer, and R.R. Engleman. 1994. *Fuel Cells: A Handbook*. Department of Energy Technical Report DOE/METC-94/1006 (DE94004072).

9. Kornosky, R.M. 1992. Liquid phase methanol—fuel for the future. *PETC Review* Summer:30–35.

10. Marsden, S.S. 1983. Methanol as a viable energy source in today's world. *Ann. Rev. Energy* 8:333–354.

11. Miles, Sr., T.R., T.R. Miles, Jr., L. Baxter, B. Jenkins, and L. Oden. Alkali deposits found in biomass power plants. *Summary Rept. National Renewable Energy Laboratory,* NREL Sub-contract TZ-2-11226-1, April 15, 1995.

12. Moran, M. and H. Shapiro. 1992. *Fundamentals of Engineering Thermodynamics,* 2nd Ed. New York: Wiley.

13. Peters, M.S., and K.D. Timmerhaus. 1980. *Plant Design and Economics for Chemical engineers,* 3rd ed. New York: McGraw-Hill.

14. Reed, T.B. 1981. In *Biomass gasification,* T. B. Reed, ed. Park Ridge, NJ: Noyes Data Corp.

15. Reed, T.B. 1993. An overview of the current status of biodiesel. In *Proc. Volume II of the First Biomass Conference of the Americas.* August 30–September 2. Burlington, Vermont.

16. Tillman, D.A. 1991. *The Combustion of Solid Fuels and Wastes.* San Diego, CA: Academic Press.

17. U.S. Department of Energy. 1993. *Statistical Abstracts of the United States 1993.* 483.

18. Wayman, M. and S. Parekh. 1990. *Biotechnology of Biomass Conversion.* Philadelphia, PA: Open University Press.

19. Williams, R.H. and E.D. Larson. 1993. Biomass-gasifier gas turbine power generating technology. In *Strategic benefits of biomass and waste fuels.* Washington, D.C.: EPRI.

INTRODUCTION TO SOLAR ECONOMIC ANALYSIS (12)

Although solar radiant energy is free, the equipment required to convert it to a useful form—thermal or electrical—is not free. Therefore, a cost must be assigned to solar-thermal or solar-electrical energy that reflects the conversion equipment cost prorated on the number of kilowatt-hours or Btus delivered by the solar system. If the solar cost is less than the cost of other energy sources that can perform the same task, there is an economic incentive to use solar energy. The purpose of an economic optimization is to maximize the savings resulting from use of solar energy. In this and following sections the method of calculating the cost of solar energy in various economic scenarios is described. The methods for calculating the amount of solar energy delivered by a given system are given in detail in Chapters 5 to 8.

12.1 OVERVIEW OF SOLAR ECONOMICS

It is rarely cost effective to provide all the energy requirements of a thermal or mechanical system by means of solar energy. If this were to be done, the solar system would be required to be capable of providing 100 percent of the energy demand for the worst set of operating conditions ever expected—inclement weather, maximum demand, and no sunshine. A solar system capable of providing the peak demand for a lengthy period would be oversized for all less severe operating conditions; it would thus be greatly oversized most of the time. An oversized system delivers more energy than can be used, thereby requiring energy rejection in some manner. However, the oversized system must be paid for even though it delivers below its capability nearly continuously. A solar system with such a low load factor is uneconomical and impractical.

The best use of solar energy is in conjunction with conventional fuels, which are used as an auxiliary source for special high-demand situations. It is then possible to design a base-load solar system that operates at nearly full capacity (high load factor) most of the time, although it provides less than 100 percent of the annual demands. Such a system is more economical, since it delivers more energy per unit system cost than an oversized full-capacity solar system. Peak loads, for which it is uneconomical to size a solar system, are carried by the auxiliary system. It will be shown shortly that the solar-plus-auxiliary method is nearly always the least-cost method. The determination of the fraction of the annual demand to be provided by solar energy and what solar system size is required for that energy delivery is the key problem addressed in this portion of the chapter.

Several indices or figures of merit can be used to select the best solar-auxiliary mix depending on what national economic system applies. In the Western economies it is the annual cost—dollars, marks, pounds, francs, etc.—of providing energy to a given load that is the figure of merit. The lower the cost, the more desirable the configuration. The least-cost approach is used in this book. Other approaches could include minimization of entropy production or minimization of fossil energy usage.

12.2 LIFE CYCLE COSTING

A solar system differs fundamentally from a fossil fuel system in a manner that requires an economic analysis reflecting the benefits accrued by solar usage throughout the lifetime of the system. Nonsolar systems usually have relatively small initial costs and relatively large annual operating costs reflecting raw energy purchases. Solar systems, however, are relatively expensive initially but have negligible nonsolar energy cost during their lifetime. If the selection of a system (solar versus nonsolar) were made on the basis of initial cost alone, the solar system would rarely be selected. However, the naive initial-cost judgment cannot account for the principal reasons a solar system is considered—to reduce the usage of expensive and scarce fuels, and because of the environmental benefits associated with solar systems.

The concept of life cycle costing includes both the initial capital cost and the year-to-year operating cost in making economic decisions. The life cycle cost of any energy system is defined as the total of the following cost components over the life of a system:

1. Capital equipment cost,
2. Acquisition costs,
3. Operating costs—fuels, etc.,
4. Interest charges if money for capital is borrowed,
5. Maintenance, insurance, and miscellaneous charges,
6. Taxes—sales, local, state, federal,
7. Other recurring or one-time costs associated with the system, and
8. Salvage value (usually a negative cost).

For example, the life cycle cost of a conventional forced-air, gas-heating system would include the cost of the furnace, distribution system and controls, the total cost of gas burned throughout the period of analysis, the portion of property tax associated with the heating system, any repair charges and replacement parts (filters, motors, dampers), and the mortgage interest charge on the heating system. It is clear that the life cycle cost is a much more useful cost index than is the initial cost of the gas furnace.

In the following sections of this chapter the parameters affecting the life cycle cost of solar systems will be described in analytical detail. The important parameters include:

Initial cost, including hardware, transportation, installation, system designer's fee, special building features, and value of space used by system,
Down payment and mortgage interest rate,
Life cycle time,
Depreciation rate and final salvage value,
Repair and replacement costs,
Maintenance costs,
Nonsolar fuel costs,
Inflation rate of cost of nonsolar fuel,
General inflation rate, and
Local, state, and federal taxes.

It will be shown that residential and commercial (income-producing) life cycle cost analyses differ somewhat in the treatment of taxes. Inflation, mortgage, and tax parameters can be significant for the homeowner, whereas repair and maintenance costs for a well-designed solar system can be made quite small.

12.3 PRINCIPLES OF DISCOUNTED CASH-FLOW ANALYSIS

12.3.1 Annualized Present Worth

Since it is convenient to conduct economic analyses in a unit of fixed monetary value, the concept of present worth is used. The *present* worth of a *future* cash flow or payment is the future value of that flow with the time value of money factored out. The time value of money represents the opportunity for investment yielding a future return; it is an effect separate from inflation. For example, the future value X of a sum of money P invested at an annual interest rate i dollars per dollars per year is

$$X = P(1 + i)^t, \tag{12.1}$$

in which t is the future time expressed in years (or other time unit corresponding to the time basis of the interest rate i). Equation (12.1) states that an initial sum of money P appreciates by a multiplier factor $(1 + i)$ each year (disregarding general inflation).
Stated alternatively, a future sum of money X has a present worth P given by

$$P = \frac{X}{(1 + i)^t}. \tag{12.2}$$

Equation (12.2) indicates that the present worth of a given amount of money in the future is discounted, in constant dollars, by a factor $(1 + i)^{-1}$ for each year in the future. In this context, i is generally called the discount rate. The factor $(1 + i)^{-t}$ is called the present-worth factor $PWF(i, t)$ given by

$$PWF(i,t) \equiv \frac{1}{(1 + i)^t}. \tag{12.3}$$

The present-worth factor can be multiplied by any cash flow in the future at time t to give its present value. Table 12.1 is a tabulation of present-worth factors.

The present-worth concept is the key idea in the use of discounted cash flow analysis. Its use permits all calculations to be made in present, discounted monies. In life cycle cost analyses in this book, the present-worth approach is used instead of forming the total of all life cycle costs. This approach seems more useful, since the sums of money are more intuitively familiar when expressed on an annual, not total, basis.

Example 12.1. Assume that opportunity A generates benefits equal to $100, $150, and $200 at the end of years 1, 2, and 3, respectively. Assume that opportunity B yields benefits of $225 in year 2 and $225 in year 3. Therefore over 3 years, both opportunities A and B yield benefits of $450. However, the timing of the benefits received is different in each case. By using the present-worth technique, their two benefit flows can be viewed in terms of today's dollar value.

Solution.

Step 1
Compute the present-worth factor PWF using Eq. (12.3). With an interest rate of 10 percent, the calculation of present-worth factors is as follows:

$$\text{Year 1} = \frac{1}{(1 + 0.10)^1} = \frac{1}{(1.10)^1} = 0.9091$$

$$\text{Year 2} = \frac{1}{(1 + 0.10)^2} = \frac{1}{(1.10)^2} = 0.8264$$

$$\text{Year 3} = \frac{1}{(1 + 0.10)^3} = \frac{1}{(1.10)^3} = 0.7513$$

Step 2
Compute the present worth of each opportunity benefit flow by multiplying the present-worth factor by the annual benefit amount.

Table 12.1. Present-worth factors[a]

n	0%	2%	4%	6%	8%	10%	12%	15%	20%	25%
1	1.0000	0.9804	0.9615	0.9434	0.9259	0.9091	0.8929	0.8696	0.8333	0.8000
2	1.0000	0.9612	0.9246	0.8900	0.8573	0.8264	0.7972	0.7561	0.6944	0.6400
3	1.0000	0.9423	0.8890	0.8396	0.7938	0.7513	0.7118	0.6575	0.5787	0.5120
4	1.0000	0.9238	0.8548	0.7921	0.7350	0.6830	0.6355	0.5718	0.4823	0.4096
5	1.0000	0.9057	0.8219	0.7473	0.6806	0.6209	0.5674	0.4972	0.4019	0.3277
6	1.0000	0.8880	0.7903	0.7050	0.6302	0.5645	0.5066	0.4323	0.3349	0.2621
7	1.0000	0.8706	0.7599	0.6651	0.5835	0.5132	0.4523	0.3759	0.2791	0.2097
8	1.0000	0.8535	0.7307	0.6274	0.5403	0.4665	0.4039	0.3269	0.2326	0.1678
9	1.0000	0.8368	0.7026	0.5919	0.5002	0.4241	0.3606	0.2843	0.1938	0.1342
10	1.0000	0.8203	0.6756	0.5584	0.4632	0.3855	0.3220	0.2472	0.1615	0.1074
11	1.0000	0.8043	0.6496	0.5268	0.4289	0.3505	0.2875	0.2149	0.1346	0.0859
12	1.0000	0.7885	0.6246	0.4970	0.3971	0.3186	0.2567	0.1869	0.1122	0.0687
13	1.0000	0.7730	0.6006	0.4688	0.3677	0.2897	0.2292	0.1625	0.0935	0.0550
14	1.0000	0.7579	0.5775	0.4423	0.3405	0.2633	0.2046	0.1413	0.0779	0.0440
15	1.0000	0.7430	0.5553	0.4173	0.3152	0.2394	0.1827	0.1229	0.0649	0.0352
16	1.0000	0.7284	0.5339	0.3936	0.2919	0.2176	0.1631	0.1069	0.0541	0.0281
17	1.0000	0.7142	0.5134	0.3714	0.2703	0.1978	0.1456	0.0929	0.0451	0.0225
18	1.0000	0.7002	0.4936	0.3503	0.2502	0.1799	0.1300	0.0808	0.0376	0.0180
19	1.0000	0.6864	0.4746	0.3305	0.2317	0.1635	0.1161	0.0703	0.0313	0.0144
20	1.0000	0.6730	0.4564	0.3118	0.2145	0.1486	0.1037	0.0611	0.0261	0.0115
25	1.0000	0.6095	0.3751	0.2330	0.1460	0.0923	0.0588	0.0304	0.0105	0.0038
30	1.0000	0.5521	0.3083	0.1741	0.0994	0.0573	0.0334	0.0151	0.0042	0.0012
40	1.0000	0.4529	0.2083	0.0972	0.0460	0.0221	0.0107	0.0037	0.0007	0.0001
50	1.0000	0.3715	0.1407	0.0543	0.0213	0.0085	0.0035	0.0009	0.0001	—
100	1.0000	0.1380	0.0198	0.0029	0.0005	0.0001	—	—	—	—

[a]For interest rates i from 0 to 25 percent and for periods of analysis n from 1 to 100 yr.

		Annual benefit ($)		Present worth ($)	
Year	Present-worth factor	A	B	A	B
1	0.9091	100	0	90.91	0
2	0.8264	150	225	123.96	185.94
3	0.7513	200	225	150.26	169.04
Total		450	450	365.13	354.98

The effect of the time value of money on the investment opportunity is seen by comparing the total annual benefit with the total present worth for the two investment opportunities.

The discount rate is a measure of an investor's preference for the time value of money. A discount rate of 8 percent means that an investor regards a sum of money as worth 8 percent less next year than it is worth this year. Alternatively, the investor is indifferent when offered the option of $100 this year or $108 next year (apart from inflation).

Large businesses have well-established values of discount rate or required rates of return on investments. Homeowners do not. Most analysts agree that the lower bound of a homeowner's discount rate is his return on personal savings or other investments. Likewise, the upper bound is the interest rate on personal loans or mortgages. An individual would not borrow a sum of money at a higher interest rate than that expected for the investment of the borrowed sum in an income-producing venture. Since the discount rates of homeowners and small business owners are somewhat subjective, both high and low limits should be considered in a life cycle cost analysis.

The effect of inflation on the future value of an invested sum is to reduce the future value by a factor $(1 + j)$ per year, where j is the inflation rate per year (for example, dollars per dollars per year). The future value X from Eq. (12.1), taking inflation into account, is

$$X = P\left(\frac{1 + i}{1 + j}\right)^t. \tag{12.4}$$

Alternatively, the present worth P of a future sum X under inflation is

$$P = \frac{X}{[(1 + i)/(1 + j)]^t}. \tag{12.5}$$

If an *effective* interest rate i' is defined to include both interest and inflation effects such that

$$P = \frac{X}{(1 + i')^t} = X\,PWF(i', t), \tag{12.6}$$

it is easy to show that the effective rate is

$$i' = \frac{1 + i}{1 + j} - 1 = \frac{i - j}{1 + j}. \tag{12.7}$$

The concept of the present-worth factor can be modified to account for inflation by replacing the interest rate i with the effective or real interest rate i'. For small inflation rates j the effective rate i' is approximately the difference between the interest rate and the inflation rate:

$$i' \approx i - j \quad \text{for small } j. \tag{12.8}$$

The general inflation rate is usually measured by the consumer price index in the United States. See Hirst [1] for a survey of the usefulness of this index.

12.3.2 Series of Payments

In many economic analyses a *series* of annual or monthly sums is invested or used to pay off a loan. The method above can easily be extended from a single payment to a series of payments. If a sum P_{ann} is invested each year at interest rate i, the present worth of the *sum* of these payments S is

$$S = \frac{P_{ann}}{1 + i} + \frac{P_{ann}}{(1 + i)^2} + \frac{P_{ann}}{(1 + i)^3} + \dots, \tag{12.9}$$

or

$$S = P_{ann}[(1 + i)^{-1} + (1 + i)^{-2} + (1 + i)^{-3} + \dots]. \tag{12.10}$$

The expression in brackets is a geometric series with first term $P_{ann}/(1 + i)$ and ratio $(1 + i)^{-1}$. It follows from the expression for the sum of such a series [3] that

$$S = P_{ann}\left[\frac{1 - (1 + i)^{-t}}{i}\right]. \tag{12.11}$$

Alternatively, the annual payment P_{ann} required to form a total amount S, for example, to purchase a solar system, is

$$P_{ann} = S\left[\frac{i}{1 - (1 + i)^{-t}}\right] \tag{12.12}$$

The term in brackets in Eq. (12.12) is called the *capital-recovery factor CRF* (i, t):

$$CRF(i, t) \equiv P_{ann}/S = \frac{i}{1 - (1 + i)^{-t}} \tag{12.13}$$

Table 12.2 is a summary tabulation of capital-recovery factors; Appendix 8 contains detailed tables of capital-recovery factors. The term P_{ann} in Eq. (12.12) is the annual payment on a self-amortizing loan of value S. Each payment is a mix of interest and principal repayment. Early payments are mostly interest because of the large outstanding balance; later payments are primarily principal repayment. Most solar systems owned by private firms or individuals are purchased with self-amortizing loans.

Example 12.2. Compute the annual cost of a solar energy system with the characteristics tabulated below.

Factor	Specification
Expected system lifetime t (yr)	20
Discount rate (%)	8
Collector area A_c (m²)	20
Collector cost ($/m²)	100
Storage cost ($/m²)	6.25
Cost of control system ($)	100
Miscellaneous costs (for example, pipes, pumps, motors, heat exchangers) ($)	$200 + (5A_c)$

Solution. To obtain the total cost of this solar energy system, add the costs for the controls, the collector, the storage, and miscellaneous items, or,

$$S = \$100 + (\$100 \times 20) + (\$6.25 \times 20) + [\$200 + (5.00 \times 20)]$$

$$S = \$2525.$$

From Table 12.2 a capital-recovery factor CRF for 20 yr at 8 percent is

$$CRF(20, 0.08) = 0.102.$$

The annual cost P_{ann} is then

$$P_{ann} = \$2525 \times 0.102 = \$257/yr.$$

The effect of inflation on a series of payments is analyzed in the same manner as for a single payment [see Eqs. (12.4 to 12.8)]. Inflation effects are included by replacing the basic interest rate i with the effective or real rate i'.

Example 12.3. The owner of a four-unit apartment building of masonry construction is considering the retrofit installation of a solar heater. Under the existing climatic conditions it is expected that the solar unit can supply 87.5×10^6 kJ/yr and will thus save 5530 liters of fuel oil (with an effective heating value of 15,800 kJ/liter and an efficiency of 43 percent in the existing heating plant). In present dollars, this savings resulting from the solar unit represents a sum of $510 (9.2¢/liter of fuel oil saved). If the owner can borrow money for the solar retrofit at 10 percent per year and wants

Table 12.2. Capital-recovery factors[a]

i / n	0%	2%	4%	6%	8%	10%	12%	15%	20%	25%
1	1.00000	1.02000	1.04000	1.06000	1.08000	1.10000	1.12000	1.15000	1.20000	1.25000
2	0.50000	0.51505	0.53020	0.54544	0.56077	0.57619	0.59170	0.61512	0.65455	0.69444
3	0.33333	0.34675	0.36035	0.37411	0.38803	0.40211	0.41635	0.43798	0.47473	0.51230
4	0.25000	0.26262	0.27549	0.28859	0.30192	0.31547	0.32923	0.35027	0.38629	0.42344
5	0.20000	0.21216	0.22463	0.23740	0.25046	0.26380	0.27741	0.29832	0.33438	0.37184
6	0.16667	0.17853	0.19076	0.20336	0.21632	0.22961	0.24323	0.26424	0.30071	0.33882
7	0.14286	0.15451	0.16661	0.17914	0.19207	0.20541	0.21912	0.24036	0.27742	0.31634
8	0.12500	0.13651	0.14853	0.16101	0.17401	0.18744	0.20130	0.22285	0.26061	0.30040
9	0.11111	0.12252	0.13449	0.14702	0.16008	0.17364	0.18768	0.20957	0.24808	0.28876
10	0.10000	0.11133	0.12329	0.13587	0.14903	0.16275	0.17698	0.19925	0.23852	0.28007
11	0.09091	0.10218	0.11415	0.12679	0.14008	0.15396	0.16842	0.19107	0.23110	0.27349
12	0.08333	0.09156	0.10655	0.11928	0.13270	0.14676	0.16144	0.18148	0.22526	0.26845
13	0.07692	0.08812	0.10014	0.11296	0.12652	0.14078	0.15568	0.17911	0.22062	0.26454
14	0.07143	0.08260	0.09467	0.10758	0.12130	0.13575	0.15087	0.17469	0.21689	0.26150
15	0.06667	0.07783	0.08994	0.10296	0.11683	0.13147	0.14682	0.17102	0.21388	0.25912
16	0.06250	0.07365	0.08582	0.09895	0.11298	0.12782	0.14339	0.16795	0.21144	0.25724
17	0.05882	0.06997	0.08220	0.09544	0.10963	0.12466	0.14046	0.16537	0.20944	0.25576
18	0.05556	0.06670	0.07899	0.09236	0.10670	0.12193	0.13794	0.16319	0.20781	0.25459
19	0.05263	0.06378	0.07614	0.08962	0.10413	0.11955	0.13576	0.16134	0.20646	0.25366
20	0.05000	0.06116	0.07358	0.08718	0.10185	0.11746	0.13388	0.15976	0.20536	0.25292
25	0.04000	0.05122	0.06401	0.07823	0.09368	0.11017	0.12750	0.15470	0.20212	0.25095
30	0.03333	0.04465	0.05783	0.07265	0.08883	0.10608	0.12414	0.15230	0.20085	0.25031
40	0.02500	0.03656	0.05052	0.06646	0.08386	0.10226	0.12130	0.15056	0.20014	0.25003
50	0.02000	0.03182	0.04655	0.06344	0.08174	0.10086	0.12042	0.15014	0.20002	0.25000
100	0.01000	0.02320	0.04081	0.06018	0.08004	0.10001	0.12000	0.15000	0.20000	0.25000
∞		0.02000	0.04000	0.06000	0.08000	0.10000	0.12000	0.15000	0.20000	0.25000

[a]For interest rates i from 0 to 25 percent and for periods of analysis n from 1 to 100 yr.

to repay the loan in 7.8 yr, how much should the owner invest in the solar-heating system?

Solution. If

$$P_{ann} = \$510 \text{ (savings resulting from of solar energy for fuel oil)},$$
$$j = 0.12 \text{ (expected increase in fuel cost)},$$
$$i = 0.10 \text{ (interest rate for loan), and}$$
$$t = 7.8 \text{ (time to repay loan, in years)},$$

then

$$i = \frac{1 + 0.10}{1 + 0.12} - 1 = -0.0179 \text{ (effective interest rate)},$$

and

$$S = 510 \times \frac{0.982^{7.8} - 1}{-0.0179 \, (0.982)^{7.8}} = \$4,300$$

The owner can pay \$4,300 or less for the retrofit installations.

Repeat the problem, assuming the repayment time can be extended to 15 yr and government-backed, solar-improvement loans of 5 percent interest are available. Observe the large effect of interest rate on the calculations.

12.4 SOLAR SYSTEM LIFE CYCLE COSTS

The principal costs of a solar system are the capital and interest required to pay for equipment. In the life cycle cost analysis of a solar system, it is the *extra cost* of the solar system in excess of the basic auxiliary system that is considered. It is assumed that a nonsolar system is present whether solar energy is used or not. The savings in conventional fuels pay for the extra investment in solar equipment, along with the recurring expenses of maintenance, taxes, and insurance.

Although a life-cycle cost analysis normally deals with totals of annual costs during the life cycle period, it is more convenient to deal with life cycle costs on an annual basis. Both approaches are equivalent, as shown by the duality of annual costs (or payments P_{ann}) and the total costs S in the preceding analysis.

12.4.1 Solar System Initial Costs

The initial extra cost of a solar system consists of a number of components. For a solar-heating system for a building, for example, the following initial costs will usually be present:

Solar collector—cost and delivery; special siting requirements,
Thermal storage and insulation; special siting requirements,
Pumps or fans, piping, valves, and insulation; working fluids,
Heat exchangers,
Tanks—expansion, etc.,
Controls,
Wiring,
Labor,
Value of any building floor space used by solar components,
Extra collector support structure,
Special distribution system features,
Testing and checkout, and
Profit and overhead of installer.

Solar system costs are of two types. If the costs are dependent on system size, they are called *variable;* examples are costs of collector, storage, heat exchangers, pumps, etc. Costs independent of system size are called *fixed*; examples are controls, some labor, building floor space used for solar components, etc. Numerous experimental and theoretical analyses of solar systems for buildings have shown that the sizes of most components are related to collector area by specific rules as described in Chapters 4 to 6. Consequently, *most variable costs in a solar system can be related to collector size or area, which, therefore, becomes the obvious index of size of a solar system* over a limited range.

Since solar collector area is a measure of system size and therefore of each component's size, variable costs are usually expressed as cost per unit area of collector. For example, storage, while measured in physical units of kg or m^3, is usually priced as \$/m$_c^2$, the subscript c denoting collector. Of course, fixed costs do not depend on system size and are usually expressed in units such as dollars. Fixed costs can be considered as those required to purchase and install a solar system with a very small collector area, that is, small energy delivery.

In most cases the initial costs listed above are paid off by the owner in monthly or annual installments. The size of the periodic payments is determined by applying the appropriate value of *CRF* to the initial sum. These periodic costs, combined with recurring future costs described below, comprise the total annual cost of solar energy.

12.4.2 Solar System Future Costs

The components of life cycle costs that recur throughout the useful life of a solar system are called future costs and include:

Maintenance—personnel, materials,
Repair—personnel, materials,
Replacement—personnel, materials,
Power costs for operating pumps, fans, and controls,
Taxes—local, state, and federal, and
Insurance.

These costs generally vary throughout the life cycle of a system subject to inflationary pressures and political decisions. They must each be converted to their present value with appropriate values of the present-worth factor.

Maintenance and repair costs vary with the solar application. High-temperature systems, using tracking concentrating collectors, will require more maintenance and repair than low-temperature systems for building heating. Likewise, power costs will vary. For example, because of the relatively low heat capacitance of air vis-à-vis water and the consequent higher air flow rates, a heating system for a building using air as the working fluid will require more motive power than a system using water.

Taxes and insurance are widely variable. Local or state decisions on solar system tax policy have resulted in situations where solar systems are tax-free in one location but are taxed at the prevailing rate in another. Assessments of solar systems can range from 0 to 50 percent of their market value. Likewise, insurance rates for solar systems vary widely depending on the insurer's guidelines, location of the solar building, and location of the collector on the building. Since no general rules apply, a determination of these costs is required for each solar project in each location.

12.4.3 Annualized Solar Costs—Residential Applications

The total cost of solar energy on an annual basis C_y is the total of amortized initial costs and distributed costs expressed on a present-worth basis. Since most residences are not income producing, the only federal and state tax credits against these costs are for local taxes and interest. Income-producing installations will have more extensive tax credits and are described in the next section. Annualized solar costs C_y can be expressed in constant dollars as

$$C_y = (C_{s,\text{tot}} - ITC)CRF(i', t) \qquad \text{Initial investment,} \qquad (12.14)$$

$$- C_{s,\text{salv}} PWF(i', t) \, CRF(i', t) \qquad \text{Salvage value,}$$

$$+ \left[\sum_{k=1}^{t} R_k \, PWF(i', t_k) \right] CRF(i', t) \qquad \text{Replacements,}$$

$$+ C_e \frac{CRF(i', t)}{CRF(i'', t)} \qquad \text{Energy,}$$

$$+ T_{\text{prop}} \, C_{s,\text{ass}} \qquad \text{Property tax,}$$

$$- T_{\text{inc}} \, T_{\text{prop}} \, C_{s,\text{ass}} \qquad \text{Property tax, tax deduction,}$$

$$- T_{\text{inc}} \, i_m \left[\sum_k \frac{P_k}{(1 + i)^k} \right] CRF(i', t) \qquad \text{Interest, tax deduction,}$$

$$+ M \qquad \text{Maintenance, and}$$

$$+ I \qquad \text{Insurance,}$$

where $C_{s,tot}$ = the total initial solar investment including sales tax,
$C_{s,salv}$ = solar system salvage value at end of period of analysis,
$C_{s,ass}$ = assessed value of solar system,
C_e = energy cost to operate solar system in year one,
i' = $(i-j)/(1+j)$ = effective discount rate,
i'' = $(i-j_e)/(1+j_e)$ = effective discount rate for energy,
i = discount rate, that is, bounded below by opportunity cost to homeowner or the rate of return foregone on the next best alternative investment in market value; bounded above by the cost of borrowing (the rate i, strictly speaking, is the marginal discount rate, that is, the rate applicable to the solar unit purchaser's next set of investment decisions. It can be different from the average discount rate on previous investments),
M = maintenance ($/yr),
j = general inflation rate,
i_m = market mortgage rate (real mortgage rate + general inflation rate),
j_e = energy inflation rate ($/$ · yr),
k = years at which replacements or repairs are made (k can denote any time increments if i, j, etc. correspond),
I = insurance charges ($/yr),
ITC = investment tax credit,
R_k = replacement costs in year k (some R_k = 0)($/yr) in constant $,
P_k = outstanding principal (unpaid balance) of $C_{s,tot}$ in year k,
t = life cycle time or period of analysis (yr),
T_{prop} = property tax rate ($/$ assessed · year),
T_{inc} = state tax rate + federal tax rate − state tax rate × federal tax rate, where the rates are based on the last dollar earned

The summation in the seventh term of Eq. (12.14) can be evaluated by noting that the remaining principal P_k during the year k is

$$P_k = C_{s,tot}\left[(1+i_m)^{k-1} + \frac{(1+i_m)^{k-1}-1}{(1+i_m)^{-t}-1}\right]. \tag{12.15}$$

The summation is then given by

$$i_m\sum_k\frac{P_k}{(1+i)^k} = C_{s,tot}\left\{\frac{CRF(i_m, t)}{CRF(i, t)}\right.$$
$$\left. + \frac{1}{1+i_m}\frac{1}{CRF\left(\frac{i-i_m}{1+i_m}, t\right)}[i_m - CRF(i_m, t)]\right\} \tag{12.16}$$

The terms in Eq. (12.14) represent two types of real payments—those that remain constant in time and those that vary. For example, maintenance, insurance, property tax, and property tax credits are usually the same, in constant dollars, for the period of analysis t. If a real escalation is expected, these terms can readily be treated as the energy (fourth) term of Eq. (12.14). The initial investment (first) term of Eq. (12.14) represents the amortized cost of the solar system on an annual basis. Tending to offset this cost is the salvage value (second) term, which acts as a credit when reduced to its present worth by use of the PWF value.

Repair and replacement R_k are considered to be a series of charges, small and infrequent for a well-designed system, which are reduced to their respective present values by the use of the PWF for year k. Energy requirements (usually only electricity) for solar system operation may escalate in price at a rate j_e differing from the general inflation rate j. Therefore, the ratio of capital-recovery factors is required to determine the present value of operating energy.

Interest tax deductions diminish during a life cycle as the principal of the solar loan is reduced year by year. This is reflected by the expression in Eq. (12.15), which, when multiplied by the market mortgage rate i_m, is the interest payment for year k. A fraction of the interest cost is deductible from state and federal returns depending on the tax bracket of the building owner. Property taxes based on the extra assessment of property value resulting from the solar system $C_{s,ass}$ are, likewise, partially deductible. During a period of general inflation, a price index must be applied to annual tax deductions to reduce them to constant dollars. Tending to offset this inflation effect, however, is the increase in tax rate T_{inc} (that is, higher tax bracket) because of income increases that reflect cost-of-living increases frequently applied to many homeowners' wages.

Local and federal governments can provide tax incentives for the adoption of solar systems by eliminating property taxes, initiating special tax credits, subsidizing solar equipment manufacturers, offering low-interest loans, or causing grants to be made for the purchase and installation of solar systems. The efficacy of these measures can be evaluated by the use of Eq. (12.14). Although an annual time scale has been used in the analysis, any time scale (months, days, continuous cash flow) compatible with the time scale of the interest and discount rates can be used.

Example 12.4. Determine the annualized cost of a solar system with the following specifications over a 15-yr period ($t = 15$):

Solar system cost	$8,000,
Salvage value	$0,
Assessed value	$4,000,
Effective discount rate	$i' = 0.02$,
Effective discount rate for energy	$i'' = -0.02$ (4 percent differential energy inflation),
Market mortgage rate	$i_m = 0.10$,
Repairs: one of $200 at year 10	$R_{10} = \$200; R_{k \neq 10} = \$0$,
Maintenance and insurance	neglect,
Property tax rate (local law exempts solar systems)	$T_{prop} = 0$ percent,
Income tax bracket	$T_{inc} = 25$ percent, and
Electric power	$C_e = \$20/\text{yr}$.

Solution. For clarity, each term in Eq. (12.14) is evaluated in the table below. The required capital-recovery factors and present-worth factors are calculated first.

$$CRF(i', t) = CRF(0.02, 15) = 0.0778,$$

$$CRF(i'', t) = CRF(-0.02, 15) = 0.0565, \text{ and}$$

$$PWF(i', t_k) = PWF(0.02, 10) = 0.820.$$

Term type	Analytical expression	Value($)
Annualized extra cost	$C_{s,tot}\, CRF(i', t)$	622.40
Salvage value	$C_{s,salv}\, (= 0)$	0
Repair and replacement	$R_{10} PWF(i', 10)\, CRF(i', t)$	12.76
Electric energy	$C_e \dfrac{CRF\,(i', t)}{CRF(i'', t)}$	27.53
Property tax	(Tax exempt)	0
Property tax credit	(Tax exempt)	0
Interest credit [use Eq. (12.16)]	$T_{inc} i_m\, CRF(i', t)\sum \dfrac{P_k}{(1 + i')^k}$	-124.16
Maintenance and insurance	$I = M = 0$	0
		538.53/yr

The annualized solar system cost C_y is $538.53. Note that the interest tax credit is quite significant in reducing the annualized cost.

In many cases a down payment D_s is made on a solar system to reduce the mortgage amount. This payment reflects the increase in down payment for a solar-equipped building required by a banking institution above that required for a nonsolar building. The annual solar cost equation can be modified to include the down payment D_s by replacing the first term with two terms, that is,

$$(C_{s,tot} - D_s)\, CRF(i', t) + \frac{D_s i'^*}{(1 + i')^t - 1} \rightarrow C_{s,tot}\, CRF(i', t). \quad (12.17)$$

Likewise, the tax deduction (seventh term) in Eq. (12.14) replaces $C_{s,tot}$ with $(C_{s,tot} - D_s)$. An investment tax credit is treated as a negative cost occurring in year one.

Example 12.5. If the homeowner in the preceding example makes a $1,000 down payment, how much is the annualized solar cost reduced because of interest savings on the mortgage?

Solution. The difference between the two terms in Eq. (12.17) represents the annual cost difference. From the preceding example

---

*It is assumed here that the time value of the down payment amount is i', the same as the real discount rate.

$$C_{s,\text{tot}} \, CRF(i', t) = \$622.40.$$

If a \$1,000 down payment is made, this sum is replaced by

$$(C_{s,\text{tot}} - D_s) \, CRF(i', t) + \frac{D_s i'}{(1 + i')^t - 1} = \$544.60 + \$57.83 = \$602.43.$$

The difference in the two values is \$19.97. In addition, the annualized tax credit is reduced to \$108.64. The net annual payment is then \$534.08.

Equation (12.14) is based on the effective discount rate i', net of inflation. An equivalent method of analysis could be based on simple discount rates by inflating costs at the general inflation rate j. The results of the two analytical modes are identical.

The complete annual cost equation [Eq. (12.14)] does not consider certain indirect costs and benefits. For example, the use of solar energy to displace fossil fuels or electricity for heating would be expected to have the effect of reducing air pollution, conserving scarce fossil fuels for use as petrochemical feedstocks, and increasing industrial energy usage in the sectors providing materials for solar collectors. It is not anticipated that those factors would be considered in most microeconomic analyses. However, on the national macroeconomic scale, such factors are quite important.

12.4.4 Annualized Solar Costs—Commercial Applications

The calculation of annualized solar costs for an income-producing application is similar in most respects to that for noncommercial usages. However, in addition to the deductions described above, additional deductions apply for solar buildings. These include depreciation, operating costs, maintenance costs, and insurance. There will be an offsetting effect in these deductions because of the reduced deduction for fuels, since solar-plus-fossil fuel systems consume less fuel than nonsolar systems. The method of calculating all additional deductions except depreciation for commercial systems is identical to that used in Eq. (12.14) for property tax deductions and need not be described in detail.

There are three widely used methods for depreciating investments—straight line, sum-of-years digits, and declining balance. Each method results in a different depreciation schedule, as shown in Fig. 12.1. The basic annualized solar cost equation [Eq. (12.14)] can be modified to account for depreciation by subtracting the term

$$\text{Annualized depreciation tax deduction} = T_{\text{inc}} \, CRF(i', t) \sum_k \frac{D_k}{(1 + i')^k}, \quad (12.18)$$

where D_k is the annual depreciation amount in current dollars.

The annual depreciation amount D_k is given below for the three common write-off procedures.

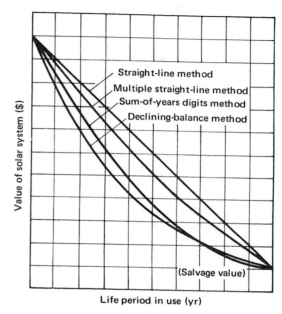

Figure 12.1. Comparison of straight-line, multiple straight-line, sum-of-years digits, and declining-balance methods for calculating solar system depreciation for income-producing property. Ref. [2] with permission.

Straight line (uniform):

$$D_k = \frac{C_{s,\text{tot}} - C_{s,\text{salv}}}{t}. \tag{12.19}$$

Sum-of-years digits (based on remaining service life):

$$D_k = \frac{2(t - k + 1)}{t(t + 1)} (C_{s,\text{tot}} - C_{s,\text{salv}}). \tag{12.20}$$

Declining balance (based on fixed percent of value in each year):

$$D_k = C_{s,\text{tot}} \left(\frac{C_{s,\text{salv}}}{C_{s,\text{tot}}}\right)^{k/t} \left[1 - \left(\frac{C_{s,\text{salv}}}{C_{s,\text{tot}}}\right)^{1/t}\right]. \tag{12.21}$$

Of the three methods, the straight-line approach is easiest to use and is recommended for preliminary studies. The declining-balance method is particularly sensitive to the salvage value, a parameter difficult to evaluate a priori. In addition, this method cannot be used for an assumed zero salvage value. The sum-of-years digits method has the advantage of accelerated early depreciation and simplicity of use. As salvage value estimates improve in accuracy through the life of a system, a multiple straight-line method can be used to refine the depreciation schedule.

A special tax situation exists for commercial rental property. Rents on solar buildings are expected to be higher than on nonsolar buildings, since the owner must recover solar system capital costs. This is not a disincentive for renting solar buildings, since conventional fuel savings would normally exceed the extra rental costs. However, the owner of a solar building realizes a larger income on the property because of the increased rentals. This increased benefit after taxes acts as an incentive to use solar energy; it is a negative cost to be subtracted from the annualized solar cost, Eq. (12.14).

Sales taxes on solar equipment can be partially recovered by an income tax deduction. To calculate the deduction, the credit on the one-time tax payment during the purchase in the first year can be annualized as a negative cost in the same way as the one-time down payment is annualized in Eq. (12.17).

12.4.5 Continuous Cash Flows

In some economic analyses, cash flows occur with such regularity and frequency that they may be assumed to occur continuously. Although this situation is uncommon in small-scale solar applications, it is common in industrial practice. The capital-recovery factors and present-worth factors for continuous cash flows differ from those for discrete cash flows and they are given in Eqs. (12.22) and (12.23).

$$CRF(i', t) = \frac{i'}{1 - e^{-i't}}, \tag{12.22}$$

$$PWF(i', t) = e^{-i't}, \tag{12.23}$$
$$\text{where } i' \equiv i - j.$$

The annualized solar cost equation can be used directly by substituting the above factors for their discrete-cash-flow counterparts in Eq. (12.14).

12.5 COST-BENEFIT ANALYSIS AND OPTIMIZATION

The optimization of a solar energy system requires that all components be sized to provide the *least-cost mix of solar energy and conventional fuel* for the application and site in question. In the preceding section the methods of calculating the annual solar cost for a given system have been detailed. In this section the concept of a production function is used to evaluate the energy delivery or *benefit* resulting from a solar system investment. The values of costs and benefits are those used to determine the optimal system configuration that maximizes benefits for a given level of solar-plus-nonsolar energy demand—the constraint on the optimization process.

12.5.1 Production Functions

A production function is the technical relationship specifying the maximum amount of output capable of being produced by each and every set of specified inputs [3]. It is defined for a given state of technical knowledge. For example, suppose a solar energy system consisting of a 100-m² collector, 2 metric tons of aqueous storage and associated pumps, heat exchangers, and controls can deliver 200,000 MJ/yr of thermal energy. The *technical production function* relates the physical solar components' sizes and characteristics to the output of 200,000 MJ. The *economic production function* relates the costs of the physical inputs to the value of the energy output. The value of energy output is equal to the value of nonsolar energy displaced by solar energy. In this book, the technical production function is used rather than the economic production function. There exists a duality between the two based on energy prices and solar equipment prices so either could be used in practice. In most economic analyses the production function is taken to be a long-term average applicable for the entire period of analysis.

The calculation of a production function is not an exact science; rather, an empirical method is used to relate energy output to endogenous solar system parameters such as collector size, efficiency and orientation, storage size, control strategy, and many other factors. The exogenous variables of climate and energy demand type and profile are also system output determinants. The empirical methods used differ with each solar application—heating, cooling, steam production, direct conversion. Although a thermal demand may not be thought of as a production function input, it behaves as one, since a given solar system will behave differently when coupled to different demands. In this chapter several types of production functions are described and their use in system optimization is derived in detail.

Nearly all inputs to a solar system production function are subject to the law of diminishing returns. The basic technological law states that extra output from a system diminishes relatively when successive equal units of an input are added to fixed amounts of other inputs [3]. For example, doubling storage capacity in a solar cooling system will not double the solar cooling effect. Figure 12.2 shows qualitatively this law applied to a solar-heating system as an example. It is seen that adding extra storage, collector area, heat-exchanger area, or pump capacity gains relatively little beyond a certain point. This fundamental law necessitates the optimization method of sizing solar systems. If equal or greater returns-to-scale were possible, 100 percent solar systems could be used with no auxiliary.

12.5.1.1 Polynomial production functions. A particularly simple production function is a multiple-variable polynomial with coefficients based on regression analysis of system performance—either data from computer models or actual equipment. It is expressed in the form

$$Q_s = \sum_{i=1}^{N} P_i(X_i),$$

(12.24)

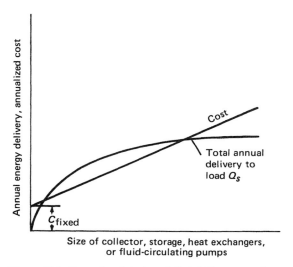

Figure 12.2. Examples of the law of diminishing returns for a solar-heating system. C_{fixed} is the cost component independent of the size of the solar system component.

where P_i is a polynomial in the parameter X_i. The X_i's include system component sizes, solar-radiation levels, weather parameters, energy-demand parameters, and the like; Q_s is the annual solar energy *delivery-to-load*. Instead of a summation of polynomials, a product could be used

$$Q_s = \prod_{i=1}^{N} P_i(X_i). \tag{12.25}$$

Polynomial production functions can be selected to have any number of terms required to ensure the best accuracy as adjudged by an F-test. However, the basic information contained in such a production function is limited to calculating Q_s. The coefficients and exponents do not provide any additional information as is the case in other production functions below. The principal advantage of the polynomial is its simplicity.

12.5.1.2 Logarithmic and exponential production functions. A solar energy production function can be formed from logarithms (or their inverse, exponents) as follows:

$$Q_s = \ln \prod_{i=1}^{N} a_i X_i^{n_i}, \tag{12.26}$$

where n_i are powers (integral or nonintergral) and a_i are coefficients. This form is equivalent to

$$Q_s = (a_1' + n_1 \ln X_1) + (a_2' + n_2 \ln X_2) + \dots$$
$$= A + n_1 \ln X_1 + n_2 \ln X_2 + \dots . \tag{12.27}$$

The coefficients n_i can be evaluated by regression methods using data relating Q_s to X_i.

Although the logarithmic form has a limited number of coefficients and possibly lower regression accuracy than the polynomial form, the coefficients n_i contain additional information. If Eq. (12.27) is differentiated, we have

$$dQ_s = n_1 \frac{dX_1}{X_1} + n_2 \frac{dX_2}{X_2} + \dots . \tag{12.28}$$

The coefficients are seen to be the change in Q_s resulting from a unit change in the independent variables X_i. The coefficients, therefore, indicate where a change in physical inputs can have the largest impact.

Example 12.6. A production function for a particular solar space-heating system in Chicago has been determined to be of the form

$$Q_s = L\left\{0.08 + \ln\left[\left(\frac{A_c}{L}\right)^{1/3}\left(\frac{S}{L}\right)^{1/20}\right]\right\}, \tag{12.29}$$

where A_c is the collector area, L is the annual energy demand of 100 GJ, and S is the storage size. Determine the expected increase in delivery Q_s for a 1 percent increase in both storage size and collector size if the load remains unchanged.

Solution. Expand Eq. (12.29) to get

$$Q_s = L\left(0.8 + \frac{1}{3}\ln A_c + \frac{1}{20}\ln S - \frac{23}{60}\ln L\right).$$

For small changes in A_c or S, the differential approximation can be used. Differentiating we have

$$dQ_s = \frac{L}{3}\frac{dA_c}{A_c} + \frac{L}{20}\frac{dS}{S}.$$

Therefore, for a 1 percent increase in collector area A_c ($dA_c/A_c = 0.01$),

$$dQ_s = \frac{100{,}000 \text{ MJ}}{3} \times 0.01 = 333 \text{ MJ}.$$

For a 1 percent increase in storage S

$$dQ_s = \frac{100{,}000 \text{ MJ}}{20} \times 0.01 = 50 \text{ MJ}.$$

In this example a unit increase in collector area A_c delivers $6\frac{2}{3}$ times more energy than the same percentage increase in storage delivers.

12.5.1.3 Power-law production functions. A production function can be formed from the product of powers of physical inputs:

$$Q_s = v_O \prod_{1=i}^{N} X_i^{v_i}. \tag{12.30}$$

As in previous cases the v_i's are determined by a regression analysis of performance data. The exponents in the power-law production function are similar to those in the logarithmic expression and represent the percentage change in output for a percentage change in input. It is easy to show that

$$\frac{dQ_s}{Q_s} = \sum_{i=1}^{N} v_i \frac{dX_i}{X_i}. \tag{12.31}$$

In addition, *the sum of the v_i's represents the return-to-scale* of a solar system. Returns-to-scale are the change in output Q_s resulting from a uniform unit change in all inputs X_i. If $\sum_{i=1}^{N} v_i$ is greater than one, returns-to-scale are increasing; if less than one, decreasing; and if equal to one, proportionate.

Numerical production functions. Instead of evaluating a closed-form empirical production function from data by regression methods, the data itself can be used as a tabular production function. Such a function does not lend itself to ready manipulation by hand, but can be used by a computer to perform optimization calculations. No information regarding returns-to-scale can be determined directly from a numerical production function, however.

12.5.2 Economic Optimization Methodology

12.5.2.1 Marginal cost analysis. *The optimal size of a solar system is that which minimizes the annual cost of solar-plus-auxiliary energy.* The total annual cost C_T can be expressed as

$$C_T = C_y(Q_s) + C_a(Q_a), \tag{12.32}$$

where $C_a(Q_a)$ is the annual cost of nonsolar energy and $C_y(Q_s)$ is the annualized solar cost—Eq. (12.14). Both costs are assumed to depend on the quantity of energy delivered. This is always true for solar systems and for nonsolar sources. Three typical total cost curves are shown in Fig. 12.3. The three generic types of cost curves can result from the following sets of conditions:

1. High nonsolar energy costs and/or high sun–high load factor location (curve 1); largest system possible is optimal.
2. Average or typical mix of energy cost, sun amount, and demand (curve 2); and
3. Low nonsolar energy costs and/or low sun–low load factor location (curve 3); the no-solar system is optimal.

A discussion of nonsolar energy prices is presented at the end of this section.

The value of the nonsolar energy cost $C_a(Q_a)$ in Eq. (12.32) represents the *annualized life cycle fuel cost*. As such it is not the cost for any single year during the life cycle. It is rather the total life cycle cost in present dollars annualized by the capital-recovery factor $CRF(i',t)$. It is given analytically by an expression of the form of the fourth (nonsolar) term in the annualized solar cost equation [Eq. (12.14)]. The quantity C_e in Eq. (12.14) can be thought of, in the present analysis, as representing the backup fuel cost in current dollars for the first year of the period of analysis. In most economic

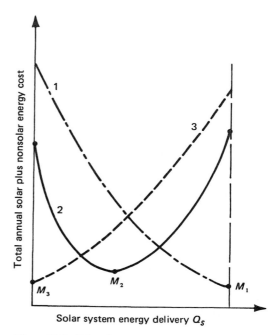

Figure 12.3. Solar-plus-nonsolar total annual cost curves; minima shown by M. Curves 1 and 3 represent boundary minima; boundaries represent maximum and minimum solar delivery for a specific application (M_3 normally represents zero collector area or the null system).

analyses it is assumed that the amount of fuel displaced by solar energy each year is the same; only its value increases year by year because of the differential inflation of fuel price.

The optimization of total cost is subject to the constraint

$$Q_T = Q_s + Q_a, \tag{12.33}$$

where Q_T is the total annual energy demand to be met, Q_s is the solar portion of load and Q_a is the nonsolar portion of load. Equation (12.32) can be unconstrained by use of the Lagrange multipler λ:

$$C_T = C_y(Q_s) + C_a(Q_a) + \lambda(Q_T - Q_s - Q_a). \tag{12.34}$$

It is desired to find a stationary point of Eq. (12.34), that is, a point where $dC_T = 0$:

$$dC_T = \frac{dC_y}{dQ_s} dQ_s + \frac{dC_a}{dQ_a} dQ_a + \lambda(-dQ_a - dQ_s) = 0. \tag{12.35}$$

Since the differentials dQ_s and dQ_a are unconstrained, they may be independently set to 0 to get

$$\frac{dC_y}{dQ_s} = \frac{dC_a}{dQ_a} = \lambda. \tag{12.36}$$

Equation (12.36) states that the minimum annual cost occurs at the point where *both marginal energy costs are equal.* This condition means that investment in any solar system should continue up to the point where the last dollar invested generates precisely $1 in nonsolar energy savings. Figure 12.4 shows the above analysis graphically. At the point where total cost curves for solar and nonsolar energy have equal slope, marginal costs are equal as required for the optimal system.*

At any point between A and B in Fig. 12.4, the solar source represents a less expensive alternative than the nonsolar source. That is, at any point between A and B the solar system owner saves both money and nonsolar energy vis-à-vis the full nonsolar option. Point M is the least-cost optimum. If the goal of economic policy were to minimize nonsolar energy use while paying the same total annual cost for energy, the design point will be B in Fig. 12.4. In this book the least-cost point M, representing rational consumer behavior in most developed economies, is adjudged the optimum. If the solar curve does not intercept or is not tangent to the nonsolar curve in Fig. 12.4a, there is no microeconomic incentive to consider solar energy.

---

*If there is a budget upper limit, the marginal cost criterion may not be achievable. In that case the full budget amount should be spent.

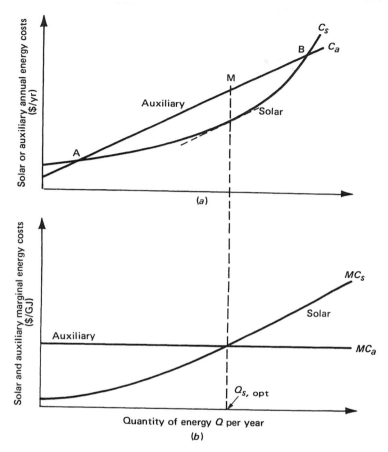

Figure 12.4. Optimum solar size determination: (*a*) total annualized costs of solar and nonsolar energy; (*b*) marginal-cost curves for solar and nonsolar energy showing optimal system size at intersection of marginal-cost curves.

Example 12.7. A single-family dwelling in St. Joseph, MS, has an annual energy demand of 70 GJ for heating. Computer model simulations of a solar system for this building have indicated that the annualized cost function for the solar system is

$$C_y = 400 - 20Q_s + 0.4\,Q_s^2 \quad (20 < Q_s < 70), \tag{12.37}$$

in which Q_s is measured in gigajoules and C_y in dollars. Fuel oil is the nonsolar energy source for this building and its cost function is

$$C_a = 5Q_a, \tag{12.38}$$

where Q_a is measured in gigajoules. Determine the optimum mix of solar energy and fuel oil.

Solution. The optimal solar fraction of load is determined by equating marginal costs of solar energy MC_s and fuel oil MC_a. From Eqs. (12.37) and (12.38),

$$MC_s = \frac{dC_y}{dQ_s} = -20 + 0.8\,Q_s \qquad (12.39)$$

$$MC_a = \frac{dC_a}{dQ_a} = 5.$$

Equating and solving for the optimal Q_s we have

$$-20 + 0.8\,Q_s = 5,$$
$$Q_s = 31 \text{ GJ}.$$

The fraction of load carried by solar energy f_s is

$$f_s = \frac{31}{70} = 0.44 \text{ or } 44 \text{ percent}.$$

12.5.2.2 Simplified analysis. In many solar analyses it is possible to simplify the above method. As mentioned in a preceding section, collector area is a reliable index by which to determine the size and cost of all ancillary components in most solar building systems. The cost function Eq. (12.32) can then be determined directly by inverting the technical production function as shown below. If the production function is given by

$$Q_s = f(X_1, X_2, \dots), \qquad (12.40)$$

then it can be simplified to

$$Q_s = f(A_c), \qquad (12.41)$$

using rules referred to earlier. The annual cost of a solar system, likewise, can be keyed to collector area by

$$C_y = c_y A_c + C_f, \qquad (12.42)$$

where c_y is the annual solar cost per unit area and C_f is the fixed cost. The nonsolar portion of the annual cost function can also be keyed to collector area. If

$$C_a = p(Q_a), \qquad (12.43)$$

then

$$C_a = p(Q_T - Q_s) = q(Q_s), \qquad (12.44)$$

where Q_T is the total annual demand. By use of Eq. (12.41) we have

$$C_a = q[f(A_c)] = h(A_c). \qquad (12.45)$$

Combining C_y and C_a to form total cost C_T

$$C_T = g(A_c) + h(A_c), \qquad (12.46)$$

or

$$C_T = F(A_c). \qquad (12.47)$$

Determination of the optimal system size can be done directly by differentiating Eq. (12.47) with respect to A_c to get the optimization criterion

$$F'(A_{c,\text{opt}}) = 0, \qquad (12.48)$$

subject to

$$F''(A_{c,\text{opt}}) > 0, \qquad (12.49)$$

and subject to

$$C_T(A_{c,\text{opt}}) < C_T(A_{c,\text{max}}), \text{ and} \qquad (12.50)$$

$$C_T(A_{c,\text{opt}}) < C_T(A_{c,\text{min}}), \qquad (12.51)$$

where the $A_{c,\text{min}}$ and $A_{c,\text{max}}$ are the extreme values of the collector area domain of interest. It is necessary to check end points, since in some cases the cost curve may have a boundary minimum as shown in Fig. 12.3. The following example shows how a technical production function can be inverted and used directly to find the optimum collector area.

Example 12.8. The simplified solar energy production function for a building in Denver, CO, is given by

$$Q_s = 35\sqrt{A_c} \qquad (12.52)$$

where A_c is the collector area in square meters ($350 > A_c > 50$ m²). If the annualized cost of the solar system c_y, per square meter, is \$10, find the optimum collector size. The total annual energy demand is 650 GJ and the cost of the nonsolar energy source c_a is \$8/GJ.

Solution. The solar cost function C_y is

$$C_y = c_y A_c = 10 A_c.$$

Note that the annualized cost $c_y = 10$ is based on an initial cost of \$150/m² and an annual multiplier [Eq. (12.14)] of 0.0667.

The nonsolar cost function is calculated from Eqs. (12.44) and (12.45):

$$C_a = c_a(Q_T - Q_s)$$
$$= 8(650 - 35\sqrt{A_c}).$$

The total cost function C_T is

$$C_T = C_y + C_a$$
$$= 10 A_c - 280\sqrt{A_c} + 5200. \qquad (12.53)$$

Equation (12.53) must be differentiated and equated to 0 to find the optimal area:

$$\frac{dC_T}{dA_c} = 10 - \frac{280}{2} \frac{1}{\sqrt{A_{c,opt}}} = 0.$$

Solving for $A_{c,opt}$,

$$A_{c,opt} = 196 m^2.$$

Equation (12.52) permits calculation of the solar energy produced:

$$Q_s = 35\sqrt{196} = 490 \text{ GJ}.$$

This represents the following portion of load f_s

$$f_s = \frac{490}{650} = 0.75 \text{ or } 75 \text{ percent}.$$

The total cost curve for the example is shown in Fig. 12.5 with the least-cost point illustrated.

It is important to recognize that the production function and cost functions used in the two preceding examples are for a specific system in a specific building in a specific location. *It is not possible to use a production function derived for one system in the analysis of another system.* Since no a priori or theoretical method exists to calculate the technical production functions, they can only be derived from simulation studies of system thermal performance as described in Chapters 5 to 8.

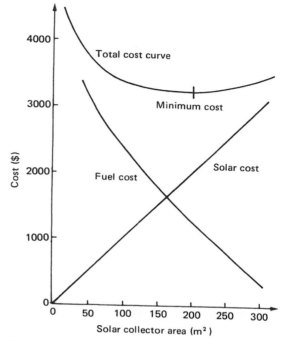

Figure 12.5. Cost curves for solar system used in example above. Total cost curve is the sum of the solar and fuel curves.

12.5.2.3 Other economic indices. Although the proper method for choosing the optimal solar configuration is the method outlined above, other indices are sometimes used. The payback-period and return-on-investment approaches are commonly used. They are easy to calculate and have an intuitive appeal because of their simplicity.

However, these indices are incomplete and therefore incorrect for two reasons. First, the payback period does not consider the time value of money. Secondly, they are not a quantitative measure of cash flow, that is, they do not have a dollar magnitude associated with them for either costs or benefits. In addition, the payback method ignores any cash flows beyond the payback period, thereby giving only a partial view of the entire time series of cash flows.

Since the investment in a solar system does require a fixed number of years for repayment, it is a long-term investment. It may not be as attractive as a shorter, more liquid investment at the same rate of return. Therefore, a somewhat larger return, that is, larger discount rate, may be needed to induce investment in a solar system. This effect is usually small because of compensating considerations such as the effect of inflation on the investment market.

If a solar system owner sells a facility before the term of the mortgage, it could be argued that a loss is experienced, because the lifetime of the system has not been realized. This is an incomplete view, since a solarized building would command a higher price in the marketplace, making it easier for the owner to recover the unamortized portion of the solar investment.

A year-by-year calculation of costs and benefits can be made to determine the profitability of a solar system for periods of time shorter than the mortgage term. The annual payment for the solar system represents a constant number of current dollars (the tax credit decreases during the life cycle, however) whereas the payment for fuel would represent a constantly increasing dollar amount in a period of inflation. Since the dollar value of fuel savings will be largest near the end of the life cycle, it may be expected that during the initial years the solar system is not cost-effective. Tending to offset this trend is the tax deduction for the solar system, which is largest in the earlier years because of large mortgage interest payments. The following example shows annual cash flows and a calculation of the cash benefit of the solar system.

Example 12.9. Calculate the annual cash flows for a solar-heating system on a residence providing an average of 100 GJ/yr. The initial cost of the system is $6,500, and the owner pays no property taxes and is in the 25 percent federal tax bracket. The mortgage interest rate i_m is 10 percent for 15 yr; for simplicity, ignore maintenance and insurance costs. Fuel costs are $5/GJ and are expected to increase by 12 percent/yr. Perform the calculation in current dollars not constant dollars. Compare the cost year by year of both solar and nonsolar systems capable of delivering 100 GJ/yr.

Solution. Cash flows will be tabulated below. The annual mortgage payment is

$$CRF(10 \text{ percent, } 15 \text{ yr}) \times 6500 = \$854.58.$$

The charge for fuel to supply the full load in the first year is

$$100 \text{ GJ } \times \$5/GJ = \$500/yr.$$

This sum inflates at a rate of 12 percent/yr. Table 12.3 summarizes cash flows for the first 10 yr of the mortgage.

The table shows that the homeowner has no positive benefit (savings for nonsolar versus solar) until the fifth year, when fuel inflation overtakes the fixed annual solar system cost. Within another 1–2 yr the total early diseconomy of the solar system has been more than eliminated because of fuel inflation. A calculation such as this can be made to show the homeowner's cash flows in current dollars. Unless extensive simplifying assumptions are made, it is not possible to carry out the calculation above in a general analytical form to determine the break-even point or the payout period.

It should be noted that savings in fuel costs by use of a solar system are a kind of nontaxable return on an investment (the solar system). As such they are equivalent to about twice their dollar value in earnings from a conventional investment subject to capital gains taxes (maximum of 50 percent in the United States).

Table 12.3. Annual cash flows for example above[a]

Annual solar cost $CRF \times C_{s,tot}$ (1)	Remaining principal P_k (2)	Interest $i_m P_k$ (3)	Principal payment [(1)–(3)] (4)	Tax credit $i_m T_{inc} P_k$ (5)	Solar cost [(1)–(5)] (6)	Fuel cost for no solar usage[b]
854.58	6500.00	650.00	204.58	162.50	692.08	500.00
854.58	6295.42	629.54	225.04	157.39	697.19	560.00
854.58	6070.38	607.04	247.54	151.76	702.82	627.20
854.58	5822.84	582.28	272.30	145.57	709.01	702.46
854.58	5550.54	555.05	239.53	138.76	715.82	786.76
854.58	5251.01	525.10	329.48	131.28	723.30	881.17
854.58	4921.53	492.15	362.43	123.04	731.54	989.91
854.58	4559.10	455.91	398.67	113.98	740.60	1105.34
854.58	4160.43	416.04	438.54	104.01	750.57	1237.98
854.58	3721.89	372.19	482.39	93.05	761.53	1386.54

[a]In current, not constant dollars.
[b]This column represents the cost of providing the solar energy portion of load (100 GJ/yr) from a non-solar source. It does not represent the total cost of heating the subject building.

In summary, economic indices such as payout period, return-on-investment, or break-even point are incomplete and incorrect indices to use for selection of the optimal solar system. They are, at best, intuitively appealing methods because of their simplicity.

12.5.2.4 Nonsolar energy costs—utility pricing strategies.
The cost of the auxiliary energy used to provide that fraction of an energy demand for which it is uneconomical to use solar energy is equal in importance to the cost of the solar system itself. Utilities that provide nonsolar energies can have a significant effect on the future viability of solar energy.

Two pricing strategies have been widely used in the United States. The first, the declining-block rate, results in charges that, on a unit-energy basis, decline with increasing energy usage rate. Part of the rationale for use of this rate by electric utilities is the higher generator load factor and increased return to the utility for a fixed capital investment.

The second strategy—used mainly by electric utilities—is called the peak-demand rate. It is frequently a declining-block rate with an additional charge related to the peak monthly demand. Often, a fraction of the peak annual demand rate (50–75 percent) is charged as the minimum utility bill for a given month even if no utility-based energy is used. This fraction is called the *billing demand* and is distinct from the *actual demand*. Part of the rationale for the demand rate is that those who use peaking power provided by extra peaking capacity should pay for the capital investment in that extra capacity.

Electric utilities have large investments in capital and therefore use the demand pricing strategy in parts of the United States to recover that investment. For solar sys-

tems in small and large buildings using electricity as the backup, the type of rate structure in force can make the difference between investment in a solar system or investment in an energy conservation system (to reduce peak demand) costing the same
amount. The following example illustrates the trade-offs.

Example 12.10. The owner of a building with a peak heat demand in January of
15.8 kW (54,000 Btu/hr) and a total demand of 5600 kWh (19×10^6 Btu) wishes to reduce the utility bill. Compare the economies of a solar system that can provide half the
January heating energy demand for $50/month (including amortization of the solar
system loan, taxes, and maintenance) with an energy conservation system (insulation,
inflation control, etc.) that can reduce the building heat loss by one-third. The extra
mortgage payment for this conservation system is the same as that for the solar system—$50/month.
   The utility uses a peak-demand pricing strategy wherein each hour of peak demand costs $2.70/month and energy costs $0.008/kWh.

Solution. The *solar system* cost for January will be

$$\text{Demand charge} = 15.8 \times \$2.70 = \$42.66,$$

$$\text{Electric energy charge} = \frac{5600}{2} \times \$.008 = \$22.40,$$

$$\text{Solar energy mortgage charge} = \$50.00,$$

$$\text{Total charge with the solar system} = \$115.06.$$

The *energy conservation system* cost for January will be

$$\text{Demand charge} = \left(\frac{2}{3} \times 15.8\right) \times \$2.70 = \$28.44,$$

$$\text{Electric energy charge} = \left(\frac{2}{3} \times 5600\right) \times \$.008 = \$29.87,$$

$$\text{Extra mortgage charge} = \$50.00,$$

$$\text{Total charge with the energy conservation system} = \$108.30.$$

For this example, in January, the energy conservation system costs less than the solar
system even though more electricity is purchased from the utility with the conservation scheme than with the solar scheme. A complete economic analysis requires that
the above calculations be made and totals determined for an entire year. In addition,
peak-shaving strategies should be analyzed.
   As an exercise, compare the economies of the energy conservation and solar systems described above for a scenario in which the auxiliary electricity is priced on a declining-block basis instead of a peak-demand basis (see previous problem.)

The example illustrates that solar systems reduce consumption, not demand, in general. It is the relative cost of reduced consumption versus reduced demands that dictates the investment size and direction.

Other utility pricing strategies have been considered. For example, as an inducement to flatten diurnal peaks, a utility could price its "valley" energy at a lower cost than its peak energy. This could be accomplished by a time-of-day pricing strategy. Presumably this would lead to a redistribution of diurnal loads in the utility grid to flatten peaks and use generating capacity at a higher load factor. It must be recognized that a solar system simply displaces the lowest-cost energy available from a utility if diurnal pricing is used. If solar energy were not used, a simple storage system that could store offpeak power for use during peak periods would be the first investment made, not the investment in a solar system. Therefore, the cost-effectiveness of a solar system must be weighed against the off-peak energy rate, not the peak rate.

Summary of Optimization Methodology

The optimization and sizing methodology described in this chapter can be summarized in a series of steps that must be followed to ensure the proper sizing of the cost-optimal solar system. These steps are

1. Establish the technical production function for a range of all important solar system component sizes. Determine the annualized cost for each configuration using discounted cash flow analysis. The energy delivery Q_s of the production function is usually established by modeling techniques or empirical data (see Chapters 5 to 8).
2. Establish the technical production function for the nonsolar auxiliary energy alternative.
3. Calculate marginal costs from the solar and nonsolar production functions. Equate to determine the optimal amount of solar energy to be supplied. The portion of load supplied by the nonsolar source is the difference between the total demand and the optimal solar fraction. Marginal costs can be calculated numerically or algebraically.

PROBLEMS

12.1. If the initial extra cost of a solar system is $5,000, what is the annual payment if the interest rate is 8 percent and the mortgage term is 15 yr?

12.2. A solar system in China Lake, CA, delivers the amounts of energy each year shown in the table. If the solar system costs $195/m² and is amortized (paid off) over 25 yr at 10 percent interest, prepare the total cost curve and specify the cost-optimal system. The backup fuel is fuel oil at a cost of $7.00/GJ; fuel price is inflating 8 percent/yr.

Collector area (m²)	Energy delivered (GJ)
100	336
150	444
200	531
250	612
300	673
400	791
500	856
600	915
(no solar)	0

The total annual energy demand is 1000 GJ/yr.

12.3. What is the present value of a $1,000 payment in 2010 if the discount rate is 7 percent?

12.4. What is the present value in constant dollars of the $1,000 payment in Problem 12.3 if the inflation rate is 4 percent/yr?

12.5. Calculate and tabulate the annual cash flows associated with a solar system for a 10-yr period of analysis if the extra initial solar cost is $6,000 and if

Interest rate   = 9 percent,
Power cost   = $30/yr escalating at 10 percent/yr,
Property tax   = 0,
Income tax bracket   = 32 percent,
Maintenance   = $\frac{1}{2}$ percent/yr,
Scrap value   = 50 percent of initial cost.

If the solar system saves $550/yr in conventional fuel (escalating at 10 percent/yr), is it cost effective? Work the problem in current dollars.

12.6. Capital-recovery factors based on monthly and yearly payments differ slightly. What is the percentage difference in these two payment schedules for an 8 percent loan over 20 yr?

12.7. What is the one-term logarithmic production function for the system in Problem 12.10? Use a regression method or a plot on semi-log graph paper.

12.8. The production function of a solar water-heating system is $Q_s = 12 A_c^{0.6}$ GJ/yr. The annual solar system cost is $20/m_c^2 \cdot$ yr. If the conventional energy is valued at $9/GJ and the annual energy demand is 140 GJ, what is cost-optimal collector area?

12.9. Rework Example 12.10 if the utility uses a declining-block rate instead of a demand rate. The declining-block rate is

0–200 kWh $7.41,
200–1000 kWh $0.03117/kWh,
over 1000 kWh $0.02130/kWh.

Is the solar system or the energy conservation system cheaper for January?

REFERENCES

1. Hirst, E. 1976. "Changes in retail energy prices and the Consumer Price Index." *Energy* 1, 33–43.
2. Peters, M. S., and R. D. Timmerhaus. 1968. *Plant design and economics for chemical engineers.* New York: McGraw-Hill Book Co.
3. Samuelson, P. 1976. *Economics,* 10th ed. New York: McGraw-Hill Book Co.
4. Weast, R. C. 1968. *Handbook of chemistry and physics.* Cleveland: Chemical Rubber Co.

THE INTERNATIONAL SYSTEM OF UNITS, FUNDAMENTAL CONSTANTS, AND CONVERSION FACTORS

Appendix

1

The International System of Units (SI) is based on seven base units. Other derived units can be related to these base units through governing equations. The base units with the recommended symbols are listed in Table A1.1. Derived units of interest in solar engineering are given in Table A1.2.

Standard prefixes can be used in the SI system to designate multiples of the basic units and thereby conserve space. The standard prefixes are listed in Table A1.3.

Table A1.4 lists some physical constants that are frequently used in solar engineering, together with their values in the SI system of units.

Conversion factors between the SI and English systems for commonly used quantities are given in Table A1.5.

Table A1.1 The seven SI base units

Quantity	Name of unit	Symbol
Length	Meter	m
Mass	Kilogram	kg
Time	Second	sec
Electric current	Ampere	A
Thermodynamic temperature	Kelvin	K
Luminous intensity	Candela	cd
Amount of a substance	Mole	mol

Table A1.2 SI derived units

Quantity	Name of unit	Symbol
Acceleration	Meters per second squared	m/sec^2
Area	Square meters	m^2
Density	Kilogram per cubic meter	kg/m^3
Dynamic viscosity	Newton-second per square meter	$N \cdot sec/m^2$
Force	Newton $(= 1 \text{ kg} \cdot m/sec^2)$	N
Frequency	Hertz	Hz
Kinematic viscosity	Square meter per second	m^2/sec
Plane angle	Radian	rad
Potential difference	Volt	V
Power	Watt $(= 1 \text{ J/s})$	W
Pressure	Pascal $(= 1 \text{ N/m}^2)$	Pa
Radiant intensity	Watts per steradian	W/sr
Solid angle	Steradian	sr
Specific heat	Joules per kilogram-Kelvin	$J/kg \cdot K$
Thermal conductivity	Watts per meter-Kelvin	$W/m \cdot K$
Velocity	Meters per second	m/sec
Volume	Cubic meter	m^3
Work, energy, heat	Joule $(= 1 \text{ N} \cdot m)$	J

Table A1.3 English prefixes

Multiplier	Symbol	Prefix	Multiplier	Symbol
10^{12}	T	Tera	10^3	M (thousand)
10^9	G	Giga	10^6	MM (million)
10^6	m	Mega		
10^3	k	Kilo		
10^2	h	Hecto		
10^1	da	Deka		
10^{-1}	d	Deci		
10^{-2}	c	Centi		
10^{-3}	m	Milli		
10^{-6}	μ	Micro		
10^{-9}	n	Nano		
10^{-12}	p	Pico		
10^{-15}	f	Femto		
10^{-18}	a	Atto		

Table A1.4 Physical constants in SI units

Quantity	Symbol	Value
Avogadro constant	N	$6.022169 \times 10^{26} \text{ kmol}^{-1}$
Boltzmann constant	k	$1.380622 \times 10^{-23} \text{ J/K}$
First radiation constant	$C_1 = 2\pi hc^2$	$3.741844 \times 10^{-16} \text{ W} \cdot m^2$
Gas constant	R	$8.31434 \times 10^3 \text{ J/kmol} \cdot K$
Planck constant	h	$6.626196 \times 10^{-34} \text{ J} \cdot sec$
Second radiation constant	$C_2 = hc/k$	$1.438833 \times 10^{-2} m \cdot K$
Speed of light in a vacuum	c	$2.997925 \times 10^8 \text{ m/sec}$
Stefan-Boltzmann constant	σ	$5.66961 \times 10^{-8} \text{ W/m}^2 \cdot K^4$

Table A1.5 Conversion factors

Physical quantity	Symbol	Conversion factor
Area	A	1 ft² = 0.0929 m²
		1 acre = 43,560 ft² = 4047 m²
		1 hectare = 10,000 m²
		1 square mile = 640 acres
Density	ρ	1 lb$_m$/ft³ = 16.018 kg/m³
Heat, energy, or work	Q or W	1 Btu = 1055.1 J
		1 kWh = 3.6 MJ
		1 Therm = 105.506 MJ
		1 cal = 4.186 J
		1 ft · lb$_f$ = 1.3558 J
Force	F	1 lb$_f$ = 4.448 N
Heat flow rate, Refrigeration	q	1 Btu/hr = 0.2931 W
		1 ton (refrigeration) = 3.517 kW
		1 Btu/sec = 1055.1 W
Heat flux	q/A	1 Btu/hr · ft² = 3.1525 W/m²
Heat-transfer coefficient	h	1 Btu/hr · ft² · F = 5.678 W/m² · K
Length	L	1 ft = 0.3048 m
		1 in = 2.54 cm
		1 mi = 1.6093 km
Mass	m	1 lb$_m$ = 0.4536 kg
		1 ton = 2240 lbm
		1 tonne (metric) = 1000 kg
Mass flow rate	$\dot{m}$	1 lb$_m$/hr = 0.000126 kg/sec
Power	$\dot{W}$	1 hp = 745.7 W
		1 kW = 3415 Btu/hr
		1 ft · lb$_f$/sec = 1.3558 W
		1 Btu/hr = 0.293 W
Pressure	p	1 lb$_f$/in² (psi) = 6894.8 Pa (N/m²)
		1 in. Hg = 3,386 Pa
		1 atm = 101,325 Pa (N/m²) = 14.696 psi
Radiation	l	1 langley = 41,860 J/m²
		1 langley/min = 697.4 W/m²
Specific heat capacity	c	1 Btu/lb$_m$ · °F = 4187 J/kg · K
Internal energy or enthalpy	e or h	1 Btu/lb$_m$ = 2326.0 J/kg
		1 cal/g = 4184 J/kg
Temperature	T	$T(°R) = (9/5)T(K)$
		$T(°F) = [T(°C)](9/5) + 32$
		$T(°F) = [T(K) - 273.15](9/5) + 32$
Thermal conductivity	k	1 Btu/hr · ft · °F = 1.731 W/m · K
Thermal resistance	R_{th}	1 hr · °F/Btu = 1.8958 K/W
Velocity	V	1 ft/sec = 0.3048 m/sec
		1 mi/hr = 0.44703 m/sec
Viscosity, dynamic	μ	1 lb$_m$/ft · sec = 1.488 N · sec/m²
		1 cP = 0.00100 N · sec/m²
Viscosity, kinematic	ν	1 ft²/sec = 0.09029 m²/sec
		1 ft²/hr = 2.581 × 10⁻⁵ m²/sec
Volume	V	1 ft³ = 0.02832 m³ = 28.32 liters
		1 barrel = 42 gal (U.S.)
		1 gal (U.S. liq.) = 3.785 liters
		1 gal (U.K.) = 4.546 liters

Table A1.5 Conversion factors *(Continued)*

Physical quantity	Symbol	Conversion factor
Volumetric flow rate	$\dot{Q}$	1 ft³/min (cfm) = 0.000472 m³/sec 1 gal/min (GPM) = 0.0631 l/sec

SOLAR RADIATION DATA

Table A2.1 Solar irradiance for different air masses[a]

| Wavelength | Air mass; $\alpha = 0.66$; $\beta = 0.085$[b] | | | | |
	0	1	4	7	10
0.290	482.0	0.0	0.0	0.0	0.0
0.295	584.0	0.0	0.0	0.0	0.0
0.300	514.0	4.1	0.0	0.0	0.0
0.305	603.0	11.4	0.0	0.0	0.0
0.310	689.0	30.5	0.0	0.0	0.0
0.315	764.0	79.4	0.1	0.0	0.0
0.320	830.0	202.6	2.9	0.0	0.0
0.325	975.0	269.5	5.7	0.1	0.0
0.330	1059.0	331.6	10.2	0.3	0.0
0.335	1081.0	383.4	17.1	0.8	0.0
0.340	1074.0	431.3	24.9	1.8	0.1
0.345	1069.0	449.2	33.3	2.5	0.2
0.350	1093.0	480.5	40.8	3.5	0.3
0.355	1083.0	498.0	48.4	4.7	0.5
0.360	1068.0	513.7	57.2	6.4	0.7
0.365	1132.0	561.3	68.4	8.3	1.0
0.370	1181.0	603.5	80.5	10.7	1.4
0.375	1157.0	609.4	89.0	13.0	1.9
0.380	1120.0	608.0	97.2	15.6	2.5
0.385	1098.0	609.8	104.5	17.9	3.1
0.390	1098.0	623.9	114.5	21.0	3.9
0.395	1189.0	691.2	135.8	26.7	5.2
0.400	1429.0	849.9	178.8	37.6	7.9
0.405	1644.0	992.8	218.7	48.2	10.6
0.410	1751.0	1073.7	247.5	57.1	13.2

Note: See page 571 for footnotes.

Table A2.1 Solar irradiance for different air masses[a] *(Continued)*

| Wavelength | Air mass; $\alpha = 0.66$; $\beta = 0.085$[b] | | | | |
	0	1	4	7	10
0.415	1774.0	1104.5	266.5	64.3	15.5
0.420	1747.0	1104.3	278.9	70.4	17.8
0.425	1693.0	1086.5	287.2	78.9	20.1
0.430	1639.0	1067.9	295.4	81.7	22.6
0.435	1663.0	1100.1	318.4	92.2	26.7
0.440	1810.0	1215.5	368.2	111.5	33.8
0.445	1922.0	1310.4	415.3	131.6	41.7
0.450	2006.0	1388.4	460.3	152.6	50.6
0.455	2057.0	1434.8	486.9	165.2	56.1
0.460	2066.0	1452.2	504.4	175.2	60.8
0.465	2048.0	1450.7	515.7	183.3	65.1
0.470	2033.0	1451.2	527.9	192.0	69.8
0.475	2044.0	1470.3	547.3	203.7	75.8
0.480	2074.0	1503.4	572.6	218.1	83.1
0.485	1976.0	1443.3	562.4	219.2	85.4
0.490	1950.0	1435.2	572.2	228.2	91.0
0.495	1960.0	1453.6	592.9	241.9	98.7
0.500	1942.0	1451.2	605.6	252.7	105.5
0.505	1920.0	1440.1	607.6	256.4	108.2
0.510	1882.0	1416.8	604.4	257.8	110.0
0.515	1833.0	1384.9	597.3	257.6	111.1
0.520	1833.0	1390.0	606.1	264.3	115.2
0.525	1852.0	1409.5	621.3	273.9	120.7
0.530	1842.0	1406.9	626.9	279.4	124.5
0.535	1818.0	1393.6	627.7	282.8	127.4
0.540	1783.0	1371.7	624.5	284.4	129.5
0.545	1754.0	1354.2	623.2	286.8	132.0
0.550	1725.0	1336.6	621.7	289.2	134.5
0.555	1720.0	1335.7	625.5	293.0	137.3
0.560	1695.0	1319.2	622.0	293.3	138.3
0.565	1705.0	1330.0	631.3	299.6	142.2
0.570	1712.0	1338.4	639.5	305.6	146.0
0.575	1719.0	1346.9	647.8	311.6	149.6
0.580	1715.0	1346.7	652.0	315.7	152.8
0.585	1712.0	1347.3	656.6	320.0	156.0
0.590	1700.0	1340.7	657.7	322.6	158.3
0.595	1682.0	1329.4	656.4	324.1	160.0
0.600	1660.0	1319.6	655.8	325.9	162.0
0.605	1647.0	1311.0	661.3	333.6	168.2
0.610	1635.0	1307.9	669.6	342.8	175.5
0.620	1602.0	1294.2	682.4	359.9	189.7
0.630	1570.0	1280.9	695.6	377.8	205.2

Table A2.1 Solar irradiance for different air masses[a] *(Continued)*

Wavelength	Air mass; $\alpha = 0.66$; $\beta = 0.085$[b]				
	0	1	4	7	10
0.640	1544.0	1272.1	711.4	397.9	222.5
0.650	1511.0	1257.1	723.9	416.9	240.1
0.660	1486.0	1244.2	730.2	428.6	251.6
0.670	1456.0	1226.8	733.8	438.9	262.5
0.680	1427.0	1209.9	737.4	449.5	273.9
0.690	1402.0	1196.2	742.9	461.3	286.5
0.698	1374.6	1010.3	546.1	311.8	181.6
0.700	1369.0	1175.3	743.7	470.6	297.7
0.710	1344.0	1157.4	739.2	472.1	301.5
0.720	1314.0	1135.1	731.7	471.6	304.0
0.728	1295.5	1003.1	582.3	351.7	212.5
0.730	1290.0	1117.8	727.1	479.0	307.7
0.740	1260.0	1095.1	718.9	471.9	309.8
0.750	1235.0	1076.6	713.2	472.4	313.0
0.762	1205.5	794.0	357.1	163.6	69.1
0.770	1185.0	1039.2	700.8	472.7	318.8
0.780	1159.0	1019.4	693.6	472.0	321.1
0.790	1134.0	1000.3	686.7	471.4	323.6
0.800	1109.0	981.2	679.4	470.5	325.8
0.806	1095.1	874.4	547.7	355.9	234.4
0.825	1048.0	931.6	654.3	459.6	322.8
0.830	1036.0	921.8	649.3	457.3	322.1
0.835	1024.5	912.4	644.4	455.2	321.5
0.846	998.1	476.2	181.0	85.9	44.2
0.860	968.0	506.4	212.0	107.4	58.3
0.870	947.0	453.8	174.7	84.0	43.8
0.875	436.5	449.2	173.4	83.6	43.7
0.887	912.5	448.6	178.3	87.7	46.7
0.900	891.0	448.9	183.7	92.3	50.0
0.907	882.8	455.2	190.9	97.6	53.7
0.915	874.5	461.5	198.5	103.2	57.5
0.925	863.5	279.0	73.6	28.0	12.1
0.930	858.0	221.8	46.9	15.4	6.0
0.940	847.0	313.4	95.0	39.6	18.5
0.950	837.0	296.5	86.3	35.0	16.0
0.955	828.5	321.1	102.3	44.1	21.2
0.965	811.5	344.4	120.4	55.1	27.8
0.975	794.0	576.9	346.0	224.6	150.1
0.985	776.0	544.6	316.1	201.2	132.4
1.018	719.2	617.5	391.0	247.5	156.7
1.082	620.0	512.9	290.4	164.4	93.1
1.094	602.0	464.1	303.1	210.8	149.9
1.098	596.0	503.7	304.1	183.6	110.9

Table A2.1 Solar irradiance for different air masses[a] *(Continued)*

Wavelength	Air mass; $\alpha = 0.66$; $\beta = 0.085$[b]				
	0	1	4	7	10
1.101	591.8	504.8	362.7	267.3	198.8
1.128	560.5	135.1	27.7	9.1	3.6
1.131	557.0	152.2	35.3	12.6	5.3
1.137	550.1	143.1	31.7	11.0	4.5
1.144	542.0	191.2	57.4	24.2	11.6
1.147	538.5	174.5	48.2	19.3	8.8
1.178	507.0	399.3	195.1	95.4	46.6
1.189	496.0	402.2	214.5	114.4	61.0
1.193	492.0	424.0	310.8	233.3	176.6
1.222	464.3	391.8	235.3	141.3	84.9
1.236	451.2	390.8	254.1	165.2	107.4
1.264	426.5	329.2	209.7	140.0	94.3
1.276	416.7	342.6	238.6	172.6	126.3
1.288	406.8	347.3	216.1	134.4	83.7
1.314	386.1	298.3	137.6	63.5	29.3
1.335	369.7	190.6	85.0	46.7	27.7
1.384	343.7	5.7	0.1	0.0	0.0
1.432	321.0	44.6	5.4	1.3	0.4
1.457	308.6	85.4	20.6	7.7	3.3
1.472	301.4	77.4	17.4	6.2	2.6
1.542	270.4	239.3	165.9	115.0	79.7
1.572	257.3	222.6	168.1	130.4	102.1
1.599	245.4	216.0	166.7	131.5	104.5
1.608	241.5	208.5	157.4	122.1	95.7
1.626	233.6	206.7	160.7	127.5	101.9
1.644	225.6	197.9	152.4	120.1	95.5
1.650	223.0	195.7	150.9	119.1	94.7
1.676	212.1	181.9	114.8	72.4	45.7
1.732	187.9	161.5	102.5	65.1	41.3
1.782	166.6	136.7	75.6	41.8	23.1
1.862	138.2	4.0	0.1	0.0	0.0
1.955	112.9	42.7	14.5	6.8	3.6
2.008	102.0	69.4	35.8	17.7	6.4
2.014	101.2	74.7	45.5	28.8	17.8
2.057	95.6	69.5	41.3	25.3	14.8
2.124	87.4	70.0	35.9	18.4	9.5
2.156	83.8	66.0	32.3	15.8	7.7
2.201	78.9	66.1	49.1	38.0	29.7
2.266	72.4	61.6	46.8	36.8	29.3
2.320	67.6	57.2	43.2	33.8	26.8
2.338	66.3	54.7	39.9	30.4	23.4
2.356	65.1	52.0	36.3	26.5	19.6
2.388	62.8	36.0	18.7	11.7	7.8
2.415	61.0	32.5	15.8	9.4	6.0
2.453	58.3	29.6	13.7	7.9	5.0

Table A2.1 Solar irradiance for different air masses[a] (Continued)

Wavelength	Air mass; $\alpha = 0.66$; $\beta = 0.085$[b]				
	0	1	4	7	10
2.494	55.4	20.3	6.8	3.2	1.7
2.537	52.4	4.6	0.4	0.1	0.0
2.900	35.0	2.9	0.2	0.0	0.0
2.941	33.4	6.0	1.0	0.3	0.1
2.954	32.8	5.7	0.9	0.3	0.1
2.973	32.1	8.7	2.2	0.9	0.4
3.005	30.8	7.8	1.8	0.7	0.3
3.045	28.8	4.7	0.7	0.2	0.1
3.056	28.2	4.9	0.8	0.2	0.1
3.097	26.2	3.2	0.4	0.1	0.0
3.132	24.9	6.8	1.7	0.7	0.3
3.156	24.1	18.7	12.6	8.9	6.3
3.204	22.5	2.1	0.2	0.0	0.0
3.214	22.1	3.4	0.5	0.1	0.0
3.245	21.1	3.9	0.7	0.2	0.1
3.260	20.6	3.7	0.6	0.2	0.1
3.285	19.7	14.2	8.5	5.1	2.8
3.317	18.8	12.9	6.9	3.5	1.3
3.344	18.1	4.2	0.9	0.3	0.1
3.403	16.5	12.3	7.8	5.1	3.2
3.450	15.6	12.5	8.9	6.7	5.0
3.507	14.5	12.5	9.9	8.1	6.7
3.538	14.2	11.8	8.8	6.9	5.5
3.573	13.8	10.9	5.4	2.6	1.3
3.633	13.1	10.8	8.3	6.7	5.5
3.673	12.6	9.1	6.1	4.6	3.5
3.696	12.3	10.4	8.2	6.7	5.6
3.712	12.2	10.9	9.0	7.6	6.5
3.765	11.5	9.5	7.2	5.9	4.8
3.812	11.0	8.9	6.7	5.4	4.4
3.888	10.4	8.1	5.6	4.0	2.9
3.923	10.1	8.0	5.6	4.2	3.1
3.948	9.9	7.8	5.5	4.0	3.0
4.045	9.1	6.7	4.1	2.6	1.5
Total $W \cdot m^2$	1353	889.2	448.7	255.2	153.8

[a]$W/m^2 \cdot \mu m$; H_2O 20 mm; O_3 3.4 mm. From Thekaekara, M. P., Data on Incident Solar Energy, "The Energy Crisis and Energy from the Sun," Institute for Environmental Sciences, 1974.

[b]The parameters α and β are measures of turbidity of the atmosphere. They are used in the atmospheric transmittance equation $\bar{\tau}_{atm} = e^{-(C_1 + C_2)m}$; C_1 includes Rayleigh and ozone attenuation; $C_2 \equiv \beta/\lambda^\alpha$.

Table A2.2 Monthly averaged, daily extraterrestrial insolation on a horizontal surface (units: Wh/m²)

Latitude (deg)	Jan.	Feb.	Mar.	Apr.	May	June	July	Aug.	Sep.	Oct.	Nov.	Dec.
20	7415	8397	9552	10422	10801	10868	10794	10499	9791	8686	7598	7076
25	6656	7769	9153	10312	10936	11119	10988	10484	9494	8129	6871	6284
30	5861	7087	8686	10127	11001	11303	11114	10395	9125	7513	6103	5463
35	5039	6359	8153	9869	10995	11422	11172	10233	8687	6845	5304	4621
40	4200	5591	7559	9540	10922	11478	11165	10002	8184	6129	4483	3771
45	3355	4791	6909	9145	10786	11477	11099	9705	7620	5373	3648	2925
50	2519	3967	6207	8686	10594	11430	10981	9347	6998	4583	2815	2100
55	1711	3132	5460	8171	10358	11352	10825	8935	6325	3770	1999	1320
60	963	2299	4673	7608	10097	11276	10657	8480	5605	2942	1227	623
65	334	1491	3855	7008	9852	11279	10531	8001	4846	2116	544	97

Table A2.3a Worldwide global horizontal average solar radiation (units: MJ/sq. m-day)

Position	Lat	Long	Jan.	Feb.	Mar.	Apr.	May	June	July	Aug.	Sep.	Oct.	Nov.	Dec.
Argentina														
Buenos Aires	34.58 S	58.48 W	24.86	21.75	18.56	11.75	8.71	7.15	7.82	8.75	14.49	16.66	24.90	21.93
Australia														
Adelaide	34.93 S	138.52 E	20.99	17.50	20.15	18.27	17.98	—	18.81	19.64	20.11	20.88	20.57	20.72
Brisbane	27.43 S	153.08 E	25.36	22.22	13.25	16.61	12.23	11.52	9.70	15.10	17.61	19.89	—	—
Canberra	35.30 S	148.18 E	28.20	24.68	20.56	14.89	10.29	6.62	—	12.33	16.88	24.06	26.00	25.77
Darwin	12.47 S	130.83 E	26.92	23.40	18.13	13.62	9.30	7.89	9.41	11.15	14.85	18.87	23.43	22.34
Hobart	42.88 S	147.32 E	—	—	—	10.09	7.26	6.04	5.72	9.21	13.54	18.12	—	—
Laverton	37.85 S	114.08 E	22.96	20.42	15.59	13.40	7.48	6.10	6.54	10.43	13.24	18.76	—	—
Sydney	33.87 S	151.20 E	21.09	21.75	17.63	13.63	9.78	8.79	7.62	12.84	16.93	22.10	—	—
Austria														
Wien	48.20 N	16.57 E	3.54	7.10	8.05	14.72	16.79	20.87	19.89	17.27	12.55	8.45	3.51	2.82
Innsbruck	47.27 N	11.38 E	5.57	9.28	10.15	15.96	14.57	17.65	18.35	17.26	12.98	9.08	4.28	3.50
Barbados														
Husbands	13.15 N	59.62 W	19.11	20.23	—	21.80	19.84	20.86	21.55	22.14	—	—	18.30	16.56
Belgium														
Ostende	51.23 N	2.92 E	2.82	5.75	9.93	15.18	16.74	16.93	18.21	18.29	11.71	6.15	2.69	1.97
Melle	50.98 N	3.83 E	2.40	4.66	8.41	13.55	14.23	13.28	15.71	15.61	10.63	5.82	2.40	1.59
Brunei														
Brunei	4.98 N	114.93 E	19.46	20.12	22.71	20.54	19.74	18.31	19.38	20.08	20.83	17.51	17.39	18.12
Bulgaria														
Chirpan	42.20 N	25.33 E	6.72	6.79	8.54	13.27	17.25	17.39	19.85	14.61	12.53	8.52	5.08	5.09
Sofia	42.65 N	23.38E	4.05	6.23	7.93	9.36	12.98	19.73	19.40	17.70	14.71	6.44	—	3.14
Canada														
Montreal	45.47 N	73.75 E	4.74	8.33	11.84	10.55	15.05	22.44	21.08	18.67	14.83	9.18	4.04	4.01
Ottawa	45.32 N	75.67 E	5.34	9.59	13.33	13.98	20.18	20.34	19.46	17.88	13.84	7.38	4.64	5.04
Toronto	43.67 N	79.38 E	4.79	8.15	11.96	14.00	18.16	24.35	23.38	—	15.89	9.40	4.72	3.79
Vancouver	49.18 N	123.17 E	3.73	4.81	12.14	16.41	20.65	24.04	22.87	19.08	12.77	7.39	4.29	1.53
Chile														
Pascua	27.17 S	109.43 W	19.64	16.65	—	11.12	9.52	8.81	10.90	12.29	17.19	20.51	21.20	22.44
Santiago	33.45 S	70.70 W	18.61	16.33	13.44	8.32	5.07	3.66	3.35	5.65	8.15	13.62	20.14	23.88

Table A2.3a Worldwide global horizontal average solar radiation (units: MJ/sq. m-day) (Continued)

Position	Lat	Long	Jan.	Feb.	Mar.	Apr.	May	June	July	Aug.	Sep.	Oct.	Nov.	Dec.
China														
Beijing	39.93 N	116.28 W	7.73	10.59	13.87	17.93	20.18	18.65	15.64	16.61	15.52	11.29	7.25	6.89
Guangzhou	23.13 N	113.32 E	11.01	6.32	4.04	7.89	10.53	12.48	16.14	16.02	15.03	15.79	11.55	9.10
Harbin	45.75 N	126.77 E	5.15	9.54	17.55	20.51	20.33	17.85	19.18	16.09	13.38	14.50	10.50	6.98
Kunming	25.02 N	102.68 E	9.92	11.26	14.38	18.00	18.53	17.37	11.95	18.47	15.94	12.45	11.96	13.62
Lanzhou	36.05 N	103.88 E	7.30	12.47	10.62	18.91	17.40	20.40	20.23	17.37	13.23	10.21	8.22	6.43
Shanghai	31.17 N	121.43 E	7.44	10.31	11.78	14.36	14.23	16.79	14.63	11.85	15.96	12.03	7.73	8.70
Columbia														
Bogota	4.70 N	74.13 W	17.89	—	19.37	16.58	14.86	—	15.42	18.20	17.05	14.58	14.20	16.66
Cuba														
Havana	23.17 N	82.35 W	—	14.70	18.94	20.95	22.63	18.83	21.40	20.19	16.84	16.98	13.19	13.81
Czech														
Kucharovice	48.88 N	16.08 E	3.03	5.85	9.88	14.06	20.84	19.24	21.18	19.41	13.61	6.11	3.47	2.12
Churanov	49.07 N	13.62 E	2.89	5.82	9.24	13.18	21.32	15.68	20.51	19.49	12.84	5.68	3.36	2.99
Hradec Kralov	50.25 N	15.85 E	3.51	5.94	10.58	15.95	20.42	18.43	17.17	17.92	11.86	6.27	2.45	1.89
Denmark														
Copenhagen	55.67 N	12.30 E	1.83	3.32	7.09	11.12	21.39	24.93	—	13.92	10.10	5.20	2.81	1.23
Egypt														
Cairo	30.08 N	31.28 E	10.06	12.96	18.49	23.04	21.91	26.07	25.16	23.09	21.01	—	11.74	9.85
Mersa Matruh	31.33 N	27.22 E	8.38	11.92	18.47	24.27	24.17	—	26.67	26.27	21.92	18.28	11.71	8.76
Ethiopia														
Addis Ababa	8.98 N	38.80 E	—	11.39	—	12.01	—	—	—	6.33	9.35	11.71	11.69	11.50
Fiji														
Nandi	17.75 S	177.45 E	20.82	20.65	20.25	18.81	15.68	14.18	15.08	16.71	19.37	20.11	21.78	25.09
Suva	48.05 S	178.57 E	20.37	17.74	16.22	13.82	10.81	12.48	11.40	—	—	18.49	19.96	20.99
Finland														
Helsinki	60.32 N	24.97 E	1.13	2.94	5.59	11.52	17.60	16.81	20.66	15.44	8.44	3.31	0.97	0.63
France														
Agen	44.18 N	0.60 E	4.83	7.40	10.69	17.12	19.25	20.42	21.63	20.64	15.56	8.41	5.09	5.01
Nice	43.65 N	7.20 E	6.83	—	11.37	17.79	20.74	24.10	24.85	24.86	15.04	10.99	7.08	6.73
Paris	48.97 N	2.45 E	2.62	5.08	7.21	12.90	14.84	13.04	15.54	16.30	10.17	5.61	3.14	2.20

Table A2.3a Worldwide global horizontal average solar radiation (units: MJ/sq. m-day) (Continued)

Position	Lat	Long	Jan.	Feb.	Mar.	Apr.	May	June	July	Aug.	Sep.	Oct.	Nov.	Dec.
Germany														
Bonn	50.70 N	7.15 E	2.94	5.82	8.01	14.27	15.67	14.41	18.57	17.80	11.70	6.15	3.42	1.90
Nuremberg	53.33 N	13.20 E	3.23	6.92	9.08	15.69	15.71	18.21	21.14	17.98	12.43	8.15	2.79	2.51
Bremen	53.05 N	8.80 E	2.36	4.93	8.53	14.52	14.94	14.52	19.40	15.02	10.48	6.27	2.80	1.66
Hamburg	53.63 N	10.00 E	1.97	3.96	7.59	12.32	14.11	12.69	19.00	14.11	10.29	6.45	2.33	1.43
Stuttgart	48.83 N	9.20 E	3.59	7.18	9.22	15.81	17.72	17.44	22.21	19.87	12.36	7.81	3.19	2.54
Ghana														
Bole	9.03 N	2.48 W	18.29	19.76	19.71	19.15	16.61	—	—	13.68	16.29	17.27	17.33	15.93
Accra	5.60 N	0.17 W	14.82	16.26	18.27	16.73	18.15	13.96	13.86	13.49	15.32	19.14	18.16	14.23
Great Britain														
Belfast	54.65 N	6.22 W	2.00	3.60	6.85	12.00	15.41	15.09	15.46	13.56	11.49	4.63	2.34	1.24
Jersey	49.22 N	2.20 W	2.76	5.65	9.51	14.98	18.51	17.83	18.14	18.62	12.98	6.16	3.26	2.83
London	51.52 N	0.12 W	2.24	3.87	7.40	12.01	12.38	13.24	16.59	16.23	12.59	5.67	2.87	1.97
Greece														
Athens	37.97 N	23.72 E	9.11	10.94	15.70	20.91	23.85	25.48	24.21	23.08	19.03	13.29	5.98	6.64
Sikiwna	37.98 N	22.73 E	7.60	8.16	11.99	21.06	22.62	24.32	23.56	21.73	17.30	11.75	9.45	6.35
Guadeloupe														
Le Raizet	16.27 N	61.52 W	14.88	18.10	20.55	19.69	20.26	20.65	20.65	20.24	18.47	17.79	13.49	14.38
Guyana														
Cayenne	4.83 N	52.37 W	14.46	14.67	16.28	17.57	—	14.92	17.42	18.24	20.52	—	22.69	17.04
Hong Kong														
King's Park	22.32 N	114.17 W	12.34	7.39	6.94	9.50	11.38	13.60	16.70	17.06	15.91	16.52	14.19	10.00
Hungary														
Budapest	47.43 N	19.18 E	2.61	7.46	11.14	14.46	20.69	19.47	21.46	19.72	12.88	7.96	2.95	2.47
Iceland														
Reykjavik	64.13 N	21.90 W	0.52	2.02	6.25	11.77	13.07	14.58	16.83	11.35	9.70	3.18	1.00	0.65
India														
Bombay	19.12 N	72.85 E	18.44	21.00	22.72	24.52	24.86	19.75	15.84	16.00	18.19	20.38	19.18	17.81
Calcutta	22.53 N	88.33 E	15.69	18.34	20.09	22.34	22.37	17.55	17.07	16.55	16.52	16.90	16.35	15.00
Madras	13.00 N	80.18 E	19.09	22.71	25.14	24.88	23.89	—	18.22	19.68	19.81	16.41	14.76	15.79

Table A2.3a Worldwide global horizontal average solar radiation (units: MJ/sq. m-day) (Continued)

Position	Lat	Long	Jan.	Feb.	Mar.	Apr.	May	June	July	Aug.	Sep.	Oct.	Nov.	Dec.
India														
Nagpur	21.10 N	79.05 E	18.08	21.01	22.25	24.08	24.79	19.84	15.58	15.47	17.66	20.10	18.98	17.33
New Delhi	28.58 N	77.20 E	14.62	18.25	20.15	23.40	23.80	19.16	20.20	19.89	20.08	19.74	16.95	14.22
Ireland														
Dublin	53.43 N	6.25 W	2.51	4.75	7.48	11.06	17.46	19.11	15.64	13.89	9.65	5.77	2.93	—
Israel														
Jerusalem	31.78 N	35.22 E	10.79	13.01	18.08	23.79	29.10	31.54	31.83	28.79	25.19	20.26	12.61	10.71
Italy														
Milan	45.43 N	9.28 E	—	6.48	10.09	13.17	17.55	16.32	18.60	16.86	11.64	5.40	3.52	2.41
Rome	41.80 N	12.55 E	—	9.75	13.38	15.82	15.82	18.89	22.27	21.53	16.08	8.27	6.41	4.49
Japan														
Fukuoka	33.58 N	130.38 E	8.11	8.72	10.95	13.97	14.36	12.81	13.84	16.75	13.92	11.86	10.05	7.30
Tateno	36.05 N	140.13 E	9.06	12.17	11.00	15.78	16.52	15.26	—	—	—	9.60	8.55	8.26
Yonago	35.43 N	133.35 E	6.25	7.16	10.87	17.30	16.72	15.44	17.06	19.93	12.41	10.82	7.50	5.51
Kenya														
Mombasa	4.03 S	39.62 E	22.30	22.17	22.74	18.49	18.31	17.41	—	18.12	21.03	22.97	21.87	21.25
Nairobi	1.32 S	36.92 E	—	24.10	21.20	18.65	14.83	15.00	13.44	14.12	19.14	19.38	16.90	18.27
Lithuania														
Kaunas	54.88 N	23.88 E	1.89	4.43	7.40	12.97	18.88	18.74	21.41	15.79	10.40	5.64	1.80	1.10
Madagascar														
Antananarivo	18.80 S	47.48 E	15.94	13.18	13.07	11.53	9.25	8.21	9.32	—	—	16.43	15.19	15.62
Malaysia														
Kualalumpur	3.12 N	101.55 E	15.36	17.67	18.48	16.87	15.67	16.24	15.32	15.89	14.62	14.13	13.54	11.53
Piang	5.30 N	100.27 E	19.47	21.35	23.24	20.52	18.63	19.32	17.17	16.96	15.93	16.01	18.35	17.37
Martinique														
Le Lamentin	14.60 N	61.00 W	17.76	20.07	22.53	21.95	22.42	21.23	20.86	21.84	20.23	19.87	14.08	16.25
Mexico														
Chihuahua	28.63 N	106.08 W	14.80	—	—	—	26.94	26.28	24.01	24.22	20.25	19.55	10.57	15.79
Orizabita	20.58 N	99.20 E	19.49	23.07	27.44	27.35	26.04	25.05	—	27.53	21.06	17.85	15.48	12.93

Table A2.3a Worldwide global horizontal average solar radiation (units: MJ/sq. m-day) (Continued)

Position	Lat	Long	Jan.	Feb.	Mar.	Apr.	May	June	July	Aug.	Sep.	Oct.	Nov.	Dec.
Mongolia														
Ulan Bator	47.93 N	106.98 E	6.28	9.22	14.34	18.18	20.50	19.34	16.34	16.65	14.08	11.36	7.19	5.35
Uliasutai	47.75 N	96.85 E	6.43	10.71	14.83	20.32	23.86	20.46	21.66	17.81	15.97	10.92	7.32	5.08
Morocco														
Casablanca	33.57 N	7.67 E	11.46	12.70	15.93	21.25	24.45	25.27	25.53	23.60	19.97	14.68	11.61	9.03
Mozambique														
Maputo	25.97 S	32.60 E	26.35	23.16	19.33	20.54	16.33	14.17	—	—	—	22.55	25.48	26.19
Netherlands														
Maastricht	50.92 N	5.78 E	3.20	5.43	8.48	14.82	14.97	14.32	18.40	17.51	11.65	6.51	3.01	1.72
New Caledonia														
Koumac	20.57 S	164.28 E	24.89	21.15	16.96	18.98	15.67	14.55	15.75	17.62	22.48	15.83	27.53	26.91
New Zealand														
Wilmington	41.28 S	174.77 E	22.59	19.67	14.91	9.52	6.97	4.37	5.74	7.14	12.50	16.34	19.07	24.07
Christchurch	43.48 S	172.55 E	23.46	19.68	13.98	8.96	6.47	4.74	5.38	6.94	13.18	17.45	18.91	24.35
Nigeria														
Benin City	6.32 N	5.60 E	14.89	17.29	19.15	17.21	16.97	15.04	10.24	12.54	14.37	15.99	17.43	15.75
Norway														
Bergen	60.40 N	5.32 E	0.46	1.33	3.18	8.36	19.24	16.70	16.28	10.19	6.53	3.19	1.36	0.35
Oman														
Seeb	23.58 N	58.28 E	12.90	14.86	21.22	22.22	25.30	24.02	23.46	21.66	20.07	18.45	15.49	13.12
Salalah	17.03 N	54.08 E	16.52	16.92	18.49	20.65	21.46	16.92	8.52	11.41	17.14	18.62	16.42	—
Pakistan														
Karachi	24.90 N	67.13 E	13.84	—	—	19.69	20.31	16.62	—	—	—	—	12.94	11.07
Multan	30.20 N	71.43 E	12.29	15.86	18.33	22.35	22.57	21.65	20.31	20.44	20.57	15.91	12.68	10.00
Islamabad	33.62 N	73.10 E	10.38	12.42	16.98	22.65	—	25.49	20.64	18.91	14.20	15.30	10.64	8.30
Peru														
Puno	15.83 S	70.02 W	14.98	12.92	16.08	20.03	17.45	17.42	15.74	15.32	16.11	16.18	14.24	13.90
Poland														
Warszawa	52.28 N	20.97 E	1.73	3.83	7.81	10.53	19.22	17.11	20.18	15.00	10.65	4.95	2.39	1.68
Kolobrzeg	54.18 N	15.58 E	2.50	3.25	8.86	15.21	20.79	20.50	17.19	16.46	7.95	5.75	1.78	1.18

Table A2.3a Worldwide global horizontal average solar radiation (units: MJ/sq. m-day) (Continued)

Position	Lat	Long	Jan.	Feb.	Mar.	Apr.	May	June	July	Aug.	Sep.	Oct.	Nov.	Dec.
Portugal														
Evora	38.57 N	7.90 W	9.92	12.43	17.81	18.69	23.57	29.23	28.75	23.77	20.17	—	6.81	4.57
Lisbon	38.72 N	9.15 W	9.24	11.60	17.52	18.49	24.64	29.02	28.14	22.20	19.76	13.56	7.18	4.83
Romania														
Bucuresti	44.50 N	26.13 E	7.05	10.22	12.04	16.53	18.97	22.16	23.19	—	17.17	9.55	4.82	—
Constania	44.22 N	28.63 E	5.62	9.28	14.31	20.59	23.23	25.80	27.98	24.22	16.91	11.89	6.19	5.10
Galati	45.50 N	28.02 E	6.09	9.33	14.31	17.75	21.77	22.74	25.55	19.70	14.05	11.26	6.32	5.38
Russia														
Alexandovsko	60.38 N	77.87 E	1.34	4.17	9.16	17.05	21.83	21.34	20.26	13.05	10.16	4.68	1.71	0.68
Moscow	55.75 N	37.57 E	1.45	3.96	8.09	11.69	18.86	18.12	17.51	14.17	10.92	4.03	2.28	1.29
St. Petersburg	59.97 N	30.30 E	1.03	3.11	4.88	12.24	20.59	21.55	20.43	13.27	7.83	2.93	1.16	0.59
Verkhoyansk	67.55 N	133.38 E	0.21	2.25	7.61	15.96	19.64	—	—	14.12	7.59	3.51	0.54	—
St. Pierre & Miquelon														
St. Pierre	46.77 N	56.17 W	4.43	6.61	12.50	17.57	18.55	17.84	19.95	16.46	12.76	8.15	3.69	3.33
Singapore														
Singapore	1.37 N	103.98 E	19.08	20.94	20.75	18.20	14.89	15.22	13.92	16.66	16.51	15.82	13.81	12.67
South Korea														
Seoul	37.57 N	126.97 E	6.24	9.40	10.34	13.98	16.35	17.49	10.65	12.94	11.87	10.35	6.47	5.14
South Africa														
Cape Town	33.98 S	18.60 E	27.47	25.57	—	15.81	11.44	9.08	8.35	13.76	17.30	22.16	26.37	27.68
Port Elizabeth	33.98 S	25.60 E	27.22	22.06	19.01	15.29	11.79	11.13	10.73	13.97	18.52	23.09	23.15	27.26
Pretoria	25.73 S	28.18 E	26.06	22.43	20.52	16.09	15.67	13.67	15.19	18.65	21.62	21.75	24.82	23.43
Spain														
Madrid	40.45 N	3.72 W	7.73	10.53	15.35	21.74	22.81	22.05	26.27	22.90	18.89	10.21	8.69	5.56
Sudan														
Wad Madani	14.40 N	33.48 E	21.92	24.01	23.43	25.17	23.92	23.51	22.40	22.85	21.75	20.47	20.19	19.21
Elfasher	13.62 N	25.33 E	21.56	21.84	24.54	25.29	24.31	24.15	22.87	21.19	22.58	23.85	—	—
Shambat	15.67 N	32.53 E	23.90	27.38	—	27.45	23.21	26.15	23.55	25.46	24.05	23.51	23.82	22.53
Sweden														
Karlstad	59.37 N	13.47 E	1.26	3.13	5.02	14.01	19.90	16.70	20.92	14.14	10.52	3.98	1.47	0.94

Table A2.3a Worldwide global horizontal average solar radiation (units: MJ/sq. m-day) (Continued)

Position	Lat	Long	Jan.	Feb.	Mar.	Apr.	May	June	July	Aug.	Sep.	Oct.	Nov.	Dec.
Sweden														
Lund	55.72 N	13.22 E	1.97	3.47	6.66	12.48	17.83	13.38	18.74	14.99	10.39	5.45	1.82	1.21
Stockholm	59.35 N	18.07 E	1.32	2.69	4.75	13.21	15.58	14.79	20.52	14.48	10.50	4.04	1.19	0.83
Switzerland														
Geneva	46.25 N	6.13 E	2.56	7.21	9.46	17.07	20.98	19.78	22.38	20.50	13.62	8.44	3.31	2.87
Zurich	47.48 N	8.53 E	2.31	7.02	7.54	15.04	16.33	16.73	20.28	18.32	12.52	7.18	2.64	2.29
Thailand														
Bangkok	13.73 N	100.57 E	16.67	19.34	23.00	22.48	20.59	17.71	18.02	16.04	16.23	16.81	18.60	16.43
Trinidad & Tobago														
Crown Point	11.15 N	60.83 W	13.05	15.61	15.17	16.96	17.61	15.37	13.16	13.08	12.24	8.76	—	—
Tunisia														
Sidi Bouzid	36.87 N	10.35 E	7.88	10.38	13.20	17.98	25.12	26.68	27.43	24.33	18.87	12.11	9.37	6.72
Tunis	36.83 N	10.23 E	7.64	9.88	14.79	31.61	25.31	26.03	26.60	20.37	19.58	12.91	9.35	7.16
Ukraine														
Kiev	50.40 N	30.45 E	2.17	4.87	11.15	12.30	20.49	—	18.99	18.55	9.72	9.84	3.72	2.52
Uzbekistan														
Tashkent	41.27 N	69.27 E	7.27	10.81	15.93	23.60	25.21	29.53	28.50	26.68	20.76	13.25	8.61	4.59
Venezuela														
Caracas	10.50 N	66.88 W	14.25	13.56	16.30	15.56	15.69	15.56	16.28	17.11	17.04	15.14	14.74	13.50
St. Antonio	7.85 N	72.45 W	11.78	10.54	10.65	12.07	12.65	21.20	14.68	15.86	16.62	15.32	12.28	11.28
St. Fernando	7.90 N	67.42 W	14.92	16.82	16.89	—	—	14.09	13.78	14.42	14.86	15.27	14.25	13.11
Vietnam														
Hanoi	21.03 N	105.85 E	5.99	7.48	8.73	13.58	19.10	21.26	19.85	19.78	20.67	14.78	12.44	13.21

Table A2.3a Worldwide global horizontal average solar radiation (units: MJ/sq. m-day) (Continued)

Position	Lat	Long	Jan.	Feb.	Mar.	Apr.	May	June	July	Aug.	Sep.	Oct.	Nov.	Dec.
Yogoslavia														
Beograd	44.78 N	20.53 E	4.92	6.27	10.64	14.74	20.95	22.80	22.09	20.27	15.57	11.24	6.77	4.99
Kopaonik	43.28 N	20.80 E	7.03	10.93	14.75	12.78	13.54	20.43	22.48	—	20.14	11.61	6.26	4.64
Portoroz	45.52 N	13.57 E	5.11	7.84	13.75	17.30	23.66	22.31	25.14	21.34	13.40	8.98	6.04	3.92
Zambia														
Lusaka	15.42 S	28.32 W	16.10	18.02	20.24	19.84	17.11	16.37	19.45	20.72	21.68	23.83	23.85	20.52
Zimbabwe														
Bulawayo	20.15 S	28.62 N	20.03	22.11	21.03	18.09	17.15	15.36	16.46	19.49	21.55	23.44	25.08	23.46
Harare	17.83 S	31.02 N	19.38	19.00	19.22	17.67	18.35	16.10	14.55	17.87	21.47	23.98	19.92	21.88

(Source: Voeikov Main Geophysical Observatory, Russia: Internet address: http://wrdc-mgo.nrel.gov/html/get_data-ap.html)

Note: Data for 872 locations is available from these sources in 68 countries.

*Source for Canadian Data: Environment Canada: Internet address: http://www.ec.gc.ca./envhome.html.

Table A2.3b Average daily solar radiation on a horizontal surface in U.S.A. (units: MJ/sq. m-day)

Position	Jan.	Feb.	Mar.	Apr.	May	June	July	Aug.	Sep.	Oct.	Nov.	Dec.	Average
Alabama													
Birmingham	9.20	11.92	15.67	19.65	21.58	22.37	21.24	20.21	17.15	14.42	10.22	8.40	16.01
Montgomery	9.54	12.49	16.24	20.33	22.37	23.17	21.80	20.56	17.72	14.99	10.90	8.97	16.58
Alaska													
Fairbanks	0.62	2.77	8.31	14.66	17.98	19.65	16.92	12.36	7.02	3.20	1.01	0.23	8.74
Anchorage	1.02	3.41	8.18	13.06	15.90	17.72	16.69	12.72	8.06	3.97	1.48	0.56	8.63
Nome	0.51	2.95	8.29	15.22	18.97	19.65	16.69	11.81	7.72	3.63	0.99	0.09	8.86
St. Paul Island	1.82	4.32	8.52	12.72	14.08	14.42	12.83	10.33	7.84	4.54	2.16	1.25	7.95
Yakutat	1.36	3.63	7.72	12.61	14.76	15.79	14.99	12.15	7.95	3.97	1.82	0.86	8.18
Arizona													
Phoenix	11.58	15.33	19.87	25.44	28.85	30.09	27.37	25.44	21.92	17.60	12.95	10.56	20.56
Tucson	12.38	15.90	20.21	25.44	28.39	29.30	25.44	24.08	21.58	17.94	13.63	11.24	20.44
Arkansas													
Little Rock	9.09	11.81	15.56	19.19	21.80	23.51	23.17	21.35	17.26	14.08	9.77	8.06	16.24
Fort Smith	9.31	12.15	15.67	19.31	21.69	23.39	23.85	24.46	17.26	13.97	9.88	8.29	16.35
California													
Bakersfield	8.29	11.92	16.69	22.15	26.57	28.96	28.73	26.01	21.35	15.90	10.33	7.61	18.74
Fresno	7.61	11.58	16.81	22.49	27.14	29.07	28.96	25.89	21.12	15.56	9.65	6.70	18.62
Long Beach	9.99	12.95	17.03	21.60	23.17	24.19	26.12	24.08	19.31	14.99	11.24	9.31	17.83
Sacramento	6.93	10.68	15.56	21.24	25.89	28.28	28.62	25.32	20.56	14.54	8.63	6.25	17.72
San Diego	11.02	13.97	17.72	21.92	22.49	23.28	24.98	23.51	19.53	15.79	12.26	10.22	18.06
San Francisco	7.72	10.68	15.22	20.44	24.08	25.78	26.46	23.39	19.31	13.97	8.97	7.04	16.92
Los Angeles	10.11	13.06	17.26	21.80	23.05	23.74	25.67	23.51	18.97	14.99	11.36	9.31	17.72
Santa Maria	10.22	13.29	17.49	22.26	25.10	26.57	26.91	24.42	20.10	15.67	11.47	9.54	18.62
Colorado													
Boulder	7.84	10.45	15.64	17.94	17.94	20.47	20.28	17.12	16.07	12.09	8.66	7.10	14.31
Colorado Springs	9.09	12.15	16.13	20.33	22.26	24.98	23.96	21.69	18.51	14.42	9.99	8.18	16.81
Connecticut													
Hartford	6.70	9.65	13.17	16.69	19.53	21.24	21.12	18.51	14.76	10.68	6.59	5.45	13.74

Table A2.3b Average daily solar radiation on a horizontal surface in U.S.A. (units: MJ/sq. m-day) (*Continued*)

Position	Jan.	Feb.	Mar.	Apr.	May	June	July	Aug.	Sep.	Oct.	Nov.	Dec.	Average
Delaware													
Wilmington	7.27	10.22	13.97	17.60	20.33	22.49	21.80	19.65	15.79	11.81	7.84	6.25	14.65
Florida													
Daytona Beach	11.24	13.85	17.94	22.15	23.17	22.03	21.69	20.44	17.72	14.99	12.15	10.33	17.38
Jacksonville	10.45	13.17	17.03	21.12	22.03	21.58	21.01	19.42	16.69	14.20	11.47	9.65	16.47
Tallahassee	10.33	13.29	16.92	21.24	22.49	22.03	20.90	19.65	17.72	15.56	11.92	9.77	16.81
Miami	12.72	15.22	18.51	21.58	21.46	20.10	21.10	20.10	17.60	15.67	13.17	11.81	17.38
Key West	13.17	16.01	19.65	22.71	22.83	22.03	22.03	21.01	18.74	16.47	13.85	15.79	18.40
Tampa	11.58	14.42	18.17	22.26	23.05	21.92	20.90	19.65	17.60	16.01	12.83	11.02	17.49
Georgia													
Athens	9.43	12.38	16.01	20.21	22.03	22.83	21.80	20.21	17.26	14.42	10.45	8.40	16.29
Atlanta	9.31	12.26	16.13	20.33	22.37	23.17	22.15	20.56	17.49	14.54	10.56	8.52	16.43
Columbus	9.77	12.72	16.47	20.67	22.37	22.83	21.58	20.33	17.60	14.99	11.02	9.09	16.62
Macon	9.54	12.61	16.35	20.56	22.37	22.83	21.58	20.21	17.26	14.88	10.90	8.86	16.50
Savanna	9.99	12.72	16.81	21.01	22.37	22.60	21.80	19.76	16.92	14.65	11.13	9.20	16.58
Hawaii													
Honolulu	14.08	16.92	19.42	21.24	22.83	23.51	23.74	23.28	21.35	18.06	14.88	13.40	19.42
Idaho													
Boise	5.79	8.97	13.63	18.97	23.51	26.01	27.37	23.62	18.40	12.26	6.70	5.11	15.90
Illinois													
Chicago	6.47	9.31	12.49	16.47	20.44	22.60	22.03	19.31	15.10	10.79	6.47	5.22	13.85
Rockford	6.70	9.77	12.72	16.58	20.33	22.49	22.15	19.42	15.22	10.79	6.59	5.34	14.08
Springfield	7.50	10.33	13.40	17.83	21.46	23.51	23.05	20.56	16.58	12.26	7.72	6.13	15.10
Indiana													
Indianapolis	7.04	9.99	13.17	17.49	21.24	23.28	22.60	20.33	16.35	11.92	7.38	5.79	14.76
Iowa													
Mason City	6.70	9.77	13.29	16.92	20.78	22.83	22.71	19.76	15.33	10.90	6.59	5.45	14.31
Waterloo	6.81	9.77	13.06	16.92	20.56	22.83	22.60	19.76	15.33	10.90	6.70	5.45	14.20

Table A2.3b Average daily solar radiation on a horizontal surface in U.S.A. (units: MJ/sq. m-day) (Continued)

Position	Jan.	Feb.	Mar.	Apr.	May	June	July	Aug.	Sep.	Oct.	Nov.	Dec.	Average
Kansas													
Dodge City	9.65	12.83	16.69	21.01	23.28	25.78	25.67	22.60	18.40	14.42	10.11	8.40	17.49
Goodland	8.97	11.92	16.13	20.44	22.71	25.78	25.55	22.60	18.28	14.08	9.65	7.84	17.03
Kentucky													
Lexington	7.27	9.88	13.51	17.60	20.56	22.26	21.46	19.65	16.01	12.38	7.95	6.25	14.54
Louisville	7.27	10.22	13.63	17.83	20.90	22.71	22.03	20.10	16.35	12.38	7.95	6.25	14.76
Louisiana													
New Orleans	9.77	12.83	16.01	19.87	21.80	22.03	20.67	19.65	17.60	15.56	11.24	9.31	16.35
Lake Charles	9.77	12.83	16.13	19.31	21.58	22.71	21.58	20.33	18.06	15.56	11.47	9.31	16.58
Maine													
Portland	6.70	9.99	13.78	16.92	19.99	21.92	21.69	19.31	15.22	10.56	6.47	5.45	13.97
Maryland													
Baltimore	7.38	10.33	13.97	17.60	20.21	22.15	21.69	19.19	15.79	11.92	8.06	6.36	14.54
Massachusetts													
Boston	6.70	9.65	13.40	16.92	20.21	22.03	21.80	19.31	15.33	10.79	6.81	5.45	14.08
Michigan													
Detroit	5.91	8.86	12.38	16.47	20.33	22.37	21.92	18.97	14.76	10.11	6.13	4.66	13.63
Lansing	5.91	8.86	12.49	16.58	20.21	22.26	21.92	18.85	14.54	9.77	5.91	4.66	13.51
Minnesota													
Duluth	5.68	9.31	13.74	17.38	20.10	21.46	21.80	18.28	13.29	8.86	5.34	4.43	13.29
Minneapolis	6.36	9.77	13.51	16.92	20.56	22.49	22.83	19.42	14.65	9.99	6.13	4.88	13.97
Rochester	6.36	9.65	13.17	16.58	20.10	22.15	22.15	19.08	14.54	10.11	6.25	5.11	13.74
Mississippi													
Jackson	9.43	12.38	16.13	19.87	22.15	23.05	22.15	19.08	14.54	10.11	6.25	5.11	13.74
Missouri													
Columbia	8.06	10.90	14.31	18.62	21.58	23.62	23.85	21.12	16.69	12.72	8.29	6.70	15.56
Kansas City	7.95	10.68	14.08	18.28	21.24	23.28	23.62	20.78	16.58	12.72	8.40	6.70	15.44
Springfield	8.52	11.02	14.65	18.62	21.24	23.05	23.62	21.24	16.81	13.17	8.86	7.27	15.67
St. Louis	7.84	10.56	13.97	18.06	21.12	23.05	22.94	20.44	16.58	12.49	8.18	6.59	15.22

Table A2.3b Average daily solar radiation on a horizontal surface in U.S.A. (units: MJ/sq. m-day) *(Continued)*

Position	Jan.	Feb.	Mar.	Apr.	May	June	July	Aug.	Sep.	Oct.	Nov.	Dec.	Average
Montana													
Helena	5.22	8.29	12.61	17.15	20.67	23.28	25.21	21.24	15.79	10.45	6.02	4.43	14.20
Lewistown	5.22	8.40	12.72	17.15	20.33	23.05	24.53	20.78	15.10	10.22	5.91	4.32	13.97
Nebraska													
Omaha	7.50	10.33	13.97	18.06	21.24	2.40	23.51	20.56	16.01	11.81	7.61	6.13	15.10
Lincoln	7.33	10.10	13.65	16.22	19.26	21.21	22.15	18.87	15.44	11.54	7.76	6.20	14.16
Nevada													
Elko	7.61	10.56	14.42	18.85	22.71	25.67	26.69	23.62	19.31	13.63	8.29	6.70	16.58
Las Vegas	10.79	14.42	19.42	24.87	28.16	30.09	28.28	25.89	22.15	17.03	12.15	9.88	20.33
Reno	8.29	11.58	16.24	21.24	25.10	27.48	28.16	24.98	20.56	14.88	9.31	7.38	17.94
New Hampshire													
Concord	6.81	10.11	13.97	16.92	20.21	21.80	21.80	19.08	14.99	10.45	6.47	5.45	14.08
New Jersey													
Atlantic City	7.38	10.22	13.97	17.49	20.21	21.92	21.24	19.19	15.79	11.92	8.06	6.36	14.54
Newark	6.93	9.77	13.51	17.26	19.76	21.35	21.01	18.85	15.33	11.36	7.27	5.68	13.97
New Mexico													
Albuquerque	11.47	14.99	19.31	24.53	27.60	29.07	27.03	24.76	21.12	17.03	12.49	10.33	19.99
New York													
Albany	6.36	9.43	12.95	16.69	19.53	21.46	21.58	18.51	14.65	10.11	6.13	5.00	13.51
Buffalo	5.68	8.40	12.15	16.35	19.76	22.03	21.69	18.62	14.08	9.54	5.68	4.54	13.29
New York City	6.93	9.88	13.85	17.72	20.44	22.03	21.69	19.42	15.56	11.47	7.27	5.79	14.31
Rochester	5.68	8.52	12.26	16.58	19.87	21.92	21.69	18.51	14.20	9.54	5.68	4.54	13.29
North Carolina													
Charlotte	8.97	11.81	15.67	19.76	21.58	22.60	21.92	19.99	16.92	13.97	9.99	8.06	16.01
Wilmington	9.31	12.15	16.24	20.44	21.92	22.60	21.58	19.53	16.69	14.08	10.56	8.52	16.13
North Dakota													
Fargo	5.79	9.09	13.17	16.92	20.56	22.37	23.17	19.87	14.31	9.54	5.68	4.54	13.74
Bismarck	6.12	9.75	13.88	17.43	21.45	23.01	24.06	20.12	15.21	10.61	6.28	4.84	14.39

Table A2.3b Average daily solar radiation on a horizontal surface in U.S.A. (units: MJ/sq. m-day) (Continued)

Position	Jan.	Feb.	Mar.	Apr.	May	June	July	Aug.	Sep.	Oct.	Nov.	Dec.	Average
Ohio													
Cleveland	5.79	8.63	12.04	16.58	20.10	22.15	21.92	18.97	14.76	10.22	6.02	4.66	13.51
Columbus	6.47	9.09	12.49	16.58	19.76	21.58	21.12	18.97	15.44	11.24	6.81	5.34	13.74
Dayton	6.81	9.43	12.83	17.03	20.33	22.37	22.37	19.65	15.90	11.47	7.04	5.45	14.20
Youngstown	5.79	8.40	11.92	15.90	19.19	21.24	20.78	18.06	14.31	10.11	6.02	4.77	13.06
Oklahoma													
Oklahoma City	9.88	1.25	16.47	20.33	22.26	24.42	24.98	22.49	18.17	14.54	10.45	8.74	17.15
Oregon													
Eugene	4.54	7.04	11.24	15.79	19.99	22.37	24.19	21.01	15.90	9.65	5.11	3.75	13.40
Medford	5.34	8.52	13.17	18.62	23.39	26.23	27.82	23.96	18.62	11.92	6.02	4.43	15.67
Portland	4.20	6.70	10.68	15.10	18.97	21.24	22.60	19.53	14.88	9.20	4.88	3.52	12.61
Pacific Islands													
Guam	16.35	17.38	19.65	20.78	20.56	19.76	18.28	17.49	17.49	16.58	15.79	15.10	17.94
Pennsylvania													
Philadelphia	7.04	9.88	13.63	17.26	19.99	22.03	21.46	19.42	15.67	11.58	7.72	6.02	14.31
Pittsburgh	6.25	8.97	12.61	16.47	19.65	21.80	21.35	18.85	15.10	10.90	6.59	5.00	13.63
Rhode Island													
Providence	6.70	9.65	13.40	16.92	19.99	21.58	21.24	18.85	15.22	11.02	6.93	5.56	13.97
South Carolina													
Charleston	9.77	12.72	16.81	21.12	22.37	22.37	21.92	19.65	16.92	14.54	11.02	9.09	16.58
Greenville	9.20	12.04	15.90	19.99	21.58	22.60	21.58	19.87	16.81	14.08	10.22	8.18	16.01
South Dakota													
Pierre	6.47	9.54	13.85	17.94	21.46	24.08	24.42	21.46	16.35	11.24	7.04	5.45	14.99
Rapid City	6.70	9.88	14.20	18.28	21.46	24.19	24.42	21.80	16.92	11.81	7.50	5.79	15.33
Tennessee													
Memphis	8.86	11.58	15.22	19.42	22.03	23.85	23.39	21.46	17.38	14.20	9.65	7.84	16.24
Nashville	8.29	11.13	14.65	19.31	21.69	23.51	22.49	20.56	16.81	13.51	8.97	7.15	15.67

Table A2.3b Average daily solar radiation on a horizontal surface in U.S.A. (units: MJ/sq. m-day) (Continued)

Position	Jan.	Feb.	Mar.	Apr.	May	June	July	Aug.	Sep.	Oct.	Nov.	Dec.	Average
Texas													
Austin	10.68	13.63	17.03	19.53	21.24	23.74	24.42	22.83	18.85	15.67	11.92	9.99	17.49
Brownsville	10.33	13.17	16.47	19.08	20.78	22.83	23.28	21.58	18.62	16.13	12.38	9.88	17.03
El Paso	12.38	16.24	20.90	25.44	28.05	28.85	26.46	24.30	21.12	17.72	13.63	11.47	20.56
Houston	9.54	12.26	15.22	18.06	20.21	21.69	21.35	20.21	17.49	15.10	11.02	8.97	15.90
San Antonio	10.88	13.53	16.26	17.35	21.10	23.87	24.92	22.81	19.22	15.52	11.50	9.98	17.24
Utah													
Salt Lake City	6.93	10.45	14.76	19.42	23.39	26.46	26.35	23.39	18.85	13.29	8.06	6.02	16.47
Vermont													
Burlington	5.79	9.20	13.06	16.47	19.87	21.69	21.80	18.74	14.42	9.43	5.56	4.43	13.40
Virginia													
Norfolk	8.06	10.90	14.65	18.51	20.78	22.15	21.12	19.42	16.13	12.49	9.09	7.27	15.10
Richmond	8.06	10.90	14.76	18.62	20.90	22.49	21.58	19.53	16.24	12.61	8.97	7.15	15.22
Washington													
Olympia	3.63	6.02	9.99	14.20	18.06	20.10	21.12	18.17	13.63	7.95	4.32	3.07	11.70
Seattle	3.52	5.91	10.11	14.65	19.08	20.78	21.80	18.51	13.51	7.95	4.20	2.84	11.92
Yakima	4.88	7.95	12.83	17.83	22.49	24.87	25.89	22.26	16.92	10.68	5.56	4.09	17.76
West Virginia													
Charleston	7.04	9.65	13.40	17.15	20.21	21.69	20.90	18.97	15.56	11.81	7.72	6.02	14.20
Elkins	6.93	9.43	12.83	16.35	19.08	20.56	19.99	18.06	14.88	11.13	7.27	5.79	13.51
Wisconsin													
Green Bay	6.25	9.31	13.17	16.81	20.56	22.49	22.03	18.85	14.20	9.65	5.79	4.88	13.74
Madison	6.59	9.88	13.29	16.92	20.67	22.83	22.37	19.42	14.76	3.41	6.25	5.22	14.08
Milwaukee	6.47	9.31	12.72	16.69	20.78	22.94	22.60	19.42	14.88	10.22	6.25	5.11	13.97
Wyoming													
Rock Springs	7.61	10.90	15.10	19.42	23.17	26.01	25.78	22.94	18.62	13.40	8.40	6.70	16.58
Seridan	6.47	9.77	13.97	17.94	20.90	23.85	24.64	21.69	16.47	11.24	7.15	5.56	14.99

(Source: National Renewable Energy Laboratory, USA; Internet Address: http://rredc.nrel.gov/solar)

Table A2.4 Regression coefficients a and b for page model for solar radiation (Eq. 2.52) for worldwide locations*

Country	Station	Lat.	Alt. (m)	Climate [34]	a	b	Ref.**
Egypt	Bahtim	30.13°N	200	BWh	0.220	0.550	[26]
	Cairo	30.08°N	112	BWh	0.140	0.610	[26]
	Giza	30.05°N	19	BWh	0.230	0.540	[26]
	Kharogaosis	25.45°N	78	BWh	0.520	0.230	[26]
	Mersa M.	31.33°N	20	BWh	0.170	0.590	[26]
	Tahrir	30.65°N	16	BWh	0.290	0.460	[26]
Ghana	Accra	5.33°N	~20	BS	0.290	0.470	[3]
	Bole	9.02°N	~350	Aw	0.280	0.440	[3]
	Ho	6.35°N	~300	Aw	0.210	0.460	[3]
	Kumasi	6.41°N	~400	Aw	0.250	0.440	[3]
	Saltpond	5.12°N	~50	BS	0.260	0.450	[3]
	Takoradi	4.59°N	~20	BS	0.250	0.470	[3]
	Tamale	9.25°N	~250	Aw	0.270	0.470	[3]
	Wenchi	7.42°N	~300	Aw	0.280	0.360	[3]
	Yendi	9.26°N	~300	Aw	0.320	0.410	[3]
Greece	Agrino	38.63°N	60	Csa	0.240	0.520	[83]
	Athens	37.97°N	~300	Csa	0.230	0.460	[83]
	Chania	35.50°N	50	Csa	0.220	0.580	[83]
	Kavala	40.93°N	150	Csa	0.250	0.460	[83]
	Larissa	39.63°N	65	Csa	0.230	0.560	[83]
	Mytiline	39.15°N	65	Csa	0.240	0.510	[83]
	Rhodes	36.37°N	5	Csa	0.260	0.520	[83]
Hong Kong	Hong Kong	22.37°N	~5	Caf	0.214	0.514	[40]
India	Ahmedabad	22.37°N	55	Aw	0.302	0.464	[17]
	Bombay (Mumbai)	19.12°N	14	Aw	0.292	0.464	[17]
	Calcutta	22.65°N	6	Aw	0.327	0.399	[17]
	Goa	15.48°N	55	Am	0.279	0.514	[17]
	Jodhpur	26.30°N	224	BSh	0.309	0.439	[17]
	Madras	13.00°N	16	Aw	0.340	0.399	[17]
	Nagpur	21.10°N	310	Aw	0.293	0.460	[17]
	New Delhi	28.63°N	216	Caw	0.341	0.446	[17]
	Poona	18.53°N	559	Bs	0.330	0.453	[17]
	Trivandrum	8.48°N	64	Am	0.393	0.357	[17]
	Vizagapatnam	17.72°N	3	Aw	0.286	0.467	[17]
Italy	Alghero	40.63°N	40	Csa	0.118	0.765	[30]
	Amendola	41.53°N	56	Csa	0.212	0.635	[30]
	Ancona	43.63°N	105	Csa	0.197	0.679	[30]
	Balzano	46.47°N	241	Csa	0.192	0.719	[30]
	Bologna	44.53°N	49	Csa	0.216	0.639	[30]
	Brindisi	40.65°N	10	Csa	0.183	0.706	[30]
	Cagliari	39.25°N	18	Csa	0.175	0.679	[30]
	Cape Mele	43.95°N	221	Csa	0.129	0.779	[30]
	Cape Palinus	40.02°N	185	Csa	0.213	0.604	[30]
	Crotone	39.07°N	158	Csa	0.266	0.546	[30]
	Gela	37.09°N	33	Csa	0.105	0.851	[30]
	Genova	44.42°N	3	Csa	0.089	0.821	[30]
	M. Cimone	44.20°N	2137	H	0.104	0.755	[30]
	M. Terminillo	42.47°N	1875	H	0.246	0.482	[30]

Table A2.4 Regression coefficients a and b for page model for solar radiation (Eq. 2.52) for world wide location* *(Continued)*

Country	Station	Lat.	Alt. (m)	Climate [34]	a	b	Ref.**
Italy	Messina	38.20°N	59	Csa	0.199	0.689	[30]
	Milano	45.43°N	103	Csa	0.148	0.775	[30]
	Napoli	40.85°N	72	Csa	0.181	0.709	[30]
	Olbia	40.93°N	2	Csa	0.183	0.633	[30]
	Pantelleria	36.82°N	170	Csa	0.191	0.679	[30]
	Pescara	42.43°N	18	Csa	0.188	0.669	[30]
	Pianosa	42.58°N	27	Csa	0.143	0.771	[30]
	Pisa	43.67°N	1	Csa	0.205	0.614	[30]
	Roma	41.80°N	131	Csa	0.148	0.722	[30]
	Torino	45.18°N	282	Daf	0.161	0.780	[30]
	Trapani	37.92°N	14	Csa	0.204	0.662	[30]
	Trieste	45.65°N	20	Csa	0.157	0.701	[30]
	Udine	46.03°N	92	Csa	0.167	0.666	[30]
	Ustica	38.70°N	251	Csa	0.254	0.586	[30]
	Venezia	45.50°N	6	Csa	0.144	0.782	[30]
	Vigna Divali	42.08°N	270	Csa	0.154	0.689	[30]
Malaysia	Kota Bahru	6.17°N	5	Af	0.340	0.490	[9]
	Kuala Lumpur	3.12°N	19	Af	0.340	0.490	[9]
	Penang	5.33°N	35	Af	0.350	0.570	[9]
Nigeria	Enugu	6.47°N	137	Af	0.228	0.492	[14]
	Makurdi	7.70°N	76	Aw	0.288	0.472	[14]
	Nsukka	6.80°N	147	Af	0.217	0.490	[14]
	Port Harcourt	4.85°N	6	Af	0.246	0.488	[14]
Sierre L.	Lungi Free.	8.61°N	25	Am	0.260	0.440	[49]
Sudan	Abu Naama	12.73°N	445	BShw	0.433	0.271	[33]
	Aroma	15.83°N	430	BWh	0.460	0.208	[33]
	Dongola	19.17°N	225	BWh	0.211	0.572	[33]
	El Fesher	13.63°N	773	BShw	0.361	0.366	[33]
	El Showak	14.22°N	380	BWh	0.325	0.423	[33]
	G. Ghawazat	11.47°N	480	Aw	0.350	0.353	[33]
	Hudeiba	17.57°N	350	BWh	0.208	0.544	[33]
	Jubo	4.87°N	460	Aw	0.402	0.234	[33]
	Kadugli	11.00°N	501	Aw	0.237	0.463	[33]
	Malakai	9.55°N	387	Aw	0.339	0.359	[33]
	Port Sudan	19.58°N	5	BWh	0.315	0.402	[33]
	Shambat	15.67°N	376	BWh	0.278	0.467	[33]
	Wadi Medani	14.38°N	405	BWh	0.357	0.374	[33]
	Zalingei	12.90°N	900	BShw	0.325	0.456	[33]
Yemen	El Boun	15.73°N	2100	H	0.331	0.385	[34]
	El Khabar	14.38°N	2100	H	0.342	0.372	[34]
	El Macca	13.25°N	10	BWh	0.358	0.346	[34]
	Hodeidah	14.75°N	33	BWh	0.374	0.321	[34]
	Sana	15.52°N	2210	BSh	0.347	0.364	[34]
	Taiz	13.58°N	1400	H	0.364	0.335	[34]
Zambia	Kasama	10.22°S	1384	Aw	0.268	0.454	[31]
	Livingstone	17.82°S	986	Aw	0.187	0.613	[31]
	Luangwa	13.27°S	570	Aw	0.253	0.588	[31]
	Lusaka	15.32°S	1154	Aw	0.198	0.551	[31]

Table A2.4 Regression coefficients a and b for page model for solar radiation (Eq. 2.52) for world wide location* *(Continued)*

Country	Station	Lat.	Alt. (m)	Climate [34]	*a*	*b*	Ref.**
Zambia	Mansa	11.10°S	1259	Aw	0.265	0.476	[31]
	Mbala	8.37°S	1673	Aw	0.245	0.505	[31]
	Mongu	15.25°S	1053	Aw	0.188	0.556	[31]
	Ndolo	13.00°S	1270	Aw	0.288	0.386	[31]

*Data compiled from Ref. [1], chapter 2.
**Reference numbers refer to references in chapter 2.

Table A2.5 Solar collector tilt factor $(\overline{R_b})^a$

Month	$L = 20°$ $\beta = 20°$	$\beta = 40°$	$L = 30°$ $\beta = 30°$	$\beta = 50°$	$L = 40°$ $\beta = 40°$	$\beta = 60°$	$L = 50°$ $\beta = 50°$	$\beta = 70°$
Jan	1.36	1.52	1.68	1.88	2.28	2.56	3.56	3.94
Feb	1.22	1.28	1.44	1.52	1.80	1.90	2.49	2.62
Mar	1.08	1.02	1.20	1.15	1.36	1.32	1.65	1.62
Apr	1.00	0.83	1.00	0.84	1.05	0.90	1.16	1.00
May	0.92	0.70	0.87	0.66	0.88	0.66	0.90	0.64
Jun	0.87	0.63	0.81	0.58	0.79	0.60	0.80	0.56
Jul	0.89	0.66	0.83	0.62	0.82	0.64	0.84	0.62
Aug	0.95	0.78	0.93	0.76	0.96	0.78	1.02	0.83
Sep	1.04	0.95	1.11	1.00	1.24	1.12	1.44	1.32
Oct	1.17	1.20	1.36	1.36	1.62	1.64	2.10	2.14
Nov	1.30	1.44	1.60	1.76	2.08	2.24	3.16	3.32
Dec	1.39	1.60	1.76	1.99	2.48	2.80	4.04	4.52

^{a}The solar collector tilt factor is the ratio of monthly beam insolation on a tilted surface to monthly beam insolation on a horizontal surface. Here β = collector tilt angle and L = collector latitude. From Kreider, J. F., and F. Kreith, "Solar Heating and Cooling," revised 1st ed., Hemisphere Publ. Corp., 1977.

Table A2.6a Solar position and insolation values for 24 degrees north latitude[a]

Date	Solar time AM	PM	Solar position Alt	Azm	BTUH/sq. ft. total insolation on surface[b] Normal[c]	Horiz.	South facing surface angle with horiz. 14	24	34	44	90
Jan 21	7	5	4.8	65.6	71	10	17	21	25	28	31
	8	4	16.9	58.3	239	83	110	126	137	145	127
	9	3	27.9	48.8	288	151	188	207	221	228	176
	10	2	37.2	36.1	308	204	246	268	282	287	207
	11	1	43.6	19.6	317	237	283	306	319	324	226
	12		46.0	0.0	320	249	296	319	332	336	232
	Surface daily totals				2766	1622	1984	2174	2300	2360	1766
Feb 21	7	5	9.3	74.6	158	35	44	49	53	56	46
	8	4	22.3	67.2	263	116	135	145	150	151	102
	9	3	34.4	57.6	298	187	213	225	230	228	141
	10	2	45.1	44.2	314	241	273	286	291	287	168
	11	1	53.0	25.0	321	276	310	324	328	323	185
	12		56.0	0.0	324	288	323	337	341	335	191
	Surface daily totals				3036	1998	2276	2396	2436	2424	1476
Mar 21	7	5	13.7	83.3	194	60	63	64	62	59	27
	8	4	27.2	76.8	267	141	150	152	149	142	64
	9	3	40.2	67.9	295	212	226	229	225	214	95
	10	2	52.3	54.8	309	266	285	288	283	270	120
	11	1	61.9	33.4	315	300	322	326	320	305	135
	12		66.0	0.0	317	312	334	339	333	317	140
	Surface daily totals				3078	2270	2428	2456	2412	2298	1022
Apr 21	6	6	4.7	100.6	40	7	5	4	4	3	2
	7	5	18.3	94.9	203	83	77	70	62	51	10
	8	4	32.0	89.0	256	160	157	149	137	122	16
	9	3	45.6	81.9	280	227	227	220	206	186	46
	10	2	59.0	71.8	292	278	282	275	259	237	61
	11	1	71.1	51.6	298	310	316	309	293	269	74
	12		77.6	0.0	299	321	328	321	305	280	79
	Surface daily totals				3036	2454	2458	2374	2228	2016	488
May 21	6	6	8.0	108.4	86	22	15	10	9	9	5
	7	5	21.2	103.2	203	98	85	73	59	44	12
	8	4	34.6	98.5	248	171	159	145	127	106	15
	9	3	48.3	93.6	269	233	224	210	190	165	16
	10	2	62.0	87.7	280	281	275	261	239	211	22
	11	1	75.5	76.9	286	311	307	293	270	240	34
	12		86.0	0.0	288	322	317	304	281	250	37
	Surface daily totals				3032	2556	2447	2286	2072	1800	246
Jun 21	6	6	9.3	111.6	97	29	20	12	12	11	7
	7	5	22.3	106.8	201	103	87	73	58	41	13
	8	4	35.5	102.6	242	173	158	142	122	99	16
	9	3	49.0	98.7	263	234	221	204	182	155	18
	10	2	62.6	95.0	274	280	269	253	229	199	18
	11	1	76.3	90.8	279	309	300	283	259	227	19
	12		89.4	0.0	281	319	310	294	269	236	22
	Surface daily totals				2994	2574	2422	2230	1992	1700	204

Table A2.6*a* Solar position and insolation values for 24 degrees north latitude[a] *(Continued)*

Date	Solar time AM	Solar time PM	Solar position Alt	Solar position Azm	BTUH/sq. ft. total insolation on surface[b] Normal[c]	Horiz.	South facing surface angle with horiz. 14	24	34	44	90
Jul 21	6	6	8.2	109.0	81	23	16	11	10	9	6
	7	5	21.4	103.8	195	98	85	73	59	44	13
	8	4	34.8	99.2	239	169	157	143	125	104	16
	9	3	48.4	94.5	261	231	221	207	187	161	18
	10	2	62.1	89.0	272	278	270	256	235	206	21
	11	1	75.7	79.2	278	307	302	287	265	235	32
	12		86.6	0.0	280	317	312	298	275	245	36
	Surface daily totals				2932	2526	2412	2250	2036	1766	246
Aug 21	6	6	5.0	101.3	35	7	5	4	4	4	2
	7	5	18.5	95.6	186	82	76	69	60	50	11
	8	4	32.2	89.7	241	158	154	146	134	118	16
	9	3	45.9	82.9	265	223	222	214	200	181	39
	10	2	59.3	73.0	278	273	275	268	252	230	58
	11	1	71.6	53.2	284	304	309	301	285	261	71
	12		78.3	0.0	286	315	320	313	296	272	75
	Surface daily totals				2864	2408	2402	2316	2168	1958	470
Sep 21	7	5	13.7	83.8	173	57	60	60	59	56	26
	8	4	27.2	76.8	248	136	144	146	143	136	62
	9	3	40.2	67.9	278	205	218	221	217	206	93
	10	2	52.3	54.8	292	258	275	278	273	261	116
	11	1	61.9	33.4	299	291	311	315	309	295	131
	12		66.0	0.0	301	302	323	327	321	306	136
	Surface daily totals				2878	2194	2342	2366	2322	2212	992
Oct 21	7	5	9.1	74.1	138	32	40	45	48	50	42
	8	4	22.0	66.7	247	111	129	139	144	145	99
	9	3	34.1	57.1	284	180	206	217	223	221	138
	10	2	44.7	43.8	301	234	265	277	282	279	165
	11	1	52.5	24.7	309	268	301	315	319	314	182
	12		55.5	0.0	311	279	314	328	332	327	188
	Surface daily totals				2868	1928	2198	2314	2364	2346	1442
Nov 21	7	5	4.9	65.8	67	10	16	20	24	27	29
	8	4	17.0	58.4	232	82	108	123	135	142	124
	9	3	28.0	48.9	282	150	186	205	217	224	172
	10	2	37.3	36.3	303	203	244	265	278	283	204
	11	1	43.8	19.7	312	236	280	302	316	320	222
	12		46.2	0.0	315	247	293	315	328	332	228
	Surface daily totals				2706	1610	1962	2146	2268	2324	1730
Dec 21	7	5	3.2	62.6	30	3	7	9	11	12	14
	8	4	14.9	55.3	225	71	99	116	129	139	130
	9	3	25.5	46.0	281	137	176	198	214	223	184
	10	2	34.3	33.7	304	189	234	258	275	283	217
	11	1	40.4	18.2	314	221	270	295	312	320	236
	12		42.6	0.0	317	232	282	308	325	332	243
	Surface daily totals				2624	1474	1852	2058	2204	2286	1808

[a]From Kreider, J. F., and F. Kreith, "Solar Heating and Cooling," revised 1st ed., Hemisphere Publ. Corp., 1977.

[b]1 Btu/hr · ft^2 = 3.152 W/m^2. Ground reflection not included on normal or horizontal surfaces.

[c]Normal insolation does not include diffuse component.

Table A2.6*b* Solar position and insolation values for 32 degrees north latitude[a]

Date	Solar time AM	PM	Solar position Alt	Azm	Normal[c]	Horiz.	South facing surface angle with horiz. 22	32	42	52	90
Jan 21	7	5	1.4	65.2	1	0	0	0	0	1	1
	8	4	12.5	56.5	203	56	93	106	116	123	115
	9	3	22.5	46.0	269	118	175	193	206	212	181
	10	2	30.6	33.1	295	167	235	256	269	274	221
	11	1	36.1	17.5	306	198	273	295	308	312	245
	12		38.0	0.0	310	209	285	308	321	324	253
	Surface daily totals				2458	1288	1839	2008	2118	2166	1779
Feb 21	7	5	7.1	73.5	121	22	34	37	40	42	38
	8	4	19.0	64.4	247	95	127	136	140	141	108
	9	3	29.9	53.4	288	161	206	217	222	220	158
	10	2	39.1	39.4	306	212	266	278	283	279	193
	11	1	45.6	21.4	315	244	304	317	321	315	214
	12		48.0	0.0	317	255	316	330	334	328	222
	Surface daily totals				2872	1724	2188	2300	2345	2322	1644
Mar 21	7	5	12.7	81.9	185	54	60	60	59	56	32
	8	4	25.1	73.0	260	129	146	147	144	137	78
	9	3	36.8	62.1	290	194	222	224	220	209	119
	10	2	47.3	47.5	304	245	280	283	278	265	150
	11	1	55.0	26.8	311	277	317	321	315	300	170
	12		58.0	0.0	313	287	329	333	327	312	177
	Surface daily totals				3012	2084	2378	2403	2358	2246	1276
Apr 21	6	6	6.1	99.9	66	14	9	6	6	5	3
	7	5	18.8	92.2	206	86	78	71	62	51	10
	8	4	31.5	84.0	255	158	156	148	136	120	35
	9	3	43.9	74.2	278	220	225	217	203	183	68
	10	2	55.7	60.3	290	267	279	272	256	234	95
	11	1	65.4	37.5	295	297	313	306	290	265	112
	12		69.6	0.0	297	307	325	318	301	276	118
	Surface daily totals				3076	2390	2444	2356	2206	1994	764
May 21	6	6	10.4	107.2	119	36	21	13	13	12	7
	7	5	22.8	100.1	211	107	88	75	60	44	13
	8	4	35.4	92.9	250	175	159	145	127	105	15
	9	3	48.1	84.7	269	233	223	209	188	163	33
	10	2	60.6	73.3	280	277	273	259	237	208	56
	11	1	72.0	51.9	285	305	305	290	268	237	72
	12		78.0	0.0	286	315	315	301	278	247	77
	Surface daily totals				3112	2582	2454	2284	2064	1788	469
Jun 21	6	6	12.2	110.2	131	45	26	16	15	14	9
	7	5	24.3	103.4	210	115	91	76	59	41	14
	8	4	36.9	96.8	245	180	159	143	122	99	16
	9	3	49.6	89.4	264	236	221	204	181	153	19
	10	2	62.2	79.7	274	279	268	251	227	197	41
	11	1	74.2	60.9	279	306	299	282	257	224	56
	12		81.5	0.0	280	315	309	292	267	234	60
	Surface daily totals				3084	2634	2436	2234	1990	1690	370

Table A2.6*b* Solar position and insolation values for 32 degrees north latitude*a* *(Continued)*

Date	Solar time AM	PM	Solar position Alt	Azm	Normal[c]	Horiz.	22	32	42	52	90
							South facing surface angle with horiz.				
Jul 21	6	6	10.7	107.7	113	37	22	14	13	12	8
	7	5	23.1	100.6	203	107	87	75	60	44	14
	8	4	35.7	93.6	241	174	158	143	125	104	16
	9	3	48.4	85.5	261	231	220	205	185	159	31
	10	2	60.9	74.3	271	274	269	254	232	204	54
	11	1	72.4	53.3	277	302	300	285	262	232	69
	12		78.6	0.0	279	311	310	296	273	242	74
	Surface daily totals				3012	2558	2422	2250	2030	1754	458
Aug 21	6	6	6.5	100.5	59	14	9	7	6	6	4
	7	5	19.1	92.8	190	85	77	69	60	50	12
	8	4	31.8	84.7	240	156	152	144	132	116	33
	9	3	44.3	75.0	263	216	220	212	197	178	65
	10	2	56.1	61.3	276	262	272	264	249	226	91
	11	1	66.0	38.4	282	292	305	298	281	257	107
	12		70.3	0.0	284	302	317	309	292	268	113
	Surface daily totals				2902	2352	2388	2296	2144	1934	736
Sep 21	7	5	12.7	81.9	163	51	56	56	55	52	30
	8	4	25.1	73.0	240	124	140	141	138	131	75
	9	3	36.8	62.1	272	188	213	215	211	201	114
	10	2	47.3	47.5	287	237	270	273	268	255	145
	11	1	55.0	26.8	294	268	306	309	303	289	164
	12		58.0	0.0	296	278	318	321	315	300	171
	Surface daily totals				2808	2014	2288	2308	2264	2154	1226
Oct 21	7	5	6.8	73.1	99	19	29	32	34	36	32
	8	4	18.7	64.0	229	90	120	128	133	134	104
	9	3	29.5	53.0	273	155	198	208	213	212	153
	10	2	38.7	39.1	293	204	257	269	273	270	188
	11	1	45.1	21.1	302	236	294	307	311	306	209
	12		47.5	0.0	304	247	306	320	324	318	217
	Surface daily totals				2696	1654	2100	2208	2252	2232	1588
Nov 21	7	5	1.5	65.4	2	0	0	0	1	1	1
	8	4	12.7	56.6	196	55	91	104	113	119	111
	9	3	22.6	46.1	263	118	173	190	202	208	176
	10	2	30.8	33.2	289	166	233	252	265	270	217
	11	1	36.2	17.6	301	197	270	291	303	307	241
	12		38.2	0.0	304	207	282	304	316	320	249
	Surface daily totals				2406	1280	1816	1980	2084	2130	1742
Dec 21	8	4	10.3	53.8	176	41	77	90	101	108	107
	9	3	19.8	43.6	257	102	161	180	195	204	183
	10	2	27.6	31.2	288	150	221	244	259	267	226
	11	1	32.7	16.4	301	180	258	282	298	305	251
	12		34.6	0.0	304	190	271	295	311	318	259
	Surface daily totals				2348	1136	1704	1888	2016	2086	1794

[a]From Kreider, J. F., and F. Keith, "Solar Heating and Cooling," revised 1st ed., Hemisphere Publ. Corp., 1977.

[b]1 Btu/hr · ft² = 3.152 W/m². Ground reflection not included on normal or horizontal surfaces.

[c]Normal insolation does not include diffuse component.

Table A2.6c Solar position and insolation values for 40 degrees north latitude[a]

Date	Solar time AM	PM	Solar position Alt	Azm	Normal[c]	Horiz.	South facing surface angle with horiz. 30	40	50	60	90
Jan 21	8	4	8.1	55.3	142	28	65	74	81	85	84
	9	3	16.8	44.0	239	83	155	171	182	187	171
	10	2	23.8	30.9	274	127	218	237	249	254	223
	11	1	28.4	16.0	289	154	257	277	290	293	253
	12		30.0	0.0	294	164	270	291	303	306	263
	Surface daily totals				2182	948	1660	1810	1906	1944	1726
Feb 21	7	5	4.8	72.7	69	10	19	21	23	24	22
	8	4	15.4	62.2	224	73	114	122	126	127	107
	9	3	25.0	50.2	274	132	195	205	209	208	167
	10	2	32.8	35.9	295	178	256	267	271	267	210
	11	1	38.1	18.9	305	206	293	306	310	304	236
	12		40.0	0.0	308	216	306	319	323	317	245
	Surface daily totals				2640	1414	2060	2162	2202	2176	1730
Mar 21	7	5	11.4	80.2	171	46	55	55	54	51	35
	8	4	22.5	69.6	250	114	140	141	138	131	89
	9	3	32.8	57.3	282	173	215	217	213	202	138
	10	2	41.6	41.9	297	218	273	276	271	258	176
	11	1	47.7	22.6	305	247	310	313	307	293	200
	12		50.0	0.0	307	257	322	326	320	305	208
	Surface daily totals				2916	1852	2308	2330	2284	2174	1484
Apr 21	6	6	7.4	98.9	89	20	11	8	7	7	4
	7	5	18.9	89.5	206	87	77	70	61	50	12
	8	4	30.3	79.3	252	152	153	145	133	117	53
	9	3	41.3	67.2	274	207	221	213	199	179	93
	10	2	51.2	51.4	286	250	275	267	252	229	126
	11	1	58.7	29.2	292	277	308	301	285	260	147
	12		61.6	0.0	293	287	320	313	296	271	154
	Surface daily totals				3092	2274	2412	2320	2168	1956	1022
May 21	5	7	1.9	114.7	1	0	0	0	0	0	0
	6	6	12.7	105.6	144	49	25	15	14	13	9
	7	5	24.0	96.6	216	214	89	76	60	44	13
	8	4	35.4	87.2	250	175	158	144	125	104	25
	9	3	46.8	76.0	267	227	221	206	186	160	60
	10	2	57.5	60.9	277	267	270	255	233	205	89
	11	1	66.2	37.1	283	293	301	287	264	234	108
	12		70.0	0.0	284	301	312	297	274	243	114
	Surface daily totals				3160	2552	2442	2264	2040	1760	724
Jun 21	5	7	4.2	117.3	22	4	3	3	2	2	1
	6	6	14.8	108.4	155	60	30	18	17	16	10
	7	5	26.0	99.7	216	123	92	77	59	41	14
	8	4	37.4	90.7	246	182	159	142	121	97	16
	9	3	48.8	80.2	263	233	219	202	179	151	47
	10	2	59.8	65.8	272	272	266	248	224	194	74
	11	1	69.2	41.9	277	296	296	278	253	221	92
	12		73.5	0.0	279	304	306	289	263	230	98
	Surface daily totals				3180	2648	2434	2224	1974	1670	610

Table A2.6c Solar position and insolation values for 40 degrees north latitude[a] (Continued)

Date	Solar time AM	PM	Solar position Alt	Azm	Normal[c]	Horiz.	South facing surface angle with horiz. 30	40	50	60	90
Jul 21	5	7	2.3	115.2	2	0	0	0	0	0	0
	6	6	13.1	106.1	138	50	26	17	15	14	9
	7	5	24.3	97.2	208	114	89	75	60	44	14
	8	4	35.8	87.8	241	174	157	142	124	102	24
	9	3	47.2	76.7	259	225	218	203	182	157	58
	10	2	57.9	61.7	269	265	266	251	229	200	86
	11	1	66.7	37.9	275	290	296	281	258	228	104
	12		70.6	0.0	276	298	307	292	269	238	111
	Surface daily totals				3062	2534	2409	2230	2006	1728	702
Aug 21	6	6	7.9	99.5	81	21	12	9	8	7	5
	7	5	19.3	90.9	191	87	76	69	60	49	12
	8	4	30.7	79.9	237	150	150	141	129	113	50
	9	3	41.8	67.9	260	205	216	207	193	173	89
	10	2	51.7	52.1	272	246	267	259	244	221	120
	11	1	59.3	29.7	278	273	300	292	276	252	140
	12		62.3	0.0	280	282	311	303	287	262	147
	Surface daily totals				2916	2244	2354	2258	2104	1894	978
Sep 21	7	5	11.4	80.2	149	43	51	51	49	47	32
	8	4	22.5	69.6	230	109	133	134	131	124	84
	9	3	32.8	57.3	263	167	206	208	203	193	132
	10	2	41.6	41.9	280	211	262	265	260	247	168
	11	1	47.7	22.6	287	239	298	301	295	281	192
	12		50.0	0.0	290	249	310	313	307	292	200
	Surface daily totals				2708	1788	2210	2228	2182	2074	1416
Oct 21	7	5	4.5	72.3	48	7	14	15	17	17	16
	8	4	15.0	61.9	204	68	106	113	117	118	100
	9	3	24.5	49.8	257	126	185	195	200	198	160
	10	2	32.4	35.6	280	170	245	257	261	257	203
	11	1	37.6	18.7	291	199	283	295	299	294	229
	12		39.5	0.0	294	208	295	308	312	306	238
	Surface daily totals				2454	1348	1962	2060	2098	2074	1654
Nov 21	8	4	8.2	55.4	136	28	63	72	78	82	81
	9	3	17.0	44.1	232	82	152	167	178	183	167
	10	2	24.0	31.0	268	126	215	233	245	249	219
	11	1	28.6	16.1	283	153	254	273	285	288	248
	12		30.2	0.0	288	163	267	287	298	301	258
	Surface daily totals				2128	942	1636	1778	1870	1908	1686
Dec 21	8	4	5.5	53.0	89	14	39	45	50	54	56
	9	3	14.0	41.9	217	65	135	152	164	171	163
	10	2	20.,	29.4	261	107	200	221	235	242	221
	11	1	25.0	15.2	280	134	239	262	276	283	252
	12		26.6	0.0	285	143	253	275	290	296	263
	Surface daily totals				1978	782	1480	1634	1740	1796	1646

[a]From Kreider, J. F., and F. Kreith, "Solar Heating and Cooling," revised 1st ed., Hemisphere Publ. Corp., 1977.

[b]1 Btu/hr · ft^2 = 3.152 W/m^2. Ground reflection not included on normal or horizontal surfaces.

[c]Normal insolation does not include diffuse component.

Table A2.6*d* Solar position and insolation values for 48 degrees north latitude[a]

Date	AM	PM	Alt	Azm	Normal[c]	Horiz.	38	48	58	68	90
			Solar position				South facing surface angle with horiz.				
Jan 21	8	4	3.5	54.6	37	4	17	19	21	22	22
	9	3	11.0	42.6	185	46	120	132	140	145	139
	10	2	16.9	29.4	239	83	190	206	216	220	206
	11	1	20.7	15.1	261	107	231	249	260	263	243
	12		22.0	0.0	267	115	245	264	275	278	255
	Surface daily totals				1710	596	1360	1478	1550	1578	1478
Feb 21	7	5	2.4	72.2	12	1	3	4	4	4	4
	8	4	11.6	60.5	188	49	95	102	105	106	96
	9	3	19.7	47.7	251	100	178	187	191	190	167
	10	2	26.2	33.3	278	139	240	251	255	251	217
	11	1	30.5	17.2	290	165	278	290	294	288	247
	12		32.0	0.0	293	173	291	304	307	301	258
	Surface daily totals				2330	1080	1880	1972	2024	1978	1720
Mar 21	7	5	10.0	78.7	153	37	49	49	47	45	35
	8	4	19.5	66.8	236	96	131	132	129	122	96
	9	3	28.2	53.4	270	147	205	207	203	193	152
	10	2	35.4	37.8	287	187	263	266	261	248	195
	11	1	40.3	19.8	295	212	300	303	297	283	223
	12		42.0	0.0	298	220	312	315	309	294	232
	Surface daily totals				2780	1578	2208	2228	2182	2074	1632
Apr 21	6	6	8.6	97.8	108	27	13	9	8	7	5
	7	5	18.6	86.7	205	85	76	69	59	48	21
	8	4	28.5	74.9	247	142	149	141	129	113	69
	9	3	37.8	61.2	268	191	216	208	194	174	115
	10	2	45.8	44.6	280	228	268	260	245	223	152
	11	1	51.5	24.0	286	252	301	294	278	254	177
	12		53.6	0.0	288	260	313	305	289	264	185
	Surface daily totals				3076	2106	2358	2266	2114	1902	1262
May 21	5	7	5.2	114.3	41	9	4	4	4	3	2
	6	6	14.7	103.7	162	61	27	16	15	13	10
	7	5	24.6	93.0	219	118	89	75	60	43	13
	8	4	34.7	81.6	248	171	156	142	123	101	45
	9	3	44.3	68.3	264	217	217	202	182	156	86
	10	2	53.0	51.3	274	252	265	251	229	200	120
	11	1	59.5	28.6	279	274	296	281	258	228	141
	12		62.0	0.0	280	281	306	292	269	238	149
	Surface daily totals				3254	2482	2418	2234	2010	1728	982
Jun 21	5	7	7.9	116.5	77	21	9	9	8	7	5
	6	6	17.2	106.2	172	74	33	19	18	16	12
	7	5	27.0	95.8	220	129	93	77	59	39	15
	8	4	37.1	84.6	246	181	157	140	119	95	35
	9	3	46.9	71.6	261	225	216	198	175	147	74
	10	2	55.8	54.8	269	259	262	244	220	189	105
	11	1	62.7	31.2	274	280	291	273	248	216	126
	12		65.5	0.0	275	287	301	283	258	225	133
	Surface daily totals				3312	2626	2420	2204	1950	1644	874

Table A2.6d Solar position and insolation values for 48 degrees north latitude[a] (Continued)

Date	Solar time AM	PM	Solar position Alt	Azm	Normal[c]	Horiz.	South facing surface angle with horiz. 38	48	58	68	90
Jul 21	5	7	5.7	114.7	43	10	5	5	4	4	3
	6	6	15.2	104.1	156	62	28	18	16	15	11
	7	5	25.1	93.5	211	118	89	75	59	42	14
	8	4	35.1	82.1	240	171	154	140	121	99	43
	9	3	44.8	68.8	256	215	214	199	178	153	83
	10	2	53.5	51.9	266	250	261	246	224	195	116
	11	1	60.1	29.0	271	272	291	276	253	223	137
	12		62.6	0.0	272	279	301	286	263	232	144
	Surface daily totals				3158	2474	2386	2200	1974	1694	956
Aug 21	6	6	9.1	98.3	99	28	14	10	9	8	6
	7	5	19.1	87.2	190	85	75	67	58	47	20
	8	4	29.0	75.4	232	141	145	137	125	109	65
	9	3	38.4	61.8	254	189	210	201	187	168	110
	10	2	46.4	45.1	266	225	260	252	237	214	146
	11	1	52.2	24.3	272	248	293	285	268	244	169
	12		54.3	0.0	274	256	304	296	279	255	177
	Surface daily totals				2898	2086	2300	2200	2046	1836	1208
Sep 21	7	5	10.0	78.7	131	35	44	44	43	40	31
	8	4	19.5	66.8	215	92	124	124	121	115	90
	9	3	28.2	53.4	251	142	196	197	193	183	143
	10	2	35.4	37.8	269	181	251	254	248	236	185
	11	1	40.3	19.8	278	205	287	289	284	269	212
	12		42.0	0.0	280	213	299	302	296	281	221
	Surface daily totals				2568	1522	2102	2118	2070	1966	1546
Oct 21	7	5	2.0	71.9	4	0	1	1	1	1	1
	8	4	11.2	60.2	165	44	86	91	95	95	87
	9	3	19.3	47.4	233	94	167	176	180	178	157
	10	2	25.7	33.1	262	133	228	239	242	239	207
	11	1	30.0	17.1	274	157	266	277	281	276	237
	12		31.5	0.0	278	166	279	291	294	288	247
	Surface daily totals				2154	1022	1774	1860	1890	1866	1626
Nov 21	8	4	3.6	54.7	36	5	17	19	21	22	22
	9	3	11.2	42.7	179	46	117	129	137	141	135
	10	2	17.1	29.5	233	83	186	202	212	215	201
	11	1	20.9	15.1	255	107	227	245	255	258	238
	12		22.2	0.0	261	115	241	259	270	272	250
	Surface daily totals				1668	596	1336	1448	1518	1544	1442
Dec 21	9	3	8.0	40.9	140	27	87	98	105	110	109
	10	2	13.6	28.2	214	63	164	180	192	197	190
	11	1	17.3	14.4	242	86	207	226	239	244	231
	12		18.6	0.0	250	94	222	241	254	260	244
	Surface daily totals				1444	446	1136	1250	1326	1364	1304

[a]From Kreider, J. F., and F. Kreith, "Solar Heating and Cooling," revised 1st ed., Hemisphere Publ. Corp., 1977.

[b]1 Btu/hr · ft^2 = 3.152 W/m^2. Ground reflection not included on normal or horizontal surfaces.

[c]Normal insolation does not include diffuse component.

Table A2.6e Solar position and insolation values for 56 degrees north latitude[a]

Date	Solar time AM	Solar time PM	Solar position Alt	Solar position Azm	Normal[c]	Horiz.	South facing surface angle with horiz. 46	56	66	76	90
Jan 21	9	3	5.0	41.8	78	11	50	55	59	60	60
	10	2	9.9	28.5	170	39	135	146	154	156	153
	11	1	12.9	14.5	207	58	183	197	206	208	201
	12		14.0	0.0	217	65	198	214	222	225	217
	Surface daily totals				1126	282	934	1010	1058	1074	1044
Feb 21	8	4	7.6	59.4	129	25	65	69	72	72	69
	9	3	14.2	45.9	214	65	151	159	162	161	151
	10	2	19.4	31.5	250	98	215	225	228	224	208
	11	1	22.8	16.1	266	119	254	265	268	263	243
	12		24.0	0.0	270	126	268	279	282	276	255
	Surface daily totals				1986	740	1640	1716	1742	1716	1598
Mar 21	7	5	8.3	77.5	128	28	40	40	39	37	32
	8	4	16.2	64.4	215	75	119	120	117	111	97
	9	3	23.3	50.3	253	118	192	193	189	180	154
	10	2	29.0	34.9	272	151	249	251	246	234	205
	11	1	32.7	17.9	282	172	285	288	282	268	236
	12		34.0	0.0	284	179	297	300	294	280	246
	Surface daily totals				2586	1268	2066	2084	2040	1938	1700
Apr 21	5	7	1.4	108.8	0	0	0	0	0	0	0
	6	6	9.6	96.5	122	32	14	9	8	7	6
	7	5	18.0	84.1	201	81	74	66	57	46	29
	8	4	26.1	70.9	239	129	143	135	123	108	82
	9	3	33.6	56.3	260	169	208	200	186	167	133
	10	2	39.9	39.7	272	201	259	251	236	214	174
	11	1	44.1	20.7	278	220	292	284	268	245	200
	12		45.6	0.0	280	227	303	295	279	255	209
	Surface daily totals				3024	1892	2282	2186	2038	1830	1458
May 21	4	8	1.2	125.5	0	0	0	0	0	0	0
	5	7	8.5	113.4	93	25	10	9	8	7	6
	6	6	16.5	101.5	175	71	28	17	15	13	11
	7	5	24.8	89.3	219	119	88	74	58	41	16
	8	4	33.1	76.3	244	163	153	138	119	98	63
	9	3	40.9	61.6	259	201	212	197	176	151	109
	10	2	47.6	44.2	268	231	259	244	222	194	146
	11	1	52.3	23.4	273	249	288	274	251	222	170
	12		54.0	0.0	275	255	299	284	261	231	178
	Surface daily totals				3340	2374	2374	2188	1962	1682	1218
Jun 21	4	8	4.2	127.2	21	4	2	2	2	2	1
	5	7	11.4	115.3	122	40	14	13	11	10	8
	6	6	19.3	103.6	185	86	34	19	17	15	12
	7	5	27.6	91.7	222	132	92	76	57	38	15
	8	4	35.9	78.8	243	175	154	137	116	92	55
	9	3	43.8	64.1	257	212	211	193	170	143	98
	10	2	50.7	46.4	265	240	255	238	214	184	133
	11	1	55.6	24.9	269	258	284	267	242	210	156
	12		57.5	0.0	271	264	294	276	251	219	164
	Surface daily totals				3438	2526	2388	2166	1910	1606	1120

Table A2.6e Solar position and insolation values for 56 degrees north latitude[a] *(Continued)*

Date	Solar time AM	PM	Solar position Alt	Azm	BTUH/sq. ft. total insolation on surfaces[b] Normal[c]	Horiz.	South facing surface angle with horiz. 46	56	66	76	90
Jul 21	4	8	1.7	125.8	0	0	0	0	0	0	0
	5	7	9.0	113.7	91	27	11	10	9	8	6
	6	6	17.0	101.9	169	72	30	18	16	14	12
	7	5	25.3	89.7	212	119	88	74	58	41	15
	8	4	33.6	76.7	237	163	151	136	117	96	61
	9	3	41.4	62.0	252	201	208	193	173	147	106
	10	2	48.2	44.6	261	230	254	239	217	189	142
	11	1	52.9	23.7	265	248	283	268	245	216	165
	12		54.6	0.0	267	254	293	278	255	225	173
	Surface daily totals				3240	2372	2342	2152	1926	1646	1186
Aug 21	5	7	2.0	109.2	1	0	0	0	0	0	0
	6	6	10.2	97.0	112	34	16	11	10	9	7
	7	5	18.5	84.5	187	82	73	65	56	45	28
	8	4	26.7	71.3	225	128	140	131	119	104	78
	9	3	34.3	56.7	246	168	202	193	179	160	126
	10	2	40.5	40.0	258	199	251	242	227	206	166
	11	1	44.8	20.9	264	218	282	274	258	235	191
	12		46.3	0.0	266	225	293	285	269	245	200
	Surface daily totals				2850	1884	2218	2118	1966	1760	1392
Sep 21	7	5	8.3	77.5	107	25	36	36	34	32	28
	8	4	16.2	64.4	194	72	111	111	108	102	89
	9	3	23.3	50.3	233	114	181	182	178	168	147
	10	2	29.0	34.9	253	146	236	237	232	221	193
	11	1	32.7	17.9	263	166	271	273	267	254	223
	12		34.0	0.0	266	173	283	285	279	265	233
	Surface daily totals				2368	1220	1950	1962	1918	1820	1594
Oct 21	8	4	7.1	59.1	104	20	53	57	59	59	57
	9	3	13.8	45.7	193	60	138	145	148	147	138
	10	2	19.0	31.3	231	92	201	210	213	210	195
	11	1	22.3	16.0	248	112	240	250	253	248	230
	12		23.5	0.0	253	119	253	263	266	261	241
	Surface daily totals				1804	688	1516	1586	1612	1588	1480
Nov 21	9	3	5.2	41.9	76	12	49	54	57	59	58
	10	2	10.0	28.5	165	39	132	143	149	152	148
	11	1	13.1	14.5	201	58	179	193	201	203	196
	12		14.2	0.0	211	65	194	209	217	219	211
	Surface daily totals				1094	284	914	986	1032	1046	1016
Dec 21	9	3	1.9	40.5	5	0	3	4	4	4	4
	10	2	6.6	27.5	113	19	86	95	101	104	103
	11	1	9.5	13.9	166	37	141	154	163	167	164
	12		10.6	0.0	180	43	159	173	182	186	182
	Surface daily totals				748	156	620	678	716	734	722

[a]From Kreider, J. F., and F. Kreith, "Solar Heating and Cooling," revised 1st ed., Hemisphere Publ. Corp., 1977.

[b]1 Btu/hr · ft^2 = 3.152 W/m^2. Ground reflection not included on normal or horizontal surfaces.

[c]Normal insolation does not include diffuse component.

Table A2.6*f* Solar position and insolation values for 64 degrees north latitude[a]

Date	Solar time AM	PM	Solar position Alt	Azm	Normal[c]	Horiz.	54	64	74	84	90
Jan 21	10	2	2.8	28.1	22	2	17	19	20	20	20
	11	1	5.2	14.1	81	12	72	77	80	81	81
	12		6.0	0.0	100	16	91	98	102	103	103
	Surface daily totals				306	45	268	290	302	306	304
Feb 21	8	4	3.4	58.7	35	4	17	19	19	19	19
	9	3	8.6	44.8	147	31	103	108	111	110	107
	10	2	12.6	30.3	199	55	170	178	181	178	173
	11	1	15.1	15.3	222	71	212	220	223	219	213
	12		16.0	0.0	228	77	225	235	237	232	226
	Surface daily totals				1432	400	1230	1286	1302	1282	1252
Mar 21	7	5	6.5	76.5	95	18	30	29	29	27	25
	8	4	20.7	62.6	185	54	101	102	99	94	89
	9	3	18.1	48.1	227	87	171	172	169	160	153
	10	2	22.3	32.7	249	112	227	229	224	213	203
	11	1	25.1	16.6	260	129	262	265	259	246	235
	12		26.0	0.0	263	134	274	277	271	258	246
	Surface daily totals				2296	932	1856	1870	1830	1736	1656
Apr 21	5	7	4.0	108.5	27	5	2	2	2	1	1
	6	6	10.4	95.1	133	37	15	9	8	7	6
	7	5	17.0	81.6	194	76	70	63	54	43	37
	8	4	23.3	67.5	228	112	136	128	116	102	91
	9	3	29.0	52.3	248	144	197	189	176	158	145
	10	2	33.5	36.0	260	169	246	239	224	203	188
	11	1	36.5	18.4	266	184	278	270	255	233	216
	12		97.6	0.0	268	190	289	281	266	243	225
	Surface daily totals				2982	1644	2176	2082	1936	1736	1594
May 21	4	8	5.8	125.1	51	11	5	4	4	3	3
	5	7	11.6	112.1	132	42	13	11	10	9	8
	6	6	17.9	99.1	185	79	29	16	14	12	11
	7	5	24.5	85.7	218	117	86	72	56	39	28
	8	4	30.9	71.5	239	152	148	133	115	94	80
	9	3	36.8	56.1	252	182	204	190	170	145	128
	10	2	41.6	38.9	261	205	249	235	213	186	167
	11	1	44.9	20.1	265	219	278	264	242	213	193
	12		46.0	0.0	267	224	228	274	251	222	201
	Surface daily totals				3470	2236	2312	2124	1898	1624	1436
Jun 21	3	9	4.2	139.4	21	4	2	2	2	2	1
	4	8	9.0	126.4	93	27	10	9	8	7	6
	5	7	14.7	113.6	154	60	16	15	13	11	10
	6	6	21.0	100.8	194	96	34	19	17	14	13
	7	5	27.5	87.5	221	132	91	74	55	36	23
	8	4	34.0	73.3	239	166	150	133	112	88	73
	9	3	39.9	57.8	251	195	204	187	164	137	119
	10	2	44.9	40.4	258	217	247	230	206	177	157
	11	1	48.3	20.9	262	231	275	258	233	202	181
	12		49.5	0.0	263	235	284	267	242	211	189
	Surface daily totals				3650	2488	2342	2118	1862	1558	1356

BTUH/sq. ft. total insolation on surfaces[b]

South facing surface angle with horiz.

Table A2.6*f* Solar position and insolation values for 64 degrees north latitude[a] *(Continued)*

Date	Solar time AM	PM	Solar position Alt	Azm	Normal[c]	Horiz.	54	64	74	84	90
							South facing surface angle with horiz.				
Jul 21	4	8	6.4	125.3	53	13	6	5	5	4	4
	5	7	12.1	112.4	128	44	14	13	11	10	9
	6	6	18.4	99.4	179	81	30	17	16	13	12
	7	5	25.0	86.0	211	118	86	72	56	38	28
	8	4	31.4	71.8	231	152	146	131	113	91	77
	9	3	37.3	56.3	245	182	201	186	166	141	124
	10	2	42.2	39.2	253	204	245	230	208	181	162
	11	1	45.4	20.2	257	218	273	258	236	207	187
	12		46.6	0.0	259	223	282	267	245	216	195
	Surface daily totals				3372	2248	2280	2090	1864	1588	1400
Aug 21	5	7	4.6	108.8	29	6	3	3	2	2	2
	6	6	11.0	95.5	123	39	16	11	10	8	7
	7	5	17.6	81.9	181	77	69	61	52	42	35
	8	4	23.9	67.8	214	113	132	123	112	97	87
	9	3	29.6	52.6	234	144	190	182	169	150	138
	10	2	34.2	36.2	246	168	237	229	215	194	179
	11	1	37.2	18.5	252	183	268	260	244	222	205
	12		38.3	0.0	254	188	278	270	255	232	215
	Surface daily totals				2808	1646	2108	1008	1860	1662	1522
Sep 21	7	5	6.5	76.5	77	16	25	25	24	23	21
	8	4	12.7	72.6	163	51	92	92	90	85	81
	9	3	18.1	48.1	206	83	159	159	156	147	141
	10	2	22.3	32.7	229	108	212	213	209	198	189
	11	1	25.1	16.6	240	124	246	248	243	230	220
	12		26.0	0.0	244	129	258	260	254	241	230
	Surface daily totals				2074	892	1726	1736	1696	1608	1532
Oct 21	8	4	3.0	58.5	17	2	9	9	10	10	10
	9	3	8.1	44.6	122	26	86	91	93	92	90
	10	2	12.1	30.2	176	50	152	159	161	159	155
	11	1	14.6	15.2	201	65	193	201	203	200	195
	12		15.5	0.0	208	71	207	215	217	213	208
	Surface daily totals				1238	358	1088	1136	1152	1134	1106
Nov 21	10	2	3.0	28.1	23	3	18	20	21	21	21
	11	1	5.4	14.2	79	12	70	76	79	80	79
	12		6.2	0.0	97	17	89	96	100	101	100
	Surface daily totals				302	46	266	286	298	302	300
Dec 21	11	1	1.8	13.7	4	0	3	4	4	4	4
	12		2.6	0.0	16	2	14	15	16	17	17
	Surface daily totals				24	2	20	22	24	24	24

[a]From Kreider, J. F., and F. Keith, "Solar Heating and Cooling," revised 1st ed., Hemisphere Publ. Corp., 1977.

[b]1 Btu/hr · ft[2] = 3.152 W/m[2]. Ground reflection not included on normal or horizontal surfaces.

[c]Normal insolation does not include diffuse component.

Table A2.7 Reflectivity values for characteristic surfaces (integrated over solar spectrum and angle of incidence)[a]

Surface	Average reflectivity
Snow (freshly fallen or with ice film)	0.75
Water surfaces (relatively large incidence angles)	0.07
Soils (clay, loam, etc.)	0.14
Earth roads	0.04
Coniferous forest (winter)	0.07
Forests in autumn, ripe field crops, plants	0.26
Weathered blacktop	0.10
Weathered concrete	0.22
Dead leaves	0.30
Dry grass	0.20
Green grass	0.26
Bituminous and gravel roof	0.13
Crushed rock surface	0.20
Building surfaces, dark (red brick, dark paints, etc.)	0.27
Building surfaces, light (light brick, light paints, etc.)	0.60

[a]From Hunn, B. D., and D. O. Calafell, Determination of Average Ground Reflectivity for Solar Collectors, *Sol. Energy,* vol. 19, p. 87, 1977; see also R. J. List, "Smithsonian Meteorological Tables," 6th ed., Smithsonian Institution Press, pp. 442–443, 1949.

The altitude and azimuth of the sun are given by

$$\sin \alpha = \sin L \sin \delta_s + \cos \phi \cos \delta_s \cos h_s \qquad (1)$$

and

$$\sin a_s = - \cos \delta_s \sin h_s / \cos \alpha \qquad (2)$$

where α = altitude of the sun (angular elevation above the horizon)

  L = latitude of the observer

  δ_s = declination of the sun

  h_s = hour angle of sun (angular distance from the meridian of the observer)

  a_s = azimuth of the sun (measured eastward from north)

From Eqs. (1) and (2) it can be seen that the altitude and azimuth of the sun are functions of the latitude of the observer, the time of day (hour angle), and the date (declination).

Figure A2.1(*b–g*) provides a series of charts, one for each 5° of latitude (except 5°, 15°, 75°, and 85°) giving the altitude and azimuth of the sun as a function of the true solar time and the declination of the sun in a form originally suggested by Hand. Linear interpolation for intermediate latitudes will give results within the accuracy to which the charts can be read.

On these charts, a point corresponding to the projected position of the sun is determined from the heavy lines corresponding to declination and solar time.

To find the solar altitude and azimuth:

1. Select the chart or charts appropriate to the latitude.
2. Find the solar declination δ corresponding to the date.
3. Determine the *true solar time* as follows:
    (a) To the *local standard time* (zone time) add 4′ for each degree of longitude the station is east of the standard meridian or subtract 4′ for each degree west of the standard meridian to get the *local mean solar time*.
    (b) To the *local mean solar time* add algebraically the equation of time; the sum is the required *true solar time*.
4. Read the required altitude and azimuth at the point determined by the declination and the true solar time. Interpolate linearly between two charts for intermediate latitudes.

It should be emphasized that the solar altitude determined from these charts is the true geometric position of the center of the sun. At low solar elevations terrestrial refraction may considerably alter the apparent position of sun. Under average atmospheric refraction the sun will appear on the horizon when it actually is about 34′ below the horizon; the effect of refraction decreases rapidly with increasing solar elevation. Since sunset or sunrise is defined as the time when the upper limb of the sun appears on the horizon, and the semidiameter of the sun is 16′, sunset or sunrise occurs under average atmospheric refraction when the sun is 50′ below the horizon. In polar regions especially, unusual atmospheric refraction can make considerable variation in the time of sunset or sunrise.

The 90°N chart is included for interpolation purposes; the azimuths lose their directional significance at the pole.

Altitude and azimuth in southern latitudes. To compute solar altitude and azimuth for southern latitudes, change the sign of the solar declination and proceed as above. The resulting azimuths will indicate angular distance from *south* (measured eastward) rather than from north.

(a)

Figure A2.1 Description of method for calculating true solar time, together with accompanying meteorological charts, for computing solar-altitude and azimuth angles. (*a*) Description of method; (*b*) chart, 25°N latitude; (*c*) chart, 30°N latitude; (*d*) chart, 35°N latitude; (*e*) chart, 40°N latitude; (*f*) chart, 45°N latitude; (*g*) chart, 50°N latitude. Description and charts reproduced from the "Smithsonian Meteorological Tables" with permission from the Smithsonian Institute, Washington, D.C

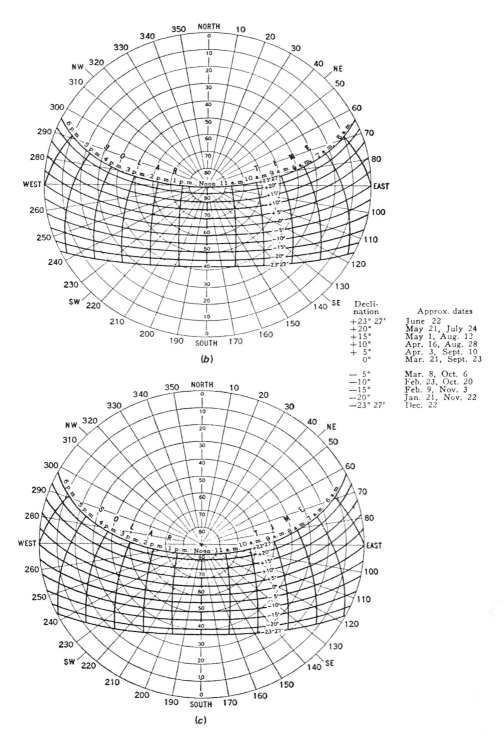

Decli- nation	Approx. dates
+23° 27′	June 22
+20°	May 21, July 24
+15°	May 1, Aug. 12
+10°	Apr. 16, Aug. 28
+ 5°	Apr. 3, Sept. 10
0°	Mar. 21, Sept. 23
— 5°	Mar. 8, Oct. 6
—10°	Feb. 23, Oct. 20
—15°	Feb. 9, Nov. 3
—20°	Jan. 21, Nov. 22
—23° 27′	Dec. 22

(b)

(c)

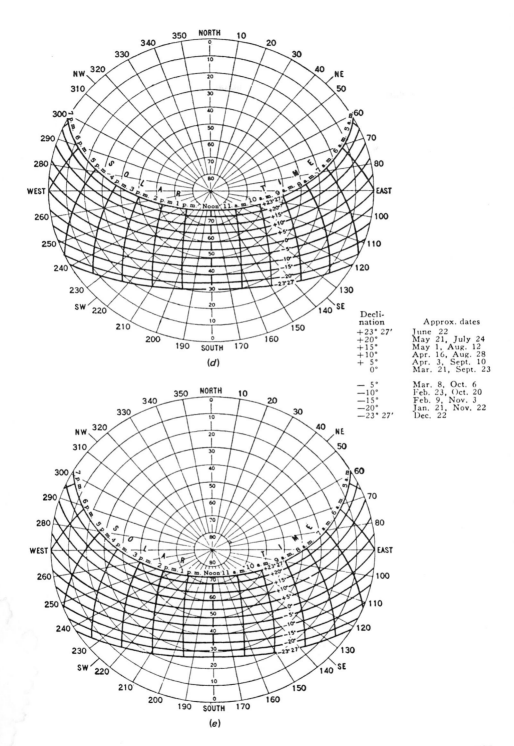

Decli- nation	Approx. dates
+23° 27′	June 22
+20°	May 21, July 24
+15°	May 1, Aug. 12
+10°	Apr. 16, Aug. 28
+ 5°	Apr. 3, Sept. 10
0°	Mar. 21, Sept. 23
— 5°	Mar. 8, Oct. 6
—10°	Feb. 23, Oct. 20
—15°	Feb. 9, Nov. 3
—20°	Jan. 21, Nov. 22
—23° 27′	Dec. 22

(d)

(e)

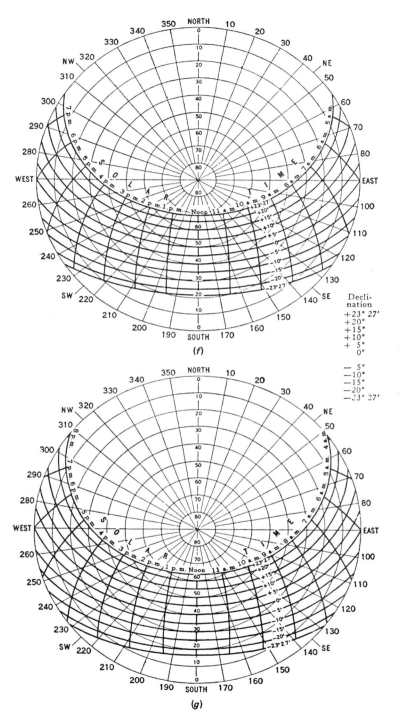

Decli- nation	Approx. dates
+23° 27′	June 22
+20°	May 21, July 24
+15°	May 1, Aug. 12
+10°	Apr. 16, Aug. 28
+ 5°	Apr. 3, Sept. 10
0°	Mar. 21, Sept. 23
— 5°	Mar. 8, Oct. 6
—10°	Feb. 23, Oct. 20
—15°	Feb. 9, Nov. 3
—20°	Jan. 21, Nov. 22
—23° 27′	Dec. 22

(f)

(g)

SUPPLEMENTARY MATERIAL FOR CHAPTER 3

Table A3.1 Properties of dry air at atmospheric pressures between 250 and 1000 K[a]

T^b (K)	ρ (kg/m³)	c_p (kJ/kg · K)	μ (kg/m · sec $\times 10^5$)	ν (m²/sec $\times 10^6$)	k (W/m · K)	α (m²/sec $\times 10^4$)	Pr
250	1.4128	1.0053	1.488	9.49	0.02227	0.13161	0.722
300	1.1774	1.0057	1.983	15.68	0.02624	0.22160	0.708
350	0.9980	1.0090	2.075	20.76	0.03003	0.2983	0.697
400	0.8826	1.0140	2.286	25.90	0.03365	0.3760	0.689
450	0.7833	1.0207	2.484	28.86	0.03707	0.4222	0.683
500	0.7048	1.0295	2.671	37.90	0.04038	0.5564	0.680
550	0.6423	1.0392	2.848	44.34	0.04360	0.6532	0.680
600	0.5879	1.0551	3.018	51.34	0.04659	0.7512	0.680
650	0.5430	1.0635	3.177	58.51	0.04953	0.8578	0.682
700	0.5030	1.0752	3.332	66.25	0.05230	0.9672	0.684
750	0.4709	1.0856	3.481	73.91	0.05509	1.0774	0.686
800	0.4405	1.0978	3.625	82.29	0.05779	1.1951	0.689
850	0.4149	1.1095	3.765	90.75	0.06028	1.3097	0.692
900	0.3925	1.1212	3.899	99.3	0.06279	1.4271	0.696
950	0.3716	1.1321	4.023	108.2	0.06525	1.5510	0.699
1000	0.3524	1.1417	4.152	117.8	0.06752	1.6779	0.702

[a]From *Natl. Bureau Standards (U.S.) Circ. 564*, 1955.

[b]Symbols: K = absolute temperature, degrees Kelvin; $v = \mu/\rho$; ρ = density; c_p = specific heat capacity; $\alpha = c_p\rho/k$; μ = viscosity; k = thermal conductivity; Pr = Prandtl number, dimensionless. The values of μ, k, c_p, and Pr are not strongly pressure-dependent and may be used over a fairly wide range of pressures.

Table A3.2 Properties of water (saturated liquid) between 273 and 533 K[a]

K	°F	°C	c_p (kJ/kg · °C)	ρ (kg/m³)	μ (kg/m · sec)	k (W/m · °C)	Pr	$\dfrac{g\beta\rho^2 c_p}{\mu k}$ (m⁻³ · °C⁻¹)
273	32	0	4.225	999.8	1.79×10^{-3}	0.566	13.25	
277.4	40	4.44	4.208	999.8	1.55	0.575	11.35	1.91×10^9
283	50	10	4.195	999.2	1.31	0.585	9.40	6.34×10^9
288.6	60	15.56	4.186	998.6	1.12	0.595	7.88	1.08×10^{10}
294.1	70	21.11	4.179	997.4	9.8×10^{-4}	0.604	6.78	1.46×10^{10}
299.7	80	26.67	4.179	995.8	8.6	0.614	5.85	1.91×10^{10}
302.2	90	32.22	4.174	994.9	7.65	0.623	5.12	2.48×10^{10}
310.8	100	37.78	4.174	993.0	6.82	0.630	4.53	3.3×10^{10}
316.3	110	43.33	4.174	990.6	6.16	0.637	4.04	4.19×10^{10}
322.9	120	48.89	4.174	988.8	5.62	0.644	3.64	4.89×10^{10}
327.4	130	54.44	4.179	985.7	5.13	0.649	3.30	5.66×10^{10}
333.0	140	60	4.179	983.3	4.71	0.654	3.01	6.48×10^{10}
338.6	150	65.55	4.183	980.3	4.3	0.659	2.73	7.62×10^{10}
342.1	160	71.11	4.186	977.3	4.01	0.665	2.53	8.84×10^{10}
349.7	170	76.67	4.191	973.7	3.72	0.668	2.33	9.85×10^{10}
355.2	180	82.22	4.195	970.2	3.47	0.673	2.16	1.09×10^{11}
360.8	190	87.78	4.199	966.7	3.27	0.675	2.03	
366.3	200	93.33	4.204	963.2	3.06	0.678	1.90	
377.4	220	104.4	4.216	955.1	2.67	0.684	1.66	
388.6	240	115.6	4.229	946.7	2.44	0.685	1.51	
399.7	260	126.7	4.250	937.2	2.19	0.685	1.36	
410.8	280	137.8	4.271	928.1	1.98	0.685	1.24	
421.9	300	148.9	4.296	918.0	1.86	0.684	1.17	
449.7	350	176.7	4.371	890.4	1.57	0.677	1.02	
477.4	400	204.4	4.467	859.4	1.36	0.665	1.00	
505.2	450	232.2	4.585	825.7	1.20	0.646	0.85	
533.0	500	260	4.731	785.2	1.07	0.616	0.83	

[a]Adapted from Brown, A. I., and S. M. Marco, "Introduction to Heat Transfer," 3d ed., McGraw-Hill Book Company, New York, 1958.

Table A3.3 Emittances and absorptances of materials[a]

Substance	Short-wave absorptance	Long-wave emittance	$\dfrac{\alpha}{\varepsilon}$
Class I substances: Absorptance to emittance ratios less than 0.5			
Magnesium carbonate, $MgCO_3$	0.025–0.04	0.79	0.03–0.05
White plaster	0.07	0.91	0.08
Snow, fine particles, fresh	0.13	0.82	0.16
White paint, 0.017 in, on aluminum	0.20	0.91	0.22
Whitewash on galvanized iron	0.22	0.90	0.24
White paper	0.25–0.28	0.95	0.26–0.29
White enamel on iron	0.25–0.45	0.9	0.28–0.5
Ice, with sparse snow cover	0.31	0.96–0.97	0.32
Snow, ice granules	0.33	0.89	0.37

Note: see page 618 for footnote.

Table A3.3 Emittances and absorptances of materials[a] (Continued)

Substance	Short-wave absorptance	Long-wave emittance	α / ε
Class I substances: Absorptance to emittance ratios less than 0.5			
Aluminum oil base paint	0.45	0.90	0.50
White powdered sand	0.45	0.84	0.54
Class II substances: Absorptance to emittance ratios between 0.5 and 0.9			
Asbestos felt	0.25	0.50	0.50
Green oil base paint	0.5	0.9	0.56
Bricks, red	0.55	0.92	0.60
Asbestos cement board, white	0.59	0.96	0.61
Marble, polished	0.5–0.6	0.9	0.61
Wood, planed oak	—	0.9	—
Rough concrete	0.60	0.97	0.62
Concrete	0.60	0.88	0.68
Grass, green, after rain	0.67	0.98	0.68
Grass, high and dry	0.67–0.69	0.9	0.76
Vegetable fields and shrubs, wilted	0.70	0.9	0.78
Oak leaves	0.71–0.78	0.91–0.95	0.78–0.82
Frozen soil	—	0.93–0.94	—
Desert surface	0.75	0.9	0.83
Common vegetable fields and shrubs	0.72–0.76	0.9	0.82
Ground, dry plowed	0.75–0.80	0.9	0.83–0.89
Oak woodland	0.82	0.9	0.91
Pine forest	0.86	0.9	0.96
Earth surface as a whole (land and sea, no clouds)	0.83	—	—
Class III substances: Absorptance to emittance ratios between 0.8 and 1.0			
Grey paint	0.75	0.95	0.79
Red oil base paint	0.74	0.90	0.82
Asbestos, slate	0.81	0.96	0.84
Asbestos, paper		0.93–0.96	—
Linoleum, red-brown	0.84	0.92	0.91
Dry sand	0.82	0.90	0.91
Green roll roofing	0.88	0.91–0.97	0.93
Slate, dark grey	0.89	—	—
Old grey rubber	—	0.86	—
Hard black rubber	—	0.90–0.95	—
Asphalt pavement	0.93	—	—
Black cupric oxide on copper	0.91	0.96	0.95
Bare moist ground	0.9	0.95	0.95
Wet sand	0.91	0.95	0.96
Water	0.94	0.95–0.96	0.98
Black tar paper	0.93	0.93	1.0
Black gloss paint	0.90	0.90	1.0
Small hole in large box, furnace, or enclosure	0.99	0.99	1.0
"Hohlraum," theoretically perfect black body	1.0	1.0	1.0

Table A3.3 Emittances and absorptances of materials[a] *(Continued)*

Substance	Short-wave absorptance	Long-wave emittance	α ε
Class IV substances: Absorptance to emittance ratios greater than 1.0			
Black silk velvet	0.99	0.97	1.02
Alfalfa, dark green	0.97	0.95	1.02
Lampblack	0.98	0.95	1.03
Black paint, 0.017 in, on aluminum	0.94–0.98	0.88	1.07–1.11
Granite	0.55	0.44	1.25
Graphite	0.78	0.41	1.90
High ratios, but absorptances less than 0.80			
Dull brass, copper, lead	0.2–0.4	0.4–0.65	1.63–2.0
Galvanized sheet iron, oxidized	0.8	0.28	2.86
Galvanized iron, clean, new	0.65	0.13	5.0
Aluminum foil	0.15	0.05	3.00
Magnesium	0.3	0.07	4.3
Chromium	0.49	0.08	6.13
Polished zinc	0.46	0.02	23.0
Deposited silver (optical reflector) untarnished	0.07	0.01	
Class V substances: Selective surfaces[b]			
Plated metals:[c]			
Black sulfide on metal	0.92	0.10	9.2
Black cupric oxide on sheet aluminum	0.08–0.93	0.09–0.21	
Copper (5×10^{-5} cm thick) on nickel or silver-plated metal			
Cobalt oxide on platinum			
Cobalt oxide on polished nickel	0.93–0.94	0.24–0.40	3.9
Black nickel oxide on aluminum	0.85–0.93	0.06–0.1	14.5–15.5
Black chrome	0.87	0.09	9.8
Particulate coatings:			
Lampblack on metal			
Black iron oxide, 47 μm grain size, on aluminum			
Geometrically enhanced surfaces:[d]			
Optimally corrugated greys	0.89	0.77	1.2
Optimally corrugated selectives	0.95	0.16	5.9
Stainless-steel wire mesh	0.63–0.86	0.23–0.28	2.7–3.0
Copper, treated with $NaClO_2$ and NaOH	0.87	0.13	6.69

[a]From Anderson, B., "Solar Energy," McGraw-Hill Book Company, 1977, with permission.

[b]Selective surfaces absorb most of the solar radiation between 0.3 and 1.9 μm, and emit very little in the 5–15 μm range—the infrared.

[c]For a discussion of plated selective surfaces, see Daniels, "Direct Use of the Sun's Energy," especially chapter 12.

[d]For a discussion of how surface selectivity can be enhanced through surface geometry, see K. G. T. Hollands, Directional Selectivity Emittance and Absorptance Properties of Vee Corrugated Specular Surfaces, *J. Sol. Energy Sci. Eng.,* vol. 3, July 1963.

Table A3.4 Thermal properties of metals and alloys[a]

Material	k, Btu/(hr)(ft)(°F) 32°F	212°F	572°F	932°F	c, Btu/(lb_m)(°F) 32°F	ρ, lb_m/ft³ 32°F	α, ft²/hr 32°F
Metals							
Aluminum	117	119	133	155	0.208	169	3.33
Bismuth	4.9	3.9	. . .	. . .	0.029	612	0.28
Copper, pure	224	218	212	207	0.091	558	4.42
Gold	169	170	. . .	. . .	0.030	1203	4.68
Iron, pure	35.8	36.6	. . .	. . .	0.104	491	0.70
Lead	20.1	19	18	. . .	0.030	705	0.95
Magnesium	91	92	. . .	. . .	0.232	109	3.60
Mercury	4.8	. . .	. . .	. . .	0.033	849	0.17
Nickel	34.5	34	32	. . .	0.103	555	0.60
Silver	242	238	. . .	. . .	0.056	655	6.6
Tin	36	34	. . .	. . .	0.054	456	1.46
Zinc	65	64	59	. . .	0.091	446	1.60
Alloys							
Admiralty metal	65	64					
Brass, 70% Cu, 30% Zn	56	60	66	. . .	0.092	532	1.14
Bronze, 75% Cu, 25% Sn	15	. . .	. . .	. . .	0.082	540	0.34
Cast iron							
Plain	33	31.8	27.7	24.8	0.11	474	0.63
Alloy	30	28.3	27	. . .	0.10	455	0.66
Constantan, 60% Cu, 40% Ni	12.4	12.8	. . .	. . .	0.10	557	0.22
18-8 Stainless steel, Type 304	8.0	9.4	10.9	12.4	0.11	488	0.15
Type 347	8.0	9.3	11.0	12.8	0.11	488	0.15
Steel, mild, 1% C	26.5	26	25	22	0.11	490	0.49

[a]From Kreith, F., "Principles of Heat Transfer," PWS Publishing Co., Boston, 1997.

Table A3.5 Thermal properties of some nonmetals[a]

Material	Average temperature, °F	$k,$ Btu/(hr)(ft)(°F)	$c,$ Btu/(lb$_m$)(°F)	$\rho,$ lb$_m$/ft³	$\alpha,$ ft²/hr
Insulating materials					
Asbestos	32	0.087	0.25	36	~0.01
	392	0.12	. . .	36	~0.01
Cork	86	0.025	0.04	10	~0.006
Cotton, fabric	200	0.046			
Diatomaceous earth,					
powdered	100	0.030	0.21	14	~0.01
	300	0.036	. . .		
	600	0.046	. . .		
Molded pipe covering	400	0.051	. . .	26	
	1600	0.088	. . .		
Glass wool					
Fine	20	0.022	. . .		
	100	0.031	. . .	1.5	
	200	0.043	. . .		
Packed	20	0.016	. . .		
	100	0.022	. . .	6.0	
	200	0.029	. . .		
Hair felt	100	0.027	. . .	8.2	
Kaolin insulating					
brick	932	0.15	. . .	27	
	2102	0.26	. . .		
Kaolin insulating					
firebrick	392	0.05	. . .	19	
	1400	0.11	. . .		
85% magnesia	32	0.032	. . .	17	
	200	0.037	. . .	17	
Rock wool	20	0.017	. . .	8	
	200	0.030	. . .		
Rubber	32	0.087	0.48	75	0.0024
Building materials					
Brick					
Fire-clay	392	0.58	0.20	144	0.02
	1832	0.95			
Masonry	70	0.38	0.20	106	0.018
Zirconia	392	0.84	. . .	304	
	1832	1.13	. . .		
Chrome brick	392	0.82	. . .	246	
	1832	0.96	. . .		
Concrete					
Stone	~70	0.54	0.20	144	0.019
10% Moisture	~70	0.70	. . .	140	~0.025
Glass, window	~70	~0.45	0.2	170	0.013
Limestone, dry	70	0.40	0.22	105	0.017
Sand					
Dry	68	0.20	. . .	95	
10% H$_2$O	68	0.60	. . .	100	

Table A3.5 Thermal properties of some nonmetals[a]

Material	Average temperature, °F	k, Btu/(hr)(ft)(°F)	c, Btu/(lb$_m$)(°F)	ρ, lb$_m$/ft^3	α, ft^2/hr
Building materials					
Soil					
Dry	70	~0.20	0.44	. . .	~0.01
Wet	70	~1.5	. . .	. . .	~0.03
Wood					
Oak ⊥ to grain	70	0.12	0.57	51	0.0041
‖ to grain	70	0.20	0.57	51	0.0069
Pine ⊥ to grain	70	0.06	0.67	31	0.0029
‖ to grain	70	0.14	0.67	31	0.0067
Ice	32	1.28	0.46	57	0.048

[a]From Kreith, F., "Principles of Heat Transfer," PWS Publishing Co., 1997.

Table A3.6 Thermal and radiative properties of collector cover materials[h]

Material name	Index of refraction (n)	τ (solar)[x] (%)	τ (solar)[y] (%)	τ (infrared)[b] (%)	Expansion coefficient (in/in · °F)	Temperature limits (°F)	Weatherability (comment)	Chemical resistance (comment)
Lexan (polycarbonate)	1.586 (D 542)[c]	125 mil 64.1 (±0.8)	125 mil 72.6 (±0.1)	125 mil 2.0 (est)[d]	$3.75 (10^{-5})$ (H 696)	250–270 service temperature	Good: 2 yr exposure in Florida caused yellowing; 5 yr caused 5% loss in τ	Good: comparable to acrylic
Plexiglas (acrylic)	1.49 (D 542)	125 mil 89.6 (±0.3)	125 mil 79.6 (±0.8)	125 mil 2.0 (est)[c]	$3.9 (10^{-9})$ at 60°F; $4.6 (10^{-6})$ at 100°F	180–200 service temperature	Average to good: based on 20 yr testing in Arizona, Florida, and Pennsylvania	Good to excellent: resists most acids and alkalis
Teflon F.E.P. (fluorocarbon)	1.343 (D 542)	5 mil 92.3 (±0.2)	5 mil 89.8 (±0.4)	5 mil 25.6 (±0.5)	$5.9 (10^{-5})$ at 160°F; $9.0 (10^{-5})$ at 212°F	400 continuous use; 475 short-term use	Good to excellent: based on 15 yr exposure in Florida environment	Excellent: chemically inert
Tedlar P.V.F. (fluorocarbon)	1.46 (D 542)	4 mil 92.2 (±0.1)	4 mil 88.3 (±0.9)	4 mil 20.7 (±0.2)	$2.8 (10^{-5})$ (D 696)	225 continuous use; 350 short-term use	Good to excellent: 10 yr exposure in Florida with slight yellowing	Excellent: chemically inert
Mylar (polyester)	1.64–1.67 (D 542)	5 mil 86.9 (±0.3)	5 mil 80.1 (±0.1)	5 mil 17.8 (±0.5)	$0.94 (10^{-5})$ (D 696-44)	300 continuous use; 400 short-term use	Poor: ultraviolet degradation great	Good to excellent: comparable to Tedlar
Sunlite[f] (fiberglass)	1.54 (D 542)	25 mil (P) 86.5 (±0.2) 25 mil (R) 87.5 (±0.2)	25 mil (P) 75.4 (±0.1) 25 mil (R) 77.1 (±0.7)	25 mil (P) 7.6 (±0.1) 25 mil (R) 3.3 (±0.3)	$1.4 (10^{-5})$ (D 696)	200 continuous use causes 5% loss in τ	Fair to good: regular, 7 yr solar life; premium, 20 yr solar life	Good: inert to chemical atmospheres
Float glass (glass)	1.518 (D 542)	125 mil 84.3 (±0.1)	125 mil 78.6 (±0.2)	125 mil 2.0 (est)[d]	$4.8 (10^{-5})$ (D 696)	1350 softening point; 100 thermal shock	Excellent: time proved	Good to excellent: time proved

Note: See page 615 for footnotes.

Table A3.7 **Thermal and radiative properties of collector cover materials**[h]

Material name	Index of refraction (n)	τ (solar)[g] (%)	τ (solar)[a] (%)	τ (infrared)[b] (%)	Expansion coefficient (in/in · °F)	Temperature limits (°F)	Weatherability (comment)	Chemical resistance (comment)
Temper glass (glass)	1.518 (D 542)	125 mil 84.3 (±0.1)	125 mil 78.6 (±0.2)	125 mil 2.0 (est)[d]	4.8 (10⁻⁶) (D 696)	450–500 continuous use; 500–550 short-term use	Excellent: time proved	Good to excellent: time proved
Clear lime sheet glass (low iron oxide glass)	1.51 (D 542)	Insufficient data provided by ASG	125 mil 87.5 (±0.5)	125 mil 2.0 (est)	5.0 (10⁻⁶) (D 696)	400 for continuous operation	Excellent: time proved	Good to excellent: time proved
Clear lime temper glass (low iron oxide glass)	1.51 (D 542)	Insufficient data provided by ASG	125 mil 87.5 (±0.5)	125 mil 2.0 (est)	5.0 (10⁻⁶) (D 696)	400 for continuous operation	Excellent: time proved	Good to excellent: time proved
Sunadex white crystal glass (0.01% iron oxide glass)	1.50 (D 542)	Insufficient data provided by ASG	125 mil 91.5 (±0.2)	125 mil 2.0 (est)	4.7 (10⁻⁶) (D 696)	400 for continuous operation	Excellent: time proved	Good to excellent: time proved

[a]Numerical integration ($\Sigma \tau_{avg} F_{\lambda_1 T - \lambda_2 T}$) for $\lambda = 0.2$–4.0μM.
[b]Numerical integration ($\Sigma \tau_{avg} F_{\lambda_1 T - \lambda_2 T}$) for $\lambda = 3.0$–50.0μM.
[c]All parenthesized numbers refer to ASTM test codes.
[d]Data not provided; estimate of 2% to be used for 125 mil samples.
[e]Degrees differential to rupture $2 \times 2 \times \frac{1}{4}$ in samples. Glass specimens heated and then quenched in water bath at 70°F.
[f]Sunlite premium data denoted by (P); Sunlite regular data denoted by (R).
[g]Compiled data based on ASTM Code E 424 Method B.
[h]Abstracted from Ratzel, A. C., and R. B. Bannerot, Optimal Material Selection for Flat-Plate Solar Energy Collectors Utilizing Commercially Available Materials, presented at ASME–AIChE Natl. Heat Transfer Conf., 1976.

Table A3.8 Saturated steam and water—SI units[a]

Temperature (K)	Pressure (MN/m²)	Specific volume (m³/kg)		Specific internal energy (kJ/kg)		Specific enthalpy (kJ/kg)			Specific entropy (kJ/kg · K)	
		v_f	v_g	u_f	u_g	h_f	h_{fg}	h_g	s_f	s_g
273.15	0.0006109	0.0010002	206.278	−0.03	2375.3	−0.02	2501.4	2501.3	−0.0001	9.1565
273.16	0.0006113	0.0010002	206.136	0	2375.3	+0.01	2501.3	2501.4	0	9.1562
278.15	0.0008721	0.0010001	147.120	+20.97	2382.3	20.98	2489.6	2510.6	+0.0761	9.0257
280.13	0.0010000	0.0010002	129.208	29.30	2385.0	29.30	2484.9	2514.2	0.1059	8.975
283.15	0.0012276	0.0010004	106.379	42.00	2389.2	42.01	2477.7	2519.8	0.1510	8.9008
286.18	0.0015000	0.0010007	87.980	54.71	2393.3	54.71	2470.6	2525.3	0.1957	8.8279
288.15	0.0017051	0.0010009	77.926	62.99	2396.1	62.99	2465.9	2528.9	0.2245	8.7814
290.65	0.0020000	0.0010013	67.004	73.48	2399.5	73.48	2460.0	2533.5	0.2607	8.7237
293.15	0.002339	0.0010018	57.791	83.95	2402.9	83.96	2454.1	2538.1	0.2966	8.6672
297.23	0.0030000	0.0010027	45.665	101.04	2408.5	101.05	2444.5	2545.5	0.3545	8.5776
298.15	0.003169	0.0010029	43.360	104.88	2409.8	104.89	2442.3	2547.2	0.3674	8.5580
302.11	0.004000	0.0010040	34.800	121.45	2415.2	121.46	2432.9	2554.4	0.4226	8.4746
303.15	0.004246	0.0010043	32.894	125.78	2416.6	125.79	2430.5	2556.3	0.4369	8.4533
306.03	0.005000	0.0010053	28.192	137.81	2420.5	137.82	2423.7	2561.5	0.4764	8.3951
308.15	0.005628	0.0010060	25.216	146.67	2423.4	146.68	2418.6	2565.3	0.5053	8.3531
309.31	0.006000	0.0010064	23.739	151.53	2425.0	151.53	2415.9	2567.4	0.5210	8.3304
312.15	0.007000	0.0010074	20.530	163.39	2428.8	163.40	2409.1	2572.5	0.5592	8.2758
313.15	0.007384	0.0010078	19.523	167.56	2430.1	167.57	2406.7	2574.3	0.5725	8.2570
314.66	0.008000	0.0010084	18.103	173.87	2432.2	173.88	2403.1	2577.0	0.5926	8.2287
316.91	0.009000	0.0010094	16.203	183.27	2435.2	183.29	2397.7	2581.0	0.6224	8.1872
318.15	0.009593	0.0010099	15.258	188.44	2436.8	188.45	2394.8	2583.2	0.6387	8.1648
318.96	0.010000	0.0010102	14.674	191.82	2437.9	191.83	2392.8	2584.7	0.6493	8.1502
323.15	0.012349	0.0010121	12.032	209.32	2443.5	209.33	2382.7	2592.1	0.7038	8.0763

Table A3.8 Saturated steam and water—SI units[a] (*Continued*)

Temperature (K)	Pressure (MN/m²)	Specific volume (m³/kg)		Specific internal energy (kJ/kg)		Specific enthalpy (kJ/kg)			Specific entropy (kJ/kg · K)	
		v_f	v_g	u_f	u_g	h_f	h_{fg}	h_g	s_f	s_g
327.12	0.015000	0.0010141	10.022	225.92	2448.7	225.94	2373.1	2599.1	0.7549	8.0085
328.15	0.015758	0.0010146	9.568	230.21	2450.1	230.23	2370.7	2600.9	0.7679	7.9913
333.15	0.019940	0.0010172	7.671	251.11	2456.6	251.13	2358.5	2609.6	0.8312	7.9096
333.21	0.020000	0.0010172	7.649	251.38	2456.7	251.40	2358.3	2609.7	0.8320	7.9085
338.15	0.025030	0.0010199	6.197	272.02	2463.1	272.06	2346.2	2618.3	0.8935	7.8310
342.25	0.030000	0.0010223	5.229	289.20	2468.4	289.23	2336.1	2625.3	0.9439	7.7686
343.15	0.031190	0.0010228	5.042	292.95	2469.6	292.98	2333.8	2626.8	0.9549	7.7553
348.15	0.038580	0.0010259	4.131	313.90	2475.9	313.93	2221.4	2635.3	1.0155	7.6824
349.02	0.040000	0.0010265	3.993	317.53	2477.0	317.58	2319.2	2636.8	1.0259	7.6700
353.15	0.047390	0.0010291	3.407	334.86	2482.2	334.91	2308.8	2643.7	1.0753	7.6122
354.48	0.050000	0.0010300	3.240	340.44	2483.9	340.49	2305.4	2645.9	1.0910	7.5939
358.15	0.057830	0.0010325	2.828	355.84	2488.4	355.90	2296.0	2651.9	1.1343	7.5445
359.09	0.060000	0.0010331	2.732	359.79	2489.6	359.86	2293.6	2653.5	1.1453	7.5320
363.10	0.070000	0.0010360	2.365	376.63	2494.5	376.70	2283.3	2660.0	1.1919	7.4797
363.15	0.070140	0.0010360	2.361	376.85	2494.5	376.92	2283.2	2660.1	1.1925	7.4791
366.65	0.080000	0.0010386	2.087	391.58	2498.8	391.66	2274.1	2665.8	1.2329	7.4346
368.15	0.084550	0.0010397	1.9819	397.88	2500.6	397.96	2270.2	2668.1	1.2500	7.4159

[a]Subscripts: f refers to a property of liquid in equilibrium with vapor; g refers to a property of vapor in equilibrium with liquid; fg refers to a change by evaporation.
Table from Bolz, R. E., and G. L. Tuve, eds., "CRC Handbook of Tables for Applied Engineering Science," 2nd ed., Chemical Rubber Co., Cleveland, Ohio, 1973.

Table A3.9 Superheated steam—SI units[a]

		Temperature									
Pressure (MN/m²) (saturation temperature)[b]		50°C 323.15 K	100°C 373.15 K	150°C 423.15 K	200°C 473.15 K	300°C 573.15 K	400°C 673.15 K	500°C 773.15 K	700°C 973.15 K	1000°C 1273.15 K	1300°C 1573.15 K
0.001 (6.98°C) (280.13 K)	v	149.093	172.187	195.272	218.352	264.508	310.661	356.814	449.117	587.571	726.025
	u	2445.4	2516.4	2588.4	2661.6	2812.2	2969.0	3132.4	3479.6	4053.0	4683.7
	h	2594.5	2688.6	2783.6	2880.0	3076.8	3279.7	3489.2	3928.7	4640.6	5409.7
	s	9.2423	9.5129	9.7520	9.9671	10.3443	10.6705	10.9605	11.4655	12.1019	12.6438
0.002 (17.50°C) (290.65 K)	v	74.524	86.081	97.628	109.170	132.251	155.329	178.405	224.558	293.785	363.012
	u	2445.2	2516.3	2588.3	2661.6	2812.2	2969.0	3132.4	3479.6	4053.0	4683.7
	h	2594.3	2688.4	2793.6	2879.9	3076.7	3279.7	3489.2	3928.7	4640.6	5409.7
	s	8.9219	9.1928	9.4320	9.6471	10.0243	10.3506	10.6406	11.1456	11.7820	12.3239
0.004 (28.96°C) (302.11 K)	v	37.240	43.028	48.806	54.580	66.122	77.662	89.201	112.278	146.892	181.506
	u	2444.9	2516.1	2588.2	2661.5	2812.2	2969.0	3132.3	3479.6	4053.0	4683.7
	h	2593.9	2688.2	2783.4	2879.8	3076.7	3279.6	3489.2	3928.7	4640.6	5409.7
	s	8.6009	8.8724	9.1118	9.3271	9.7044	10.0307	10.3207	10.8257	11.4621	12.0040
0.006 (36.16°C) (309.31 K)	v	24.812	28.676	32.532	36.383	44.079	51.774	59.467	74.852	97.928	121.004
	u	2444.6	2515.9	2588.1	2661.4	2812.2	2969.0	3132.3	3479.6	4053.0	4683.7
	h	2593.4	2688.0	2783.3	2879.7	3076.6	3279.6	3489.1	3928.7	4640.6	5409.7
	s	8.4128	8.6847	8.9244	9.1398	9.5172	9.8435	10.1336	10.6386	11.2750	11.8168
0.008 (41.51°C) (314.66 K)	v	18.598	21.501	24.395	27.284	33.058	38.829	44.599	56.138	73.446	90.753
	u	2444.2	2515.7	2588.0	2661.4	2812.1	2969.0	3132.3	3479.6	4053.0	4683.7
	h	2593.0	2687.7	2783.1	2879.6	3076.6	3279.6	3489.1	3928.7	4640.6	5409.7
	s	8.2790	8.5514	8.7914	9.0069	9.3844	9.7107	10.0008	10.5058	11.1422	11.6841
0.010 (45.81°C)	v	14.869	17.196	19.512	21.825	26.445	31.063	35.679	44.911	58.757	72.602
	u	2443.9	2515.5	2587.9	2661.3	2812.1	2968.9	3132.3	3479.6	4053.0	4683.7

Note: See page 620 for footnotes.

Table A3.9 Superheated steam—SI units^a (Continued)

Pressure (MN/m²) (saturation temperature)ᵇ		50°C 323.15 K	100°C 373.15 K	150°C 423.15 K	200°C 473.15 K	300°C 573.15 K	400°C 673.15 K	500°C 773.15 K	700°C 973.15 K	1000°C 1273.15 K	1300°C 1573.15 K
										Temperature	
(318.96 K)	h	2592.6	2687.5	2783.0	2879.5	3076.5	3279.6	3489.1	3928.7	4640.6	5409.7
	s	8.1749	8.4479	8.6882	8.9038	9.2813	9.6077	9.8978	10.4028	11.0393	11.5811
0.020	v	7.412	8.585	9.748	10.907	13.219	15.529	17.838	22.455	29.378	36.301
(60.06°C)	u	2442.2	2514.6	2587.3	2660.9	2811.9	2968.8	3132.2	3479.5	4053.0	4683.7
(333.21 K)	h	2590.4	2686.2	2782.3	2879.1	3076.3	3279.4	3489.0	3928.6	4640.6	5409.7
	s	7.8498	8.1255	8.3669	8.5831	8.9611	9.2876	9.5778	10.0829	10.7193	11.2612
0.040	v	3.683	4.279	4.866	5.448	6.606	7.763	8.918	11.227	14.689	18.151
(75.87°C)	u	2438.8	2512.6	2586.2	2660.2	2811.5	2968.6	3132.1	3479.4	4052.9	4683.6
(349.02 K)	h	2586.1	2683.8	2780.8	2878.1	3075.8	3279.1	3488.8	3928.5	4640.5	5409.6
	s	7.5192	7.8003	8.0444	8.2617	8.6406	8.9674	9.2577	9.7629	10.3994	10.9412
0.060	v	2.440	2.844	3.238	3.628	4.402	5.174	5.944	7.484	9.792	12.100
(85.94°C)	u	2435.3	2510.6	2585.1	2659.5	2811.2	2968.4	3131.9	3479.4	4052.9	4683.6
(359.09 K)	h	2581.7	2681.3	2779.4	2877.2	3075.3	3278.8	3488.6	3928.4	4640.4	5409.6
	s	7.3212	7.6079	7.8546	8.0731	8.4528	8.7799	9.0704	9.5757	10.2122	10.7541
0.080	v	1.8183	2.127	2.425	2.718	3.300	3.879	4.458	5.613	7.344	9.075
(93.50°C)	u	2431.7	2508.7	2583.9	2658.8	2810.8	2968.1	3131.7	3479.3	4052.8	4683.5
(366.65 K)	h	2577.2	2678.8	2777.9	2876.2	3074.8	3278.5	3488.3	3928.3	4640.4	5409.5
	s	7.1775	7.4698	7.7191	7.9388	8.3194	8.6468	8.9374	9.4428	10.0794	10.6213
0.100	v	1.4450	1.6958	1.9364	2.172	2.639	3.103	3.565	4.490	5.875	7.260
(99.63°C)	u	2428.2	2506.7	2582.8	2658.1	2810.4	2967.9	3131.6	3479.2	4052.8	4683.5
(372.78 K)	h	2572.7	2676.2	2776.4	2875.3	3074.3	3278.2	3488.1	3928.2	4640.3	5409.5
	s	7.0633	7.3614	7.6134	7.8343	8.2158	8.5435	8.8342	9.3398	9.9764	10.5183

Table A3.9 Superheated steam—SI units[a] (Continued)

Pressure (MN/m²) (saturation temperature)[b]		Temperature									
		50°C 323.15 K	100°C 373.15 K	150°C 423.15 K	200°C 473.15 K	300°C 573.15 K	400°C 673.15 K	500°C 773.15 K	700°C 973.15 K	1000°C 1273.15 K	1300°C 1573.15 K
0.200 (120.23°C) (393.38 K)	v	0.6969	0.8340	0.9596	1.0803	1.3162	1.5493	1.7814	2.244	2.937	3.630
	u	2409.5	2496.3	2576.9	2654.4	2808.6	2966.7	3130.8	3478.8	4052.5	4683.2
	h	2548.9	2663.1	2768.8	2870.5	3071.8	3276.6	3487.1	3927.6	4640.0	5409.3
	s	6.6844	7.0135	7.2795	7.5066	7.8926	8.2218	8.5133	9.0194	9.6563	10.1982
0.300 (133.55°C) (406.70 K)	v	0.4455	0.5461	0.6339	0.7163	0.8753	1.0315	1.1867	1.4957	1.9581	2.4201
	u	2389.1	2485.4	2570.8	2650.7	2806.7	2965.6	3130.0	3478.4	4052.3	4683.0
	h	2522.7	2649.2	2761.0	2865.6	3069.3	3275.0	3486.0	3927.1	4639.7	5409.0
	s	6.4319	6.7965	7.0778	7.3115	7.7022	8.0330	8.3251	8.8319	9.4690	10.0110
0.400 (143.63°C) (416.78 K)	v	0.3177	0.4017	0.4708	0.5342	0.6548	0.7726	0.8893	1.1215	1.4685	1.8151
	u	2366.3	2473.8	2564.5	2646.8	2804.8	2964.4	3129.2	3477.9	4052.0	4682.8
	h	2493.4	2634.5	2752.8	2860.5	3066.8	3273.4	3484.9	3926.5	4639.4	5408.8
	s	6.2248	6.6319	6.9299	7.1706	7.5662	7.8985	8.1913	8.6987	9.3360	9.8780
0.500 (151.86°C) (425.01 K)	v		0.3146	0.3729	0.4249	0.5226	0.6173	0.7109	0.8969	1.1747	1.4521
	u		2461.5	2557.9	2642.9	2802.9	2963.2	3128.4	3477.5	4051.8	4682.5
	h		2618.7	2744.4	2855.4	3064.2	3271.9	3483.9	3925.9	4639.1	5408.6
	s		6.4945	6.8111	7.0592	7.4599	7.7938	8.0873	8.5952	9.2328	9.7749

[a]From Bolz, R. E., and G. L. Tuve, eds., "CRC Handbook of Tables for Applied Engineering Science," 2nd ed., Chemical Rubber Co., Cleveland, Ohio, 1973.
[b]Symbols: v = specific volume, m³/kg; u = specific internal energy, kJ/kg; h = specific enthalpy, kJ/kg; s = specific entropy, kJ/K · kg.

Table A3.10 Normal distribution function

$$F(z) = \frac{1}{\sqrt{2\pi}} \int_{-\infty}^{z} e^{-1/2 t^2} \, dt$$

z	0.00	0.01	0.02	0.03	0.04	0.05	0.06	0.07	0.08	0.09
0.0	0.5000	0.5040	0.5080	0.5120	0.5160	0.5199	0.5239	0.5279	0.5319	0.5359
0.1	0.5398	0.5438	0.5478	0.5517	0.5557	0.5596	0.5636	0.5675	0.5714	0.5753
0.2	0.5793	0.5832	0.5871	0.5910	0.5948	0.5987	0.6026	0.6064	0.6103	0.6141
0.3	0.6179	0.6217	0.6255	0.6293	0.6331	0.6368	0.6406	0.6443	0.6480	0.6517
0.4	0.6554	0.6591	0.6628	0.6664	0.6700	0.6736	0.6772	0.6808	0.6844	0.6879
0.5	0.6915	0.6950	0.6985	0.7019	0.7054	0.7088	0.7123	0.7157	0.7190	0.7224
0.6	0.7257	0.7291	0.7324	0.7357	0.7389	0.7422	0.7454	0.7486	0.7517	0.7549
0.7	0.7580	0.7611	0.7642	0.7673	0.7704	0.7734	0.7764	0.7794	0.7823	0.7852
0.8	0.7881	0.7910	0.7939	0.7967	0.7995	0.8023	0.8051	0.8078	0.8106	0.8133
0.9	0.8159	0.8186	0.8212	0.8238	0.8264	0.8289	0.8315	0.8340	0.8365	0.8389
1.0	0.8413	0.8438	0.8461	0.8485	0.8508	0.8531	0.8554	0.8577	0.8599	0.8621
1.1	0.8643	0.8665	0.8686	0.8708	0.8729	0.8749	0.8770	0.8790	0.8810	0.8830
1.2	0.8849	0.8869	0.8888	0.8907	0.8925	0.8944	0.8962	0.8980	0.8997	0.9015
1.3	0.9032	0.9049	0.9066	0.9082	0.9099	0.9115	0.9131	0.9147	0.9162	0.9177
1.4	0.9192	0.9207	0.9222	0.9236	0.9251	0.9265	0.9279	0.9292	0.9306	0.9319
1.5	0.9332	0.9345	0.9357	0.9370	0.9382	0.9394	0.9406	0.9418	0.9429	0.9441
1.6	0.9452	0.9463	0.9474	0.9484	0.9495	0.9505	0.9515	0.9525	0.9535	0.9545
1.7	0.9554	0.9564	0.9573	0.9582	0.9591	0.9599	0.9608	0.9616	0.9625	0.9633
1.8	0.9641	0.9649	0.9656	0.9664	0.9671	0.9678	0.9686	0.9693	0.9699	0.9706
1.9	0.9713	0.9719	0.9726	0.9732	0.9738	0.9744	0.9750	0.9756	0.9761	0.9767
2.0	0.9772	0.9778	0.9783	0.9788	0.9793	0.9798	0.9803	0.9808	0.9812	0.9817
2.1	0.9821	0.9826	0.9830	0.9834	0.9838	0.9842	0.9846	0.9850	0.9854	0.9857
2.2	0.9861	0.9864	0.9868	0.9871	0.9875	0.9878	0.9881	0.9884	0.9887	0.9890
2.3	0.9893	0.9896	0.9898	0.9901	0.9904	0.9906	0.9909	0.9911	0.9913	0.9916
2.4	0.9918	0.9920	0.9922	0.9925	0.9927	0.9929	0.9931	0.9932	0.9934	0.9936
2.5	0.9938	0.9940	0.9941	0.9943	0.9945	0.9946	0.9948	0.9949	0.9951	0.9952
2.6	0.9953	0.9955	0.9956	0.9957	0.9959	0.9960	0.9961	0.9962	0.9963	0.9964
2.7	0.9965	0.9966	0.9967	0.9968	0.9969	0.9970	0.9971	0.9972	0.9973	0.9974
2.8	0.9974	0.9975	0.9976	0.9977	0.9977	0.9978	0.9979	0.9979	0.9980	0.9981
2.9	0.9981	0.9982	0.9982	0.9983	0.9984	0.9984	0.9985	0.9985	0.9986	0.9986
3.0	0.9987	0.9987	0.9987	0.9988	0.9988	0.9989	0.9989	0.9989	0.9990	0.9990
3.1	0.9990	0.9991	0.9991	0.9991	0.9992	0.9992	0.9992	0.9992	0.9993	0.9993
3.2	0.9993	0.9993	0.9994	0.9994	0.9994	0.9994	0.9994	0.9995	0.9995	0.9995
3.3	0.9995	0.9995	0.9995	0.9996	0.9996	0.9996	0.9996	0.9996	0.9996	0.9997
3.4	0.9997	0.9997	0.9997	0.9997	0.9997	0.9997	0.9997	0.9997	0.9997	0.9998

SUPPLEMENTARY MATERIAL FOR CHAPTER 4

Table A4.1 Toxicological properties of common glycols used as antifreezes in solar systems[a]

	Single oral LD50 dose in rats (ml/kg)	Repeated oral feeding in rats acceptable level in diet and duration	Single skin penetration LD50 dose in rabbits (ml/kg)	Single inhalation concentrated vapor (8 hr) in rats	Primary skin irritation in rabbits	Eye injury in rabbits
Ethylene glycol	7.40[b]	0.18 g/kg/day (30 days)	>20	Killed none of 6	None	None
Diethylene glycol	28.3	0.18 g/kg/day (30 days)	11.9	Killed none of 6	None	None
Triethylene glycol	28.2	0.83 g/kg/day (30 days)	>20	Killed none of 6	None	None
Tetraethylene glycol	28.9	1.88 g/kg/day (2 yr)	>20	Killed none of 6	None	None
Propylene glycol	34.6	2.0 g/kg/day (2 yr)[c]	>20	Killed none of 6	None	Trace
Dipropylene glycol	14.8	—	>20	Killed none of 6	None	Trace
Hexylene glycol	4.06	0.31 g/kg/day (90 days)	8.56	Killed none of 6	Trace	Minor
2-Ethyl-1,3-hexanediol	6.50	0.48 g/kg/day (90 days)	15.2	Killed none of 6	Trace	Moderate
1,5-Pentanediol	5.89[d]	—	>20	Killed none of 6	None	Trace

[a]The term LD50 refers to that quantity of chemical that kills 50 percent of dosed animals within 14 days. For uniformity, dosage is expressed in grams or milliliters per kilogram of body weight. Single skin penetration refers to a 24-hr covered skin contact with the liquid chemical. Single inhalation refers to the continuous breathing of a certain concentration of chemical for the stated period of time. Primary irritation refers to the skin response 24 hr after application of 0.01-ml amounts to uncovered skin. Eye injury refers to surface damage produced by the liquid chemical. Table from Union Carbide, Glycols, F-41515A 7/71-12M, p. 68, 1971.

[b]Single dose oral toxicity to humans is greater.

[c]Dogs.

[d]g/kg.

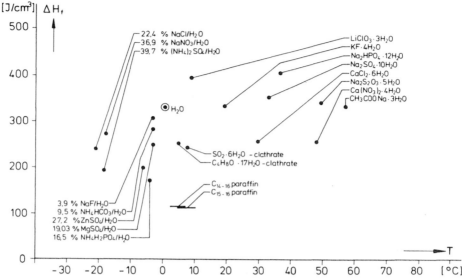

Figure A4.1 Enthalpy of fusion and phase change temperatures of low-temperature phase-change storage media. From Schröder, J., Philips GmbH Forschungslaboratorium (Aachen) Report, 1977.

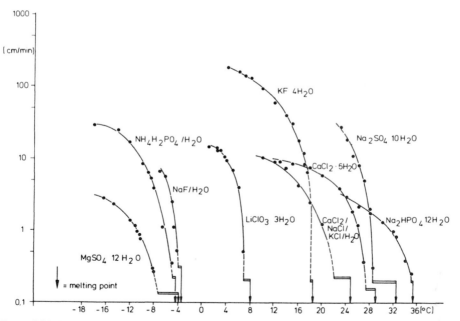

Figure A4.2 Rate of crystallization of several low-temperature hydrates and eutectics. Note that $Na_2SO_4 \cdot 10H_2O$ and $CaCl_2/NaCl/KCl/H_2O$ have very inadequate rates without substantial subcooling. From Schröder, J., Philips GmbH Forschungslaboratorium (Aachen) Report, 1977.

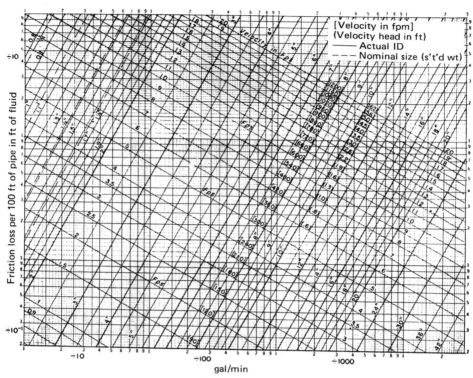

Figure A4.3 Pressure drop of water in turbulent flow for standard pipe sizes. (1 gal/min = 3.79 × 10⁻³ m³/min; 1 ft = 0.305 m; 1 in = 2.54 cm.) From Potter, Philip J., "Power Plant Theory and Design," 2nd ed. Copyright © 1959 The Ronald Press Company, New York.

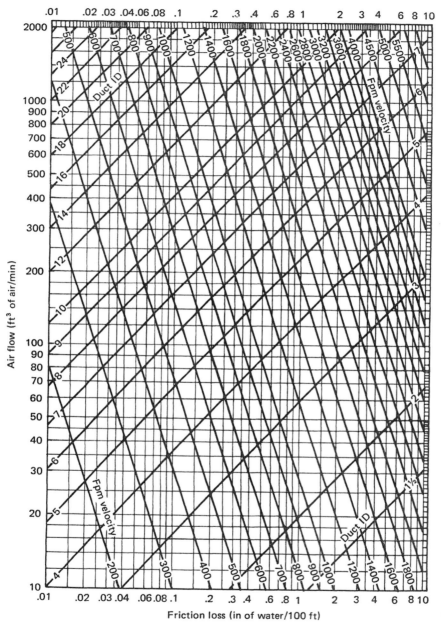

Figure A4.4 Pressure drop of air flowing in ducts; velocities shown for circular ducts. (1 in H₂O = 249 N/m²; 1 ft³ = 2.83 × 10⁻² m³; 1 in = 2.54 cm; 1 ft - 0.305 m.) From ASHRAE, "Handbook of Fundamentals," American Society of Heating, Refrigerating, and Air Conditioning Engineers, New York, 1972.

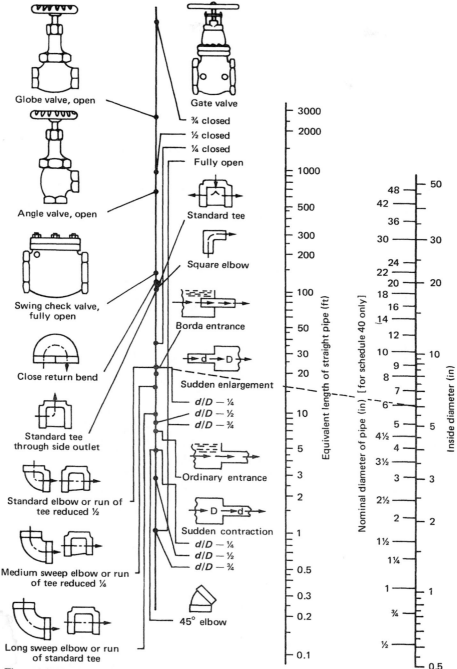

Figure A4.5 Equivalent lengths of pipe, standard pipe fittings. (1 ft = 0.305 m; 1 in = 2.54 cm.) For sudden enlargements and sudden contractions the equivalent length is in feet of pipe of the smaller diameter *d*. Dashed line shows determination of equivalent length of a 6-in standard elbow. From the Crane Co. Technical Paper 410, "Flow of Fluids," 1957.

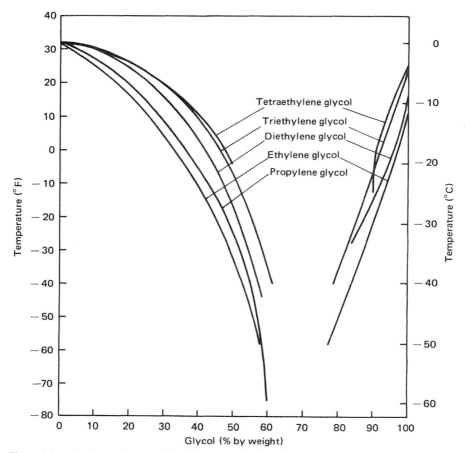

Figure A4.6 Freezing-point depression of aqueous solutions of common glycol antifreezes. From Union Carbide, Glycols, F-41515A 7/71-12M, p. 17, 1971.

SUPPLEMENTARY MATERIAL FOR CHAPTER 5

Appendix **5**

Table A5.1 Conductance and resistance values for external air surfaces[a]

Wind condition / Position of surface	Direction of heat flow	Foil		Aluminum-coated paper		Nonreflective building materials	
		Conductance C, Btu/(hr)(ft²)(°F)	Resistance R, 1/[Btu/(hr)(ft²)(°F)]	Conductance C, Btu/(hr)(ft²)(°F)	Resistance R, 1/[Btu/(hr)(ft²)(°F)]	Conductance C, Btu/(hr)(ft²)(°F)	Resistance R, 1/[Btu/(hr)(ft²)(°F)]
Still air:							
Horizontal	Up	0.76	1.32	0.91	1.10	1.63	0.61
45° slope	Up	0.73	1.37	0.88	1.14	1.60	0.62
Vertical	Horizontal	0.59	1.70	0.74	1.35	1.46	0.68
45° slope	Down	0.45	2.22	0.60	1.67	1.32	0.76
Horizontal	Down	0.22	4.55	0.37	2.70	1.08	0.92
7.5-mph wind							
Any position	(for summer calculations)	...	...	...	...	4.00	0.25
15-mph wind							
Any position	(for winter calculations)	...	...	...	...	6.00	0.17

[a]Adapted from Johns-Mansville, Denver, Colorado.

Table A5.2 Conductance and resistance values for internal air surfaces[a]

				Type of surface					
				Foil and nonreflective building materials		Aluminum-coated paper and nonreflective bldg. materials		Both surfaces non-reflective building materials	
Position of air space	Direction of heat flow[b]	Thickness, in.	Temp. cond.[c]	Conductance C, Btu/(hr) (ft²)(°F)	Resistance R, 1/[Btu/(hr) (ft²)(°F)]	Conductance C, Btu/(hr) (ft²)(°F)	Resistance R, 1/[Btu/(hr) (ft²)(°F)]	Conductance C, Btu/(hr) (ft²)(°F)	Resistance R, 1/[Btu/(hr) (ft²)(°F)]
Horizontal	Up	$\frac{3}{4}$	W	0.45	2.23	0.59	1.71	1.15	0.87
		$\frac{3}{4}$	S	0.44	2.26	0.61	1.63	1.32	0.76
		4	W	0.37	2.73	0.50	1.99	1.07	0.94
		4	S	0.36	2.75	0.53	1.87	1.24	0.80
45° slope	Up	$\frac{3}{4}$	W	0.36	2.78	0.50	2.02	1.06	0.94
		$\frac{3}{4}$	S	0.36	2.81	0.53	1.90	1.24	0.81
		4	W	0.33	3.00	0.47	2.13	1.04	0.96
		4	S	0.33	3.00	0.51	1.98	1.21	0.82
Vertical	Horiz.	$\frac{3}{4}$	W	0.29	3.48	0.42	2.36	0.99	1.01
		$\frac{3}{4}$	S	0.31	3.28	0.48	2.10	1.19	0.84
		4	W	0.29	3.45	0.43	2.34	0.99	1.01
		4	S	0.29	3.44	0.46	2.16	1.17	0.91
45° slope	Down	$\frac{3}{4}$	W	0.28	3.57	0.42	2.40	0.98	1.02
		$\frac{3}{4}$	S	0.31	3.24	0.48	2.09	1.19	0.84
		4	W	0.23	4.41	0.36	2.75	0.93	1.08
		4	S	0.23	4.36	0.40	2.50	1.11	0.90
Horizontal	Down	$\frac{3}{4}$	W	0.28	3.55	0.42	2.39	0.98	1.02
		$1\frac{1}{2}$	W	0.17	5.74	0.31	3.21	0.88	1.14

Table A5.2 Conductance and resistance values for internal air surfaces[a] (Continued)

Position of air space	Direction of heat flow[b]	Thickness, in.	Temp. cond.[c]	Foil and nonreflective building materials		Aluminum-coated paper and nonreflective bldg. materials		Both surfaces nonreflective building materials	
				Conductance C, Btu/(hr) (ft²)(°F)	Resistance R, 1/[Btu/(hr) (ft²)(°F)]	Conductance C, Btu/(hr) (ft²)(°F)	Resistance R, 1/[Btu/(hr) (ft²)(°F)]	Conductance C, Btu/(hr) (ft²)(°F)	Resistance R, 1/[Btu/(hr) (ft²)(°F)]
		4	W	0.11	8.94	0.25	4.02	0.81	1.23
		$\frac{3}{4}$	S	0.31	3.25	0.48	2.08	1.19	0.84
		$1\frac{1}{2}$	S	0.19	5.24	0.36	2.76	1.07	0.93
		4	S	0.12	8.08	0.30	3.38	1.01	0.99

Type of surface

[a]Adapted from Johns-Manville, Denver, Colorado.

[b]Heat flows from hot to cold. For ceiling instillation the direction of heat flow would normally be "up" for winter and "down" for summer. In a floor the direction of heat flow would be "down" in winter and "up" in summer. Heat flow in walls would be in a horizontal direction.

[c]W = winter; S = summer.

Table A5.3 Conductance and resistance values for exterior siding materials[a]

Material	Description	Conductivity k, Btu/(hr)(ft²) (°F/in.)	Thickness, in.	Conductance C, Btu/(hr) (ft²)(°F)	Resistance R, 1/[Btu/(hr) (ft²)(°F)]
Brick	Common	5.0	4	1.25	0.80
Brick	Face	9.0	4	2.27	0.44
Stucco		5.0	1	5.0	0.20
Asbestos cement shingles				4.76	0.21
Wood shingles	$16-7\frac{1}{2}$-in. exposure			1.15	0.87
Wood shingles	Double 16–12 in. exposure			0.84	1.19
Wood shingles	Plus $\frac{5}{16}$ in. Insulated backerboard			0.71	1.40
Asbestos cement siding	$\frac{1}{4}$ in. lapped			4.76	0.21
Asphalt roll siding				6.50	0.15
Asphalt insulating siding			$\frac{1}{2}$	0.69	1.46
Wood	Drop siding, 1 × 8 in.			1.27	0.79
Wood	Bevel, $\frac{1}{2}$ × 8 in. lapped			1.23	0.81
Wood	Bevel, $\frac{3}{4}$ × 10 in. lapped			0.95	1.05
Wood	Plywood, $\frac{3}{8}$ in. lapped			1.59	0.59
Hardboard	Medium density	0.73	$\frac{1}{4}$	2.94	0.34
	Tempered	1.00	$\frac{1}{4}$	4.00	0.25
Plywood lap siding			$\frac{3}{8}$	1.79	0.56
Plywood flat siding			$\frac{3}{8}$	2.33	0.43

[a]Adapted from Johns-Mansville, Denver, Colorado.

Table A5.4 Conductance and resistance values for sheathing and building paper[a]

Material	Description	Conductivity k, Btu/(hr)(ft²)(°F/in.)	Thickness, in.	Conductance C, Btu/(hr)(ft²)(°F)	Resistance R, 1/[Btu/(hr)(ft²)(°F)]
Gypsum	. . .	1.11	$\frac{3}{8}$	3.10	0.32
			$\frac{1}{2}$	2.25	0.45
			$\frac{5}{8}$	1.75	0.57
Plywood	. . .	0.80	$\frac{1}{4}$	3.20	0.31
			$\frac{3}{8}$	2.13	0.47
			$\frac{1}{2}$	1.60	0.62
			$\frac{5}{8}$	1.28	0.78
			$\frac{3}{4}$	1.07	0.93
Nail-base sheathing	. . .	0.44	$\frac{1}{2}$	0.88	1.14
Wood sheathing	Fir or pine	0.80	$\frac{3}{4}$	1.06	0.94
Sheathing paper	Vapor-permeable			16.70	0.06
Vapor barrier	2 layers mopped 15-lb felt			8.35	0.12
	Plastic film			Negl.	Negl.

[a]Adapted from Johns-Mansville, Denver, Colorado.

Table A5.5 Conductance and resistance values for masonry materials[a]

Material	Description	Conductivity k, Btu/(hr) (ft²)(°F/in.)	Thickness, in.	Conductance C, Btu/(hr) (ft²)(°F)	Resistance R, 1/[Btu/(hr) (ft²)(°F)]
Concrete blocks, three-oval core	Sand and gravel aggregate	. . .	4	1.40	0.71
			8	0.90	1.11
			12	0.78	1.28
	Cinder aggregate	. . .	4	0.90	1.11
			8	0.58	1.72
			12	0.53	1.89
	Lightweight aggregate	. . .	4	0.67	1.50
			8	0.50	2.00
			12	0.44	2.27
Hollow clay tile	1 cell deep	. . .	4	0.90	1.11
	2 cells deep	. . .	8	0.54	1.85
	3 cells deep	. . .	12	0.40	2.50
Gypsum partition tile	3 × 12 × 30 in. solid	. . .	3	0.79	1.26
	3 × 12 × 30 in. 4-cell	. . .	3	0.74	1.35
	4 × 12 × 30 in. 3-cell	. . .	4	0.60	1.67
Cement mortar		5.0	1	5.0	0.20
Stucco		5.0	1	5.0	0.20
Gypsum	Poured	1.66	1	1.66	0.60
	Precast	2.80	2	1.40	0.71
Concrete	Sand and gravel or stone	12.0	1	12.0	0.08
Lightweight concrete	Perlite or zonolite mixture				
	1 : 4 mix, 36 lb/ft³	0.72–0.75	1	0.74	1.35
	1 : 5 mix, 30 lb/ft³	0.61–0.72	1	0.67	1.49
	1 : 6 mix, 27 lb/ft³	0.54–0.61	1	0.58	1.72
	1 : 8 mix, 22 lb/ft³	0.47–0.54	1	0.51	1.96
Stone	. . .	12.5	1	12.5	0.08

[a]Adapted from Johns-Mansville, Denver, Colorado.

Table A5.6 Conductance and resistance values for woods[a]

Material	Description	Conductivity k, Btu/(hr) (ft²)(°F/in.)	Thickness, in.	Conductance C, Btu/(hr) (ft²)(°F)	Resistance R, 1/[Btu/(hr) (ft²)(°F)]
Maple, oak and similar hardwoods	45 lb/ft³	1.10	$\frac{3}{4}$	1.47	0.68
Fir, pine and similar softwoods	32 lb/ft³	C.80	$\frac{3}{4}$	1.06	0.94
			$1\frac{1}{2}$	0.53	1.89
			$2\frac{1}{2}$	0.32	3.12
			$3\frac{1}{2}$	0.23	4.35

[a]Adapted from Johns-Mansville, Denver, Colorado.

Table A5.7 Conductance and resistance values for wall-insulation materials[a]

Material	Description	Conductivity k, Btu/(hr) (ft²)(°F/in.)	Thickness, in.	Conductance C, Btu/(hr) (ft²)(°F)	Resistance R, 1/[Btu/(hr) (ft²)(°F)]
Fiber glass roof insulation	. . .	. . .	$\frac{15}{16}$	0.27	3.70
			$1\frac{1}{16}$	0.24	4.17
			$1\frac{5}{16}$	0.19	5.26
			$1\frac{5}{8}$	0.15	6.67
			$1\frac{7}{8}$	0.13	7.69
			$2\frac{1}{4}$	0.11	9.09
Urethane roof insulation	Thickness includes membrane roofing on both sides	0.13	$\frac{4}{5}$	0.19	5.26
			1	0.15	6.67
			$1\frac{1}{5}$	0.12	8.33
Styrofoam SM & TG	2.1 lb/ft³	0.19	$\frac{3}{4}$	0.25	3.93
			1	0.19	5.26
			$1\frac{1}{2}$	0.13	7.89
			2	0.95	10.52
Wood shredded	Cemented in preformed slabs	0.60	1	0.60	1.67
Insulating board	Building and service board, decorative ceiling panels	0.38	$\frac{3}{8}$	1.01	0.99
			$\frac{1}{2}$	0.76	1.32
			$\frac{9}{16}$	0.68	1.48
			$\frac{3}{4}$	0.51	1.98
Thermal, acoustical fiber glass	. . .	0.39	$2\frac{3}{4}$	0.14	7.00
		0.36	4	0.09	11.00
		0.34	$6\frac{1}{2}$	0.05	19.00
Corkboard	6.4 lb/ft³	0.26	1	0.26	3.85
Expanded polystyrene					
Extruded	1.8 lb/ft³	0.25	1	0.25	4.00
Molded beads	1.0 lb/ft³	0.26	1	0.26	3.85
Urethane foam					
Thurane (Dow Chemical)	1.9 lb/ft³	0.17	$\frac{3}{4}$	0.23	4.41
			$1\frac{1}{2}$	0.11	8.82
			2	0.09	11.76
Fiberglas perimeter insulation	. . .	. . .	1	0.23	4.30
			$1\frac{1}{4}$	0.19	5.40
Fiberglas form board	. . .	. . .	1	0.25	4.00

[a]Adapted from Johns-Mansville, Denver, Colorado.

Table A5.8 Conductance and resistance values for roofing materials[a]

Material	Description	Conductivity k, Btu/(hr) (ft²)(°F/in.)	Thickness, in.	Conductance C, Btu/(hr) (ft²)(°F)	Resistance R, 1/[Btu/(hr) (ft²)(°F)]
Asbestos cement shingles	120 lb/ft³	. . .	. . .	4.76	0.21
Asphalt shingles	70 lb/ft³	. . .	. . .	2.27	0.44
Wood shingles	. . .	. . .	. . .	1.06	0.94
Slate	. . .	. . .	$\frac{1}{2}$	20.0	0.05
Asphalt roll roofing	70 lb/ft³	. . .	. . .	6.50	0.15
Built-up roofing	Smooth or gravel surface	. . .	$\frac{3}{8}$	3.00	0.33
Sheet metal	. . .	. . .	. . .	Negl.	Negl.

[a]Adapted from Johns-Mansville, Denver, Colorado.

Table A5.9 Conductance and resistance values for flooring materials[a]

Material	Description	Conductivity k, Btu/(hr) (ft²)(°F/in.)	Thickness, in.	Conductance C, Btu/(hr) (ft²)(°F)	Resistance R, 1/[Btu/(hr) (ft²)(°F)]
Asphalt, vinyl, rubber, or linoleum tile	. . .	. . .	. . .	20.0	0.05
Cork tile	. . .	0.45	$\frac{1}{8}$	3.60	0.28
Terrazzo	. . .	12.5	1	12.50	0.08
Carpet and fibrous pad	. . .	. . .	. . .	0.48	2.08
Carpet and rubber pad	. . .	. . .	. . .	0.81	1.23
Plywood subfloor	. . .	0.80	$\frac{5}{8}$	1.28	0.78
Wood subfloor	. . .	0.80	$\frac{3}{4}$	1.06	0.94
Wood, hardwood finish	. . .	1.10	$\frac{3}{4}$	1.47	0.68

[a]Adapted from Johns-Mansville, Denver, Colorado.

Table A5.10 Conductance and resistance values for interior finishes[a]

Material	Description	Conductivity k, Btu/(hr) (ft²)(°F/in.)	Thickness, in.	Conductance C, Btu/(hr) (ft²)(°F)	Resistance R, 1/[Btu/(hr) (ft²)(°F)]
Gypsum board		1.11	$\frac{3}{8}$	3.10	0.32
			$\frac{1}{2}$	2.25	0.45
Cement plaster	Sand aggregate	5.0	$\frac{1}{2}$	10.00	0.10
			$\frac{3}{4}$	6.66	0.15
Gypsum plaster	Sand aggregate	5.6	$\frac{1}{2}$	11.10	0.09
			$\frac{5}{8}$	9.10	0.11
Gypsum plaster	Lightweight aggregate	1.6	$\frac{1}{2}$	3.12	0.32
			$\frac{5}{8}$	2.67	0.39
Gypsum plaster on					
Metal lath	Sand aggregate		$\frac{3}{4}$	7.70	0.13
Metal lath	Lightweight aggregate		$\frac{3}{4}$	2.13	0.47
Gypsum board, $\frac{3}{8}$ in.	Sand aggregate		$\frac{7}{8}$	2.44	0.41
Insulating board		0.38	$\frac{1}{2}$	0.74	1.35
Plywood		0.80	$\frac{3}{8}$	2.13	0.47

[a]Adapted from Johns-Mansville, Denver, Colorado.

Table A5.11 Conductance and resistance values for glass[a]

Material	Description	Conductivity k, Btu/(hr)(ft²)(°F/in.)	Thickness, in.	Conductance C (U), Btu/(hr)(ft²)(°F)	Resistance R ($1/U$), 1/[Btu/(hr)(ft²)(°F)]
Single-plate				1.13	0.88
Double-plate	Air space, $\frac{3}{16}$ in.	. . .	. . .	0.69	1.45
Storm windows	Air space, 1–4 in.	. . .	. . .	0.56	1.78
Solid-wood door	Actual thickness, $1\frac{1}{2}$ in.	. . .	. . .	0.49	2.04
Storm door, wood and glass	With wood and glass storm door	. . .	. . .	0.27	3.70
Storm door, metal and glass	With metal and glass storm door	. . .	. . .	0.33	3.00

[a]Adapted from Johns-Mansville, Denver, Colorado.

Table A5.12 Internet sources of world climatic data

The following climatic data, useful for solar energy system design, are available for various locations in the countries listed below.

- *Temperature (Max., Min., Mean)*
- *Dew Point (Mean)*
- *Precipitation*
- *Pressure*
- *Wind Speed*

Additional information about national weather data for some countries are available from the internet. For websites, see the National Oceanic and Atmospheric Administration (NOAA) website [http://www.ncdc.noaa.gov/cgi-bin/res40?page=climvisgsod.html]

Additional data available for U.S.A. from National Oceanic and Atmospheric Administration (NOAA) [http://www.ncdc.noaa.gov/ol/documentlibrary/hc/hcs.html] includes:

- *Heating Degree Days*
- *Cooling Degree Days*

Countries (Numbers in parentheses denote the number of sites for which data are available):

Africa

Algeria (55)	Angola (19)
Benin (6)	Botswana (17)
Bouvet Island (1)	Burkina Faso (11)
Burundi (2)	Cameroon (10)
Cape Verde (3)	Central African Republic (8)
Chad/Tchad (12)	Comoros (4)
Congo (14)	Cote D'Ivoire/Ivory Coast (12)
Democratic Republic of the Congo (73)	Djibouti (2)
Egypt (39)	Equatorial Guinea (2)
Eritrea (3)	Ethiopia (14)
Gabon (9)	Gambia (12)
Ghana (23)	Guinea (7)
Guniea-Bissau (4)	Kenya (11)
Lesotho (3)	Liberia (2)
Libyan Arab Jamahiriya (17)	Madagascar (9)
Maderia (2)	Malawi (7)
Mali (6)	Mauritania (6)
Morocco (24)	Mozambique (16)
Namibia (11)	Niger (7)
Nigeria (21)	Ocean Islands (12)
Rwanda (5)	Senegal (6)
Seychelles (5)	Sierra Leone (6)
Somalia (7)	South Africa (101)
Spain (Canary Islands) (7)	Sudan (8)
Swaziland (1)	Uganda (9)
Tunisia (17)	Togo (5)
United Republic of Tanzania (12)	Western Sahara (2)
Zambia (31)	Zimbabwe (27)

Asia

Afghanistan (17)	Bahrain (1)
Bangladesh (14)	Cambodia (9)
China (414)	North Korea (15)
Hong Kong (18)	India (540)
Iran (81)	Iraq (31)
Japan (315)	Kazakhstan (76)
Kuwait (19)	Kyrgyzstan (6)
Lao (19)	Macau (4)
Maldives (5)	Mongolia (38)
Myanmar (51)	Nepal (9)

Table A5.12 Internet sources of world climatic data *(Continued)*

Countries (Numbers in parentheses denote the number of sites for which data are available):

Asia

Oman (12)
Qatar (3)
Russian Federation (in Asia) (576)
Sri Lanka (26)
Thailand (73)
United Arab Emirates (5)
Viet Nam (21)

Pakistan (72)
South Korea (68)
Saudi Arabia (87)
Tajikistan (9)
Turkmenistan (29)
Uzbekistan (36)
Yemen (17)

Europe

Albania (8)
Austria (174)
Belarus (23)
Bosnia and Herzegovina (10)
Croatia (47)
Czech Republic (32)
Estonia (27)
France (180)
Germany (306)
Greece (52)
Hungary (33)
Ireland (17)
Italy (140)
Kazakhstan (Europe) (3)
Lebanon (6)
Luxembourg (3)
Netherlands (54)
Poland (72)
Republic of Moldova (3)
Russian Federation (in Europe) (342)
Slovenia (23)
Sweden (96)
Syrian Arab Republic (35)
Turkey (102)
United Kingdom and No. Ireland (344)

Armenia (36)
Azerbaijan (72)
Belgium (19)
Bulgaria (34)
Cyprus (4)
Denmark (70)
Finland (49)
Georgia (17)
Gibraltar (2)
Greenland (43)
Iceland (58)
Israel (8)
Jordan (9)
Latvia (22)
Lithuania (17)
Malta (1)
Norway (198)
Portugal (33)
Romania (42)
Slovakia (25)
Spain (79)
Switzerland (70)
Macedonia (22)
Ukraine (117)
Yugoslavia (47)

North and South America

Anguilla (1)
Argentina (90)
Bahamas (7)
Belize (4)
Brazil (342)
Canada (324)
Chile (35)
Columbia (54)
Cuba (69)
Dominica (18)
El Salvador (6)
Grenada (1)
Guyana (8)

Antigua and Barbuda (2)
Boliva (36)
Barbados (2)
Bermuda (1)
British Virgin Islands (1)
Cayman Islands (2)
Clipperton (1)
Costa Rica (5)
Curaçao (3)
Equador (37)
French Guyana (5)
Guatemala (7)
Haiti (6)

Table A5.12 Internet sources of world climatic data *(Continued)*

Countries (Numbers in parentheses denote the number of sites for which data are available):

North and South America

Honduras (7)	Islands (7)
Jamaica (4)	Martinique (2)
Mexico (89)	Paraguay (49)
Peru (54)	Suriname (7)
Uruguay (17)	Venezuela (33)
Nicaragua (13)	Panama (7)
Puerto Rico (5)	Saint Kitts (2)
Saint Lucia (2)	Saint Vincent (1)
St. Marten, St. Eustatius (5)	St. Pierre (1)
Trinidad and Tobago (2)	Turks and Caicos Islands (2)
United States of America (756)	Venezuela (1)

South-West Pacific

Australia (Additional Islands) (828)	Brunei (2)
Cook Islands (5)	Detached Islands (6)
East Timor/Timor Oriental (7)	Fiji (15)
French Polynesia (Austral Islands) (16)	Indonesia (198)
Islands (157)	Kiribati (6)
Malaysia (18)	Nauru (2)
New Caledonia (8)	New Zealand (126)
Niue (2)	Papua New Guinea (54)
Philippines (72)	Samoa and American Samoa (5)
Singapore (5)	Solomon Islands (7)
Southern Line Islands (3)	Tokelau (4)
Tonga (9)	Tuvalu (5)
Vanuatu (7)	

THERMODYNAMIC DATA FOR COOLING SYSTEMS

Appendix **6**

Table A6.1a Saturated refrigerant-134a—Temperature table

Temp., $T°C$	Press., P_{sat} MPa	Specific volume, m³/kg		Enthalpy, kJ/kg			Entropy, kJ/(kg · K)	
		Sat. liquid, v_f	Sat. vapor, v_g	Sat. liquid, h_f	Evap., h_{fg}	Sat. vapor, h_g	Sat. liquid, s_f	Sat. vapor, s_g
−40	0.05164	0.0007055	0.3569	0.00	222.88	222.88	0.0000	0.9560
−36	0.06332	0.0007113	0.2947	4.73	220.67	225.40	0.0201	0.9506
−32	0.07704	0.0007172	0.2451	9.52	218.37	227.90	0.0401	0.9456
−28	0.09305	0.0007233	0.2052	14.37	216.01	230.38	0.0600	0.9411
−26	0.10199	0.0007265	0.1882	16.82	214.80	231.62	0.0699	0.9390
−24	0.11160	0.0007296	0.1728	19.29	213.57	232.85	0.0798	0.9370
−22	0.12192	0.0007328	0.1590	21.77	212.32	234.08	0.0897	0.9351
−20	0.13299	0.0007361	0.1464	24.26	211.05	235.31	0.0996	0.9332
−18	0.14483	0.0007395	0.1350	26.77	209.76	236.53	0.1094	0.9315
−16	0.15748	0.0007428	0.1247	29.30	208.45	237.74	0.1192	0.9298
−12	0.18540	0.0007498	0.1068	34.39	205.77	240.15	0.1388	0.9267
−8	0.21704	0.0007569	0.0919	39.54	203.00	242.54	0.1583	0.9239
−4	0.25274	0.0007644	0.0794	44.75	200.15	244.90	0.1777	0.9213
0	0.29282	0.0007721	0.0689	50.02	197.21	247.23	0.1970	0.9190
4	0.33765	0.0007801	0.0600	55.35	194.19	249.53	0.2162	0.9169
8	0.38756	0.0007884	0.0525	60.73	191.07	251.80	0.2354	0.9150
12	0.44294	0.0007971	0.0460	66.18	187.85	254.03	0.2545	0.9132
16	0.50416	0.0008062	0.0405	71.69	184.52	256.22	0.2735	0.9116
20	0.57160	0.0008157	0.0358	77.26	181.09	258.35	0.2924	0.9102
24	0.64566	0.0008257	0.0317	82.90	177.55	260.45	0.3113	0.9089
26	0.68530	0.0008309	0.0298	85.75	175.73	261.48	0.3208	0.9082
28	0.72675	0.0008362	0.0281	88.61	173.89	262.50	0.3302	0.9076
30	0.77006	0.0008417	0.0265	91.49	172.00	263.50	0.3396	0.9070
32	0.81528	0.0008473	0.0250	94.39	170.09	264.48	0.3490	0.9064
34	0.86247	0.0008530	0.0236	97.31	168.14	265.45	0.3584	0.9058

Table A6.1a Saturated refrigerant-134a—Temperature table *(Continued)*

Temp., $T°C$	Press., P_{sat} MPa	Specific volume, m³/kg		Enthalpy, kJ/kg			Entropy, kJ/(kg · K)	
		Sat. liquid, v_f	Sat. vapor, v_g	Sat. liquid, h_f	Evap., h_{fg}	Sat. vapor, h_g	Sat. liquid, s_f	Sat. vapor, s_g
36	0.91168	0.0008590	0.0223	100.25	166.15	266.40	0.3678	0.9053
38	0.96298	0.0008651	0.0210	103.21	164.12	267.33	0.3772	0.9047
40	1.0164	0.0008714	0.0199	106.19	162.05	268.24	0.3866	0.9041
42	1.0720	0.0008780	0.0188	109.19	159.94	269.14	0.3960	0.9035
44	1.1299	0.0008847	0.0177	112.22	157.79	270.01	0.4054	0.9030
48	1.2526	0.0008989	0.0159	118.35	153.33	271.68	0.4243	0.9017
52	1.3851	0.0009142	0.0142	124.58	148.66	273.24	0.4432	0.9004
56	1.5278	0.0009308	0.0127	130.93	143.75	274.68	0.4622	0.8990
60	1.6813	0.0009488	0.0114	137.42	138.57	275.99	0.4814	0.8973
70	2.1162	0.0010027	0.0086	154.34	124.08	278.43	0.5302	0.8918
80	2.6324	0.0010766	0.0064	172.71	106.41	279.12	0.5814	0.8827
90	3.2435	0.0011949	0.0046	193.69	82.63	276.32	0.6380	0.8655
100	3.9742	0.0015443	0.0027	224.74	34.40	259.13	0.7196	0.8117

From Wilson, D. P., and R. S. Basu. 1988. Thermodynamic Properties of a New Stratospherically Safe Working Fluid-Refrigerant-134a. *ASHRAE Trans.* 94(2):2095–2118.

Table A6.1b Saturated refrigerant-134a—Pressure table

Press., PMPa	Temp., T_{sat} °C	Specific volume, m³/kg		Enthalpy, kJ/kg			Entropy, kJ/(kg · K)	
		Sat. liquid, v_f	Sat. vapor, v_g	Sat. liquid, h_f	Evap., h_{fg}	Sat. vapor, h_g	Sat. liquid, s_f	Sat. vapor, s_g
0.06	−37.07	0.0007097	0.3100	3.46	221.27	224.72	0.0147	0.9520
0.08	−31.21	0.0007184	0.2366	10.47	217.92	228.39	0.0440	0.9447
0.10	−26.43	0.0007258	0.1917	16.29	215.06	231.35	0.0678	0.9395
0.12	−22.36	0.0007323	0.1614	21.32	212.54	233.86	0.0879	0.9354
0.14	−18.80	0.0007381	0.1395	25.77	210.27	236.04	0.1055	0.9322
0.16	−15.62	0.0007435	0.1229	29.78	208.18	237.97	0.1211	0.9295
0.18	−12.73	0.0007485	0.1098	33.45	206.26	239.71	0.1352	0.9273
0.20	−10.09	0.0007532	0.0993	36.84	204.46	241.30	0.1481	0.9253
0.24	−5.37	0.0007618	0.0834	42.95	201.14	244.09	0.1710	0.9222
0.28	−1.23	0.0007697	0.0719	48.39	198.13	246.52	0.1911	0.9197
0.32	2.48	0.0007770	0.0632	53.31	195.35	248.66	0.2089	0.9177
0.36	5.84	0.0007839	0.0564	57.82	192.76	250.58	0.2251	0.9160
0.4	8.93	0.0007904	0.0509	62.00	190.32	252.32	0.2399	0.9145
0.5	15.74	0.0008056	0.0409	71.33	184.74	256.07	0.2723	0.9117
0.6	21.58	0.0008196	0.0341	79.48	179.71	259.19	0.2999	0.9097
0.7	26.72	0.0008328	0.0292	86.78	175.07	261.85	0.3242	0.9080
0.8	31.33	0.0008454	0.0255	93.42	170.73	264.15	0.3459	0.9066
0.9	35.53	0.0008576	0.0226	99.56	166.62	266.18	0.3656	0.9054
1.0	39.39	0.0008695	0.0202	105.29	162.68	267.97	0.3838	0.9043
1.2	46.32	0.0008928	0.0166	115.76	155.23	270.99	0.4164	0.9023
1.4	52.43	0.0009159	0.0140	125.26	148.14	273.40	0.4453	0.9003
1.6	57.92	0.0009392	0.0121	134.02	141.31	275.33	0.4714	0.8982
1.8	62.91	0.0009631	0.0105	142.22	134.60	276.83	0.4954	0.8959
2.0	67.49	0.0009878	0.0093	149.99	127.95	277.94	0.5178	0.8934
2.5	77.59	0.0010562	0.0069	168.12	111.06	279.17	0.5687	0.8854
3.0	86.22	0.0011416	0.0053	185.30	92.71	278.01	0.6156	0.8735

From Wilson, D. P., and R. S. Basu. 1988. Thermodynamic Properties of a New Stratospherically Safe Working Fluid-Refrigerant-134a. *ASHRAE Trans.* 94(2):2095–2118.

Table A6.2 Superheated refrigerant-134a

T °C	P = 0.06 MPa (T_sat = −37.07°C) v m³/kg	h kJ/kg	s kJ/(kg·K)	P = 0.10 MPa (T_sat = −26.43°C) v m³/kg	h kJ/kg	s kJ/(kg·K)	P = 0.14 MPa (T_sat = −18.80°C) v m³/kg	h kJ/kg	s kJ/(kg·K)
Sat.	0.31003	224.72	0.9520	0.19170	231.35	0.9395	0.13945	236.04	0.9322
−20	0.33536	237.98	1.0062	0.19770	236.54	0.9602			
−10	0.34992	245.96	1.0371	0.20686	244.70	0.9918	0.14549	243.40	0.9606
0	0.36433	254.10	1.0675	0.21587	252.99	1.0227	0.15219	251.86	0.9922
10	0.37861	262.41	1.0973	0.22473	261.43	1.0531	0.15875	260.43	1.0230
20	0.39279	270.89	1.1267	0.23349	270.02	1.0829	0.16520	269.13	1.0532
30	0.40688	279.53	1.1557	0.24216	278.76	1.1122	0.17155	277.97	1.0828
40	0.42091	288.35	1.1844	0.25076	287.66	1.1411	0.17783	286.96	1.1120
50	0.43487	297.34	1.2126	0.25930	296.72	1.1696	0.18404	296.09	1.1407
60	0.44879	306.51	1.2405	0.26779	305.94	1.1977	0.19020	305.37	1.1690
70	0.46266	315.84	1.2681	0.27623	315.32	1.2254	0.19633	314.80	1.1969
80	0.47650	325.34	1.2954	0.28464	324.87	1.2528	0.20241	324.39	1.2244
90	0.49031	335.00	1.3224	0.29302	334.57	1.2799	0.20846	334.14	1.2516
100							0.21449	344.04	1.2785

T °C	P = 0.18 MPa (T_sat = −12.73°C) v m³/kg	h kJ/kg	s kJ/(kg·K)	P = 0.20 MPa (T_sat = −10.09°C) v m³/kg	h kJ/kg	s kJ/(kg·K)	P = 0.24 MPa (T_sat = −5.37°C) v m³/kg	h kJ/kg	s kJ/(kg·K)
Sat.	0.10983	239.71	0.9273	0.09933	241.30	0.9253	0.08343	244.09	0.9222
−10	0.11135	242.06	0.9362	0.09938	241.38	0.9256			
0	0.11678	250.69	0.9684	0.10438	250.10	0.9582	0.08574	248.89	0.9399
10	0.12207	259.41	0.9998	0.10922	258.89	0.9898	0.08993	257.84	0.9721
20	0.12723	268.23	1.0304	0.11394	267.78	1.0206	0.09339	266.85	1.0034
30	0.13230	277.17	1.0604	0.11856	276.77	1.0508	0.09794	275.95	1.0339
40	0.13730	286.24	1.0898	0.12311	285.88	1.0804	0.10181	285.16	1.0637
50	0.14222	295.45	1.1187	0.12758	295.12	1.1094	0.10562	294.47	1.0930
60	0.14710	304.79	1.1472	0.13201	304.50	1.1380	0.10937	303.91	1.1218
70	0.15193	314.28	1.1753	0.13639	314.02	1.1661	0.11307	313.49	1.1501
80	0.15672	323.92	1.2030	0.14073	323.68	1.1939	0.11674	323.19	1.1780
90	0.16148	333.70	1.2303	0.14504	333.48	1.2212	0.12037	333.04	1.2055
100	0.16622	343.63	1.2573	0.14932	343.43	1.2483	0.12398	343.03	1.2326

Table A6.2 Superheated refrigerant-134a (*Continued*)

T °C	v m³/kg	h kJ/kg	s kJ/(kg·K)	v m³/kg	h kJ/kg	s kJ/(kg·K)	v m³/kg	h kJ/kg	s kJ/(kg·K)
	$P = 0.28$ MPa ($T_{sat} = -1.23°C$)			$P = 0.32$ MPa ($T_{sat} = 2.48°C$)			$P = 0.40$ MPa ($T_{sat} = 8.93°C$)		
Sat.	0.07193	246.52	0.9197	0.06322	248.66	0.9177	0.05089	252.32	0.9145
0	0.07240	247.64	0.9238						
10	0.07613	256.76	0.9566	0.06576	255.65	0.9427	0.05119	253.35	0.9182
20	0.07972	265.91	0.9883	0.06901	264.95	0.9749	0.05397	262.96	0.9515
30	0.08320	275.12	1.0192	0.07214	274.28	1.0062	0.05662	272.54	0.8937
40	0.08660	284.42	1.0494	0.07518	283.67	1.0367	0.05917	282.14	1.0148
50	0.08992	293.81	1.0789	0.07815	293.15	1.0665	0.06164	291.79	1.0452
60	0.09319	303.32	1.1079	0.08106	302.72	1.0957	0.06405	301.51	1.0748
70	0.09641	312.95	1.1364	0.08392	312.41	1.1243	0.06641	311.32	1.1038
80	0.09960	322.71	1.1644	0.08674	322.22	1.1525	0.06873	321.23	1.1322
90	0.10275	332.60	1.1920	0.08953	332.15	1.1802	0.07102	331.25	1.1602
100	0.10587	342.62	1.2193	0.09229	342.21	1.1076	0.07327	341.38	1.1878
110	0.10897	352.78	1.2461	0.09503	352.40	1.2345	0.07550	351.64	1.2149
120	0.11205	363.08	1.2727	0.09774	362.73	1.2611	0.07771	362.03	1.2417
130							0.07991	372.54	1.2681
140							0.08208	383.18	1.2941

T °C	v m³/kg	h kJ/kg	s kJ/(kg·K)	v m³/kg	h kJ/kg	s kJ/(kg·K)	v m³/kg	h kJ/kg	s kJ/(kg·K)
	$P = 0.50$ MPa ($T_{sat} = 15.74°C$)			$P = 0.60$ MPa ($T_{sat} = 21.58°C$)			$P = 0.70$ MPa ($T_{sat} = 26.72°C$)		
Sat.	0.04086	256.07	0.9117	0.03408	259.19	0.9097	0.02918	261.85	0.9080
20	0.04188	260.34	0.9264						

Table A6.2 Superheated refrigerant-134a (Continued)

T °C	v m³/kg	h kJ/kg	s kJ/(kg·K)	v m³/kg	h kJ/kg	s kJ/(kg·K)	v m³/kg	h kJ/kg	s kJ/(kg·K)
	$P = 0.50$ MPa ($T_{sat} = 15.74°C$)			$P = 0.60$ MPa ($T_{sat} = 21.58°C$)			$P = 0.70$ MPa ($T_{sat} = 26.72°C$)		
30	0.04416	270.28	0.9597	0.03581	267.89	0.9388	0.02979	265.37	0.9197
40	0.04633	280.16	0.9918	0.03774	278.09	0.9719	0.03157	275.93	0.9539
50	0.04842	290.04	1.0229	0.03958	288.23	1.0037	0.03324	286.35	0.9867
60	0.05043	299.95	1.0531	0.04134	298.35	1.0346	0.03482	296.69	1.0182
70	0.05240	309.92	1.0825	0.04304	308.48	1.0645	0.03634	307.01	1.0487
80	0.05432	319.96	1.1114	0.04469	318.67	1.0938	0.03781	317.35	1.0784
90	0.05620	330.10	1.1397	0.04631	328.93	1.1225	0.03924	327.74	1.1074
100	0.05805	340.33	1.1675	0.04790	339.27	1.1505	0.04064	338.19	1.1358
110	0.05988	350.68	1.1949	0.04946	349.70	1.1781	0.04201	348.71	1.1637
120	0.06168	361.14	1.2218	0.05099	360.24	1.2053	0.04335	359.33	1.1910
130	0.06347	371.72	1.2484	0.05251	370.88	1.2320	0.04468	370.04	1.2179
140	0.06524	382.42	1.2746	0.05402	381.64	1.2584	0.04599	380.86	1.2444
150				0.05550	392.52	1.2844	0.04729	391.79	1.2706
160				0.05698	403.51	1.3100	0.04857	402.82	1.2963

T °C	v m³/kg	h kJ/kg	s kJ/(kg·K)	v m³/kg	h kJ/kg	s kJ/(kg·K)	v m³/kg	h kJ/kg	s kJ/(kg·K)
	$P = 0.80$ MPa ($T_{sat} = 31.33°C$)			$P = 0.90$ MPa ($T_{sat} = 35.53°C$)			$P = 1.00$ MPa ($T_{sat} = 39.39°C$)		
Sat.	0.02547	264.15	0.9066	0.02255	266.18	0.9054	0.02020	267.97	0.9043
40	0.02691	273.66	0.9374	0.02325	271.25	0.9217	0.02029	268.68	0.9066
50	0.02846	284.39	0.9711	0.02472	282.34	0.9566	0.02171	280.19	0.9428
60	0.02992	294.98	1.0034	0.02609	293.21	0.9897	0.02301	291.36	0.9768
70	0.03131	305.50	1.0345	0.02738	303.94	1.0214	0.02423	302.34	1.0093
80	0.03264	316.00	1.0647	0.02861	314.62	1.0521	0.02538	313.20	1.0405
90	0.03393	326.52	1.0940	0.02980	325.28	1.0819	0.02649	324.01	1.0707
100	0.03519	337.08	1.1227	0.03095	335.96	1.1109	0.02755	334.82	1.1000
110	0.03642	347.71	1.1508	0.03207	346.68	1.1392	0.02858	345.65	1.1286
120	0.03762	358.40	1.1784	0.03316	357.47	1.1670	0.02959	356.52	1.1567

Table A6.2 Superheated refrigerant-134a (*Continued*)

T °C	v m³/kg	h kJ/kg	s kJ/(kg·K)	v m³/kg	h kJ/kg	s kJ/(kg·K)	v m³/kg	h kJ/kg	s kJ/(kg·K)
	$P = 0.80$ MPa ($T_{sat} = 31.33°C$)			$P = 0.90$ MPa ($T_{sat} = 35.53°C$)			$P = 1.00$ MPa ($T_{sat} = 39.39°C$)		
130	0.03881	369.19	1.2055	0.03423	368.33	1.1943	0.03058	367.46	1.1841
140	0.03997	380.07	1.2321	0.03529	379.27	1.2211	0.03154	378.46	1.2111
150	0.04113	391.05	1.2584	0.03633	390.31	1.2475	0.03250	389.56	1.2376
160	0.04227	402.14	1.2843	0.03736	401.44	1.2735	0.03344	400.74	1.2638
170	0.04340	413.33	1.3098	0.03838	412.68	1.2992	0.03436	412.02	1.2895
180	0.04452	424.63	1.3351	0.03939	424.02	1.3245	0.03528	423.40	1.3149

T °C	v m³/kg	h kJ/kg	s kJ/(kg·K)	v m³/kg	h kJ/kg	s kJ/(kg·K)	v m³/kg	h kJ/kg	s kJ/(kg·K)
	$P = 1.20$ MPa ($T_{sat} = 46.32°C$)			$P = 1.40$ MPa ($T_{sat} = 52.43°C$)			$P = 1.60$ MPa ($T_{sat} = 57.92°C$)		
Sat.	0.01663	270.99	0.9023	0.01405	273.40	0.9003	0.01208	275.33	0.8982
50	0.01712	275.52	0.9164						
60	0.01835	287.44	0.9527	0.01495	283.10	0.9297	0.01233	278.20	0.9069
70	0.01947	298.96	0.9868	0.01603	295.31	0.9658	0.01340	291.33	0.9457
80	0.02051	310.24	1.0192	0.01701	307.10	0.9997	0.01435	303.74	0.9813
90	0.02150	321.39	1.0503	0.01792	318.63	1.0319	0.01521	315.72	1.0148
100	0.02244	332.47	1.0804	0.01878	330.32	1.0628	0.01601	327.46	1.0467
110	0.02335	343.52	1.1096	0.01960	341.32	1.0927	0.01677	339.04	1.0773
120	0.02423	354.58	1.1381	0.02039	352.59	1.1218	0.01750	350.53	1.1069
130	0.02508	365.68	1.1660	0.02115	363.86	1.1501	0.01820	361.99	1.1357
140	0.02592	376.83	1.1933	0.02189	375.15	1.1777	0.01887	373.44	1.1638
150	0.02674	388.04	1.2201	0.02262	386.49	1.2048	0.01953	384.91	1.1912
160	0.02754	399.33	1.2465	0.02333	397.89	1.2315	0.02017	396.43	1.2181
170	0.02834	410.70	1.2724	0.02403	409.36	1.2576	0.02080	407.99	1.2445
180	0.02912	422.16	1.2980	0.02472	420.90	1.2834	0.02142	419.62	1.2704
190				0.02541	432.53	1.3088	0.02203	431.33	1.2960
200				0.02608	444.24	1.3338	0.02263	443.11	1.3212

From Wilson, D. P., and R. S. Basu. 1988. Thermodynamic Properties of a New Stratospherically Safe Working Fluid-Refrigerant-134a. *ASHRAE Trans.* 94(2):2095–2118.

Table A6.3 Properties of Ammonia, $NH_3{}^a$

			Saturated				100°F		200°F	
							Degrees of superheat			
t	p	v_g	h_f	h_g	s_f	s_g	h	s	h	s
−70	3.94	61.65	−31.1	584.4	−0.0771	1.5026	635.6	1.6214	686.2	1.7151
−65	4.69	52.34	−26.0	586.6	−0.0642	1.4833	638.0	1.6058	688.7	1.6990
−60	5.55	44.73	−20.9	588.8	−0.0514	1.4747	640.3	1.5907	691.1	1.6834
−55	6.54	38.38	−15.7	591.0	−0.0381	1.4614	642.6	1.5761	693.6	1.6683
−50	7.67	33.08	−10.5	593.2	−0.0254	1.4487	644.9	1.5620	696.1	1.6537
−45	8.95	28.62	−5.3	595.4	−0.0128	1.4363	647.2	1.5484	698.6	1.6395
−40	10.41	24.86	0	597.6	0	1.4242	649.4	1.5353	701.0	1.6260
−35	12.05	21.68	5.3	599.5	0.0126	1.4120	651.7	1.5226	703.4	1.6129
−30	13.90	18.97	10.7	601.4	0.0250	1.4001	653.9	1.5103	705.9	1.6002
−25	15.98	16.66	16.0	603.2	0.0374	1.3886	656.1	1.4983	708.3	1.5878
−20	18.30	14.68	21.4	605.0	0.0497	1.3774	658.3	1.4868	710.8	1.5759
−15	20.88	12.97	26.7	606.7	0.0618	1.3664	660.5	1.4756	713.2	1.5644
−10	23.74	11.50	32.1	608.5	0.0738	1.3558	662.6	1.4647	715.5	1.5531
−5	26.92	10.23	37.5	610.1	0.0857	1.3454	664.7	1.4541	717.9	1.5423
0	30.42	9.116	42.9	611.8	0.0975	1.3352	666.8	1.4438	720.3	1.5317
5	34.27	8.150	48.3	613.3	0.1092	1.3253	668.9	1.4338	722.7	1.5214
10	38.51	7.304	53.8	614.9	0.1208	1.3157	670.9	1.4241	725.0	1.5115
15	43.14	6.562	59.2	616.3	0.1323	1.3062	672.9	1.4147	727.3	1.5018
20	48.21	5.910	64.7	617.8	0.1437	1.2969	675.0	1.4055	729.6	1.4924
25	53.73	5.334	70.2	619.1	0.1551	1.2879	677.0	1.3966	731.9	1.4833
30	59.74	4.825	75.7	620.5	0.1663	1.2790	678.9	1.3879	734.2	1.4744
35	66.26	4.373	81.2	621.7	0.1775	1.2704	680.8	1.3794	736.4	1.4658
40	73.32	3.971	86.8	623.0	0.1885	1.2618	682.7	1.3711	738.6	1.4575
45	80.96	3.614	92.3	624.1	0.1996	1.2535	684.5	1.3630	740.9	1.4493
50	89.19	3.294	97.9	625.2	0.2105	1.2453	686.4	1.3551	743.1	1.4413
55	98.06	3.008	103.5	626.3	0.2214	1.2373	688.2	1.3474	745.2	1.4335
60	107.6	2.751	109.2	627.3	0.2322	1.2294	689.9	1.3399	747.4	1.4260
65	117.8	2.520	114.8	628.2	0.2430	1.2216	691.6	1.3326	749.5	1.4186
70	128.8	2.312	120.5	629.1	0.2537	1.2140	693.3	1.3254	751.6	1.4114
75	140.5	2.125	126.2	629.9	0.2643	1.2065	694.9	1.3184	753.7	1.4044
80	153.0	1.955	132.0	630.7	0.2749	1.1991	696.6	1.3115	755.8	1.3976
85	166.4	1.801	137.8	631.4	0.2854	1.1918	698.1	1.3048	757.9	1.3909
90	180.6	1.661	143.5	632.0	0.2958	1.1846	699.7	1.2982	759.9	1.3843
95	195.8	1.534	149.4	632.6	0.3062	1.1775	701.2	1.2918	761.8	1.3779
100	211.9	1.419	155.2	633.0	0.3166	1.1705	702.7	1.2855	763.8	1.3717
105	228.9	1.313	161.1	633.4	0.3269	1.1635	704.2	1.2793	765.7	1.3656
110	247.0	1.217	167.0	633.7	0.3372	1.1566	705.6	1.2732	767.6	1.3596
115	266.2	1.128	173.0	633.9	0.3474	1.1497	706.9	1.2672	769.5	1.3538

Table A6.3 Properties of Ammonia, $NH_3{}^a$ *(Continued)*

			Saturated				Degrees of superheat			
							100°F		200°F	
t	p	v_g	h_f	h_g	s_f	s_g	h	s	h	s
120	286.4	1.047	179.0	634.0	0.3576	1.1427	708.2	1.2613	771.3	1.3480
125	307.8	0.973	185.1	634.0	0.3679	1.1358	709.5	1.2555	773.1	1.3423

^{a}Symbols: t, °F; p, psia; v_g, ft³/lb; h_f, Btu/lb; h_g, Btu/lb; s_f, Btu/lb · °R; s_g, Btu/lb · °R; h, Btu/lb; s, Btu/lb · °R.

Table A6.4 Psychrometric table—SI units. Properties of moist air at 101,325 N/m^{2a}

| Temperature | | | Properties^b | | | | | | |
C	K	F	P_s	W_s	V_a	V_s	h_a	h_s	s_s
−40	233.15	−40	12.838	0.00007925	0.65961	0.65968	−22.35	−22.16	−90.659
−30	243.15	−22	37.992	0.0002344	0.68808	0.68833	−12.29	−11.72	−46.732
−25	248.15	−13	63.248	0.0003903	0.70232	0.70275	−7.265	−6.306	−24.706
−20	253.15	−4	103.19	0.0006371	0.71649	0.71724	−2.236	−0.6653	−2.2194
−15	258.15	+5	165.18	0.001020	0.73072	0.73191	+2.794	5.318	21.189
−10	263.15	14	259.72	0.001606	0.74495	0.74683	7.823	11.81	46.104
−5	268.15	23	401.49	0.002485	0.75912	0.76218	12.85	19.04	73.365
0	273.15	32	610.80	0.003788	0.77336	0.77804	17.88	27.35	104.14
5	278.15	41	871.93	0.005421	0.78759	0.79440	22.91	36.52	137.39
10	283.15	50	1227.2	0.007658	0.80176	0.81163	27.94	47.23	175.54
15	288.15	59	1704.4	0.01069	0.81600	0.82998	32.97	59.97	220.22
20	293.15	68	2337.2	0.01475	0.83017	0.84983	38.00	75.42	273.32
25	298.15	77	3167.0	0.02016	0.84434	0.87162	43.03	94.38	337.39
30	303.15	86	4242.8	0.02731	0.85851	0.89609	48.07	117.8	415.65
35	308.15	95	5623.4	0.03673	0.87274	0.92406	53.10	147.3	512.17
40	313.15	104	7377.6	0.04911	0.88692	0.95665	58.14	184.5	532.31
45	318.15	113	9584.8	0.06536	0.90115	0.99535	63.17	232.0	783.06
50	323.15	122	12339	0.08678	0.91532	1.0423	68.21	293.1	975.27
55	328.15	131	15745	0.1152	0.92949	1.1007	73.25	372.9	1221.5
60	333.15	140	19925	0.1534	0.94372	1.1748	78.29	478.5	1543.5
65	338.15	149	25014	0.2055	0.95790	1.2721	83.33	621.4	1973.6
70	343.15	158	31167	0.2788	0.97207	1.4042	88.38	820.5	2564.8
75	348.15	167	38554	0.3858	0.98630	1.5924	93.42	1110	3412.8
80	353.15	176	47365	0.5519	1.0005	1.8791	98.47	1557	4710.9

Table A6.4 Psychrometric table—SI units. Properties of moist air at 101,325 N/m²[a] (*Continued*)

Temperature			Properties[b]						
C	K	F	P_s	W_s	V_a	V_s	h_a	h_s	s_s
85	358.15	185	57809	0.8363	1.0146	2.3632	103.5	2321	6892.6
90	363.15	194	70112	1.416	1.0288	3.3409	108.6	3876	11281

[a]Symbols and units: P_s = pressure of water vapor at saturation, N/m²; W_s = humidity ratio at saturation, mass of water vapor associated with unit mass of dry air; V_a = specific volume of dry air, m³/kg; V_s = specific volume of saturated mixture, m³/kg dry air; s_s = specific enthalpy of saturated mixture, kJ/kg dry air; s_s = specific entropy of saturated mixture, J/K · kg dry air; h_a = specific enthalpy of dry air, kJ/kg; h_s = specific enthalpy of saturated mixture, kJ/kg dry air. Abstracted from Bolz, R. E., and G. L. Tuve, eds., "CRC Handbook of Tables for Applied Engineering Science," 2nd ed., Chemical Rubber Co., Cleveland, Ohio, 1973.

[b]The P_s column gives the vapor pressure of pure water at temperature intervals of 5°C. For the latest data on vapor pressures at intervals of 0.1°C, from 0 to 100°C, see Wexler, A., and L. Greenspan, Vapor Pressure Equation for Water, *J. Res. Natl. Bur. Stand. Sect. A*, 75(3):213–229, May–June 1971. For very low barometric pressures and wet-bulb temperatures, the values of h_a here are somewhat low; for corrections see the "Handbook of Fundamentals," American Society of Heating, Refrigerating, and Air-Conditioning Engineers, 1972.

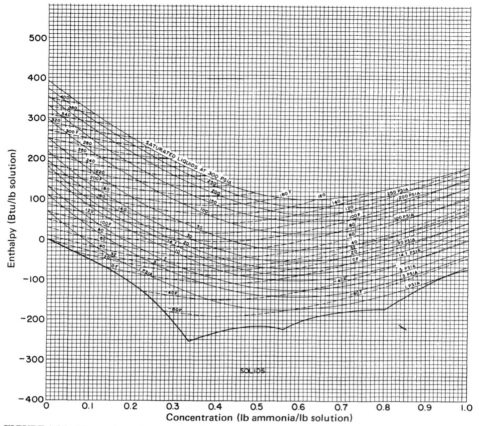

FIGURE A6.1 Thermodynamic properties of ammonia-water mixtures used for absorption air conditioning.

Table A6.5 Refrigerant temperature ($t' = °C$) and enthalpy ($h = kJ/kg$) of lithium bromide solutions

		Percent LiBr										
Temp. ($t = °C$)		0	10	20	30	40	45	50	55	60	65	70
20	t'	20	19.1	17.7	15.0	9.8	5.8	−0.4	−7.7	−15.8	−23.4#	−29.3#
	h	84.0	67.4	52.6	40.4	33.5	33.5	38.9	53.2	78.0	111.0#	145.0#
30	t'	30.0	29.0	27.5	24.6	19.2	15.0	8.6	1.0	−7.3	−15.2#	−21.6#
	h	125.8	103.3	84.0	68.6	58.3	56.8	60.5	73.5	96.8	128.4#	161.7#
40	t'	40.0	38.9	37.3	34.3	28.5	24.1	17.5	9.8	1.3	−7.0#	−14.0#
	h	167.6	139.5	115.8	96.0	82.5	79.7	82.2	93.5	115.4	145.0#	178.3#
50	t'	50.0	48.8	47.2	44.0	37.9	33.3	26.5	18.5	9.9	1.3	−6.3#
	h	209.3	175.2	147.0	123.4	106.7	102.6	103.8	114.0	134.5	163.5	195.0#
60	t'	60.0	58.8	57.0	53.6	47.3	42.5	35.5	27.3	18.4	9.5	1.4#
	h	251.1	211.7	179.1	151.4	131.7	125.8	125.8	134.7	153.7	181.4	211.9#
70	t'	70.0	68.7	66.8	63.3	56.6	51.6	44.4	36.1	27.0	17.7	9.0#
	h	293.0	247.7	210.5	178.8	155.7	148.9	148.0	155.6	173.2	199.4	228.8#
80	t'	80.0	78.6	76.7	73.0	66.0	60.8	53.4	44.8	35.6	26.0	16.7#
	h	334.9	287.8	243.6	207.3	181.0	172.8	170.0	176.2	192.6	217.2	245.7#
90	t'	90.0	88.6	86.5	82.6	75.4	70.0	62.3	53.6	44.1	34.2	24.3#
	h	376.9	321.1	275.6	235.4	206.1	195.8	192.3	197.1	212.2	235.6	262.9#

Table A6.5 Refrigerant temperature ($t' = °C$) and enthalpy ($h = $ kJ/kg) of lithium bromide solutions (*Continued*)

Temp. ($t = °C$)		Percent LiBr										
		0	10	20	30	40	45	50	55	60	65	70
100	t'	100.0	98.5	96.3	92.3	84.7	79.1	71.3	62.4	52.7	42.4	32.0
	h	419.0	357.6	307.9	263.8	231.0	219.9	214.6	218.2	231.5	253.5	279.7
110	t'	110.0	108.4	106.2	101.9	94.1	88.3	80.2	71.1	61.3	50.6	39.7
	h	461.3	394.3	340.1	292.4	255.9	243.3	236.8	239.1	251.0	271.4	296.3
120	t'	120.0*	118.3*	116.0*	111.6	103.4	97.5	89.2	79.9	69.8	58.9	47.3
	h	503.7*	431.0*	372.5*	320.9	281.0	267.0	259.0	260.0	270.2	289.5	313.4
130	t'	130.0*	128.3*	125.8*	121.3*	112.8	106.7	92.8	88.7	78.4	67.1	55.0
	h	546.5*	468.4*	404.5*	349.6*	306.2	290.7	281.0	280.4	289.1	306.9	330.2
140	t'	140.0*	138.2*	135.7*	130.9*	122.2*	115.8	107.1	97.4	87.0	75.3	62.7
	h	589.1*	505.6*	437.8*	377.9*	331.3*	314.2	303.2	301.1	308.1	324.7	346.9
150	t'	150.0*	148.1*	145.5*	140.6*	131.5*	125.0*	116.1*	106.2	95.5	83.5	70.3
	h	632.2*	542.7*	470.5*	406.8*	356.6*	337.8*	325.5*	321.6	327.3	342.7	363.6
160	t'	160.0*	158.1*	155.3*	150.3*	140.9*	134.2*	125.0*	115.0	104.1	91.8	78.9
	h	675.6*	580.8*	503.1*	435.4*	381.9*	361.2*	347.7*	342.2	346.1	360.3	380.1
170	t'	170.0*	168.0*	165.2*	159.9*	150.3*	143.3*	134.0*	123.7	112.7	100.0	85.7
	h	719.2*	618.9*	536.1*	464.3*	406.8*	384.9*	369.9*	362.9	365.4	378.3	396.0
180	t'	180.0*	177.9*	175.0*	169.6*	159.6*	152.5*	142.9*	132.5*	121.2*	108.2	93.3
	h	763.2*	657.1*	569.4*	493.4*	432.1*	408.8*	392.1*	383.4*	384.3*	395.8	411.3

*Extensions of data above 115°C are well above the original data and should be used with care.
#Supersaturated solution.

SUPPLEMENTARY MATERIAL FOR CHAPTER 7

Appendix

7

Table A7.1 Designations and characteristics for 94 reference passive systems

(a) Overall System Characteristics

Masonry properties

thermal conductivity (k)	
sunspace floor	0.5 Btu/hr/ft/°F
all other masonry	1.0 Btu/hr/ft/°F
density (Q)	150 lb/ft³
specific heat (c)	0.2 Btu/lb/°F
infrared emittance of normal surface	0.9
infrared emittance of selective surface	0.1

Solar absorptances

waterwall	1.0
masonry, Trombe wall	1.0
direct gain and sunspace	0.8
sunspace: water containers	0.9
lightweight common wall	0.7
other lightweight surfaces	0.3

Glazing properties

transmission characteristics	diffuse
orientation	due south
index of refraction	1.526
extinction coefficient	0.5 inch⁻¹
thickness of each pane	one-eighth inch
gap between panes	one-half inch
ared emittance	0.9

Control range

room temperature	65 to 75°F
sunspace temperature	45 to 95°F
internal heat generation	0

Table A7.1 Designations and characteristics for 94 reference passive systems (*Continued*)

(a) Overall System Characteristics

Thermocirculation vents (when used)

vent area/projected area (sum of both upper and lower vents)	0.06
height between vents	8 ft
reverse flow	none

Nighttime insulation (when used)

thermal resistance	R9
in place, solar time	5:30 P.M. to 7:30 A.M.

Solar radiation assumptions

shading	none
ground diffuse reflectance	0.3

(b) Direct-Gain (DG) System Types

Designation	Thermal Storage Capacity* (in Btu/ft²/°F)	Mass Thickness* (inches)	Mass-Area-to-Glazing-Area Ratio	No. of Glazings	Nighttime Insulation
A1	30	2	6	2	no
A2	30	2	6	3	no
A3	30	2	6	2	yes
B1	45	6	3	2	no
B2	45	6	3	3	no
B3	45	6	3	2	yes
C1	60	4	6	2	no
C2	60	4	6	3	no
C3	60	4	6	2	yes

(c) Vented Trombe-Wall (TW) System Types

Designation	Thermal Storage Capacity* (Btu/ft²/°F)	Wall Thickness* (inches)	ρck Btu²/hr/ft⁴/°F²)	No. of Glazings	Wall Surface	Nighttime Insulation
A1	15	6	30	2	normal	no
A2	22.5	9	30	2	normal	no
A3	30	12	30	2	normal	no
A4	45	18	30	2	normal	no
B1	15	6	15	2	normal	no
B2	22.5	9	15	2	normal	no
B3	30	12	15	2	normal	no
B4	45	18	15	2	normal	no
C1	15	6	7.5	2	normal	no
C2	22.5	9	7.5	2	normal	no
C3	30	12	7.5	2	normal	no
C4	45	18	7.5	2	normal	no

Table A7.1 Designations and characteristics for 94 reference passive systems *(Continued)*

(c) Vented Trombe-Wall (TW) System Types

D1	30	12	30	1	normal	no
D2	30	12	30	3	normal	no
D3	30	12	30	1	normal	yes
D4	30	12	30	2	normal	yes
D5	30	12	30	3	normal	yes
E1	30	12	30	1	selective	no
E2	30	12	30	2	selective	no
E3	30	12	30	1	selective	yes
E4	30	12	30	2	selective	yes

(d) Unvented Trombe-Wall (TW) System Types

Designation	Thermal Storage Capacity* (Btu/ft²/°F)	Wall Thickness* (inches)	ρck Btu²/hr/ft⁴/°F²)	No. of Glazings	Wall Surface	Nighttime Insulation
F1	15	6	30	2	normal	no
F2	22.5	9	30	2	normal	no
F3	30	12	30	2	normal	no
F4	45	18	30	2	normal	no
G1	15	6	15	2	normal	no
G2	22.5	9	15	2	normal	no
G3	30	12	15	2	normal	no
G4	45	18	15	2	normal	no
H1	15	6	7.5	2	normal	no
H2	22.5	9	7.5	2	normal	no
H3	30	12	7.5	2	normal	no
H4	45	18	7.5	2	normal	no
I1	30	12	30	1	normal	no
I2	30	12	30	3	normal	no
I3	30	12	30	1	normal	yes
I4	30	12	30	2	normal	yes
I5	30	12	30	3	normal	yes
J1	30	12	30	1	selective	no
J2	30	12	30	2	selective	no
J3	30	12	30	1	selective	yes
J4	30	12	30	2	selective	yes

(e) Waterwall (WW) System Types

Designation	Thermal Storage Capacity* (in Btu/ft²/°F)	Wall Thickness (inches)	No. of Glazings	Wall Surface	Nighttime Insulation
A1	15.6	3	2	normal	no
A2	31.2	6	2	normal	no
A3	46.8	9	2	normal	no
A4	62.4	12	2	normal	no

Table A7.1 Designations and characteristics for 94 reference passive systems (*Continued*)

(e) Waterwall (WW) System Types

A5	93.6	18	2	normal	no
A6	124.8	24	2	normal	no
B1	46.8	9	1	normal	no
B2	46.8	9	3	normal	no
B3	46.8	9	1	normal	yes
B4	46.8	9	2	normal	yes
B5	46.8	9	3	normal	yes
C1	46.8	9	1	selective	no
C2	46.8	9	2	selective	no
C3	46.8	9	1	selective	yes
C4	46.8	9	2	selective	yes

(f) Sunspace (SS) System Types

Designation	Type	Tilt (degrees)	Common Wall	End Walls	Nighttime Insulation
A1	attached	50	masonry	opaque	no
A2	attached	50	masonry	opaque	yes
A3	attached	50	masonry	glazed	no
A4	attached	50	masonry	glazed	yes
A5	attached	50	insulated	opaque	no
A6	attached	50	insulated	opaque	yes
A7	attached	50	insulated	glazed	no
A8	attached	50	insulated	glazed	yes
B1	attached	90/30	masonry	opaque	no
B2	attached	90/30	masonry	opaque	yes
B3	attached	90/30	masonry	glazed	no
B4	attached	90/30	masonry	glazed	yes
B5	attached	90/30	insulated	opaque	no
B6	attached	90/30	insulated	opaque	yes
B7	attached	90/30	insulated	glazed	no
B8	attached	90/30	insulated	glazed	yes
C1	semienclosed	90	masonry	common	no
C2	semienclosed	90	masonry	common	yes
C3	semienclosed	90	insulated	common	no
C4	semienclosed	90	insulated	common	yes
D1	semienclosed	50	masonry	common	no
D2	semienclosed	50	masonry	common	yes
D3	semienclosed	50	insulated	common	no
D4	semienclosed	50	insulated	common	yes
E1	semienclosed	90/30	masonry	common	no
E2	semienclosed	90/30	masonry	common	yes
E3	semienclosed	90/30	insulated	common	no
E4	semienclosed	90/30	insulated	common	yes

*The thermal storage capacity is per unit of projected area, or, equivalently, the quantity ρck. The wall thickness is listed only as an appropriate guide by assuming $\rho c = 30$ Btu/ft^3/°F.

Source: PSDH, 1984.

Table A7.2 LCR tables for six representative cities (Albuquerque, Boston, Madison, Medford, Nashville, Santa Maria)

SSF	.10	.20	.30	.40	.50	.60	.70	.80	.90
Santa Maria, California									3053 DD
WW A1	1776	240	119	73	50	35	25	18	12
WW A2	617	259	154	103	74	54	39	28	19
WW A3	523	261	164	114	82	61	45	33	22
WW A4	482	260	169	119	87	65	48	35	24
WW A5	461	263	175	125	92	69	52	38	26
WW A6	447	263	177	128	95	72	54	40	27
WW B1	556	220	128	85	60	43	32	23	15
WW B2	462	256	168	119	88	66	49	36	25
WW B3	542	315	211	151	112	85	64	47	32
WW B4	455	283	197	144	109	83	63	47	32
WW B5	414	263	184	136	103	79	60	45	31
WW C1	569	330	221	159	118	89	67	49	33
WW C2	478	288	197	143	107	81	61	45	31
WW C3	483	318	228	170	130	100	77	57	40
WW C4	426	280	200	149	114	88	68	51	35
TW A1	1515	227	113	70	48	34	24	17	11
TW A2	625	234	134	89	63	46	33	24	16
TW A3	508	231	140	95	68	50	37	27	18
TW A4	431	217	137	95	69	51	38	28	19
TW B1	859	212	112	71	49	35	25	18	12
TW B2	502	209	124	83	59	43	32	23	15
TW B3	438	201	123	84	60	44	33	24	16
TW B4	400	184	112	76	55	40	30	22	14
TW C1	568	188	105	69	48	35	25	18	12
TW C2	435	178	105	70	50	36	27	19	13
TW C3	413	165	97	64	46	33	25	18	12
TW C4	426	146	82	54	38	27	20	14	10
TW D1	403	170	101	67	48	35	25	18	12
TW D2	488	242	152	105	76	57	42	31	21
TW D3	509	271	175	123	90	67	50	36	25
TW D4	464	266	177	127	94	71	53	39	27
TW D5	425	250	169	122	91	69	52	38	26
TW E1	581	309	199	140	102	76	57	42	28
TW E2	512	283	186	132	97	73	55	40	27
TW E3	537	328	225	164	123	94	71	53	36
TW E4	466	287	199	145	109	83	63	47	32
TW F1	713	198	107	68	47	34	25	18	12
TW F2	455	199	120	81	58	42	31	22	15
TW F3	378	190	120	83	60	45	33	24	16
TW F4	311	169	110	77	57	42	32	23	16
TW G1	450	170	98	65	46	33	24	17	12
TW G2	331	163	102	70	51	38	28	20	14
TW G3	278	147	94	66	48	36	27	20	13
TW G4	222	120	78	55	40	30	22	16	11
TW H1	295	137	84	57	41	30	22	16	11
TW H2	226	118	75	52	38	28	21	15	10
TW H3	187	99	64	44	33	24	18	13	9

Table A7.2 LCR tables for six representative cities (Albuquerque, Boston, Madison, Medford, Nashville, Santa Maria) *(Continued)*

SSF	.10	.20	.30	.40	.50	.60	.70	.80	.90
TW H4	143	75	48	33	24	18	14	10	7
TW I1	318	144	88	59	42	31	23	16	11
TW I2	377	203	132	93	68	51	38	28	19
TW I3	404	226	149	106	78	58	44	32	22
TW I4	387	230	156	113	84	64	48	36	24
TW I5	370	226	155	113	85	65	49	36	25
TW J1	483	271	179	127	94	71	53	39	26
TW J2	422	246	165	119	88	67	50	37	25
TW J3	446	283	199	146	111	85	65	48	33
TW J4	400	254	178	132	100	77	58	43	30
DG A1	392	188	117	79	55	38	26	16	7
DG A2	389	190	121	85	61	45	32	22	14
DG A3	443	220	142	102	77	58	44	31	19
DG B1	384	191	122	86	64	48	35	24	13
DG B2	394	196	127	91	69	53	40	29	19
DG B3	445	222	145	105	80	62	49	37	25
DG C1	451	225	146	104	78	61	47	34	21
DG C2	453	226	148	106	80	63	49	37	25
DG C3	509	254	167	121	92	73	58	45	31
SS A1	1171	396	220	142	98	69	49	34	22
SS A2	1028	468	283	190	135	98	71	50	33
SS A3	1174	380	209	133	91	64	45	31	20
SS A4	1077	481	289	193	136	98	71	50	32
SS A5	1896	400	204	127	86	60	42	29	18
SS A6	1030	468	283	190	135	97	71	50	32
SS A7	2199	359	178	109	72	50	35	24	15
SS A8	1089	478	285	190	133	96	69	48	31
SS B1	802	298	170	111	77	55	40	28	18
SS B2	785	366	224	152	108	79	57	41	27
SS B3	770	287	163	106	74	52	37	26	17
SS B4	790	368	224	152	108	78	57	40	26
SS B5	1022	271	144	91	62	44	31	22	14
SS B6	750	356	219	149	106	77	56	40	26
SS B7	937	242	127	80	54	38	27	19	12
SS B8	750	352	215	146	103	75	55	39	25
SS C1	481	232	144	99	71	52	39	28	19
SS C2	482	262	170	120	88	66	49	36	24
SS C3	487	185	107	71	50	36	27	19	13
SS C4	473	235	147	102	74	55	41	30	20
SS D1	1107	477	282	188	132	95	68	48	31
SS D2	928	511	332	232	169	125	92	66	43
SS D3	1353	449	248	160	110	78	56	39	25
SS D4	946	500	319	222	160	117	86	61	40
SS E1	838	378	227	153	108	78	56	40	26
SS E2	766	419	272	190	138	102	75	54	36
SS E3	973	322	178	115	79	56	40	28	18
SS E4	780	393	247	170	122	89	65	47	31

Table A7.2 LCR tables for six representative cities (Albuquerque, Boston, Madison, Medford, Nashville, Santa Maria) *(Continued)*

SSF	.10	.20	.30	.40	.50	.60	.70	.80	.90
Albuquerque, New Mexico									4292 DD
WW A1	1052	130	62	38	25	18	13	9	6
WW A2	354	144	84	56	39	29	21	15	10
WW A3	300	146	90	62	45	33	24	18	12
WW A4	276	146	93	65	47	35	26	19	13
WW A5	264	148	97	69	50	38	28	21	14
WW A6	256	148	99	70	52	39	30	22	15
WW B1	293	111	63	41	28	20	15	11	7
WW B2	270	147	96	67	49	37	28	20	14
WW B3	314	179	119	84	62	47	35	26	18
WW B4	275	169	116	85	64	49	37	28	19
WW B5	252	159	110	81	61	47	36	27	19
WW C1	333	190	126	89	66	50	38	28	19
WW C2	287	171	115	83	62	47	36	27	18
WW C3	293	191	136	101	77	59	46	34	24
WW C4	264	172	122	91	69	54	41	31	22
TW A1	900	124	60	37	25	17	12	9	6
TW A2	361	130	73	48	33	24	18	13	8
TW A3	293	129	77	52	37	27	20	15	10
TW A4	249	123	76	52	38	28	21	15	10
TW B1	502	117	60	38	26	18	13	9	6
TW B2	291	118	68	45	32	23	17	12	8
TW B3	254	114	68	46	33	24	18	13	9
TW B4	233	104	63	42	30	22	16	12	8
TW C1	332	106	58	37	26	19	14	10	6
TW C2	255	101	58	39	27	20	15	11	7
TW C3	243	94	54	36	25	18	13	10	7
TW C4	254	84	46	30	21	15	11	8	5
TW D1	213	86	50	33	23	17	12	9	6
TW D2	287	139	86	59	43	32	24	17	12
TW D3	294	153	97	68	49	37	27	20	14
TW D4	281	158	104	74	55	41	31	23	16
TW D5	260	151	101	73	54	41	31	23	16
TW E1	339	177	113	78	57	43	32	23	16
TW E2	308	168	109	77	56	42	32	23	16
TW E3	323	195	133	96	72	55	42	31	21
TW E4	287	175	120	88	66	50	38	28	20
TW F1	409	108	57	36	24	17	13	9	6
TW F2	260	110	65	43	31	22	17	12	8
TW F3	216	106	66	45	33	24	10	13	9
TW F4	178	95	61	42	31	23	17	13	9
TW G1	256	93	53	34	24	17	13	9	6
TW G2	189	91	56	38	27	20	15	11	7
TW G3	159	82	52	36	26	20	15	11	7
TW G4	128	68	43	30	22	16	12	9	6
TW H1	168	76	45	31	22	16	12	9	6
TW H2	130	66	41	29	21	15	11	8	6

Table A7.2 LCR tables for six representative cities (Albuquerque, Boston, Madison, Medford, Nashville, Santa Maria) *(Continued)*

SSF	.10	.20	.30	.40	.50	.60	.70	.80	.90
TW H3	108	56	35	25	8	13	10	7	5
TW H4	83	42	27	19	13	10	7	5	4
TW I1	166	73	43	29	20	15	11	8	5
TW I2	221	117	75	52	30	28	21	16	11
TW I3	234	128	83	59	43	32	24	10	12
TW I4	234	137	92	66	49	37	28	21	14
TW I5	226	136	93	67	50	38	29	22	15
TW J1	282	156	102	72	53	40	30	22	15
TW J2	254	146	97	69	51	39	29	22	15
TW J3	269	169	118	86	65	50	38	29	20
TW J4	247	155	106	80	60	46	35	26	18
DG A1	211	97	57	36	22	13	5	—	—
DG A2	227	107	67	46	32	23	16	10	5
DG A3	274	131	83	59	44	34	25	18	10
DG B1	210	97	60	42	30	21	13	6	—
DG B2	232	110	69	49	37	28	21	14	8
DG B3	277	134	85	61	47	37	28	21	14
DG C1	253	120	74	53	39	30	22	14	—
DG C2	271	130	82	59	45	35	26	19	12
DG C3	318	155	96	71	54	43	34	26	18
SS A1	591	187	101	64	44	31	22	16	10
SS A2	531	232	137	92	65	47	34	25	16
SS A3	566	170	90	56	38	27	19	13	8
SS A4	537	230	135	89	63	45	33	23	15
SS A5	980	187	92	56	37	26	18	13	8
SS A6	529	231	136	91	64	47	34	24	16
SS A7	1103	158	74	44	29	20	14	10	6
SS A8	540	226	131	87	61	44	32	23	15
SS B1	403	141	78	50	35	25	18	13	8
SS B2	412	186	111	75	53	39	28	20	14
SS B3	372	130	71	46	31	22	16	11	7
SS B4	403	181	106	72	51	37	27	20	13
SS B5	518	127	65	40	27	19	13	9	6
SS B6	390	179	106	73	52	38	28	20	13
SS B7	457	108	54	33	22	16	11	8	5
SS B8	379	171	102	69	49	35	26	19	12
SS C1	270	126	77	52	37	27	20	15	10
SS C2	282	150	97	68	49	37	28	20	14
SS C3	276	101	57	37	26	19	14	10	7
SS C4	277	135	83	57	41	31	23	17	11
SS D1	548	225	130	85	59	43	31	22	14
SS D2	474	253	162	113	82	61	45	33	22
SS D3	683	212	113	72	49	35	25	17	11
SS D4	484	248	156	107	77	57	42	30	20
SS E1	410	176	103	68	48	35	25	18	12
SS E2	390	208	133	92	67	50	37	27	18
SS E3	487	151	80	51	35	25	18	12	8
SS E4	400	195	120	82	59	43	32	23	15

Table A7.2 LCR tables for six representative cities (Albuquerque, Boston, Madison, Medford, Nashville, Santa Maria) *(Continued)*

SSF	.10	.20	.30	.40	.50	.60	.70	.80	.90
Nashville, Tennessee									3696 DD
WW A1	588	60	24	13	8	5	3	2	1
WW A2	192	70	38	23	15	11	7	5	3
WW A3	161	72	42	27	18	13	9	6	4
WW A4	148	72	43	29	20	14	10	7	5
WW A5	141	74	46	31	22	16	11	8	5
WW A6	137	74	47	32	22	16	12	8	5
WW B1	135	41	19	10	6	3	2	—	—
WW B2	152	78	48	33	23	17	12	9	6
WW B3	179	97	61	42	30	22	16	12	8
WW B4	164	97	65	46	34	25	19	14	9
WW B5	153	93	63	45	33	25	19	14	9
WW C1	193	105	67	46	33	24	18	13	8
WW C2	169	97	63	44	32	24	18	13	8
WW C3	181	115	79	58	43	33	25	18	12
WW C4	164	104	72	53	39	30	23	17	11
TW A1	509	59	25	13	8	5	3	2	1
TW A2	199	64	33	20	13	9	6	4	3
TW A3	160	65	36	23	15	11	8	5	3
TW A4	136	62	36	23	16	11	8	6	4
TW B1	282	57	26	15	9	6	4	3	2
TW B2	161	59	32	20	13	9	6	4	3
TW B3	141	58	32	21	14	10	7	5	3
TW B4	131	54	30	19	13	9	7	5	3
TW C1	188	53	27	16	10	7	5	3	2
TW C2	144	52	28	18	12	8	6	4	2
TW C3	139	49	27	17	11	8	5	4	2
TW C4	149	45	23	14	9	7	5	3	2
TW D1	99	33	16	9	5	3	2	1	—
TW D2	164	75	44	29	20	14	10	7	5
TW D3	167	82	49	33	23	17	12	8	5
TW D4	168	91	58	40	29	21	15	11	7
TW D5	160	89	58	40	29	22	16	12	8
TW E1	198	98	59	40	28	20	15	10	7
TW E2	182	95	59	40	29	21	15	11	7
TW E3	197	115	76	54	39	29	22	16	11
TW E4	178	105	70	50	37	27	20	15	10
TW F1	221	50	23	13	8	5	4	2	1
TW F2	139	53	29	18	12	8	6	4	2
TW F3	116	52	30	19	13	9	7	5	3
TW F4	96	47	28	19	13	9	7	5	3
TW G1	137	44	22	13	9	6	4	3	2
TW G2	101	44	25	16	11	8	5	4	2
TW G3	86	41	24	16	11	8	6	4	2
TW G4	69	34	21	14	10	7	5	3	2
TW H1	89	36	20	13	8	6	4	3	2
TW H2	69	33	19	12	9	6	4	3	2
TW H3	59	28	17	11	8	5	4	3	2

Table A7.2 LCR tables for six representative cities (Albuquerque, Boston, Madison, Medford, Nashville, Santa Maria) *(Continued)*

SSF	.10	.20	.30	.40	.50	.60	.70	.80	.90
TW H4	46	22	13	9	6	4	3	2	1
TW I1	74	26	13	7	4	2	1	—	—
TW I2	125	62	38	25	18	13	9	7	4
TW I3	133	69	43	29	20	15	11	8	5
TW I4	139	78	51	35	26	19	14	10	7
TW I5	137	80	53	37	27	20	15	11	7
TW J1	164	86	54	36	26	19	14	10	6
TW J2	150	82	53	36	26	19	14	10	7
TW J3	165	101	68	49	36	27	20	15	10
TW J4	153	93	63	46	34	25	19	14	10
DG A1	98	34	—	—	—	—	—	—	—
DG A2	130	55	31	19	11	6	—	—	—
DG A3	173	78	47	32	23	16	11	7	2
DG B1	100	36	17	—	—	—	—	—	—
DG B2	134	58	33	22	15	10	6	—	—
DG B3	177	81	49	33	24	18	14	10	6
DG C1	131	52	28	17	9	—	—	—	—
DG C2	161	71	42	28	20	14	10	6	—
DG C3	205	94	57	39	29	22	17	12	8
SS A1	351	100	50	29	19	13	9	6	4
SS A2	328	135	76	49	33	24	17	12	8
SS A3	330	87	41	24	15	10	6	4	2
SS A4	331	133	74	47	32	22	16	11	7
SS A5	595	98	43	24	15	10	7	4	2
SS A6	324	132	75	48	32	23	16	11	7
SS A7	668	79	32	17	10	6	4	2	1
SS A8	330	129	71	45	30	21	15	10	6
SS B1	236	74	38	23	15	10	7	5	3
SS B2	258	110	63	41	28	20	14	10	6
SS B3	212	65	32	19	12	8	5	3	2
SS B4	251	105	60	39	27	19	13	9	6
SS B5	307	65	30	17	10	7	4	3	2
SS B6	241	104	60	39	27	19	14	10	6
SS B7	264	52	23	12	7	5	3	2	—
SS B8	233	98	56	36	25	17	12	9	5
SS C1	141	60	33	21	14	10	7	5	3
SS C2	161	81	50	33	23	17	12	9	6
SS C3	149	48	25	15	10	7	4	3	2
SS C4	160	73	43	28	19	14	10	7	5
SS D1	317	119	64	39	26	18	13	8	5
SS D2	287	147	90	61	43	31	23	16	10
SS D3	405	113	55	33	21	14	10	6	4
SS D4	295	144	87	58	40	29	21	15	10
SS E1	229	89	48	29	19	13	9	6	4
SS E2	233	118	72	48	34	24	18	12	8
SS E3	283	77	37	22	14	9	6	4	2
SS E4	242	111	65	43	29	21	15	11	7

Table A7.2 LCR tables for six representative cities (Albuquerque, Boston, Madison, Medford, Nashville, Santa Maria) *(Continued)*

SSF	.10	.20	.30	.40	.50	.60	.70	.80	.90
Medford, Oregon									4930 DD
WW A1	708	64	24	11	—	—	—	—	—
WW A2	212	73	38	22	13	7	3	—	—
WW A3	174	75	41	25	16	9	5	2	—
WW A4	158	74	43	27	17	11	6	3	1
WW A5	149	75	45	29	19	12	7	4	2
WW A6	144	75	46	30	20	13	8	4	2
WW B1	154	43	16	—	—	—	—	—	—
WW B2	162	80	48	31	21	14	9	6	3
WW B3	190	100	62	41	28	19	13	8	5
WW B4	171	99	65	45	32	23	16	11	7
WW B5	160	95	63	45	32	23	17	12	7
WW C1	205	108	67	45	31	21	15	10	6
WW C2	178	99	63	43	30	22	15	10	6
WW C3	189	117	80	57	42	31	23	16	10
WW C4	170	106	72	52	38	28	21	15	9
TW A1	607	63	25	12	5	—	—	—	—
TW A2	222	68	33	19	11	6	2	—	—
TW A3	175	67	36	21	13	8	4	2	—
TW A4	147	64	36	22	14	9	5	3	1
TW B1	327	61	27	14	7	3	—	—	—
TW B2	178	62	32	19	12	7	4	2	—
TW B3	154	60	33	20	12	8	4	2	1
TW B4	143	56	31	19	12	8	5	2	1
TW C1	212	56	27	15	9	5	2	—	—
TW C2	159	55	28	17	11	7	4	2	—
TW C3	154	52	27	16	10	6	4	2	1
TW C4	167	48	24	14	9	5	3	2	—
TW D1	112	34	14	—	—	—	—	—	—
TW D2	177	77	44	28	18	12	8	5	3
TW D3	180	85	50	32	21	14	9	6	3
TW D4	177	93	58	39	27	19	13	9	5
TW D5	168	92	58	40	28	20	14	10	6
TW E1	213	101	60	39	26	18	12	8	4
TW E2	194	98	59	39	27	19	13	9	5
TW E3	208	118	77	53	38	27	20	13	8
TW E4	186	108	71	49	36	26	19	13	8
TW F1	256	53	23	12	5	—	—	—	—
TW F2	153	56	29	17	10	5	2	—	—
TW F3	125	54	30	18	11	7	3	1	—
TW F4	102	48	28	18	11	7	4	2	1
TW G1	153	46	22	12	7	—	—	—	—
TW G2	109	46	25	15	9	5	3	1	—
TW G3	92	42	24	15	9	6	3	2	—
TW G4	74	35	20	13	8	5	3	2	—
TW H1	97	38	20	12	7	4	1	—	—
TW H2	75	34	19	12	7	5	3	1	—

Table A7.2 LCR tables for six representative cities (Albuquerque, Boston, Madison, Medford, Nashville, Santa Maria) *(Continued)*

SSF	.10	.20	.30	.40	.50	.60	.70	.80	.90
TW H3	63	29	17	10	7	4	3	1	—
TW H4	49	23	13	8	5	3	2	1	—
TW I1	83	27	10	—	—	—	—	—	—
TW I2	133	64	38	24	16	11	7	4	2
TW I3	142	71	43	28	19	13	9	5	3
TW I4	146	80	51	35	25	17	12	8	5
TW I5	144	82	53	37	26	19	13	9	6
TW J1	175	89	54	36	24	17	11	7	4
TW J2	158	85	53	36	25	18	12	8	5
TW J3	173	103	69	48	35	26	18	13	8
TW J4	160	96	64	45	33	24	17	12	8
DG A1	110	35	—	—	—	—	—	—	—
DG A2	142	58	32	18	9	—	—	—	—
DG A3	187	82	48	32	22	15	9	5	—
DG B1	110	40	15	—	—	—	—	—	—
DG B2	146	61	35	21	13	7	—	—	—
DG B3	193	84	51	34	24	17	12	7	3
DG C1	144	57	29	13	—	—	—	—	—
DG C2	177	75	44	28	19	12	6	—	—
DG C3	224	98	60	41	29	21	14	10	5
SS A1	415	110	51	28	16	9	4	2	—
SS A2	372	146	79	48	31	21	14	8	5
SS A3	397	96	42	21	10	—	—	—	—
SS A4	379	144	76	46	29	19	12	7	4
SS A5	732	111	45	23	12	5	—	—	—
SS A6	368	143	77	47	30	20	13	8	4
SS A7	846	90	33	14	—	—	—	—	—
SS A8	379	140	73	44	27	17	11	6	3
SS B1	274	81	38	21	12	6	3	—	—
SS B2	288	117	65	40	26	18	12	7	4
SS B3	249	71	33	17	8	—	—	—	—
SS B4	282	113	62	38	25	16	11	7	4
SS B5	368	72	30	15	7	—	—	—	—
SS B6	269	111	62	30	25	17	11	7	4
SS B7	323	58	23	10	—	—	—	—	—
SS B8	262	106	57	35	23	15	9	6	3
SS C1	153	62	33	19	11	5	—	—	—
SS C2	172	83	50	32	22	15	10	6	3
SS C3	166	51	24	13	7	3	—	—	—
SS C4	173	76	43	27	18	12	8	5	3
SS D1	367	129	65	37	22	13	7	3	1
SS D2	318	156	92	60	40	27	18	12	7
SS D3	480	124	57	31	18	10	5	2	—
SS D4	328	153	89	57	38	26	17	11	6
SS E1	262	95	48	27	15	7	—	—	—
SS E2	257	124	73	47	31	21	14	9	5
SS E3	334	84	38	20	10	4	—	—	—
SS E4	269	118	67	42	27	18	12	7	4

Table A7.2 LCR tables for six representative cities (Albuquerque, Boston, Madison, Medford, Nashville, Santa Maria) *(Continued)*

SSF	.10	.20	.30	.40	.50	.60	.70	.80	.90
Boston, Massachusetts									5621 DD
WW A1	368	28	9	—	—	—	—	—	—
WW A2	119	41	20	12	7	5	3	2	—
WW A3	101	43	24	15	10	6	4	3	1
WW A4	93	44	26	16	11	7	5	3	2
WW A5	89	45	27	18	12	8	6	4	2
WW A6	87	46	28	19	13	9	6	4	3
WW B1	59	—	—	—	—	—	—	—	—
WW B2	103	52	31	21	15	10	7	5	3
WW B3	123	66	41	28	20	14	10	7	5
WW B4	118	70	46	33	24	18	13	9	6
WW B5	113	69	46	33	25	18	14	10	7
WW C1	135	72	46	31	22	16	12	8	5
WW C2	121	68	44	31	22	16	12	9	6
WW C3	136	86	60	44	33	25	19	14	9
WW C4	124	78	54	40	30	23	17	12	8
TW A1	324	30	11	4	—	—	—	—	—
TW A2	126	37	18	10	6	4	2	1	—
TW A3	102	39	21	13	8	5	3	2	1
TW A4	88	38	22	14	9	6	4	3	2
TW B1	180	32	13	7	4	2	—	—	—
TW B2	104	36	19	11	7	5	3	2	1
TW B3	92	36	19	12	8	5	3	2	1
TW B4	86	34	19	12	8	5	4	2	1
TW C1	122	32	15	9	5	3	2	1	—
TW C2	95	33	17	10	7	4	3	2	1
TW C3	93	31	16	10	6	4	3	2	1
TW C4	102	29	15	9	6	4	3	2	1
TW D1	45	—	—	—	—	—	—	—	—
TW D2	112	49	28	18	12	9	6	4	3
TW D3	113	54	32	21	15	10	7	5	3
TW D4	121	64	41	28	20	15	11	8	5
TW D5	118	66	42	30	21	16	12	8	6
TW E1	138	67	40	27	18	13	9	7	4
TW E2	130	66	41	28	20	14	10	7	5
TW E3	146	84	56	39	29	21	16	11	8
TW E4	133	78	52	37	27	20	15	11	7
TW F1	134	25	10	4	—	—	—	—	—
TW F2	86	30	16	9	5	3	2	1	—
TW F3	72	31	17	11	7	4	3	2	1
TW F4	61	29	17	11	7	5	3	2	1
TW G1	83	24	11	6	3	2	—	—	—
TW G2	63	26	14	9	5	4	2	1	—
TW G3	54	25	14	9	6	4	3	2	1
TW G4	45	21	12	8	5	4	3	2	1
TW H1	54	21	11	6	4	2	1	—	—
TW H2	44	20	11	7	5	3	2	1	—
TW H3	38	17	10	6	4	3	2	1	—

Table A7.2 LCR tables for six representative cities (Albuquerque, Boston, Madison, Medford, Nashville, Santa Maria) (Continued)

SSF	.10	.20	.30	.40	.50	.60	.70	.80	.90
TW H4	30	14	8	5	3	2	2	1	—
TW I1	30	—	—	—	—	—	—	—	—
TW I2	84	41	24	16	11	8	6	4	2
TW I3	91	46	28	19	13	9	7	5	3
TW I4	100	56	36	25	18	13	10	7	5
TW I5	101	58	38	27	20	15	11	8	5
TW J1	114	59	37	25	17	12	9	6	4
TW J2	107	58	37	25	18	13	10	7	4
TW J3	123	75	51	36	27	20	15	11	7
TW J4	115	70	47	34	25	19	14	10	7
DG A1	43	—	—	—	—	—	—	—	—
DG A2	85	34	18	9	—	—	—	—	—
DG A3	125	56	33	22	16	11	7	4	—
DG B1	44	—	—	—	—	—	—	—	—
DG B2	87	36	20	12	7	—	—	—	—
DG B3	129	58	35	24	17	13	9	6	—
DG C1	71	23	—	—	—	—	—	—	—
DG C2	109	47	27	17	12	8	4	—	—
DG C3	151	68	41	28	21	16	12	8	5
SS A1	230	61	29	16	10	6	4	2	1
SS A2	231	93	52	33	22	15	11	7	5
SS A3	205	48	20	10	4	—	—	—	—
SS A4	229	90	49	31	20	14	9	6	4
SS A5	389	58	23	11	6	3	—	—	—
SS A6	226	91	50	32	21	15	10	7	4
SS A7	420	40	12	—	—	—	—	—	—
SS A8	226	86	46	28	19	12	8	6	3
SS B1	151	44	21	12	7	4	2	1	—
SS B2	183	77	43	28	19	13	9	6	4
SS B3	129	36	16	8	3	—	—	—	—
SS B4	176	73	41	26	17	12	8	6	4
SS B5	193	36	15	7	3	—	—	—	—
SS B6	169	72	41	26	18	12	9	6	4
SS B7	157	25	7	—	—	—	—	—	—
SS B8	160	66	37	23	16	11	7	5	3
SS C1	84	33	17	10	6	4	2	1	—
SS C2	110	54	33	22	15	11	8	5	3
SS C3	91	26	12	7	4	2	—	—	—
SS C4	109	48	28	18	12	9	6	4	3
SS D1	206	73	38	22	14	9	5	3	2
SS D2	203	103	63	42	29	21	15	10	6
SS D3	264	69	32	18	10	6	4	2	1
SS D4	208	100	60	39	27	19	14	9	6
SS E1	140	51	25	14	8	4	2	—	—
SS E2	161	80	48	32	22	15	11	7	5
SS E3	177	44	19	10	5	2	—	—	—
SS E4	166	75	43	28	19	13	9	6	4

Table A7.2 LCR tables for six representative cities (Albuquerque, Boston, Madison, Medford, Nashville, Santa Maria) *(Continued)*

SSF	.10	.20	.30	.40	.50	.60	.70	.80	.90
Madison, Wisconsin									7730 DD
WW A1	278	—	—	—	—	—	—	—	—
WW A2	91	27	12	—	—	—	—	—	—
WW A3	77	30	15	8	3	—	—	—	—
WW A4	72	32	17	10	5	—	—	—	—
WW A5	69	33	19	11	7	4	—	—	—
WW A6	67	34	19	12	7	4	2	—	—
WW B1	—	—	—	—	—	—	—	—	—
WW B2	84	41	24	15	10	7	5	3	2
WW B3	102	53	32	21	15	10	7	5	3
WW B4	101	59	39	27	19	14	10	7	5
WW B5	98	59	39	28	20	15	11	8	5
WW C1	113	59	37	25	17	12	8	6	3
WW C2	103	57	37	25	18	13	9	6	4
WW C3	119	75	51	37	28	21	15	11	7
WW C4	109	68	47	34	25	19	14	10	7
TW A1	249	16	—	—	—	—	—	—	—
TW A2	97	26	11	4	—	—	—	—	—
TW A3	79	28	13	7	3	—	—	—	—
TW A4	69	28	15	9	5	3	—	—	—
TW B1	139	20	5	—	—	—	—	—	—
TW B2	81	26	12	6	3	—	—	—	—
TW B3	72	27	13	7	4	2	—	—	—
TW B4	69	26	13	8	5	3	1	—	—
TW C1	96	23	10	4	—	—	—	—	—
TW C2	76	25	12	7	4	2	—	—	—
TW C3	75	24	12	7	4	2	1	—	—
TW C4	84	23	11	6	4	2	1	—	—
TW D1	—	—	—	—	—	—	—	—	—
TW D2	91	39	22	13	9	6	4	2	1
TW D3	93	43	25	16	10	7	5	3	1
TW D4	103	54	34	23	16	12	8	6	4
TW D5	102	56	36	25	18	13	10	7	4
TW E1	115	54	32	21	14	10	7	4	3
TW E2	110	55	34	22	16	11	8	5	3
TW E3	126	72	47	33	24	18	13	9	6
TW E4	116	68	45	32	23	17	13	9	6
TW F1	99	13	—	—	—	—	—	—	—
TW F2	65	20	8	—	—	—	—	—	—
TW F3	55	22	11	5	—	—	—	—	—
TW F4	47	21	11	7	4	2	—	—	—
TW G1	61	14	—	—	—	—	—	—	—
TW G2	47	18	8	4	—	—	—	—	—
TW G3	42	18	9	5	3	—	—	—	—
TW G4	35	16	9	5	3	2	—	—	—
TW H1	41	13	6	—	—	—	—	—	—
TW H2	34	14	7	4	2	—	—	—	—

Table A7.2 LCR tables for six representative cities (Albuquerque, Boston, Madison, Medford, Nashville, Santa Maria) *(Continued)*

SSF	.10	.20	.30	.40	.50	.60	.70	.80	.90
TW H3	29	13	7	4	2	1	—	—	—
TW H4	24	10	6	3	2	1	—	—	—
TW I1	—	—	—	—	—	—	—	—	—
TW I2	68	32	18	12	8	5	3	2	1
TW I3	75	37	22	14	10	7	4	3	2
TW I4	85	47	30	21	15	11	8	5	3
TW I5	87	50	33	23	16	12	9	6	4
TW J1	95	48	29	19	13	9	6	4	3
TW J2	91	48	30	21	14	10	7	5	3
TW J3	106	65	43	31	23	17	12	9	6
TW J4	100	61	41	29	21	16	12	9	6
DG A1	—	—	—	—	—	—	—	—	—
DG A2	68	25	11	—	—	—	—	—	—
DG A3	109	47	28	18	12	8	5	—	—
DG B1	—	—	—	—	—	—	—	—	—
DG B2	70	27	14	6	—	—	—	—	—
DG B3	114	50	30	20	14	10	7	4	—
DG C1	47	—	—	—	—	—	—	—	—
DG C2	91	37	21	13	7	—	—	—	—
DG C3	133	59	35	24	17	13	9	6	3
SS A1	192	47	20	9	3	—	—	—	—
SS A2	200	78	42	26	17	12	8	5	3
SS A3	166	32	—	—	—	—	—	—	—
SS A4	197	74	39	23	15	10	6	4	2
SS A5	329	42	13	—	—	—	—	—	—
SS A6	195	75	40	25	16	11	7	5	3
SS A7	349	22	—	—	—	—	—	—	—
SS A8	192	69	36	21	13	8	5	3	2
SS B1	122	32	13	5	—	—	—	—	—
SS B2	158	64	36	22	15	10	7	5	3
SS B3	100	22	—	—	—	—	—	—	—
SS B4	150	60	33	29	13	9	6	4	2
SS B5	156	24	—	—	—	—	—	—	—
SS B6	145	59	33	20	13	9	6	4	2
SS B7	122	—	—	—	—	—	—	—	—
SS B8	136	54	29	18	11	7	5	3	2
SS C1	61	20	7	—	—	—	—	—	—
SS C2	90	43	25	16	11	7	5	3	2
SS C3	67	16	—	—	—	—	—	—	—
SS C4	90	38	22	13	9	6	4	2	1
SS D1	169	56	26	13	6	—	—	—	—
SS D2	175	86	51	34	23	16	11	7	5
SS D3	221	52	21	10	—	—	—	—	—
SS D4	179	84	49	32	21	15	10	7	4
SS E1	108	34	12	—	—	—	—	—	—
SS E2	135	65	38	24	16	11	7	5	3
SS E3	141	29	8	—	—	—	—	—	—
SS E4	140	61	34	21	14	9	6	4	2

Source: PSDH, 1984.

Table A7.3 SLR Correlation Parameters for the 94 Reference Systems

Type	A	B	C	D	R	G	H	LCRs	STDV
WW A1	0.0000	1.0000	.9172	.4841	−9.0000	0.00	1.17	13.0	.053
WW A2	0.0000	1.0000	.9833	.7603	−9.0000	0.00	.92	13.0	.046
WW A3	0.0000	1.0000	1.0171	.8852	−9.0000	0.00	.85	13.0	.040
WW A4	0.0000	1.0000	1.0395	.9569	−9.0000	0.00	.81	13.0	.037
WW A5	0.0000	1.0000	1.0604	1.0387	−9.0000	0.00	.78	13.0	.034
WW A6	0.0000	1.0000	1.0735	1.0827	−9.0000	0.00	.76	13.0	.033
WW B1	0.0000	1.0000	.9754	.5518	−9.0000	0.00	.92	22.0	.051
WW B2	0.0000	1.0000	1.0487	1.0851	−9.0000	0.00	.78	9.2	.036
WW B3	0.0000	1.0000	1.0673	1.0087	−9.0000	0.00	.95	8.9	.038
WW B4	0.0000	1.0000	1.1028	1.1811	−9.0000	0.00	.74	5.8	.034
WW B5	0.0000	1.0000	1.1146	1.2771	−9.0000	0.00	.56	4.5	.032
WW C1	0.0000	1.0000	1.0667	1.0437	−9.0000	0.00	.62	12.0	.038
WW C2	0.0000	1.0000	1.0846	1.1482	−9.0000	0.00	.59	8.7	.035
WW C3	0.0000	1.0000	1.1419	1.1756	−9.0000	0.00	.28	5.5	.033
WW C4	0.0000	1.0000	1.1401	1.2378	−9.0000	0.00	.23	4.3	.032
TW A1	0.0000	1.0000	.9194	.4601	−9.0000	0.00	1.11	13.0	.048
TW A2	0.0000	1.0000	.9680	.6318	−9.0000	0.00	.92	13.0	.043
TW A3	0.0000	1.0000	.9964	.7123	−9.0000	0.00	.85	13.0	.038
TW A4	0.0000	1.0000	1.0190	.7332	−9.0000	0.00	.79	13.0	.032
TW B1	0.0000	1.0000	.9364	.4777	−9.0000	0.00	1.01	13.0	.045
TW B2	0.0000	1.0000	.9821	.6020	−9.0000	0.00	.85	13.0	.038
TW B3	0.0000	1.0000	.9980	.6191	−9.0000	0.00	.80	13.0	.033
TW B4	0.0000	1.0000	.9981	.5615	−9.0000	0.00	.76	13.0	.028
TW C1	0.0000	1.0000	.9558	.4709	−9.0000	0.00	.89	13.0	.039
TW C2	0.0000	1.0000	.9788	.4964	−9.0000	0.00	.79	13.0	.033
TW C3	0.0000	1.0000	.9760	.4519	−9.0000	0.00	.76	13.0	.029
TW C4	0.0000	1.0000	.9588	.3612	−9.0000	0.00	.73	13.0	.026
TW D1	0.0000	1.0000	.9842	.4418	−9.0000	0.00	.89	22.0	.040
TW D2	0.0000	1.0000	1.0150	.8994	−9.0000	0.00	.80	9.2	.036
TW D3	0.0000	1.0000	1.0346	.7810	−9.0000	0.00	1.08	8.9	.036
TW D4	0.0000	1.0000	1.0606	.9770	−9.0000	0.00	.85	5.8	.035
TW D5	0.0000	1.0000	1.0721	1.0718	−9.0000	0.00	.61	4.5	.033
TW E1	0.0000	1.0000	1.0345	.8753	−9.0000	0.00	.68	12.0	.037
TW E2	0.0000	1.0000	1.0476	1.0050	−9.0000	0.00	.66	8.7	.035
TW E3	0.0000	1.0000	1.0919	1.0739	−9.0000	0.00	.61	5.5	.034
TW E4	0.0000	1.0000	1.0971	1.1429	−9.0000	0.00	.47	4.3	.033
TW F1	0.0000	1.0000	.9430	.4744	−9.0000	0.00	1.09	13.0	.047
TW F2	0.0000	1.0000	.9900	.6053	−9.0000	0.00	.93	13.0	.041
TW F3	0.0000	1.0000	1.0189	.6502	−9.0000	0.00	.86	13.0	.036
TW F4	0.0000	1.0000	1.0419	.6258	−9.0000	0.00	.80	13.0	.032
TW G1	0.0000	1.0000	.9693	.4714	−9.0000	0.00	1.01	13.0	.042
TW G2	0.0000	1.0000	1.0133	.5462	−9.0000	0.00	.88	13.0	.035
TW G3	0.0000	1.0000	1.0325	.5269	−9.0000	0.00	.82	13.0	.031
TW G4	0.0000	1.0000	1.0401	.4400	−9.0000	0.00	.77	13.0	.030
TW H1	0.0000	1.0000	1.0002	.4356	−9.0000	0.00	.93	13.0	.034
TW H2	0.0000	1.0000	1.0280	.4151	−9.0000	0.00	.83	13.0	.030
TW H3	0.0000	1.0000	1.0327	.3522	−9.0000	0.00	.78	13.0	.029
TW H4	0.0000	1.0000	1.0287	.2600	−9.0000	0.00	.74	13.0	.024
TW I1	0.0000	1.0000	.9974	.4036	−9.0000	0.00	.91	22.0	.038
TW I2	0.0000	1.0000	1.0386	.8313	−9.0000	0.00	.80	9.2	.034
TW I3	0.0000	1.0000	1.0514	.6886	−9.0000	0.00	1.01	8.9	.034
TW I4	0.0000	1.0000	1.0781	.8952	−9.0000	0.00	.82	5.8	.032
TW I5	0.0000	1.0000	1.0902	1.0284	−9.0000	0.00	.65	4.5	.032
TW J1	0.0000	1.0000	1.0537	.8227	−9.0000	0.00	.65	12.0	.037
TW J2	0.0000	1.0000	1.0677	.9312	−9.0000	0.00	.62	8.7	.035
TW J3	0.0000	1.0000	1.1153	.9831	−9.0000	0.00	.44	5.5	.034

Table A7.3 SLR Correlation Parameters for the 94 Reference Systems *(Continued)*

Type	A	B	C	D	R	G	H	LCRs	STDV
TW J4	0.0000	1.0000	1.1154	1.0607	−9.0000	0.00	.38	4.3	.033
DG A1	.5650	1.0090	1.0440	.7175	.3931	9.36	0.00	0.0	.046
DG A2	.5906	1.0060	1.0650	.8099	.4681	5.28	0.00	0.0	.039
DG A3	.5442	.9715	1.1300	.9273	.7068	2.64	0.00	0.0	.036
DG B1	.5739	.9948	1.2510	1.0610	.7905	9.60	0.00	0.0	.042
DG B2	.6180	1.0000	1.2760	1.1560	.7528	5.52	0.00	0.0	.035
DG B3	.5601	.9839	1.3520	1.1510	.8879	2.38	0.00	0.0	.032
DG C1	.6344	.9887	1.5270	1.4380	.8632	9.60	0.00	0.0	.039
DG C2	.6763	.9994	1.4000	1.3940	.7604	5.28	0.00	0.0	.033
DG C3	.6182	.9859	1.5660	1.4370	.8990	2.40	0.00	0.0	.031
SS A1	0.0000	1.0000	.9587	.4770	−9.0000	0.00	.83	18.6	.027
SS A2	0.0000	1.0000	.9982	.6614	−9.0000	0.00	.77	10.4	.026
SS A3	0.0000	1.0000	.9552	.4230	−9.0000	0.00	.83	23.6	.030
SS A4	0.0000	1.0000	.9956	.6277	−9.0000	0.00	.80	12.4	.026
SS A5	0.0000	1.0000	.9300	.4041	−9.0000	0.00	.96	18.6	.031
SS A6	0.0000	1.0000	.9981	.6660	−9.0000	0.00	.86	10.4	.028
SS A7	0.0000	1.0000	.9219	.3225	−9.0000	0.00	.96	23.6	.035
SS A8	0.0000	1.0000	.9922	.6173	−9.0000	0.00	.90	12.4	.028
SS B1	0.0000	1.0000	.9683	.4954	−9.0000	0.00	.84	16.3	.028
SS B2	0.0000	1.0000	1.0029	.6802	−9.0000	0.00	.74	8.5	.026
SS B3	0.0000	1.0000	.9689	.4685	−9.0000	0.00	.82	19.3	.029
SS B4	0.0000	1.0000	1.0029	.6641	−9.0000	0.00	.76	9.7	.026
SS B5	0.0000	1.0000	.9408	.3866	−9.0000	0.00	.97	16.3	.030
SS B6	0.0000	1.0000	1.0068	.6778	−9.0000	0.00	.84	8.5	.028
SS B7	0.0000	1.0000	.9395	.3363	−9.0000	0.00	.95	19.3	.032
SS B8	0.0000	1.0000	1.0047	.6469	−9.0000	0.00	.87	9.7	.027
SS C1	0.0000	1.0000	1.0087	.7683	−9.0000	0.00	.76	16.3	.025
SS C2	0.0000	1.0000	1.0412	.9281	−9.0000	0.00	.78	10.0	.027
SS C3	0.0000	1.0000	.9699	.5106	−9.0000	0.00	.79	16.3	.024
SS C4	0.0000	1.0000	1.0152	.7523	−9.0000	0.00	.81	10.0	.025
SS D1	0.0000	1.0000	.9889	.6643	−9.0000	0.00	.84	17.8	.028
SS D2	0.0000	1.0000	1.0493	.8753	−9.0000	0.00	.70	9.9	.028
SS D3	0.0000	1.0000	.9570	.5285	−9.0000	0.00	.90	17.8	.029
SS D4	0.0000	1.0000	1.0356	.8142	−9.0000	0.00	.73	9.9	.028
SS E1	0.0000	1.0000	.9968	.7004	−9.0000	0.00	.77	19.6	.027
SS E2	0.0000	1.0000	1.0468	.9054	−9.0000	0.00	.76	10.8	.027
SS E3	0.0000	1.0000	.9565	.4827	−9.0000	0.00	.81	19.6	.028
SS E4	0.0000	1.0000	1.0214	.7694	−9.0000	0.00	.79	10.8	.027

Source: PSDH, 1984.

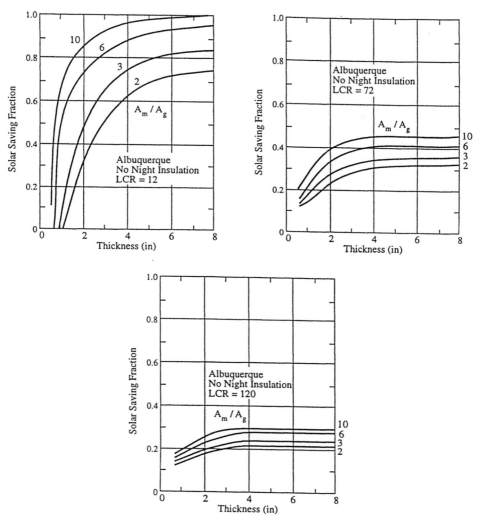

FIGURE A7.1 (a) Storage Wall: Mass Thickness Sensitivity of SSF to off-reference conditions. *Source:* PSDH, 1984.

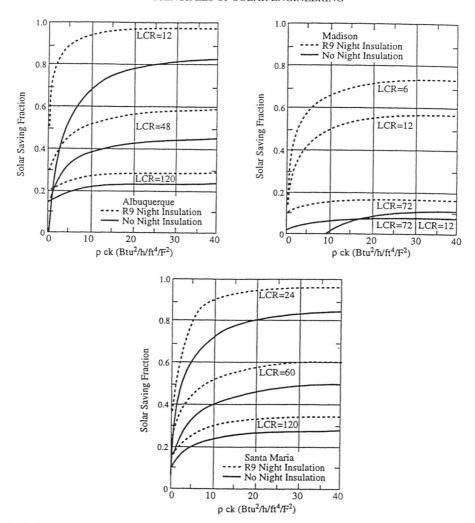

FIGURE A7.1 (b) Storage Wall: ρ cK Product

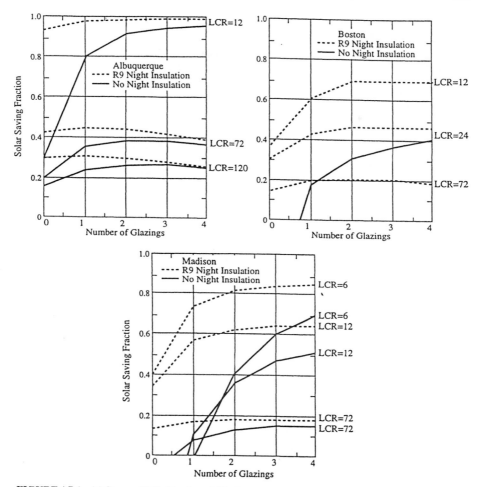

FIGURE A7.1 (c) Storage Wall: Number of Glazings

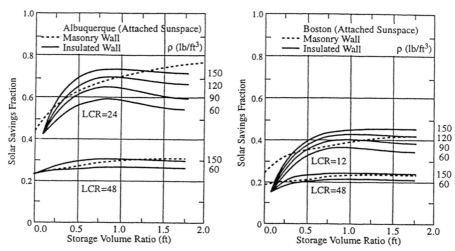

FIGURE A7.1 (d) Sunspace: Storage Volume to Projected Area Ratio

ECONOMIC AND SYSTEMS' COMPONENT DATA

Table A8.1 Capital-recovery factors[a]

n	7	7½	8	8½	9	9½	10	10½	11	11½	12
					Annual mortgage interest rate						
1	1.070	1.075	1.080	1.085	1.090	1.095	1.100	1.105	1.110	1.115	1.120
2	0.553	0.557	0.561	0.565	0.568	0.572	0.576	0.580	0.584	0.588	0.592
3	0.381	0.385	0.388	0.392	0.395	0.399	0.402	0.406	0.409	0.413	0.416
4	0.295	0.299	0.302	0.305	0.309	0.312	0.315	0.319	0.322	0.326	0.329
5	0.244	0.247	0.250	0.254	0.257	0.260	0.264	0.267	0.271	0.274	0.277
6	0.210	0.213	0.216	0.220	0.223	0.226	0.230	0.233	0.236	0.240	0.243
7	0.186	0.189	0.192	0.195	0.199	0.202	0.205	0.209	0.212	0.216	0.219
8	0.167	0.171	0.174	0.177	0.181	0.184	0.187	0.191	0.194	0.198	0.201
9	0.153	0.157	0.160	0.163	0.167	0.170	0.174	0.177	0.181	0.184	0.188
10	0.142	0.146	0.149	0.152	0.156	0.159	0.163	0.166	0.170	0.173	0.177
11	0.133	0.137	0.140	0.143	0.147	0.150	0.154	0.158	0.161	0.165	0.168
12	0.126	0.129	0.133	0.136	0.140	0.143	0.147	0.150	0.154	0.158	0.161
13	0.120	0.123	0.127	0.130	0.134	0.137	0.141	0.144	0.148	0.152	0.156
14	0.114	0.118	0.121	0.125	0.128	0.132	0.136	0.139	0.143	0.147	0.151
15	0.110	0.113	0.117	0.120	0.124	0.128	0.131	0.135	0.139	0.143	0.147
16	0.106	0.109	0.113	0.117	0.120	0.124	0.128	0.132	0.136	0.139	0.143
17	0.102	0.106	0.110	0.113	0.117	0.121	0.125	0.129	0.132	0.136	0.140
18	0.099	0.103	0.107	0.110	0.114	0.118	0.122	0.126	0.130	0.134	0.138
19	0.097	0.100	0.104	0.108	0.112	0.116	0.120	0.124	0.128	0.132	0.136
20	0.094	0.098	0.102	0.106	0.110	0.113	0.117	0.121	0.126	0.130	0.134

[a] n is the mortgage term in years.

Table A8.2 Interest fraction of mortgage payment

Yr left on mortgage	Annual mortgage interest rate										
	7	7½	8	8½	9	9½	10	10½	11	11½	12
20	0.742	0.765	0.785	0.804	0.822	0.837	0.851	0.864	0.876	0.887	0.896
19	0.723	0.747	0.768	0.788	0.806	0.822	0.836	0.850	0.862	0.874	0.884
18	0.704	0.728	0.750	0.770	0.788	0.805	0.820	0.834	0.847	0.859	0.870
17	0.683	0.708	0.730	0.750	0.769	0.786	0.802	0.817	0.830	0.843	0.854
16	0.661	0.686	0.708	0.729	0.748	0.766	0.782	0.798	0.812	0.825	0.837
15	0.638	0.662	0.685	0.706	0.725	0.744	0.761	0.776	0.791	0.805	0.817
14	0.612	0.637	0.660	0.681	0.701	0.719	0.737	0.753	0.768	0.782	0.795
13	0.585	0.609	0.632	0.654	0.674	0.693	0.710	0.727	0.742	0.757	0.771
12	0.556	0.580	0.603	0.624	0.644	0.663	0.681	0.698	0.714	0.729	0.743
11	0.525	0.549	0.571	0.592	0.612	0.631	0.650	0.667	0.683	0.698	0.713
10	0.492	0.515	0.537	0.558	0.578	0.596	0.614	0.632	0.648	0.663	0.678
9	0.456	0.478	0.500	0.520	0.540	0.558	0.576	0.593	0.609	0.625	0.639
8	0.418	0.439	0.460	0.479	0.498	0.516	0.533	0.550	0.566	0.581	0.596
7	0.377	0.397	0.417	0.435	0.453	0.470	0.487	0.503	0.518	0.533	0.548
6	0.334	0.352	0.370	0.387	0.404	0.420	0.436	0.451	0.465	0.480	0.493
5	0.287	0.303	0.319	0.335	0.350	0.365	0.379	0.393	0.407	0.420	0.433
4	0.237	0.251	0.265	0.278	0.292	0.304	0.317	0.329	0.341	0.353	0.364
3	0.184	0.195	0.206	0.217	0.228	0.238	0.249	0.259	0.269	0.279	0.288
2	0.127	0.135	0.143	0.151	0.158	0.166	0.174	0.181	0.188	0.196	0.203
1	0.065	0.070	0.074	0.078	0.083	0.087	0.091	0.095	0.099	0.103	0.107

Table A8.3 Present-worth factors

Year	Discount rate (%)													
	0	1	2	3	4	5	6	7	8	9	10	11	12	13
0	1.000													
1	1.000	0.990	0.980	0.971	0.962	0.952	0.943	0.935	0.926	0.917	0.909	0.901	0.893	0.885
2	1.000	0.980	0.961	0.943	0.925	0.907	0.890	0.873	0.857	0.842	0.826	0.812	0.797	0.783
3	1.000	0.971	0.942	0.915	0.889	0.864	0.840	0.816	0.794	0.772	0.751	0.731	0.712	0.693
4	1.000	0.961	0.924	0.888	0.855	0.823	0.792	0.763	0.735	0.708	0.683	0.659	0.636	0.613
5	1.000	0.951	0.906	0.863	0.822	0.784	0.747	0.713	0.681	0.650	0.621	0.593	0.567	0.543
6	1.000	0.942	0.888	0.837	0.790	0.746	0.705	0.666	0.630	0.596	0.564	0.535	0.507	0.480
7	1.000	0.933	0.871	0.813	0.760	0.711	0.665	0.623	0.583	0.547	0.513	0.482	0.452	0.425
8	1.000	0.923	0.853	0.789	0.731	0.677	0.627	0.582	0.540	0.502	0.467	0.434	0.404	0.376
9	1.000	0.914	0.837	0.766	0.703	0.645	0.592	0.544	0.500	0.460	0.424	0.391	0.361	0.333
10	1.000	0.905	0.820	0.744	0.676	0.614	0.558	0.508	0.463	0.422	0.386	0.352	0.322	0.295
11	1.000	0.896	0.804	0.722	0.650	0.585	0.527	0.475	0.429	0.388	0.350	0.317	0.287	0.261
12	1.000	0.887	0.788	0.701	0.625	0.557	0.497	0.444	0.397	0.356	0.319	0.286	0.257	0.231
13	1.000	0.879	0.773	0.681	0.601	0.530	0.469	0.415	0.368	0.326	0.290	0.258	0.229	0.204
14	1.000	0.870	0.758	0.661	0.577	0.505	0.442	0.388	0.340	0.299	0.263	0.232	0.205	0.181
15	1.000	0.861	0.743	0.642	0.555	0.481	0.417	0.362	0.315	0.275	0.239	0.209	0.183	0.160
16	1.000	0.853	0.728	0.623	0.534	0.458	0.394	0.339	0.292	0.252	0.218	0.188	0.163	0.141
17	1.000	0.844	0.714	0.605	0.513	0.436	0.371	0.317	0.270	0.231	0.198	0.170	0.146	0.125
18	1.000	0.836	0.700	0.587	0.494	0.416	0.350	0.296	0.250	0.212	0.180	0.153	0.130	0.111
19	1.000	0.828	0.686	0.570	0.475	0.396	0.331	0.277	0.232	0.194	0.164	0.138	0.116	0.098
20	1.000	0.820	0.673	0.554	0.456	0.377	0.312	0.258	0.215	0.178	0.149	0.124	0.104	0.087

INDEX